D0762249

Forest Canopies

Physiological Ecology
A Series of Monographs, Texts, and Treatises

Series Editor
Harold A. Mooney
Stanford University, Stanford, California

Editorial Board
Fakhri Bazzaz   F. Stuart Chapin   James R. Ehleringer
Robert W. Pearcy   Martyn M. Caldwell   E.-D. Schulze

T. T. KOZLOWSKI (Ed.). Growth and Development of Trees, Volumes I and II, 1971

D. HILLEL (Ed.). Soil and Water: Physical Principles and Processes, 1971

V. B. YOUNGER and C. M. McKELL (Eds.). The Biology and Utilization of Grasses, 1972

J. B. MUDD and T. T. KOZLOWSKI (Eds.). Responses of Plants to Air Pollution, 1975

R. DAUBENMIRE (Ed.). Plant Geography, 1978

J. LEVITT (Ed.). Responses of Plants to Environmental Stresses, 2nd Edition
Volume I: Chilling, Freezing, and High Temperature Stresses, 1980
Volume II: Water, Radiation, Salt, and Other Stresses, 1980

J. A. LARSEN (Ed.). The Boreal Ecosystem, 1980

S. A. GAUTHREAUX, JR. (Ed.). Animal Migration, Orientation, and Navigation, 1981

F. J. VERNBERG and W. B. VERNBERG (Eds.). Functional Adaptations of Marine Organisms, 1981

R. D. DURBIN (Ed.). Toxins in Plant Disease, 1981

C. P. LYMAN, J. S. WILLIS, A. MALAN, and L. C. H. WANG (Eds.). Hibernation and Torpor in Mammals and Birds, 1982

T. T. KOZLOWSKI (Ed.). Flooding and Plant Growth, 1984

E. L. RICE (Ed.). Allelopathy, Second Edition, 1984

M. L. CODY (Ed.). Habitat Selection in Birds, 1985

R. J. HAYNES, K. C. CAMERON, K. M. GOH, and R. R. SHERLOCK (Eds.). Mineral Nitrogen in the Plant–Soil System, 1986

T. T. KOZLOWSKI, P. J. KRAMER, and S. G. PALLARDY (Eds.). The Physiological Ecology of Woody Plants, 1991

H. A. MOONEY, W. E. WINNER, and E. J. PELL (Eds.). Response of Plants to Multiple Stresses, 1991

The list of titles in this series continues at the end of this volume.

574.52642
L954

Forest Canopies

WITHDRAWN

Edited by

Margaret D. Lowman

Marie Selby Botanical Gardens
Sarasota, Florida

Nalini M. Nadkarni

Evergreen State College
Olympia, Washington

Academic Press

San Diego   New York   Boston   London   Sydney   Tokyo   Toronto

LIBRARY ST. MARY'S COLLEGE

Cover photograph: Aerial photograph of a canopy in a dipterocarp forest in the Danum Valley, Sabah, Malaysia. See Fig. 10C of Chapter 10 by Louise H. Emmons.

This book is printed on acid-free paper. ∞

Copyright © 1995 by ACADEMIC PRESS, INC.

All Rights Reserved.
No part of this publication may be reproduced or transmitted in any form or by any means, electronic or mechanical, including photocopy, recording, or any information storage and retrieval system, without permission in writing from the publisher.

Academic Press, Inc.
A Division of Harcourt Brace & Company
525 B Street, Suite 1900, San Diego, California 92101-4495

United Kingdom Edition published by
Academic Press Limited
24-28 Oval Road, London NW1 7DX

Library of Congress Cataloging-in-Publication Data

Forest canopies / edited by Margaret D. Lowman, Nalini M. Nadkarni.
      p.   cm. -- (Physiological ecology series)
    Includes bibliographical references and index.
    ISBN 0-12-457650-8 (case)
    1. Forest canopy ecology. 2. Forest canopies. I. Lowman,
Margaret. II. Nadkarni, Nalini Moreshwar. III. Series:
  Physiological ecology.
  QH541.5.F6F66   1995
  574.5'2642--dc20   94-41251
                            CIP

PRINTED IN THE UNITED STATES OF AMERICA
95 96 97 98 99 00 BB 9 8 7 6 5 4 3 2 1

Contents

4. Structure and Microclimate of Forest Canopies
Geoffrey G. Parker

Part II
Organisms in Tree Canopies

5. Measuring Arthropod Biodiversity in the Tropical Forest Canopy
Terry L. Erwin

6. Ecology and Diversity of Tropical Forest Canopy Ants
John E. Tobin

7. Lizard Ecology in the Canopy of an Island Rain Forest
Douglas P. Reagan

8. Canopy Access Techniques and Their Importance for the Study of Tropical Forest Canopy Birds
Charles A. Munn and Bette A. Loiselle

9. Forest Structure and the Abundance and Diversity of Neotropical Small Mammals
Jay R. Malcolm

10. Mammals of Rain Forest Canopies
Louise H. Emmons

11. Vascular Epiphytes
David H. Benzing

12. The Ecology of Hemiepiphytes in Forest Canopies
Guadalupe Williams-Linera and Robert O. Lawton

13. Ecology and Population Biology of Mistletoes
Nick Reid, Mark Stafford Smith, and Zhaogui Yan

14. Vines in Treetops: Consequences of Mechanical Dependence
Francis E. Putz

15. Life on the Forest Phylloplane: Hairs, Little Houses, and Myriad Mites
David Evans Walter and Dennis J. O'Dowd

16. Nonvascular Epiphytes in Forest Canopies: Worldwide Distribution, Abundance, and Ecological Roles
Fred M. Rhoades

Part III
Processes in Tree Canopies

17. Photosynthesis in Forest Canopies
N. Michele Holbrook and Christopher P. Lund

18. Herbivory as a Canopy Process in Rain Forest Trees
Margaret D. Lowman

Part IV
Human Impacts on Canopy Research

Contributors

Numbers in parentheses indicate the pages on which the authors' contributions begin.

Bradley C. Bennett (547), Department of Biological Sciences, Florida International University, Miami, Florida 33199

David H. Benzing (225), Department of Biology, Oberlin College, Oberlin, Ohio 44074

Darwyn S. Coxson (495), Faculty of Natural Resources and Environmental Studies, University of Northern British Columbia, Prince George, British Columbia, Canada V2N 4Z9

Louise H. Emmons (199), Division of Mammals, Smithsonian Institution, Washington, DC 20560

Terry L. Erwin (109), Department of Entomology, Smithsonian Institution, Washington, DC 20560

David R. Fitzjarrald (45), Atmospheric Sciences Research Center, State University of New York at Albany, Albany, New York 12205

Francis Hallé (27), Botanical Institute of the University, 34000 Montpellier, France

N. Michele Holbrook (411), Department of Biological Sciences, Stanford University, Stanford, California 94305

Stephen W. Ingram (587), Marie Selby Botanical Gardens, Sarasota, Florida 34236

Robert O. Lawton (255), Department of Biological Sciences, University of Alabama, Huntsville, Alabama 35899

Bette A. Loiselle (165), Department of Biology and International Center for Tropical Ecology, University of Missouri, St. Louis, St. Louis, Missouri 63121

Margaret D. Lowman (3, 431, 587, 609), Marie Selby Botanical Gardens, Sarasota, Florida 34236

Christopher P. Lund (411), Department of Biological Science, Stanford University, Stanford, California 94305

Jay R. Malcolm (179), Department of Biology, Queen's University, Kingston, Ontario, Canada K7L 3N6

Mark W. Moffett (3), Museum of Comparative Zoology, Harvard University, Cambridge, Massachusetts 02138

Kathleen E. Moore (45), Atmospheric Sciences Research Center, State University of New York at Albany, Albany, New York 12205

Charles A. Munn (165), Wildlife Conservation Society, International Program, Bronx, New York 10460

Darlyne A. Murawski (457), Harvard University Herbaria, Cambridge, Massachusetts 02138

Nalini M. Nadkarni (495, 609), Evergreen State College, Olympia, Washington 98505

Dennis J. O'Dowd (325), Department of Ecology and Evolutionary Biology, Monash University, Clayton, Victoria 3168, Australia

Geoffrey G. Parker (73), Smithsonian Environmental Research Center, Edgewater, Maryland 21037

Donald Perry (605), Tropical Treetop Exploration, Branchport, New York 14418

Francis E. Putz (311), Department of Botany, University of Florida, Gainesville, Florida 32611

Douglas P. Reagan (149), Woodward Clyde Federal Services, Denver, Colorado 80237

Nick Reid (285), Department of Ecosystem Management, University of New England, Armidale, New South Wales 2351, Australia

Fred M. Rhoades (353), Department of Biology, Western Washington University, Bellingham, Washington 98225

Mark Stafford Smith (285), Division of Wildlife and Ecology, CSIRO, Alice Springs, Northern Territory 0871, Australia

John E. Tobin (129), Museum of Comparative Zoology and Harvard Law School, Harvard University, Cambridge, Massachusetts 02138

David Evans Walter (325), Department of Entomology and Centre for Tropical Pest Management, University of Queensland, St. Lucia, Queensland 4072, Australia

Guadalupe Williams-Linera (255), Instituto de Ecologia Xalapa, Veracruz 91000, Mexico

Zhaogui Yan (285), Department of Agriculture, South Perth, Western Australia 6151, Australia

Foreword

When I started working in tropical forests in the mid 1960s, the canopy rendered me the biologist's equivalent of Tantalus from the very outset. Frank Chapman's *My Tropical Air Castle* (1929, D. Appleton & Co., New York) and Marston Bates' *The Forest and the Sea* (1965, Random House, New York) had already stirred the imagination. So had Colin Pittendrigh's hypothesis that bromeliads arose in xeric environments and then invaded the canopy. Early on I met the colorful Jorge Boshell, Bates' successor as director at Villavicencio, who had solved the riddle of jungle yellow fever's ability to vault from its canopy cycle with howler monkeys to people far below on the forest floor. Boshell had noted a cloud of blue *Haemagogus* mosquitoes, normally canopy dwellers, swarming around woodsmen who had felled a canopy tree. Today there are ominous echoes of this critical observation in a tale of deforestation with disquieting portent, namely that of the Zaire Ebola virus which nearly escaped containment in suburban Washington, DC. For me, however, the canopy will always be symbolized by the exquisite but deadly metallic blue mosquitoes and the less ominous *Sabethes* mosquitoes which float on flanges rather like an entomological equivalent of a South Pacific outrigger canoe.

It is not surprising then, that even after Boshell solved the riddle of the Sylvan yellow fever, tropical epidemiologists have had a compulsion about canopy studies. At the mouth of the Amazon, the Belem Virus Laboratory erected a modest 35-meter tower where one could glimpse canopy birds close at hand aided by the same fearlessness for which Galapagos wildlife is renowned. There was also an ingenious counterweighted rope "elevator" so someone could easily zip up into the canopy to put out or collect back sentinel animals, such as chickens, to detect the presence of any arthropod borne viruses. Philip Humphrey and I set out a couple of mistnets rigged like a sail to catch birds in the upper reaches of the forest. Imagine our delight on the very first day when we showed this new technique to Helmut Sick, dean of South American ornithology, and caught two swifts, *Brachyura spinicauda,* the very species Sick had come north to collect for Brazil's Museu Nacional.

These experiences merely highlight how very little is known about the forest canopy and how challenging it is to do canopy research. While I expect no one would disagree that canopy biology is still in its infancy, the field has undergone an intellectual radiation in recent years. Physically, it may be represented by the use of cranes which the Smithsonian Tropical Research Institute has recently installed in Panama. Intellectually, it is represented by this volume, and a rich smorgasbord it is indeed.

There is no better evidence than canopy biology that the age of exploration is not over. We can anticipate a diverse panoply of discoveries emanating from this field. Some will be of serious practical benefit to ourselves as living organisms. Others will illuminate aspects of biology never before dreamed of. Yet others will astound with their beauty or be intrinsically fascinating. In all cases it will be clear that canopy biology, as a recognized field of intellectual endeavor, began with this book.

THOMAS E. LOVEJOY
Smithsonian Institution
Washington, DC

Preface

In the 1970s, many of the contributors to this volume were dangling precariously from ropes, trying to answer questions about what lives in the tops of trees. Ironically, most of their ropes were dangled throughout remote regions of tropical and temperate forests, and very few of these pioneers in canopy biology ever met one another. It came as a great surprise when two of us (Margaret Lowman and Nalini Nadkarni) published back-to-back papers in *Biotropica* in 1984, each having developed the same techniques for using slingshots and ropes to conduct ecologically replicated sampling throughout tropical tree canopies. One of us was struggling to avoid giant stinging tree hairs of the Urticaceae in the paleotropics while measuring herbivory, while the other was peering through the misty branches at epiphytes in a cloud forest in Costa Rica. If only we could have networked via E-mail, then perhaps we could have been making exciting intercontinental comparative studies of epiphytes and herbivores! Now, years later, many of us still have never met due to the isolated nature of our scientific research.

How unfortunate that the canopy biology network has been so undeveloped that very little communication among researchers has taken place prior to publication of results. As with all new fields, canopy biology has experienced the excitement of ideas, the growth of hypotheses, and the development of methodologies that are imperative to the foundations of any discipline. With the impending harmonization of a range of methods, and the establishment of sampling protocols, the real work can begin: addressing hypotheses, collecting collaborative databases, making large-scale comparisons, and figuring out the dynamics of this previously uncharted region of forests.

Within the past decade, the number of scientific publications on canopies has grown at a disproportionately rapid pace relative to other aspects of biology (Chapter 24). Other evidence for the burgeoning interest in forest canopies are recent symposia featuring canopy biota, e.g., Missouri Botanical Gardens 1987; Marie Selby Botanical Gardens 1986, 1991, 1994; and Association for Tropical Biology 1992. Two reasons for the rapid growth of published material and scientific interest in canopies are: (1) the development of more sophisticated methods to address canopy research (e.g., the dirigible, the canopy crane, remote sensing); and (2) the political and economic interests (in addition to biological interests) in tropical rain forest conservation, with the canopy emerging as a center of biodiversity on the planet.

These reasons have also led to an unprecedented media interest in this new arena of research. Climbing trees is of great interest to children, and so forest canopy ecology has become a good mechanism for inspiring young students in science. The threats to biodiversity have provided a stimulus for taxonomists to concentrate on canopy organisms. Tropical rain forests have become a "bandwagon" for conservation organizations to bring attention to environmental issues. While these initiatives may not have a solid scientific basis, the opportunities for canopy research and education are greater than ever before. The importance of accurate and careful scientific protocols is paramount.

Given the importance of and interest in canopy biology, the dearth of research on this topic is startling. There are relatively few long-term studies, and almost no large-scale comparative studies to date. It is still difficult—if not impossible—to observe bird behavior or to count orchid seedlings within the canopy. Canopy biologists are distributed throughout the fields of ecology, geography, meteorology, botany, zoology, and forestry, and the canopy literature is scattered throughout myriad journals. No central journals or institutions that specialize in this field exist, and no protocols for sampling have been formally adopted.

This book is the first synthesis of this exciting field aimed at a scientific audience. It is by no means comprehensive. Obvious gaps include microbial ecology, fungi, bats, and evolutionary aspects of diversity of organisms in the canopy. The more obvious lack is the dearth of long-term studies to understand the seasonal variability in canopy dynamics. For the first time, however, it is possible to get a holistic perspective on canopy biology—what information exists, what studies are currently in progress, and what remains to be explored.

Most authors elected to review their particular field of canopy research, using the scientific literature and personal data to illustrate concepts, and to present avenues of future study for aspiring students and researchers. This was not possible in all cases, due to the infancy of some topics. It was impossible to review other aspects due to the large volume of information that already exists (e.g., lower plants!). We avoided making all chapters homogeneous in their style and organization, to make the reading more interesting and to give the authors more opportunities to be creative. For example, Don Perry, an important pioneer explorer of tree crowns, discusses how a biologist has applied his interests to ecotourism and conservation.

It is our hope that this book will foster enthusiasm in canopy biology, and lead to the improved integration of results and to the recruitment of a plethora of students who will define sound field protocols for comparative and replicated sampling. Both primary and secondary forests require study if we are to implement sound management practices to the world's over-exploited forests. Field studies of forests are an urgent priority, and the conservation of this important resource requires comprehensive and well-executed field research in both the upper and lower regions of trees.

We pay special tribute to recently deceased biologists who have pioneered aspects of canopy research: Alan Smith, Al Gentry, and Ted Parker.

We are indebted to many scientists who generously spent time amid their busy field schedules to review sections of this book, including: James Ackerman, Joachim Adis, Robin Andrews, Peter Ashton, Henrik Balslev, Yves Basset, Kamal Bawa, David Benzing, Paul Berry, Beryl Black, Brian Boom, Bart Bouricius, Judith Bronstein, George Carroll, Paul Catling, Robert Colwell, Joe Connell, Stefan Cover, Thomas Croat, D. A. Crossley, Roman Dial, John Eisenberg, Robin Foster, J. L. Hamrick, Harold Heatwole, Mandy Heddle, E. E. Hegerty, Bert Hölldobler, Paul Jarvis, Michael Kaspari, Michael Keller, Daniel Kelly, J. Egbert Leigh, Jr., Peter Lesica, Betty Loiselle, John Longino, John Lowman, Jon Majer, Robert Marquis, Teri Matelson, Larry Mitchell, Mark Moffett, Linton Musselman, Peter Myerscough, Dan Nickrent, Roy Norton, Richard Primack, Stanley Rand, Bruce Rinker, Phillip Rundel, William Schlesinger, Timothy Schowalter, Richard Schultes, David Shaw, Steve Sillett, Benjamin Stone, Nigel Stork, Michael Sutton, Charlotte Taylor, Tony Underwood, Dennis Whigham, Mark Whitten, Henk Wolda, S. Joseph Wright, and Jess Zimmerman. And we thank the authors, whose enthusiasm and timely submissions made this volume a reality in only two years.

We thank our administrative assistants, Ellen Baskerville and Cara Stallman, for their loyal support throughout the production of this volume. Chuck Crumly, at Academic Press, is a superb and understanding editor. The staff at our respective institutions, the Marie Selby Botanical Gardens and the Evergreen State College, were also very supportive throughout our pursuit of canopy-related careers.

Finally, we dedicate this book to our children: Edward and James, Gus and Erika, potential future biologists, who have given us great enthusiasm to pursue scientific research and who know how much fun it is to climb trees.

MARGARET D. LOWMAN
NALINI M. NADKARNI

I

Structure and Function in Tree Canopies

1

Canopy Access Techniques

Mark W. Moffett and Margaret D. Lowman

To know the forest, we must study it in all aspects, as birds soaring above its roof, as earth-bound bipeds creeping slowly over its roots.
—*Alexander F. Skutch, "A Naturalist in Costa Rica" (1971)*

I. Introduction

In this chapter, we have intentionally departed from the rigorous scientific presentation of the other authors. Our chapter does not offer hypotheses or results; rather, it is a *story* of the development of one of the most exciting and innovative frontiers of ecology. The "heroes" of this tale are the scientists whose writing follows ours, and their pioneering studies are setting the stage for the young researchers and students who may become stimulated by the discoveries reported here. We have, however, highlighted three pioneering techniques in the development of canopy biology (see Boxes).

We emphasize an important underlying message with our review of canopy methods: *SAFETY.* To all who become inspired to attempt canopy research, and particularly those who have not previously worked with experienced arborists or canopy biologists, please be cautious and use all possible safety precautions. Though our descriptions are intended to entice, the concepts of canopy access require serious attention to safety measures.

In contemplating the tangled web of species interactions in forest canopies, scientists have relied so far on sparse information. As this volume

Copyright © 1995 by Academic Press, Inc.
All rights of reproduction in any form reserved.

shows, canopy biologists must dangle precariously for hours to find out such basic data as where innocuous ants horde epiphyte seeds; chart the complex pathways of lianas to sunny spots where they flower; trace the passage of nitrogen from tissue to tissue through floral arcades; and loiter for weeks in aerial blinds to observe the fruit preferences of enigmatic birds. Concentrated endeavors of this kind are cornerstones to an intimate understanding of treetop ecology (Lowman and Moffett, 1993). Nonetheless, many canopy biologists still invest as much time and effort to get into the trees as to collect data from them. This chapter reviews the practical side of the researcher's canopy experiences.

Although canopy access technologies have expanded over the last 15 years or so, many of the methods can be traced back to antecedents from decades earlier (reviews in Mitchell, 1982; Moffett, 1992). Entire volumes that report climbing techniques in the field have been written by Hingston (1932), Mitchell (1981), Perry (1986), Moffett (1993b), and Lowman (1995). Of the modern methods reviewed here, some require shrewd financial lobbying, whereas others can be managed on a shoestring budget; some are cumbersome and reach a limited area, whereas others allow scientists to touch the tips of lofty branches with the grace of an acrobat. Most are currently in use somewhere in the tropics, and every one has merit under the right circumstances.

Our review emphasizes tropical rain forest situations mainly because the widest variety of techniques has been attempted in this tall, most architecturally complex forest type. The chapter starts by addressing stratagems for gathering canopy data from the ground. Thereafter, our coverage focuses on techniques created to actually transport people into the treetops.

II. Techniques of Canopy Access

A. Ground-Based Methods of Access

It is by no means necessary to climb into the canopy to complete a canopy study. For collecting many kinds of data, climbing would be a waste of time. Ground-based methods are notably useful in studying species that are either extremely mobile or too sensitive to disturbance to be monitored from within the canopy, in gathering museum samples (such as plant specimens or bird skins), or when the sampling protocol is so demanding that it is impractical to climb so often.

Nearly every naturalist has taken advantage of a ridgetop to view the canopy at its own level, close at hand, even if only to ponder the magnificence of the treetops. Some biologists actually get better information from a good ground-based vantage point. For example, primates can be tracked with the greatest ease from the ground, binoculars or telephoto lens in hand (see Chapter 10). Day-Glo colors painted on reptiles permit the location and identification of cryptic species in the trees (Robert Henderson, personal communication). Concealed vertebrates can be located and tracked by radiotelemetry (e.g., Montgomery *et al.*, 1973). Fish-eye photographs of light flecks in the canopy can be analyzed by computer to determine the light regimes at specific understory positions (Becker *et al.*, 1988). Canopy data can be gathered photographically from above the trees with the aid of balloons, ultralight aircraft, or—from still higher up—planes and satellites (O'Neill, 1993). With the aid of binoculars, most canopy trees can be identified quickly by the experienced eye (e.g., Robin Foster, personal communication) (Fig. 1).

In collecting canopy plant specimens there are several alternatives to climbing trees (see also Chapter 23). E. J. H. Corner (1992) reported that

Figure 1 Robin Foster using binoculars to identify canopy plants.

domesticated monkeys were trained in Malaysia in the 1930s to retrieve botanical samples from the canopy:

> By means of the coconut or pig-tailed monkey (*Macaca nemestrina*) . . . I obtained, at last, a fair measure of the forest without destruction of the trees. Hitherto, baffled by the height of trees, climbers and epiphytes, I had been obliged to content myself with forest that was being felled. That was hard, hot, and commonly unsuccessful work, for the plants were often neither in flower or fruit. I grew to detest climbing over a jumble of fragments in the endeavor to piece them together, while streaming with sweat, harassed by the glare, and often assailed by irate bees and wasps. . . . [Using monkeys] I developed a rubber neck that I could lay back on my shoulders for half-an-hour at a stretch, while gazing upwards and shouting commands to the treetops.

Similar plant (or animal) sampling can be accomplished from the ground with a slingshot, rifle, or (for low branches) a pole pruner (Fig. 2). Fogging trees with insecticide is another "knock down" approach to sampling, but applied only to invertebrates (Erwin, 1989; Kitching *et al.,* 1993; and Chap-

Figure 2   Darlyne Murawski using a slingshot to knock down fruits of a cuipo tree in Panama for isozyme analysis.

Figure 3 Terry Erwin fogging for arthropods in a Peruvian tree crown as part of his biodiversity studies.

ter 5) (Fig. 3). Without climbing a tree, D. H. Murphy (personal communication) collects airborne insects with sticky traps raised to the canopy by tethered helium balloons. The normal "rain" of canopy organisms and debris can also be assessed with litter traps (e.g., Lowman, 1988).

A widely applicable ground-based technique involves hoisting an apparatus into the canopy with a pulley system, then lowering it as required to check an experiment. An initial climb into the canopy may be required to set up a pulley, or researchers may elect to shoot a line over a branch from the ground and use that line to pull up and support a second haul line on a pulley. This has been done with mist nets (Munn, 1991), light traps (Wolda, 1992; Sutton, 1979), butterfly bait traps (DeVries, 1988), mammal traps (Malcolm, 1991; Deedra McClearn, personal communication), branches harboring epiphytes (Cristian Samper, personal communication), trays that hold sprouting fig seeds (Timothy Laman, personal communication), and numerous other experiments.

Table I Methods for Canopy Access and Assessment of Their Use, Rated from 1 (Least Desirable) to 10 (Most Desirable)

Method	Cost-effectiveness	Ease of training before use	Ease for researcher	Ease to build	Ease to move to new site	Ease to avoid canopy disturbances	Assistant necessary (N = none; U = unskilled; S = skilled)	Group size at one time	Shape of accessed area	Horizontal reach between trees (N = no; Y = yes)	Access uppermost twigs (N = no; Y = yes)
Peconha[a]	10	9	1	10	10	8[b]	N	1–2[c]	Vertical line[d]	N	N
Ladders	8	10	9	8	5	8[b]	N	1–3[c]	Vertical line[d]	N	N
Single rope	8	8	6	9	9	8[b]	N	1–3[c]	Vertical line	N[e]	N
Tower	5	10	9	5	2[f]	8	N	1–4[c]	Vertical line	N	Y[g]
Rope web	8	6	4[h]	4	2[f]	9	N	1–3[c]	Cylinder	Y	Y
Boom	8	8	7	7	8	9	U	1	Cylinder	Y	Y
Cherry picker	2	10	10	10	5	8–9	U	1–2	Horizontal line[i]	Y	Y[g]
Crane	1	10	10	4	2[f]	6–9[j]	S	1–3	Cylinder	Y	Y
Raft	1	10	10	1	7	6	S	1–6	Cylinder[k]	Y	Y
Sled	1	10	10	1	10	6	S	2–3	Horizontal line[l]	Y[l]	Y[l]
Walkway or tram	5	10	10	5	1–3[f]	6	N	1–10+	Horizontal line[k, m]	Y	N

Note: Our comparisons among techniques are personal opinion, but based on firsthand experience with all the methods conducted under tropical conditions. Advocates of a given method might argue for higher scores under certain circumstances, and scores pertaining to damage to the canopy environment may be optimistic. A range of scores could often be applied. Walkways, for instance, vary tremendously in expense depending on their length, construction materials, and intended duration of use.

[a] Can only climb trees of limited width.

[b] Damage caused mostly by frequent necessity of climbing onto branches.

[c] Even more can be accommodated by building large platforms.

[d] Very restricted to branch or tree trunk surfaces.

[e] Movement restricted to climbing on branch surfaces or ascending through areas beneath major boughs.

[f] Usually intended to be permanent or long term.

[g] Access to upper strata requires sufficient apparatus size.

[h] Robotic version under testing at Rara Avis, Costa Rica, will be an improvement.

[i] Usually works only along forest margins (e.g., roadsides).

[j] Depends on canopy density; descent on rope instead of gondola reduces problems.

[k] Limited vertical range unless used in combination with climbing ropes.

[l] Only method in which only crown peripheries are accessible.

[m] Horizontal expansion possible between strong trees.

B. Climbing Methods of Access

We have compiled a brief description of many climbing techniques and summarized their attributes (Table 1). No attempt is made to describe the techniques in sufficient detail for this chapter to serve as a how-to guide. It is best to consult the primary literature (or better yet, talk with those who are currently using the method).

The peconha is a technique originated by Brazilian Indians to climb the trunks of trees up to 40 cm in diameter (Fig. 4). All that is required is a loop of webbing. Other direct trunk-climbing methods are avoided by responsible biologists whenever their use inflicts damage on trees: climbing spikes, tree surgeon's belt with spiked boots, tree bicycles or Swiss tree grippers, or boards with nails to create steps.

As an acceptable alternative, ladders can be lashed into place one above the other along the trunk with relatively few nails. Mori (1984) has used Swiss tree grippers to inch-worm up trunks, although this method should only be used on hardwoods and trees with few epiphytes on the trunk so that they inflict little damage.

Figure 4   Jay Malcolm inch-worming his way up a tree trunk with a peconha in Brazil to access his mammal traps.

Figure 5   Nalini Nadkarni assisting Jack Longino, who is climbing a tree using a mountaineering rope and ascenders.

Single rope techniques (SRT) enable researchers to sample away from the confines of tree trunks (Fig. 5). To climb mountaineering ropes, most people purchase rock-climbing ascenders (Perry, 1978; Whitacre, 1981; Padgett and Smith, 1987), but it is also possible to replace a plain rope with a rope ladder or to design block-and-tackle systems, or even motorized chairs to make the climb less arduous. Rope webs (Perry and Williams, 1981) and the boom (see Box 1) are highly modified rope-climbing methods that facilitate greater reach from the rope (Fig. 6).

Towers are free-standing structures that, like ropes, permit ascent away from tree trunks. In contrast, ropes can only be placed over strong tree limbs, but towers can be erected anywhere and may even extend above the canopy. Towers include an assortment of constructions, from narrow structures that are little more than free-standing ladders to configurations with landings every few meters that can accommodate several people.

The tower crane, developed in Panama by the Smithsonian Tropical Research Institution, is basically a tower with an arm to provide horizontal reach. It allows comprehensive access to a permanent suite of trees (Fig. 7; see Box 2). Researchers board a gondola at ground level and are driven upward (review in Parker *et al.*, 1992). To maneuver the gondola to a specific site in the canopy, it is usually necessary to ascend over the trees and then descend again; in this sense, approach to the canopy is from above. A cherry picker is a small, relatively mobile variant of the canopy crane in which the crowns are approached from below. They have been used

Box 1. Using Booms in Canopy Research

To study reproductive aspects of tree population biology in the field, it may be necessary to carry out simultaneous or near-simultaneous studies in the crowns of several trees of the same species, which can be widely scattered through the species-rich tropical canopy. This research might involve observing between-tree visits of pollinators or seed dispersers, controlled cross-pollination experiments, or precise quantitative analyses of the synchrony of phenological events. Problems that such research may impose include the need to rapidly locate and map all reproductive trees in dispersed populations, to observe the few individuals that may flower at short notice in an average year, and to gain access to the slender outermost twigs where flowers and fruits are borne (even beyond them if fine observations or manipulations of small flowers are needed). Hands must be free to use fine brushes, forceps, and a magnifying lens.

These problems were faced in a pioneer study of the reproductive biology of rain forest trees in Malaya, carried out by six Malayan graduate students under Professors Engkik Seopadmo and Peter Ashton and colleagues at the Universities of Malaya and Aberdeen during the 1970s. Early on, it was discovered that pollination of the main dipterocarps under study took place at night, that the flowers lasted one night only, and that the whole tree population completed its flowering within two weeks, once every five years. This was not an ideal subject for a doctoral thesis, but the team succeeded, as there was a mass of flowering in the middle of the day. After three years, which gave them time to devise methods to quickly get to fine twigs at 60 m and to develop experimental techniques on specimen trees in the Forest Research Institute arboretum, a massive flowering occurred in 1976.

With the assistance of a Scottish oil field engineer, a system of affordable telescopic booms was designed, several of which could be stationed simultaneously in the crowns of different emergent dipterocarps at short notice. Each boom consists of five lengths of standard-diameter aluminum alloy pipe: a central tube 20 cm in diameter and two pairs of longer, narrower pipes that slide into one another. One, three, or five segments can be used, and a series of holes in the pipes allows them to be locked together at several points of insertion by metal pins. The maximum length with all five segments is 20 m. Because standard-diameter pipes do not fit snugly, epoxy resin sleeves are fitted within the longer pipes, allowing a tight fit with smooth insertion.

continues

Box 1. *Continued*

The length and angle of crown penetration required are estimated from the ground, where the boom is assembled. The booms are lifted by a steel cable that passes through a pulley attached to one of the main branches of the tree by a sleeve that does not abrade the bark. This must first be put in place by a tree climber, using SRT. One end of the cable is attached to a steel bracket near the middle of the boom, whose position can be adjusted to achieve the inclination desired. The other end of the cable is passed through a manual lifting gear attached to the base of a nearby tree.

Two ropes are attached at the lower end of the boom and, when fully lifted, are used to swing the other end into the desired position in or through the crown. They are then tied around the trunks of two different trees to anchor the boom while in use. The entire device can be lifted and adjusted into position within one hour. The researcher is attached on a block-and-tackle harness, with both hands free for research activities, and is lifted by pulling the rope through the block-and-tackle from the ground. The equipment cannot be operated by one person, which serves as an added safety precaution.

The principal inconvenience of the method proved to be carrying the boom sections through the forest, each of which needs two carriers. Transferring a boom from one tree to another within a kilometer could be done in a day by two to four people. The booms have not been used in recent years, but they still exist and await a bidder!

Peter Ashton
Harvard Institute of International Development
Gray Herbarium
Harvard University
Cambridge, Massachusetts 02138

successfully in dry, relatively sparse forests, where it is possible to drive among the stand (e.g., Australian sclerophyll forests; Lowman and Heatwole, 1992).

The canopy raft and sled were developed by Operation Canopee in France (Hallé, 1990; Hallé and Pascal, 1992). Both the raft and sled are lowered onto the canopy surface by a dirigible (Fig. 8). The raft remains in place for several days, whereas the sled—a miniaturized version of the

Figure 6   The canopy boom. Left: Suspension bracket at midpoint of boom from which the boom is lifted into the canopy by means of a steel cable. Right: H. T. Chan in a bosun's chair suspended from one end of the boom. (Photographs by Peter Ashton, Pasoh, Malaysia, 1975.)

raft—is repeatedly dragged over the canopy surface by the dirigible, enabling scientists to sample between many tree crowns in rapid succession (Lowman *et al.*, 1993).

Once in the trees, a common challenge is to provide a stable working area and to enlarge horizontal reach in and between crowns. At minimum, a wide branch crotch or hanging rock-climber's cot may provide comfort for short durations (e.g., Nadkarni, 1988). A more substantial approach is to construct a platform (e.g., Leighton and Thomas, 1980; Lowman and Bouricius, 1993). (The raft, crane, and some towers already have a stable "platform" built in.) Horizontal reach can also be extended away from the platform with pole pruners or long-handled nets.

Suspended walkways (Fig. 9; see Box 3) can be built in conjunction with platforms, utilizing a modular concept. They are most often reached from below by a ladder, rope, or tower, but some walkways have also been built directly out into the canopy from hillsides, circumventing a vertical ascent (e.g., Muul and Lait, 1970). Trams (cable cars) supported by steel towers

Figure 7   Researchers Solby Chavarria and Mirna Samaniego aboard the gondola suspended from the arm of the Smithsonian Tropical Research Institute's canopy tower crane in Panama.

Figure 8   The raft and balloon operation, Radeau des Cimes, in Cameroon, Africa.

Box 2.   The Canopy Crane

The use of standard construction tower cranes to gain access to the forest canopy was pioneered by Dr. Alan P. Smith at the Smithsonian Tropical Research Institute (STRI) in Panama. Free-standing construction tower cranes consist of a central vertical shaft, a horizontal boom that moves through 360 degrees, and an electrical motor. A gondola carrying biologists and their equipment can be delivered to any unobstructed point below the boom. Tower cranes have unique advantages for forest canopy studies: canopy access is safe, rapid, and supported from above. The same branch, leaf, or point in space can be visited repeatedly.

Tower cranes also have unique requirements. The heaviest components must be moved into position by mobile cranes or by helicopter at remote sites. Electrical demand is substantial, and a generator is required, which creates noise disturbance in the understory. It is one of the most expensive methods in which to invest. In 1990, a prototype crane with a 40-m vertical shaft and a 30-m boom was leased commercially in Panama for $2000 per month. Costs increase rapidly with the height of the vertical shaft and the length of the boom. In 1993, STRI purchased a 60-m-tall crane with a 51-m boom for $240,000. Maintenance, security, and operator costs were about $40,000 per year in 1993.

A crane has been operating over a tropical dry forest in central Panama and has demonstrated the utility of the method. Demography, herbivory, and leaf gas exchange are monitored each month for several thousand sun and shade leaves. Behavioral observations of pollinators and herbivores are also under way. Nighttime operation is routine. The stability of the gondola permits real-time *in situ* measurements. Micropressure transducers have been inserted into leaf blades to monitor water potentials at heights of 35 m. In sum, the tower crane permits the full range of investigations possible from the ground with few of the drawbacks of other techniques.

S. Joseph Wright
Smithsonian Tropical Research Institute
Box 2072
Balboa, Republic of Panama

Figure 9   Meg Lowman and son Edward on a bridge suspended at 22 m over Blue Creek, Belize, as part of a walkway system built by Bart Bouricius.

can be suspended either in or above the treetops (e.g., Leonard and Eschner, 1968).

III. Logistic Considerations of Canopy Access

A. Cost of Different Methods

Why spend money at all? As just about anyone who can recall childhood knows, climbing trees can be done without special equipment. Although the sheer physical strength required inhibits most large-bodied adults from free-climbing, native free-climbing specialists live in many parts of the tropics and work for years without injury. Yet, aspiring arborealists take warning! A bare-handed approach to tropical canopy access has serious safety drawbacks. Tropical forests can be impressively tall, and the selection of sound climbing trees is sometimes difficult except by a professional arborist or experienced researcher. Understory species tend to be spindly with weak vertical limbs, and overstory trees seldom offer branches near ground level.

Of the methods currently in favor today, only ropes or ladders lie within the budgets of most students or small grants. Of moderate cost are platforms and walkways; they offer the additional advantage of creating a permanent structure, which can subsequently be used by many others.

Because equipment for the more expensive methods is durable and permits longer work hours with proportionally larger group sizes, we under-

Box 3.   The Aerial Walkway Technique

Aerial canopy walkways are a relatively simple, flexible, and inexpensive method of studying a broad swath of the forest canopy. For short-term studies (e.g., a few days or weeks), structures held together and supported by ropes and high-strength industrial tapes have been used (A. W. Mitchell, personal communication). Longer-term studies will be the focus of this discussion.

For over five years, I have designed and built aerial walkways that consist of platforms and bridges linked together to form pathways through the trees. The platforms are supported by stainless-steel or galvanized aircraft cable (tensile strengths of 12,000 and 14,000 kg, respectively). The bridges are suspended from above or supported from below with these cables.

The platforms and bridges have netting and handrails made of rope or steel cable. Overhead cables and other strong attachment points are provided with safety lanyards, which allow users of the walkways to remain tethered at all times. Cables supporting bridges and platforms are bolted with galvanized steel bolts through the trees and are secured from the other side by washers and nuts. Suspending the structures on cables prevents structural members from rubbing against the tree in the wind. Trees in the walkways are also guyed with rigid seven-strand cable to stabilize and counterbalance the weight of these platforms and bridges. I do not advise encircling the tree trunks or limbs with cables or ropes to avoid risk of injuring the cambium.

I have worked extensively with a canopy biologist, Meg Lowman, to integrate both structural and scientific factors in walkway design. Factors in selection of a site for an aerial walkway include:

a. appropriateness and accessibility of the site for the specific research, education, or tourism project involved;
b. nearness of the site to ridgetops, clearings, or other physical features that may increase the incidence of lightning or wind damage;
c. placement of trees of sufficient size, soundness, and proximity to facilitate efficient and economical construction of a walkway;
d. trees that would enable future expansion of the walkway;
e. access to a range of canopy levels and trees to maximize scientific use of the structure.

Several factors affect the cost and materials that are required to construct a walkway. In the temperate zone where there is moderate to light rainfall, galvanized cable may be used as guy wires and for

continues

Box 3. *Continued*

structural support. In wet tropical forests or in temperate rain forest sites, more expensive stainless steel is required, at least for the main bridge support cables. Stainless-steel cable, when used for all support cables of both bridges and platforms, raises the materials cost of an average project by 15–20%. Stainless steel generally lasts for 40–50 years; galvanized cable will generally last for 20 years in temperate areas and for half that duration in moist tropical regions.

Pressure-treated wood, rot-resistant wood, or aluminum are good choices for platform decking, support joists, and bridge treads. The greater expense of aluminum plus possible metallic noise problems must be balanced against the greater weight of wood and its tendency to deteriorate more rapidly than aluminum.

Access methods to the platform include wood or aluminum ladders bolted to the tree trunks, a block-and-tackle arrangement of polyester rope and pulleys, and counterweight systems. A counterweight system requires extreme caution and extensive training in safety procedures, and can only be used by one researcher at a time.

The cost of materials for 70 m of bridge spans, four platforms (2 m × 3 m), guy wires, and a ladder or block-and-tackle for access ranges between $11,000 and $14,000 (1994 values). Labor and other costs bring the total to $17,000 to $22,000 for sites in the United States. To date, canopy research platforms have been completed at Williams College (Massachusetts), Hampshire College (Massachusetts), Coweeta Hydrological Station (North Carolina), Blue Creek (Belize), and Selby Botanical Gardens (Florida).

Bart Bouricius
Canopy Construction Co.
32 Mt. View Circle
Amherst, Massachusetts 01002

took a comparison of costs on the basis of our rough estimation of person-hours of canopy work. In our judgment, the methods in this case would be ordered in a similar hierarchy for cost-effectiveness as indicated in Table I, but the cost differences would be less pronounced because more expensive methods often allow more researchers and longer work hours. (Obviously, this ranking may also change with the scope of any design.) It is important to recognize that different researchers and different studies will have varying constraints. For example, canopy walkways and platforms may be ideal for long-term studies in one forest by a large group of students, whereas SRT is preferable for a study that requires examination of replicate crowns throughout a forest region.

B. Safety Precautions (with Special Consideration of SRT)

Most methods require an assistant or, in the case of the canopy raft, a skilled support staff. In addition, safety lines, backup hardware, and protective clothing are necessary.

There are many potential physical hazards in trees. For example, structurally weakened wood or the presence of fungal pathogens may be difficult to detect; epiphyte mats may slide off beneath a climber's feet; termite-infected limbs can snap unexpectedly; a young vine's grip on a tree may be more tenuous than it would appear from its size; and ants, wasps, and snakes may lurk among bromeliads.

Regardless of the method, it is prudent to carry a length of rope or webbing to tie oneself to a branch upon attaining the crown. Although no climbing technique is foolproof, in general, the more expensive the technique, the greater the security it potentially can offer. Thus, the peconha is not much safer than free-climbing and hardly less stressful, whereas when used with common sense, a tower crane appears to be very secure. Cranes and towers offer the advantage of not being dependent on the structural integrity of the canopy itself, so they can be relied on to access trees with unmanageable crown architectures or fragile wood.

For long-term success as a canopy researcher, one not only needs to *be* safe, but also must *feel* safe. Developing confidence with the methods can be difficult. For example, based on our experiences, to descend a tree with the single rope technique, we attach to the rope a simple device like a rappelling rack, which bears our weight on the way down. We loop the climbing rope into the rack, then secure the rack to our waist harness. Now for the hard part: we must work up the nerve to roll or leap off the branch. When we do, we free-fall a short distance (usually 0.5–2 m) until the rack's friction on the rope slows us. Thereafter, descent is effortless and fun. We float down as the climbing rope slithers through the rack at eye level.

Glenn Prestwich of the State University of New York (Stony Brook) had first-hand experience with the problems of treetop safety. Before his first descent, he triple-checked the rappelling rack to be sure he had strung the rope through it correctly. Unfortunately, he concentrated so intently on the rope itself that he forgot to attach the rack to his harness. Glenn made a death-defying leap, leaving the rack dangling from the rope without him.

"If I hadn't had gloves on, I'd have been in bad shape," Glenn told us. "I grabbed the rope as tightly as I could and slid 120 feet, hitting bottom fast." He hobbled off without internal injuries, although afterward he went into shock. Fourteen years later Glenn still has scars down his arms from the rope burns. We advocate careful training with professional arborists, mountaineering or caving clubs, or experienced researchers (see Laman, 1994).

C. Assembling the Apparatus

Among climbing techniques, the strategies for putting a climbing rope in place are especially varied. Heavy-gauge monofilament line is typically shot over the branch from below. For low branches, it may only be necessary to tie a stone to the line and toss it up. To reach higher limbs, biologists have used crossbows, hunting bows, slingshots, or line guns (Fig. 2). Whatever the method, the line is tied to a blunt arrow or lead weight heavy enough to propel it back to earth (Tucker and Powell, 1991). The thin line is then replaced by a parachute cord that can in turn hoist the relatively heavy climbing rope. Finally, a climbing rope is pulled over and lashed to a firm support at one end. A crossbow reportedly works best for positioning horizontal lines between trees in building walkways or canopy webs. The positioning of lines *must* be based on safety precautions, not on ease of rigging the branch.

A variant of the arborist's techniques has been devised by Dial and Tobin (1994) in which neither end of the rope is secured at the ground. The rope—which should be a brand made specifically for tree climbing—is tied into the waist harness, and the climber snaps the ascenders to the rope's *opposite* side. Because the branch over which the rope was thrown serves much like a pulley, the effort required for ascent can be considerably reduced. This works best on bare, relatively narrow branches, as these offer little resistance to rope movement.

The assembly of other methods, such as booms, cranes, and walkways, is discussed in Boxes 1–3, respectively. As it rarely takes more than 15 minutes to climb in and out of the canopy once the climbing apparatus is in place, continuous treetop visits of more than a few hours are seldom obligatory, except as a personal challenge. For protracted stays, however, a roof and space for food and toiletries may be desirable (Perry, 1986).

D. Spatial Coverage and Mobility

From a research perspective, the most crucial attributes of a climbing method are the volume and shape of the space one can enter and the permanency and mobility of the hardware.

Trunk-climbing methods, single rope techniques, and towers limit canopy access to a vertical transect line through the forest. But single rope techniques are also fairly flexible; for example, one can swing on a rope over to another tree trunk or limb, or reposition the rope once in the crown. Towers, however, can be erected above the height of the canopy or where no strong tree limbs exist to support a rope.

In contrast, walkways, bridges, and trams permit work along a horizontal transect line through the forest that can be extended indefinitely as the budget permits. They also offer vertical reach during ascent to the platform, and one can tie and descend on a rope anywhere along their span.

Solidly built towers have been the longest-lasting canopy structures, with some towers operational over many years (e.g., McClure, 1977). Built in 1958, Uganda's Haddow tower (Haddow *et al.,* 1961) still seems to be in reasonable shape (M. Moffett, personal observation), despite having been moved once and then abandoned years ago when the research projects on insect vectors ended. Some walkways and trams can likewise be classified as permanent, although they may require frequent inspection.

All climbing hardware requires regular inspection and replacement of parts as needed. Climbing ropes should never be left in the forest, but can be replaced with parachute cord between climbs to reduce exposure to the elements. Actually, ropes, ladders, or other seemingly temporary apparatuses can be climbed indefinitely without rerigging a tree if they are checked regularly and replaced as needed. Compared to a peconha, rope, or ladder, walkways and towers are awkward to disassemble and move to a new site, but it can be done given appropriate modular construction.

The most versatile methods encompass a larger volume of space, typically shaped like a wide cylinder. Of these, the simplest approach may be the boom, which can also be readily moved from one site to the next. The canopy web and canopy crane cover larger areas than a boom, but can be moved only with difficulty. A cherry picker tends to be restricted to transects along forest edges (chiefly roadsides) (e.g., Smith, 1968).

Most scientists on the French canopy raft investigate the ring of treetop foliage along the raft's perimeter. It is also possible to access any vertical transect down from the outer edge of the raft and from some points within the raft's central area by descending on a rope. The raft is most appropriate for studies that require at most a few days at any given site, after which it is moved for logistical and safety reasons and to avoid excessive damage to the tree crowns.

The most troublesome canopy areas to study are branch tips, whether laterally within tree crowns or at the zenith of the topmost trees. The importance of this twig layer cannot be overstated for here exist most of the resources that support the canopy ecosystem: the majority of flowers, fruits, and leaves, and, at the very top of the canopy, high solar radiation for plant growth. Climbing methods that access cylindrical volumes of canopy space reach these areas most effectively. The uppermost leaves generally present the greatest difficulty, and in this regard the newest climbing innovations— the raft, the sled, and the crane—are especially effective: most of the projects involving them have centered around the topmost foliage or the boughs immediately below.

One fortuitous option for accessing branch tips is to use canoes and motor boats in black- and white-water flooded forests bordering the Amazon River and its tributaries. When the floods crest, however, the water currents can make it taxing to stay in one place for long (Darlyne Murawski, personal communication). Because the flooding is annual, these forest canopies have unique ecological dynamics (Goulding, 1993).

Tree trunks present their own, albeit less demanding, challenges for access and study. Certain trunks may be impossible to examine by canopy crane, for example, without snapping twigs and branches as the gondola is lowered through the upper canopy. Perhaps just as troublesome, the peconha and other trunk-based methods may physically disturb organisms on the trunk.

With careful equipment maintenance, any climbing method permits round-the-clock treetop visits so that different people can work at the same place sequentially. Methods differ somewhat in the number of people that can be accommodated in the canopy at once (often this varies with the site; for example, given a wide platform or enough branches to sit on, several people can climb the same rope in succession to work together in a tree). The "carrying capacity" of scientists for different methods is becoming increasingly critical as research priorities shift from solitary endeavors to coordinated team projects (Moffett, 1993a). The raft operation is particularly ideal for collaborative studies (Hallé and Pascal, 1992).

E. Impact of Canopy Techniques on Ecosystems

The "perfect" climbing and research technique would enable ecologists to examine a selected canopy volume without disturbing it, except as required by experimental protocol. Yet working in tree-tops alters the environment: nailed steps deface the wood; ropes slung over branches scar bark and smash epiphytes; and as a canopy raft is lowered in place, it snaps branches and twigs beneath and stirs up the surrounding vegetation. Disturbances range from massive, long-term physical damage to subtle, momentary shifts.

Certain disturbances to the canopy may have little effect on research, as when twigs are broken during construction of an arboreal blind for a clear view of bird behavior. The appropriateness of a technique will depend on the species or canopy properties under study. For lack of information on how organisms respond to the equipment, a researcher must make intuitive decisions. For example, a walkway provides a connection between trees that had once been isolated, alters air flow patterns all around it, and partially fills some of the open spaces in the forest. A crane introduces sounds that may disturb some organisms, and the raft may alter light and moisture regimes. How profoundly might these changes affect the climbing or flying vertebrate and insect communities in that area?

The obtrusiveness of canopy techniques will become an increasingly important concern as the attention of canopy biologists turns to detailed studies of intact canopy structure. Rather than attempting to judge the different techniques here in relation to the magnitude of their potential disturbance—which would be premature—we describe key research directions that will be difficult to pursue without care in minimizing human impact on the canopy environment. Consider, for example, the dispersal of plants and animals in the canopy:

1. Movement Patterns in Airborne Organisms Ideally, we need to map the distribution of open spaces within the forest that are broad enough to accommodate, say, flying animals of a given wing spread, body configuration, and flight dynamics. As wing spread grows, how do the flight path options through the forest decline? For example, are large fliers channeled along wide horizontal corridors between tree strata? Are they largely confined furthermore to these specific heights by the scarcity of vertical passages bridging one stratum to the next? Or is forest stratification so ill-defined that large species are forced to fly over the canopy or at ground level, unless they can clamber across branches to get from one open pocket to the next within the trees? And how is the dispersal of the smallest airborne objects, such as spores and some insects, influenced by air currents in a forest?

2. Movement Patterns in Climbing Organisms For animals and plants (such as vines that rely on solid supports), the number of routes available through the forest must decline with increasing weight (as well as with how well the organisms cling to horizontal or vertical faces, spread their mass over several supports like a vine or snake, or bridge gaps). How does variation in architecture and woody structure (e.g., tensile strength) of the supporting plants affect route availability? And what of the position and spacing of trees and climbing plants relative to each other (e.g., crown shyness and stratification)?

We might assume that the optimal spatial scale for examining an organism's environment would increase with increasing organism size, but given the general rarity of most species of trees and other plants, and the extent of species-specific associations of insects and plants in the tropics (Erwin, 1991), even small insects may disperse long distances as part of their life-history strategies. Generalist insect species can be mobile as well. The classic example of long-distance locomotion in rain forest insects are euglossine bees that, as Janzen (1971) showed, can routinely travel kilometers between conspecific flowering plants; pollen dispersal in tropical plant species is dictated by such insects (see Chapter 19). We need to understand how insects travel through and orient long distances in the rain forest labyrinth. But what methods can we use to document such problems in an actual forest?

Broadening our questions from static forest composition to forest dynamics magnifies methodological problems manyfold. Even in the ongoing, two-dimensional, ground-based surveys such as the 50-ha plot of tree population dynamics on Barro Colorado Island, Panama (Condit *et al.,* 1992), it may be impossible to avoid having fragile tree seedlings crushed underfoot by the surveying team and other visitors—seedlings whose survivorship determines the distributions of the older size classes under survey. Similarly, within the tree crowns, each broken twig may alter the foraging patterns of ants, and each scuffed branch—potentially removing any soil deposited over years as traces by rain and mist (Nadkarni and Matelson, 1991)—may alter the local prospects for epiphyte survival. Although permanent constructions such as canopy cranes and walkways seem to be ideal for long-term projects, regardless of the climbing technique, strict protocols must be adopted. Otherwise repeated minor disturbances will culminate in substantial (but potentially unrecognized) changes to the canopy.

IV. Canopy Access Techniques and Future Research Directions

No single method is an ultimate solution for ecological studies of the canopy. Indeed, as canopy research protocols become increasingly demanding, more ecologists should probably stop thinking of the methods individually. Instead, emphasis should be changed to combining several methods for a more flexible approach. A good field base for canopy work would stock the most mobile climbing equipment, such as ropes and booms, plus ground-based equipment such as binoculars and pole pruners. These supplies can be used in conjunction with more permanent structures like towers, walkways, and cranes, chosen based on the horizontal and vertical coverage required for the projects and arranged in the forest to take best advantage of the local landscape. For example, do researchers need to reach the topmost veneer of leaves or even the open air above the canopy?

Today's canopy access techniques have opened an unparalleled biological frontier (Wilson, 1991; Moffett, 1993b; Lowman, 1995). Before the development of these techniques, as rain forest pedestrians, people were dazzled by the silhouettes of exotic vertebrates and herbage above. Now as climbers with the canopy at our fingertips, smaller organisms materialize before us to enrich our image of the canopy's lavish ecological tapestry. And if we seek out animalcules hidden from us by size, and larger beings tucked from view within canopy soil or behind leafy veils and palisades of bark and wood—what then? We will have barely scratched the surface of the canopy as the earth's grandest expression of organic life. In the next few years the energies of arboreal biologists will hopefully shift more and more from the problems of canopy access to those of data collection within the treetop's ecological labyrinth.

Acknowledgments

We thank National Geographic (both the Committee for Research and Exploration and the magazine) (M.M.) and the Australian Research Grants Scheme plus the Sydney Spelunking Society (M.L.) for the initial support that got our canopy experiences under way. Tim Laman, Darlyne Murawski, Laurie Burnham, and Beryl Black commented on the manuscript. Special thanks to Glenn Prestwich for allowing us to describe his canopy fall. We also pay special tribute to two tragically deceased men who were among the finest of all canopy biologists, Alan Smith and Alwyn Gentry. We gratefully recognize the encouragement they gave to our efforts; their thoughtful and meticulous comments have improved our canopy literature immeasurably over the years.

References

Becker, P. F., Erhart, D. W., and Smith, A. P. (1988). Analysis of forest light environments. I. Computerized estimation of solar radiation from hemispherical photographs. *Agric. For. Meteorol.* 44, 217–232.

Condit, R., Hubbell, S. P., and Foster, R. B. (1992). Short-tern dynamics of a neotropical forest: Change within limits. *BioScience* 42, 822–828.

Corner, E. J. H. (1992). "Botanical Monkeys." Pentland Press, Edinburgh.

DeVries, P. J. (1988). Stratification of fruit-feeding nymphalid butterflies in a Costa Rican rainforest. *J. Res. Lepid.* 26, 98–108.

Dial, R., and Tobin, C. (1994). Description of arborist methods for forest canopy access and movement. *Selbyana* 15(2), 24–37.

Erwin, T. L. (1989). Canopy arthropod diversity: A chronology of sampling techniques and results. *Rev. Peru. Entomol.* 32, 71–77.

Erwin, T. L. (1991). How many species are there? Revisited. *Conserv. Biol.* 5, 330–333.

Goulding, M. (1993). Flooded forests of the Amazon. *Sci. Am.* 266, 114–120.

Haddow, A. J., Corbet, P. S., and Gillett, J. D. (1961). Entomological studies from a tower in Mpanga forest, Uganda. *Trans. R. Entomol. Soc. London* 113, 249–256.

Hallé, F. (1990). A raft atop the rain forest. *Natl. Geogr.* 178, 129–138.

Hallé, F., and Pascal, O., eds. (1992). "Biologie d'Une Canopée de Forêt Equatoriale. II. Rapport de Mission: Radeau des Cimes Octobre/Novembre 1991. Reserve de Campo, Cameroun." Fondation Elf, Paris.

Hingston, R. W. G. (1932). "A Naturalist in the Guiana Forest." Longmans, Green, New York.

Janzen, D. H. (1971). Euglossine bees as long distance pollinators of tropical plants. *Science* 171, 203–205.

Kitching, R. L., Bergelson, J. M., Lowman, M. D., McIntyre, S., and Carruthers, G. (1993). The biodiversity of arthropods from Australian rainforest canopies: General introduction, methods, sites and ordinal results. *Aust. J. Ecol.* 18, 181–191.

Laman, T. G. (1994). Safety recommendations for climbing rain forest trees with single rope technique. *Biotropica* (in press).

Leighton, M., and Thomas, B. (1980). A canopy observation platform in East Kalimantan, Indonesia. *Flora Malesiana Bull.* 33, 3432–3434.

Leonard, R. E., and Eschner, A. R. (1968). A treetop tramway system for meteorological studies. *USDA For. Serv. Pap. NE NE-92.*

Lowman, M. D. (1988). Litterfall and leaf decay in three Australian rainforest formations. *J. Ecol.* 76, 451–465.

Lowman, M. D. (1995). "In the Treetops—Observations of a Canopy Biologist." Yale Univ. Press, New Haven, CT (in press).

Lowman, M. D., and Bouricius, B. (1993). Canopy walkways—Techniques for their design and construction. *Selbyana* 14, 4.

Lowman, M. D., and Heatwole, H. (1992). Spatial and temporal variability in defoliation of Australian eucalypts. *Ecology* 73, 129–142.

Lowman, M. D., and Moffett, M. W. (1993). The ecology of tropical rain forest canopies. *Trends Ecol. Evol.* 8, 104–107.

Lowman, M. D., Moffett, M. W., and Rinker, H. B. (1993). Insect sampling in forest canopies: A new method. *Selbyana* 14, 75–79.

Malcolm, J. R. (1991). Comparative abundances of neotropical small mammals by trap height. *J. Mammal.* 72, 188–192.

McClure, H. E. (1977). The secret life of a tree. *Anim. Kingdom* 80, 16–23.

Mitchell, A. W. (1981). "Operation Drake: Voyage of Discovery." Severn House, London.

Mitchell, A. W. (1982). "Reaching the Rain Forest Roof. A Handbook on Techniques of Access and Study in the Canopy." Leeds Philosophical and Literary Society/U.N.E.P, Oxford.

Moffett, M. W. (1992). Vorstoss in die Wipfel-Welt. *Geo,* June, pp. 18–34.

Moffett, M. W. (1993a). The tropical rainforest canopy: Researching a new frontier. *Selbyana* 14, 3–4.

Moffett, M. W. (1993b). "The High Frontier: Exploring the Tropical Rainforest Canopy." Harvard Univ. Press, Cambridge, MA.

Montgomery, G. G., Cochran, W. W., and Sunquist, M. E. (1973). Radiolocating arboreal vertebrates in tropical forest. *J. Wildl. Manage.* 37, 426–428.

Mori, S. A. (1984). Use of "Swiss tree grippers" for making botanical collections of tropical trees. *Biotropica* 16, 79–80.

Munn, C. A. (1991). Tropical canopy netting and shooting lines over tall trees. *J. Field Ornithol.* 62, 454–463.

Muul, I., and Lait, L. (1970). Vertical zonation in a tropical rain forest in Malaysia: A method of study. *Science* 169, 788–789.

Nadkarni, N. M. (1988). Use of a portable platform for observations of tropical forest canopy animals. *Biotropica* 20, 350–351.

Nadkarni, N. M., and Matelson, T. J. (1991). Fine litter dynamics within canopy of a tropical cloud forest. *Ecology* 72, 2071–2082.

O'Neill, T. (1993). New sensors eye the rain forest. *Natl. Geogr.* 184, 118–130.

Padgett, A., and Smith, B. (1987). "On Rope." National Speleological Society, Huntsville, AL.

Parker, G. G., Smith, A. P., and Hogan, K. P. (1992). Access to the upper forest canopy with a large tower crane. *BioScience* 42, 664–670.

Perry, D. R. (1978). A method of access into the crowns of emergent and canopy trees. *Biotropica* 10, 155–157.

Perry, D. R. (1986). "Life above the Jungle Floor." Simon & Schuster, New York.

Perry, D. R., and Williams, J. (1981). The tropical rain forest canopy: A method providing total access. *Biotropica* 13, 283–285.

Smith, N. G. (1968). The advantages of being parasitized. *Nature (London)* 219, 690–694.

Sutton, S. L. (1979). A portable light trap for studying insects of the upper canopy. *Brunei Mus. J.* 4, 156–160.

Tucker, G. F., and Powell, J. R. (1991). An improved canopy access technique. *North. J. Appl. For.* 8, 29–32.

Whitacre, D. F. (1981). Additional techniques and safety hints for climbing tall trees, and some additional equipment and information sources. *Biotropica* 13, 286–291.

Wilson, E. O. (1991). Rain forest canopy: The high frontier. *Natl. Geogr.* 180, 78–107.

Wolda, H. (1992). Trends in abundance of tropical forest insects. *Oecologia* 89, 47–52.

2

Canopy Architecture in Tropical Trees: A Pictorial Approach

Francis Hallé

I. Introduction

The study of tree architecture is a relatively recent discipline (Hallé and Oldeman, 1970, 1975; Oldeman, 1974; Hallé *et al.*, 1978) that first focused on humid tropical forests. Its purpose is to identify the endogenous processes controlling the growth and form of the whole tree. Architectural analysis is a dynamic approach to understanding whole-tree development; it considers the growth and branching of a tree species, using successive observations of tree form from seed germination to tree death. Several trees are observed and drawn at each stage of development: how many are observed at a given stage depends on the architectural complexity and flexibility of the species. Results are summarized in a series of diagrams that symbolize successive growth stages.

II. Architectural Model–Architectural Unit

The growth pattern that determines the successive architectural phases of a tree is called its architectural model, which is a genetically based ontogenetic program that dictates both the manner in which the plant elaborates its form and the resulting architecture. Identification of the relevant model is easy by observing simple morphological features of young trees prior to reiteration (see Section III). Determining features include: (1) presence or absence of aboveground vegetative branching; (2) orien-

Forest Canopies

27

Copyright © 1995 by Academic Press, Inc.
All rights of reproduction in any form reserved.

tation of vegetative axes, either vertical, horizontal, or intermediate; (3) absence of terminal bud scars due to continuous growth, or presence of terminal bud scars usually associated with rhythmic growth; (4) monopodial or sympodial branching; and (5) lateral sexuality allowing indeterminate growth, or terminal sexuality forcing the growth of axes to become determinate.

Each architectural model is defined by a combination of these morphological features. Although the number of combinations is theoretically very high, it appears that only thirty architectural models occur in nature. These models are realized in both tropical and temperate regions and in herbaceous, climbing, and arborescent plants. A fascinating aspect of tree form is that the same architectural model appears in plants belonging to unrelated taxa.

Figure 1 illustrates some of these architectural models, each of which is represented by the saplings of thousands of species. Growth patterns defined by the architectural models are genetically determined and a few architectural mutations have already been described (Hallé, 1978). Conversely, different models can be represented in the same genus of plants. Moreover, many forms are intermediate between two or three models, proving that the disjunction between the models, although real, is not absolute. Along this architectural continuum, the models represent the forms that are likely to have the highest fitness and most complete stability (Barthélemy *et al.*, 1991).

The morphological characters used to delimit and define models are general ones. The concept of the architectural model (Hallé and Oldeman, 1970, 1975) was intended to emphasize the common features that unite the growth patterns of unrelated plants; therefore, these models are unifying concepts, but they do not provide an exact description of the actual plant growth. A more precise description is often needed. If one wishes to emphasize the morphological peculiarities of a plant species, the branching order and a precise description of every branch order should be quantified. Every branch order has its own leaf arrangement, length and diameter, longevity, share in total photosynthetic activity, and sexuality. For a given plant, the specific expression of its model is called its architectural unit (Caraglio and Edelin, 1990; Edelin, 1991). Figures 2a and 2b give an example of two plants having the same architectural model but having quite distinct architectural units.

III. Reiteration

Some trees conform to one single architectural unit throughout their life span, but this is exceptional (Fig. 3). In most dicotyledonous trees, particu-

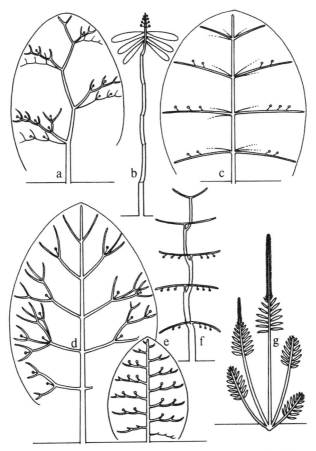

Figure 1 Architecture models. (a) Koriba's model exemplified by the balsa tree, *Ochroma lagopus* Swartz (Bombacaceae, South America). (b) Chamberlain's model in the broken-bones tree, *Oroxylum indicum* Vent. (Bignoniaceae, Southeast Asia). (c) Massart's model in *Lepionurus sylvestris* Bl. (Opiliaceae, Indonesia). (d) Scarrone's model in *Gardenia imperialis*. K. Schum. (Rubiaceae, tropical Africa). (e) Petit's model in *Leptaulus daphnoides* Benth. (Icacinaceae, tropical Africa). (f) Nozeran's model in *Geissospermum sericeum* (Sagot) Benth. (Apocynaceae, Guyanas). (g) Tomlinson's model exemplified by *Lobelia gibberoa* Hemsley (Lobeliaceae, montane forests of East Africa).

larly canopy species, the fundamental architecture is repeated (i.e., reiterated) during ontogenesis (Oldeman, 1974). Reiteration is a process through which the organism duplicates its own elementary architecture, that is, its architectural unit. As a result of this process, the tree can be considered a colony (Oldeman, 1974; Hallé *et al.*, 1978). Without examples of free-living individuals, the concept of coloniality would have no basis; the trees shown in Fig. 2 are such individuals.

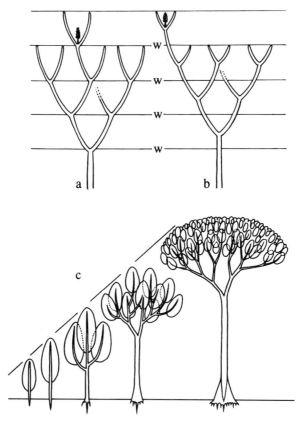

Figure 2 Architectural unit and reiteration. (a, b) Two plants having the same architecture model but having distinct architectural units. *Rhus* sp. (a) and *Cornus* sp. (b) belong to Leeuwenberg's model. In *Rhus*, apical flowering occurs at the end of the growing season and proleptic branching produces renewal shoots. In *Cornus*, apical flowering and sylleptic branching both occur in the middle of the growing season. w = winter. Prolepsis is the discontinuous development of a lateral from a terminal meristem to establish a branch, with an intervening period of rest of the lateral meristem. Branches so developed are referred to as proleptic branches. Syllepsis is the continuous development of a lateral from a terminal meristem to establish a branch, without an intervening period of rest of the lateral meristem. Branches so developed are referred to as sylleptic branches. (c) General picture of the reiteration process occurring during the growth of a large dicotyledonous tree. (Modified from Hallé *et al.*, 1978.)

Viewing a tree as a colony, rather than an individual, is not a new idea. From the beginning of the eighteenth century, botanists have known that a single plant is "not a simple unit but a collection of subunitary parts," in the words of White (1979), from which I took many of the following historical facts. As early as 1708, de la Hire compared the tree with a colony, the bud with a seed, and the flushing of this bud with germination. In his view, the root of the germinating bud worms its way between wood and bark of

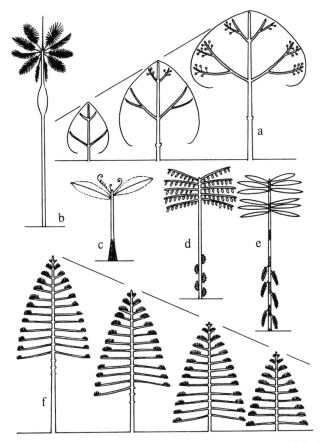

Figure 3   Trees having a strict model organization without reiteration. (a) A pioneer tree, such as *Macaranga* sp. (Euphorbiaceae, Southeast Asia). (b) A palm; in the whole Arecaceae family, most of the trees are devoid of the capability of reiteration. (c) A tree fern, 10 m high (Ivory Coast). (d) *Unonopsis stipitata* (Annonaceae); a treelet, 7 m high, from the Peruvian rain forest. (e) *Trichoscypha ferruginea* Engl. (Anacardiaceae, Gabon); a fruit tree of the forest undergrowth. (f) *Araucaria columnaris* (Forst.) Hook. (Araucariaceae); a large coniferous tree from New Caledonia.

the supporting tree. Bradley (1721) wrote: "The twigs and branches of trees are really so many plants growing upon one another"; von Goethe (1790): "The lateral branches which spring from the nodes of plants may be regarded as individual plantlets which take their stand upon the body of the mother, just as the latter is fixed in the earth"; and according to Erasmus Darwin (1800):

> If a bud be torn from the branch of a tree and cut out and planted in the earth . . . , or if it be inserted into the bark of another tree, it will grow, and become a plant in every respect like its parent. This evinces that every bud of a tree is an individual vegetable

being; and that a tree therefore is a family or swarm of individual plants like the polypus, with its growing young out of its sides, or like the branching cells of the coral-insect.

The polyp–plant analogy of Erasmus Darwin was used by his grandson, but in reverse, when he wrote of corals: "In these compound animals . . . surprising as this union of separate individuals in a common stock must always appear, every tree displays the same fact, for buds must be considered individual plants" (Charles Darwin, 1839).

The concept of reiteration, if not the word itself, appeared in Braun's (1855a, b; 1856) writings: "The sight of the most ramified plant-stocks, especially by a tree . . . excite the presentiment that this is not one single being . . . comparable with the animal or human individual, but rather a world of united individuals which have sprung from each other." In complete convergence with these opinions, the architectural studies developed since 1970 have confirmed that coloniality exists in trees and should be considered a normal corollary of the sessile way of life. Therefore, (1) most of the trees in a tropical rain forest are colonies and (2) within a tree, the elementary individual is not a bud, but an architectural unit.

A general picture of the reiteration process occurring during the growth of a large tree is given in Fig. 2c. Figure 4 illustrates specific development and crown construction in some tropical trees. Edelin (1990) has demonstrated that the reiteration process is a spontaneous event occurring after a definite threshold of differentiation during the normal development of a tree. Therefore, in a tropical rain forest canopy, most of the large trees are reiterated.

As the tree grows, the number of reiterated units tends to increase, but their size tends to diminish and ultimately only parts of the architectural unit are reiterated; this is called "partial reiteration." During the growth of trees, there is a progressive reduction of size and simplification of structure of reiterated architectural units (Fig. 5). Barthélémy (1988, 1991) described this reduction of the reiterated architectural unit during the tree growth as being merely vegetative and counterbalanced by a greater abundance of flowers. He coined the term "minimal unit" to describe the smallest and simplest reiterated units forming the peripheral portion of a mature crown.

With the canopy-raft technique (Hallé and Blanc, 1990; Hallé and Pascal, 1992), it is easy to land on forest canopies. Blanc (1992) described populations of minimal units at the very top of high tropical trees: "Any visitor reaching our platform for the first time and discovering the forest canopy is struck by an unexpected and rather 'non-exotic' landscape. Foliage is dense everywhere, the countless small twigs bear medium-sized coriaceous leaves all alike; and the hilly landscape reminds him of the dense bushy vegetation of mountains or mediterranean scrublands" (Blanc, 1992).

The following description of the minimal unit is from observations made

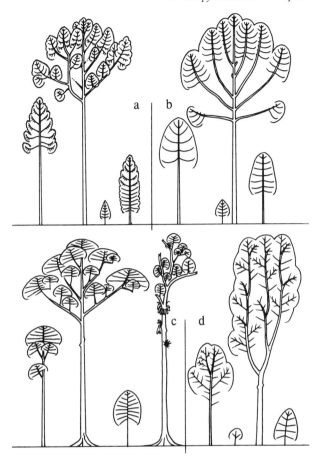

Figure 4 The reiteration process and crown construction. (a) Developmental sequence of *Agathis dammara* (Lamb.) Rich. (Araucariaceae), a large coniferous tree of Southeast Asia (Sumatra to New Guinea). (b) *Dipterocarpus costulatus* V. Sl. (Dipterocarpaceae, Southeast Asia). (c) *Shorea stenoptera* Burck. (Dipterocarpaceae, Southeast Asia); Roux's model is quite clear. (d) *Acacia auriculiformis* A. Cunn. ex Benth. (Mimosaceae, northern Australia). (Modified from Edelin, 1990. Reprinted by permission of John Wiley & Sons, Inc., from *Physiology of Trees* (A. S. Raghavendra, ed.), pp. 1–20. Copyright © 1991 John Wiley & Sons, Inc.)

and data collected during the canopy-raft missions in Guyana and Cameroon. Although based on a limited number of species, this description is a starting point for further morphological studies in tree crowns. Minimal units are small leafy twigs, 5 to 30 cm long; usually the more vertical the growth orientation of the twig, the longer the unit. Hallé (1986a, b) suggested that the simpler units could phylogenetically evolve into "growth units" of rhythmically growing axes; this could be the origin of at least some cases of endogenous rhythmic growth in the trees of the humid tropics.

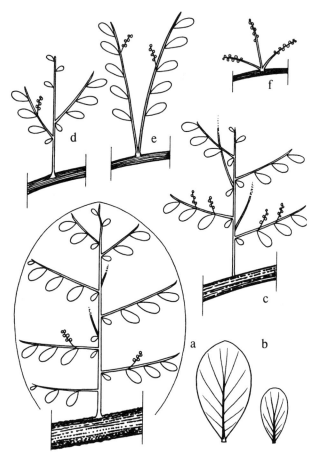

Figure 5 Miniaturization of the architectural unit (I). (a) *Combretocarpus rotundatus* (Rhizo-phoraceae) is a large tree of the swamp forest in Sumatra. On a large horizontal branch, the reiterated architectural unit clearly expresses Roux's model, with a vertical trunk and horizon-tal branches. (b) The leaves are much bigger on the branches than on the trunk, allowing easy recognition of the two kinds of axes. (c, d) On smaller horizontal limbs, the units become miniaturized; Roux's model can still be seen. (e) The trunk has vanished, and the architectural unit is made up of branches only. (f) On very thin horizontal branches, the minimal unit (Barthélémy, 1988, 1991) becomes a group of inflorescences.

Although the episodic initiation of sets of minimal units is widespread in tropical canopies (Blanc, 1992), variation exists between tree species. For example, in *Desbordesia glaucescens* (Irvingiaceae, Cameroon), the resting and growing phases are synchronized between trees within the same area. In *Dialium pachyphyllum* (Caesalpiniaceae, Cameroon), the adult tree crown is subdivided into two to four subcrowns that are not synchronized; one of

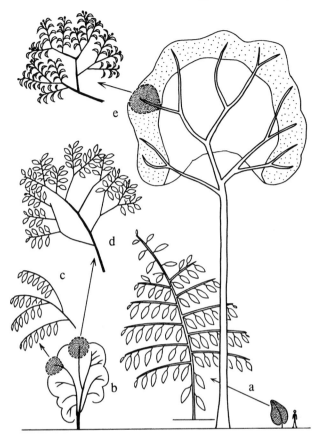

Figure 6   Miniaturization of the agricultural unit (II). (a) *Sacoglottis gabonensis* (Humiriaceae) is a large tree of the African rain forest. The young plant displays an architecture close to Roux's model, with continuous growth and sylleptic branching. (b–d) When the young tree is approximately 10 m high, sylleptic branching stops and the reduced architectural units proceed from each other by proleptic branching. (e) In the crown of large trees, architectural units are short, equivalent to each other, and bear flowers or fruits. (Modified from Blanc, 1992.)

them may be resting or flowering, while another one is growing new leaves. In *Couepia* sp. (Chrysobalanaceae, Guyana) or *Eperua falcata* (Caesalpiniaceae, Guyana), growth is generally synchronized within the crown, but minimal units at the top of the crown tend to grow more actively. Such variation between the phenology of different species is important to the biology of the tree with respect to sexual reproduction, level of herbivory, and fruit dispersal (Charles-Dominique *et al.*, 1981; Gautier-Hion and Michaloud, 1989).

Patterns of flowering are also diverse. In *Sacoglottis gabonensis* (Humiriaceae, Cameroon), minimal units bear axillary inflorescences so that flowers are produced among the leaves (Fig. 6). More often (e.g., *Taralea oppositifolia*, Fabaceae, Guyana; *Eschweilera* sp., Lecythidaceae, Guyana), the inflorescences are terminal, leafy units alternate in time with inflorescences, and the whole crown is either green or blooming. The apical meristem of the minimal unit often dries up or aborts.

Branching is proleptic; this is probably because growing meristems at the top of tall trees may desiccate, particularly in the microclimate of the canopy (relative atmospheric humidity drops from 100% at night to 40% at noon during sunny days; Salager *et al.*, 1990). The degree of branching of the minimal unit is related to its orientation. In many tree species, the more vertical the minimal unit, the more it will branch. In contrast, a horizontal unit usually does not branch. In such cases, it will certainly be shed as it has no secondary growth and is shaded by the rest of the crown. Moreover, downward shedding of branches is active and results in various levels of crown-shyness (Fig. 7).

The uniformity among the minimal units that comprise a forest canopy is striking. Not only do those units display a high degree of equivalence within one crown, but to a large extent they are also similar or even equivalent between the crowns of various species in the forest canopy (Leigh, 1990). The morphodiversity of a forest evolves thus: (1) in the tropical rain forest understory, most of the known architectural models and a wide array of foliar morphologies, life-forms, and biological adaptations are present; in this permissive environment, the morphodiversity is high; and (2) during tree ascendance into the canopy, and throughout the process of reiteration, the architectural units decrease in size, become simplified and unbranched, and turn into minimal units. Therefore the morphodiversity diminishes as the bioclimate conditions become harsh when approaching the upper canopy.

IV. Prospects for Canopy Architecture

Elucidating the growth and form of trees opens up prospects for pure and applied research, for example, crop monitoring in fruit trees (de Reffye *et al.*, 1988a; Lauri, 1991, 1992; Nicolini, 1991; Costes, 1993), pruning or use of suckers for air layering (Aumeeruddy and Pinglo, 1989), surveys of tree pathology related to architecture (de Reffye, 1976), wood mechanics (de Reffye, 1976; Fisher and Stevenson, 1981; Hallé *et al.*, 1992), agroforestry (Michon *et al.*, 1983), landscaping and horticulture (Caraglio and Edelin, 1990; Barthélémy *et al.*, 1989, 1992; Donoghue, 1981), growth quantification and computer modeling (de Reffye and Snoeck, 1976; de Reffye *et*

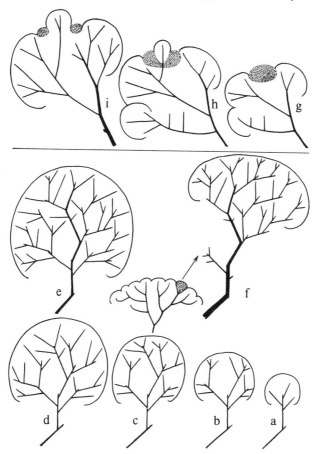

Figure 7 Canopy architectural dynamics. (a–d) The growth of a crownlet accumulates minimal units through proleptic branching. The more vertical the minimal unit, the more branched it is. (e–f) Secondary growth and downward abscission produce large branches. (g–i) Reiteration allows a crownlet to outgrow an invading liana. (Modified from Blanc, 1992.)

al., 1986, 1988b, 1989a, b, c, 1991; Lauri and Térouanne, 1991; Fisher and Hibbs, 1982; Fisher, 1992), studies of relationships between trees and lianas (Bordenave, 1990; Caballé *et al.,* 1992) or trees and strangling figs (Michaloud and Michaloud-Pelletier, 1987), herbivory (Bongers *et al.,* 1990), genetics and improvement (Nozeran *et al.,* 1982; Thiébaut, 1982; Thiébaut *et al.,* 1985, 1988), phylogeny (Banchilhon, 1969, 1971; Nozeran *et al.,* 1984), and paleobotany (Bateman and Dimichele, 1993).

Among these prospects, I discuss an ongoing study related to the water supply of reiterated architectural units. From the origin of the concept of reiteration (Oldeman, 1974), reiterated units have merely been considered as leafy branch systems; the question of their root systems had never been posed (Figs. 8 and 9). The architectural unit has a root system when it originates from the seed. Assuming that reiteration exactly copies the architectural blueprint, there is no reason why the reiterated architectural unit should not also possess its own root system. Hallé (1991) posed the hypothesis that the bole is made up of the aggregated root systems of all the reiterated units forming the tree crown. It may then be inferred that the trunk wood in a reiterated tree would be of root origin; here the term "root" has its physiological sense: an organ or a tissue providing water and mineral supply to the reiterated units. Concerning the anatomical sense of the term "root," three assumptions should be put forward, which could alleviate the somewhat heretical character of the hypothesis:

(1) Whereas root and stem are easy to differentiate on the basis of the anatomy of primary tissues, the wood of a root and the wood of a stem are sometimes identical within a single species, for example, *Ficus benjamina* L. (Moraceae) (Zimmerman *et al.,* 1968).

(2) In several tropical tree species, the trunk is irrefutably made of roots. For some tropical trees, both the diametric growth of the trunk and its mechanical stability are provided by roots (Fig. 8).

(3) The root system of a reiterated unit is sometimes visible. This often happens in the shady and wet environment of the rain forest understory, for example, on fallen trunks (Fig. 9d). Nadkarni's (1981) findings are of critical importance in this context; at the canopy level, beneath the thick mats of accumulated organic material and epiphytes covering their boles and branches, tropical trees put forth adventitious roots. Nadkarni (1981) coined the term "canopy roots" and explained that through an "extensive network of canopy roots, [trees] gain access to the arboreal nutrient sources generated and retained by the epiphytes." She reported canopy roots in the crowns of more than 20 species of tropical trees in the montane forests of Costa Rica and Hawaii, including *Metrosideros* (Myrtaceae), *Cheirodendron* and *Didymopanax* (Araliaceae), *Ocotea* (Lauraceae), *Xylosma* (Flacourtiaceae), and *Weinmannia* (Cunoniaceae). I suggest that canopy roots are part of the root systems of reiterated architectural units.

For a given architectural unit, the orientation of the parent axis controls the visibility of its root system (Fig. 9). On a vertical trunk this root system is usually not visible, but it becomes evident on horizontal branches. The most interesting cases are observed on leaning axes. In *Gnetum gnemon* (Gnetaceae, Indonesia), *Schefflera venulosa* (Araliaceae, Indonesia), *Maclura pomifera* (Moraceae, southern United States), and others, architectural units sprouting on leaning axes display a kind of "root system," hidden between the wood and the secondary phloem, or embedded within the

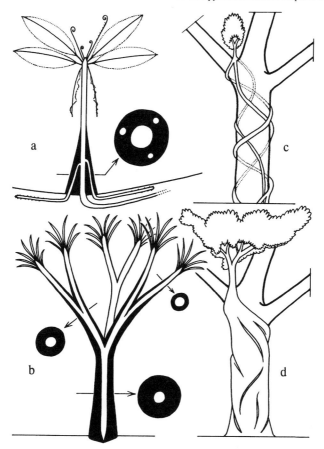

Figure 8 Some examples of trees having trunks made of roots. (a) *Cyathea manniana* Hooker (Cyatheaceae), a large tree fern from the montane forests of West Africa. Roots (in black) ensure diametric growth and mechanical stability. (b) *Vellozia flavicans* Mart. (Velloziaceae), a small monocotyledonous treelet of the Brazilian "cerrado." Roots (in black) cover the stems. (c, d) Two stages in the life history of a strangling fig (*Ficus* sp., Moraceae). The fig tree germinates in a fork of the host tree; its roots grow downward and fuse together. After the death of the host tree, the fig tree stands on a trunk that is made of roots. Reprinted from Hallé, 1991, with permission.

tissues of the bark. The developmental anatomy of this "root system" is presently being studied in several tree species (Grzeskowiak, personal communication).

In *Taralea oppositifolia* (Fabaceae, Guyana), a fallen trunk bore some large suckers, or reiterated architectural units (Blanc and Hallé, 1990; Hallé, 1991). At the base of the reiterated unit, horizontal roots were visible, to-

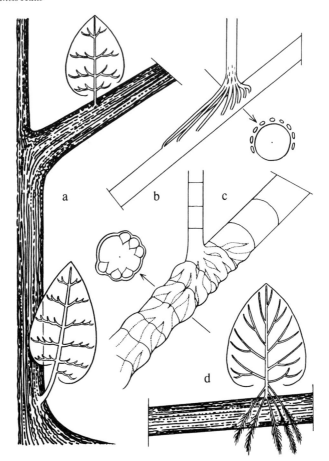

Figure 9   The root system of reiterated units may become visible under certain conditions. (a) With most tree species, the root system of reiterated units is not visible. (b) *Schefflera venulosa* (Araliaceae), an epiphytic shrub from Indonesia. A reiterated unit, arising from a leaning axis, shows a kind of "root system" embedded within the tissues of the bark. Here the bark was removed and the woody parts were retained. (c) *Gnetum gnemon* (Gnetaceae), a fruit tree from Indonesian agroforests. On a leaning trunk, the reiterated unit develops a "root system" between the wood and the bark. (d) On a fallen trunk, in the shady and wet environment of the rain forest understory, the root systems of reiterated units are often visible. Reprinted from Hallé, 1991, with permission.

gether with a thin layer of wood covering the wood of the fallen trunk. This layer originated at the same time as the sucker itself, and grew downward. Because it conferred its mechanical stability to the sucker and provided a water supply, the layer fulfilled the function of a taproot.

If the *Taralea* had not fallen, instead of producing suckers from its trunk, it may have produced many minimal units from its thin branches at the crown periphery. Nevertheless, whatever the spatial distribution of the reiterated units, the point is that these units produced a layer of downward-growing wood ensuring their water supply and having the function of taproots. Hypothetically, the trunk wood itself could be constructed by the aggregation of such layers. Although anatomical analysis and physiological experimentation are needed to validate this hypothesis, its agreement with botanical thought dating back to the early eighteenth century and its relation to recent progress in tropical botany is noteworthy.

V. Summary

This chapter comprises a brief survey, gathering what is known of canopy architecture. The concepts of architectural unit (i.e., the architecture of a particular species) and reiteration deserve attention in the canopy context and they are discussed with some detail; in contrast, architectural models are not emphasized because they are easier to observe in the forest understory than at the canopy level. The root system of reiterated architectural units is hypothetically and pictorially presented in a way that may be somewhat heretical to North American readers.

Interested and somewhat polyglot readers could plunge deeper into the subject of tree architecture by consulting the following list of references.

References

Aumeeruddy, Y., and Pinglo, F. (1989). "Phytopractices in Tropical Regions. A Preliminary Survey of Traditional Crop Improvement Techniques." UNESCO, Paris.

Bancilhon, L. (1969). Etude expérimentale de la morphogenèse et plus spécialement de la floraison d'un groupe de *Phyllanthus* (Euphorbiacées) à rameaux dimorphes. *Ann. Sci. Nat., Bot. Biol. Veg.* [10], 2, 127–224.

Bancilhon, L. (1971). Contribution à l'étude taxonomique du genre *Phyllantus*. *Boissiera* 18, 1–81.

Barthélémy, D. (1988). Architecture et sexualité chez quelques plantes tropicales: le concept de floraison automatique. Thèse de Doctorat, Université de Montpellier II, Montpellier, France.

Barthélémy, D. (1991). Levels of organization and repetition phenomena in seed plants. *Acta Biotheor.* 39, 309–323.

Barthélémy, D., Edelin, C., and Hallé, F. (1989). Some architectural aspects of tree ageing. *Ann. Sci. For.* 46, Suppl., 194–198.

Barthélémy, D., Edelin, C., and Hallé, F. (1991). Canopy architecture. *In* "Physiology of Trees" (A. S. Raghavendra, ed.), pp. 1–20. Wiley, New York.

Barthélémy, D., Caraglio, Y., Drénou, C., and Figureau, C. (1992). Architecture et sénescence des arbres., *Forêt-entreprise* 83, 15–36.

Bateman, R. M., and Dimichele, W. A. (1993). Saltational evolution of form in vascular plants: A neoGoldschmidtian synthesis. *Bot. J. Linn. Soc.* (in press).

Blanc, P. (1992). Comment poussent les couronnes d'arbres dans la canopée? *In* "Biologie d'une Canopée de Forêt Equatoriale. *II.* Rapport de Mission: Radeau des Cimes Octobre/ Novembre 1991. Réserve de Campo, Cameroun" (F. Hallé and O. Pascal, eds.), pp. 155–172. Fondation Elf, Paris.

Blanc, P., and Hallé, F. (1990). Timidité et multiplication végétative d'un arbre guyanais: *Taralea oppositifolia* Aublet. (Légumineuse—Papilionaceae). *In* "Biologie d'une Canopée de Forêt Equatoriale. Rapport de Mission: Radeau des Cimes Octobre–Novembre 1989. Guyane Française" (F. Hallé and P. Blanc, eds.), pp. 125–135. Xylochimie, Paris.

Bongers, F., van der Meer, P. J., Oldeman, R. A. A., Schalk, B., and Sterck, F. J. (1990). Levels of herbivory in a tropical rainforest canopy in french Guyana. *In* "Biologie d'une Canopée de Forêt Equatoriale. Rapport de Mission: Radeau des Cimes Octobre–Novembre 1989. Guyane Française" (F. Hallé and P. Blanc, eds.), pp. 166–174. Xylochimie, Paris.

Bordenave, B. (1990). Observations sur la dynamique de croissance d'une liane de la canopée de forêt Guyanaise. *In* "Biologie d'une Canopée de Forêt Equatoriale. Rapport de Mission: Radeau des Cimes Octobre–Novembre 1989. Guyane Française" (F. Hallé and P. Blanc, eds.), pp. 115–124. Xylochimie, Paris.

Bradley, R. (1721). "A Philosophical Account of the Works of Nature." Mears, London.

Braun, A. (1855a). The vegetable individual in its relation to species. Part I. *Am. J. Sci. Arts* 19, 233–256.

Braun, A. (1855b). The vegetable individual in its relation to species. Part II. *Am. J. Sci. Arts* 20, 181–201.

Braun, A. (1856). The vegetable individual in its relation to species. Part III. *Am. J. Sci. Arts* 21, 58–79.

Caballé, G., Coudurier, T., Berger, A., and Salager, J. L. (1992). Flux de sève et conditions climatiques chez les lianes ligneuses forestières. *In* "Biologie d'une Canopée de Forêt Equatoriale. II. Rapport de Mission: Radeau des Cimes Octobre/Novembre 1991. Réserve de Campo, Cameroun" (F. Hallé and O. Pascal, eds.), pp. 185–192. Fondation Elf, Paris.

Caraglio, Y., and Edelin, C. (1990). Architecture et dynamique de croissance du platane. *Platanus hybrida* (Platanaceae) (syn. *Platanus acerifolia* Aiton) Willd. *Bull. Soc, Bot. Fr., Lett. Bot.* 137, 279–291.

Charles-Dominique, P., Atramentowicz, M., Charles-Dominique, M., Gérard, H., Hladik, A., Hladik, C. M., and Prévost, M. F. (1981). Les mammifères frugivores arboricoles nocturnes d'une forêt guyanaise: inter-relations plantes–animaux. *Rev. Ecol. Terre Vie* 35, 341–435.

Costes, E. (1993). L'architecture aérienne de l'abricotier en développement libre. *Acta Bot. Gallica* 140, 249–261.

Darwin, C. (1839). "Journal of Researches into Geology and Natural History of the Various Countries Visited by H. M. S. Beagle, under the Command of Captain Fitz Roy from 1832 to 1836." London.

Darwin, E. (1800). "Phytologia; or the Philosophy of Agriculture and Gardening." Johnson, London.

de la Hire, M. (1708). Explication physique de la direction verticale et naturelle des tiges des plantes des branches des arbres, et de leurs racines. *Hist. Acad. R. Sci., Mém. Math. Phys.,* pp. 231–235.

de Reffye, P. (1976). Modélisation et simulation de la verse du caféier, à l'aide de la résistance des matériaux. *Café, Cacao, Thé* 20, 251–272.

de Reffye, P. and Snoeck, J. (1976). Modèle mathématique de base pour l'étude et la simulation de la croissance et l'architecture de *Coffea robusta. Café, Cacao, Thé* 20, 11–32.

de Reffye, P., Edelin, C., Jaeger, M., and Cabart, C. (1986). Simulation de l'architecture des arbres. *C. Colloq. Int. Arbre,* Montpellier, 1985, *Naturalia Monspeliensia (Montpellier, France), Hors Sér.,* pp. 223–240.

de Reffye, P., Cognée, M., Jaeger, M., and Traoré, B. (1988a). Modélisation stochastique de la croissance et de l'architecture du cotonnier. 1. Tiges principales et branches fructifères primaires. *Coton Fibres Trop.* 43, 269–291.

de Reffye, P., Edelin, C., Françon, J. F., and Jaeger, M. (1988b). Plant models faithful to botanical structure and development. *Comput. Graph.* 22, 15.

de Reffye, P., Edelin, C., and Jaeger, M. (1989a). Modélisation de la croissance des plantes. *La Recherche* 20, 158–168.

de Reffye, P., Edelin, C., and Jaeger, M. (1989b). Computer simuliert Pflanzenwachstum: Die grüne Zeitmashine. *Bild Wiss.* 8, 46–52.

de Reffye, P., Lecoustre, R., Edelin, C., and Dinouard, P. (1989c). Modelling plant growth and architecture. *In* "Cell to Cell Signalling: From Experiments to Theoretical Models" (A. Goldbeter, ed.), pp. 237–246. Academic Press, London.

de Reffye, P., Elguero, E., and Costes, E. (1991). Growth units construction in trees: A stochastic approach. *Acta Biotheor.* 39, 325–342.

Donoghue, M. (1981). Growth patterns in woody plants with examples from the genus *Viburnum. Arnoldia* 41, 2–23.

Edelin, C. (1990). "The Monopodial Architecture: The Case of Some Tree Species from Tropical Asia," Research Pamphlet N. 105. Forest Research Institute, Kuala Lumpur, Malaysia.

Edelin, C., ed. (1991). L'Arbre: Biologie et Développement, Actes du 2e Colloque International. Montpellier, 10–15 Septembre 1990. *Naturalia Monspeliensia (Montpellier, France),* Hors Sér., A7.

Fisher, J. B. (1992). How predictive are computer simulations of tree architecture? *Int. J. Plant Sci.* 153, 137–146.

Fisher, J. B., and Hibbs, D. E. (1982). Plasticity of tree architecture: Specific and ecological variations found in Aubreville's model. *Am. J. Bot.* 69, 690–702.

Fisher, J. B., and Stevenson, J. W. (1981). Occurrence of reaction wood in branches of dicotyledons and its role in tree architecture. *Bot. Gaz. (Chicago)* 142, 82–95.

Gautier-Hion, A., and Michaloud, G. (1989). Are figs always keystone resources for tropical frugivorous vertebrates? A test in Gabon. *Ecology* 70, 1826–1833.

Hallé, F. (1978). Architectural variation at the specific level in tropical trees. *In* "Tropical Trees as Living Systems" (P. B. Tomlinson and M. H. Zimmermann, eds.), pp. 209–221. Cambridge Univ. Press, Cambridge, UK.

Hallé, F. (1986a). Deux stratégies pour l'arborescence: gigantisme et répétition. *C. R. Colloq. Int. Arbre,* Montpellier, *1985, Naturalia Monspeliensia (Montpellier, France), Hors Sér.* pp. 159–170.

Hallé, F. (1986b). Modular growth in seed plants. *In* "The Growth and Form of Modular Organisms" (J. L. Harper, ed.), pp. 77–87. Philosophical Transactions of the Royal Society, London.

Hallé, F. (1991). Le bois constituant un tronc peut-il être de nature racinaire? Une hypothèse. *Actes Colloq. Int. Arbre: Biol. Dév. 2nd,* Montpellier, 1990, *Naturalia Monspeliensia (Montpellier, France), Hors Sér.,* A7, pp. 97–111.

Hallé, F., and Blanc, P., eds. (1990). "Biologie d'une Canopée de Forêt Equatoriale. Rapport de Mission: Radeau des Cimes Octobre–Novembre 1989. Guyane Française". Xylochimie, Paris.

Hallé, F., and Oldeman, R. A. A. (1970). "Essai sur l'Architecture et la Dynamique de Croissance des Arbres Tropicaux," Monogr. Bot. Biol. Vég., N. 6. Masson, Paris.

Hallé, F., and Oldeman, R. A. A. (1975). "An Essay on the Architecture and Dynamics of Growth of Tropical Trees." Penerbit Universiti Malaya, Kuala Lumpur, Malaysia.

Hallé, F., and Pascal, O., eds. (1992). "Biologie d'une Canopée de Forêt Equatoriale. II. Rapport de Mission: Radeau des Cimes Octobre/Novembre 1991. Réserve de Campo, Cameroun." Fondation Elf, Paris.

Hallé, F., Oldeman, R. A. A., and Tomlinson, P. B. (1978). "Tropical Trees and Forests: An Architectural Analysis." Springer-Verlag, Berlin.

Hallé, F., Chanson, B., Delavault, O., and Fournier, M. (1992). Déformations de maturation du bois et connexions vasculaires des complexes réitérés avec leurs axes porteurs. *In* "Biologie d'une Canopée de Forêt Equatoriale. II. Rapport de Mission: Radeau des Cimes Octobre/Novembre 1991. Réserve de Campo, Cameroun" (F. Hallé and O. Pascal, eds.), pp. 173–183. Fondation Elf, Paris.

Lauri, P. E. (1991). Données sur l'évolution de la ramification et de la floraison du pêcher (*Prunus persica* (L.) Batsch) au cours de sa croissance. *Ann. Sci. Nat., Bot. Biol. Veg.* [12] 11, 95–103.

Lauri, P. E. (1992). Données sur le contexte végétatif lié à la floraison chez le cerisier (*Prunus avium*). *Can. J. Bot.* 70, 1848–1859.

Lauri, P. E., and Térouanne, E. (1991). Eléments pour une approache morphométrique de la croissance végétale et de la floraison: le cas d'espèces tropicales du modèle de Leeuwenberg. *Can. J. Bot.* 69, 2095–2112.

Leigh, E. G., Jr. (1990). Tree shape and leaf arrangement: A quantitative comparison of montane forests with emphasis on Malaysia and South India. *In* "Conservation in Developing Countries: Problems and Prospects" (D. C. Daniel and J. S. Serrao, eds.), pp. 119–174. Oxford Univ. Press, Bombay.

Michaloud, G., and Michaloud-Pelletier, S. (1987). Ficus hémi-épiphytes (Moraceae) et arbres supports. *Biotropica* 19, 125–136.

Michon, G., Bompard, J. M., Hecketsweiler, P., and Ducatillion, C. (1983). Tropical forest architectural analysis as applied to agroforests in the humid tropics: The example of traditional village-agroforests in West Java. *Agrofor. Syst.* 1, 117–129.

Nadkarni, N. M. (1981). Canopy roots: Convergent evolution in rain forest nutrient cycles. *Science* 214, 1023–1024.

Nicolini, E. (1991). Premières données sur l'architecture des genres *Citrus et Poncirus* (Rutaceae). *Fruits* 46, 653–669.

Nozeran, R., Ducreux, G., and Rossignol-Bancilhon, L. (1982). Réflexions sur les problèmes de rajeunissement chez les végétaux. *Bull. Soc. Bot. Fr. Lett. Bot.* 129, 107–130.

Nozeran, R., Rossignol-Bancilhon, L., and Mangenot, G. (1984). Les recherches sur le genre *Phyllanthus* (Euphorbiaceae): Acquis et perspectives. *Bot. Helv.* 94, 199–233.

Oldeman, R. A. A. (1974). "L'Architecture de la Forêt Guyanaise," Mém. ORSTOM, N. 73. ORSTOM, Paris.

Salager, J. L., Salager, D., and Roy, J. (1990). Quelques données microclimatiques sur l'environnement forestier tropical. *In* "Biologie d'une Canopée de Forêt Equatoriale. Rapport de Mission: Radeau des Cimes Octobre–Novembre 1989. Guyane Française" (F. Hallé and P. Blanc, eds.), pp. 44–53. Xylochimie, Paris.

Thiébaut, B. (1982). Observations sur le développement de plantules de Hêtres (*Fagus sylvatica*) cultivées en pépinière, orthotropie et plagiotropie. *Can. J. Bot.* 60, 1292–1303.

Thiébaut, B., Cuguen, J., and Dupré, S. (1985). Architecture des jeunes hêtres *Fagus sylvatica*. *Can. J. Bot.* 63, 2100–2110.

Thiébaut, B., Comps, B., and Teissier du Cros, E. (1988). Développement des axes des arbres: pousse annuelle, syllepsie et prolepsie. *Can. J. Bot.* 68, 202–211.

von Goethe, J. W. (1790). "Die Metamorphose der Pflanzen zu Erklären."

White, J. (1979). The plant as a metapopulation. *Annu. Rev. Ecol. Syst.* 10, 109–145.

Zimmerman, M. H., Wardrop, A. B., and Tomlinson, P. B. (1968). Tension wood in aerial roots of *Ficus benjamina* L. *Wood Sci. Techn.* 2, 95–104.

3

Physical Mechanisms of Heat and Mass Exchange between Forests and the Atmosphere

David R. Fitzjarrald and Kathleen E. Moore

I. Introduction

Trees must exchange water, carbon dioxide, and energy with the atmosphere to survive. Because they are efficient sources of water vapor and can be significant sinks of carbon, forest canopies are an important component of surface control on climate from the very local to global scales. Biologists must consider aspects of canopy microclimate that are usually ignored by climatologists. Rich species diversity found in many forests is related in part to the diurnal course of temperature and humidity at different levels in the canopy. Tall canopies create a microclimate that differs from the above-canopy environment to a degree that depends on the vertical distribution of biomass. Micrometeorological techniques, inherently nonintrusive and amenable to automatic measurement, are useful to determine quantities such as water use efficiency, net carbon fixation, pollutant deposition, and spore dispersal. In this chapter, we examine some of the processes that act to control energy and mass transports within and above forest canopies from the point of view of atmospheric scientists. We examine features of the "accepted wisdom" for describing canopy–atmosphere interactions, assess current revisions, and comment on prospects for future advances. Our intent is to offer the canopy researcher a glimpse of the current state of physical understanding of transport in canopies. Those interested in the history of the topic should consult Monteith and Unsworth (1990) or Hutchinson

Copyright © 1995 by Academic Press, Inc.
All rights of reproduction in any form reserved.

and Hicks (1985). We concentrate on observations made in or just above forest canopies and illustrate with examples from our research.

The canopy appears to the meteorologist as a porous volume in which the leaf and branch structures lead to radiation and momentum absorption, processes that modulate trace gas evolution. The microclimate of the canopy includes not only its current state (e.g., temperature, humidity), but also the components of the energy budget that combine to alter its state. Profiles of temperature, humidity, and wind speed are part of the description of distinct ecological niches whose physical location may shift vertically, diurnally, or seasonally. The micrometeorological observational approach yields estimates of the heat and water vapor exchanges averaged spatially over a weighted upwind sector of influence known as the "tower footprint" and temporally over periods from 30 minutes to days to seasons. In a typical daytime situation, the peak in the weighting may lie within 300 m of a tower, but a "tail" of diminishingly important areas can extend up to several kilometers (Horst and Weil, 1992).

Ecologists have used leaf- or twig-scale measurements on mature trees to estimate physiological functions, (e.g., stomatal conductance, CO_2 exchange). They have observed stomatal aperture dependence on incident light and saturation deficit near the leaf. There are sometimes discrepancies between the extrapolated leaf-scale measurements and whole-canopy results obtained using micrometeorological techniques (Jarvis and McNaughton, 1986). The difficulty in scaling up small-scale measurements to apply to the entire canopy reflects the fact that the canopy environment is determined by the cooperative action of biological and physical influences. The near-leaf environment in part reflects the accumulated effects of turbulent air motions that mix air from above the forest into the canopy. However, these same air motions within and just above the canopy depend in part on canopy structure.

To describe bulk exchanges of mass or energy with the atmosphere above, meteorologists have treated the forest as equivalent to an ordinary flat surface that is simply displaced upward. However, some features of canopy wind flows and turbulent exchanges are not explained by this standard approach. These "anomalous" features are common to turbulent flow over many types of "rough-wall turbulent flows" originally studied in laboratory wind tunnel investigations. Deviations from the flat-plane boundary layer characteristics above the canopy can extend vertically more than three canopy thicknesses. This has important implications for how flux observations are interpreted and how canopies are represented in atmospheric models.

Many important biological processes are dependent on aspects of the physical canopy environment. Physical and biological controls on canopy microclimate are complementary. Seed and pollen dispersal depend heavily on the turbulent wind field, which is itself influenced by the biomass profile. Fungal and bacterial pathogen epidemiology depends on both the mean moisture state and the rate of canopy drying (Huber and Gillespie, 1992). Microclimate influences CO_2 exchange. Infrequent strong gusts above some threshold may be necessary for seed dispersal. Diurnal changes in the relative humidity within the canopy may relate to vertical motions of insects and their entrainment into the wind above the canopy. Seeking to understand how turbulent motions distribute heat and mass in vegetation is a goal common to canopy researchers, foresters, meteorologists, and fluid dynamicists.

II. Profiles and Fluxes within and Just above the Canopy

Radiation is absorbed or scattered when it passes through the canopy. A useful approximation is that a constant fraction is absorbed each unit length, leading to exponential light decay with increasing cumulative leaf area. Details of the canopy radiation environment are presented elsewhere (Fox, 1985; Shuttleworth, 1989; Chapter 4). The rate of decay of radiative energy with distance into the canopy (the radiative flux divergence) causes heating of the biomass and interstitial air. This pushes the upper canopy toward a stable density stratification (dense cooler air below warm lighter air) and partially isolates the lower canopy from the atmosphere above. At night, the canopy interferes with the loss of longwave radiation inside the canopy, but appreciable radiative cooling occurs at the canopy top. On clear nights, this can lead to strong stable layers near the top of the canopy. Vertical air motion is suppressed both by the presence of the canopy and by the stability of the air. Both effects must be overcome by energy in turbulent motions to ventilate the lower canopy. The vertical temperature and humidity differences that result from the balance of the stability-producing and -destroying mechanisms provide a spectrum of different living conditions for forest organisms.

The decoupling of canopy air from the atmosphere above is most pronounced for canopies with thick foliage near the canopy top, such as tropical rain forests or midlatitude deciduous forests, with leaf area indices (LAI) from 3 to 6. Examples from the Amazon rain forest (Fig. 1) illustrate the diurnal changes in density stratification, humidity, and wind speed caused by such a thick canopy. We follow meteorological practice by using the virtual potential temperature, Θ_v, as an indicator of air density. The value of Θ_v depends on the temperature, humidity, and atmospheric pressure: $\Theta_v = T(P_0/P)^{0.286} (1 + 0.61r)$, where T is the temperature, r the mixing ratio of water vapor, P_0, a reference pressure, and P the atmospheric

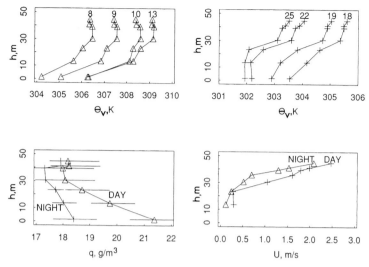

Figure 1 Mean profiles of virtual potential temperature, Θ_v (K), absolute humidity, q (g/ m^{-3}), and wind speed, U (m s^{-1}) within the Amazon rain forest canopy near Manaus, Brazil (2°57′S, 59°57′W). The Θ_v profiles are identified by the hour of day over which they were averaged (12 = 1200; 25 = 0100). From Fitzjarrald *et al.* (1990), copyright by the American Geophysical Union.

pressure (Stull, 1988). Because air density is proportional to the reciprocal of Θ_v, an inversion ($\Delta\Theta_v/\Delta z > 0$) indicates a statically stable situation. In such a situation, air is most dense at the bottom of a layer, and any air displaced vertically tends to return to its initial level. A convective situation ($\Delta\Theta_v/\Delta z < 0$) occurs when a surface is heated and buoyant plumes rise and produce convective mixing.

The seasonal course of a midlatitude deciduous forest presents an interesting annual "experiment" on the importance of canopy foliage on microclimate. We present results from unique long-term observations we are conducting with colleagues from Harvard University at a midlatitude oak– maple deciduous forest site (LAI ≈ 3) in central Massachusetts (Harvard Forest, 42°36′N, 72°14′W). The canopy is approximately 20 m high. Details of the site are in Wofsy *et al.* (1993). Winter–summer differences in the midday Θ_v profile at Harvard Forest (Fig. 2) vividly illustrate the role of leaves in determining the stability of the air in the canopy. In winter at noon, the warmest level is at the forest floor, but the warmest level is displaced to the forest top at noon in summer. Weaker temperature inversions are set up in this midlatitude forest canopy as compared to the rain forest due to the smaller LAI, different canopy architecture, and lower sun angles at the midlatitude site.

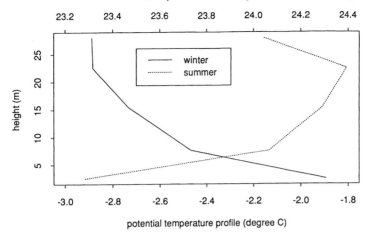

Figure 2 Typical noontime mean potential temperature (Θ, C) profiles in a midlatitude deciduous forest during summer and winter (Harvard Forest, Massachusetts, 42°36′ N, 72°14′W). From Lu and Fitzjarrald (1994). Reprinted by permission of Kluwer Academic Publishers.

Onset of the growing season alters the energy balance within and just above the canopy (Fig. 3). We present the sensible heat flux emanating from the top of the canopy on clear days (H1, 30 m), at 11 m ($\approx$ 50% of canopy height, H2), and at 6.2 m (H3). Net radiation at the canopy top on clear days follows solar elevation. Sensible heat flux from the northeastern forest peaks in late spring (a fact familiar to glider pilots). Heat flux into the soil is greatest at about the same time of year. A qualitative estimate of canopy cover from digital video images (Fig. 3) shows that leaf-out occurs during only a very few days. Leaf-out leads to an abrupt drop in sensible heat flux above the canopy and, by considering the difference between the available energy and that used for sensible heat, to the expected strong jump in evapotranspiration. The leaves shade the forest floor and the sensible heat flux there drops precipitously. These examples illustrate how the canopy structure leads to consequent atmospheric structure.

Mean wind profiles inside and just above the forest canopies are often similar to the rain forest case in Fig. 1, though a secondary wind maximum sometimes appears in the lower canopy if there is little understory. The wind profile in the upper canopy and lower atmosphere for a variety of situations exhibits an approximately sigmoid shape (Raupach *et al.*, 1991). To describe the lower part of the sigmoid, Cionco (1972) fitted the expression $U = U_h \exp[a(z/h - 1)]$ to the in-canopy horizontal wind profile, where U_h is the wind speed at canopy top h and a is an empirical "canopy index." Average vertical wind fluctuations (σ_w) inside the canopy damp out quickly below the canopy top. Work by Shaw and Zhang (1992) indicated that pres-

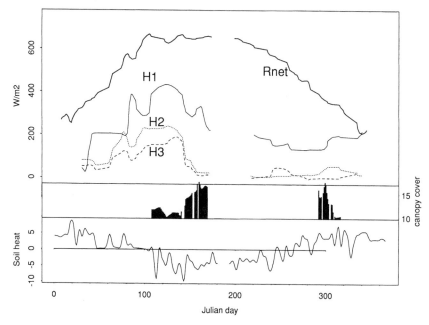

Figure 3 Seasonal course of thermal balance components (W m^{-2}) at Harvard Forest on clear days: net radiation (Rnet) above the canopy, sensible heat flux at $z \approx 1.5h$ (H1), at $z \approx 0.5h$ (H2), and at $z \approx 0.2h$ (H3) for 1992. Center panel shows qualitative estimate of canopy cover during leaf-out and leaf-fall periods made with video images. Bottom panel shows flux of heat (W m^{-2}) into the soil.

sure gradients associated with large eddies passing above the forest can induce motions deep inside the canopy that could also affect mean profiles. Well above the canopy, the wind profile returns to the logarithmic shape $[U \approx \ln(z - d) + \text{constant}, d$ a "displacement height"$]$ that is well known over flat terrain (Fazu and Schwerdtfeger, 1989).

Consistent with mean wind profile observations, the turbulent transport of horizontal momentum typically decays very rapidly as one enters the canopy. In a midlatitude deciduous forest, momentum flux (the rate at which the atmosphere loses momentum to the canopy) was found to be 10% of its above-canopy value at $z/h \approx 0.6$ (Shaw *et al.*, 1988). The degree to which the canopy is connected to the atmosphere above varies with time of day. The drag coefficient C_d (dimensionless), the ratio of vertical momentum transport to the canopy to the mean horizontal momentum in the flow, is a measure of momentum coupling efficiency at the canopy top. Typical diurnal variation of C_d above the rain forest (Fig. 4) shows that the coupling is 5–6 times more efficient during the day than at night. After it rains, the canopy can be "slick" for a time until water on leaves evaporates (Fig. 3). A

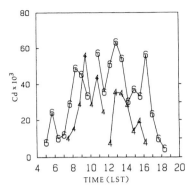

Figure 4 Diurnal variability of the drag coefficient just above the Amazon rain forest canopy on two separate days (July 24 and 26, 1986, represented by 4 and 6, respectively). Note the abrupt drop in C_d near noon on July 24 due to cooling and stability resulting from a rain shower. From Fitzjarrald *et al.* (1988), copyright by the American Geophysical Union.

similar daytime drag coefficient over short vegetation, for example, would be approximately 10 times smaller (Fitzjarrald and Moore, 1992). Turbulence properties of a plane surface layer occur first from two to three canopy heights above the sparse subarctic lichen woodland canopy (e.g., Fitzjarrald and Moore, 1994).

Many aspects of the turbulence environment just above the canopy are observed to resemble the findings of tests with small-scale models in wind tunnels (Raupach *et al.*, 1991). However, it is difficult to simulate the stability effects inside the forest canopy in a wind tunnel. Canopy researchers and modelers will continue to rely on field observations to provide new information.

Observed scalar fluxes are approximately constant with height just above the canopy, consistent with the fact that no sources or sinks are there. The fluxes can change abruptly with z/h within the canopy. The form of the flux profile depends on the canopy structure and the source or sink levels. Most within-canopy scalar flux observations are for sensible heat and water vapor. A complication is that the source and sink levels for CO_2, water vapor, momentum, and radiation are usually at different heights. This is an unavoidable difficulty that puts some limits on our ability to infer the transport of one substance by analogy with another.

Jarvis and McNaughton (1986) invented a "coupling coefficient" Ω, from 0 to 1, to express how closely the saturation deficit near the leaves is coupled to the air outside the leaf boundary layer. ($\Omega \to 1$ is the decoupled limit.) Black and Kelliher (1989) found that the understory contribution can account for 20–50% of total stand evapotranspiration in pine forests, somewhat higher than we infer from the Harvard Forest results in Fig. 1. Black and Kelliher estimated Ω in one conifer stand to be 0.5–0.6, indicating that turbulent transport was keeping the canopy dry. The detailed definition of Ω has been questioned recently by the authors themselves (McNaughton and Jarvis, 1991), but the concept is still useful. Because instrumentation limitations have only recently been overcome, relatively few studies have been done that include fluxes and mean gradients at many levels of the canopy.

For selected forest types and for limited periods of time, the environment in the canopy has been well-observed. Major unknowns include how to characterize biomass structure and how to get sufficient observations to scale up local measurements to apply to the whole forest. Indirect estimates of LAI using hemispherical photographs or light sensors (e.g., Chen *et al.*, 1991) must be complemented by detailed studies of time changes in the vertical biomass profile (Parker *et al.*, 1989). Moreover, average quantities give an incomplete picture of how canopy–atmosphere exchanges occur.

III. Relating Mean Conditions and Fluxes

How a homogeneous canopy as a whole functions is revealed by considering the fluxes of mass and energy at its upper and lower boundaries. Until recently, it has been much easier to measure mean gradients of quantities than their turbulent fluxes directly. Thus, empirical relationships between fluxes and gradients just above forests have been of great importance. Such flux–gradient relations are required to incorporate the lower boundary condition (fluxes) for a prognostic variable (mean quantity) in computer simulations of air flow over canopies. In this section, we discuss the origins and limitations of a conceptual model that linearly relates fluxes to gradients. This model is so widely used that observations are routinely compared from one site to another using its parameters.

Near a homogeneous flat surface there can be no mean vertical air transport—where would the air go? Most transport of any quantity X (e.g., CO_2, water vapor, or another scalar) must be effected by irregular turbulent gusts, known as eddies. The turbulent flux is a statistical quantity that expresses transports associated with transient gusts. Average profiles of quantity X change with the difference of the turbulent fluxes (F_x) across the volume, which we call the *turbulent flux divergence*. This divergence can occur by action of forcing such as absorption of radiation or by chemical reaction. Mathematically, we write $dX/dt = -\partial F_X/\partial z + S_X$. The first term on the right is the turbulent flux divergence and the second term represents any local source or sink of X. We have made the common assumption that the most important fluxes are in the vertical (z) direction. In many cases, however, one must go beyond a description of mean conditions within the canopy and study the eddies that accomplish transport and mixing within the canopy. Such work may be important to canopy researchers, who, for example, may be interested in spore or insect transport. These topics have usually not been the primary focus of micrometeorologists, who tend to concentrate on energy or CO_2 budgets.

Turbulent diffusion is commonly described through analogy with molecular diffusion. Such an analogy is fundamentally incorrect, as turbulent diffusion depends on the flow rather than the fluid considered, but useful empirical estimates of an equivalent turbulent diffusivity have been the mainstay of micrometeorology. This analogy can only explain direct "down-gradient" transport; fluxes go from high to low concentrations, just as would be expected in molecular diffusion. A common way to express this is $F_X = -K (X_2-X_1)/\Delta z$, where X_2 at level $z+\Delta z$ and X_1 at level z are values of any scalar quantity and K is an exchange coefficient. Linear flux–gradient relations for turbulent transport are often referred to as "K-theories." Invoking an analogy with Ohm's law, the flux–gradient relation is sometimes written as $F_X = (X_1-X_2)/r$, where r is an exchange resistance.

Above a plane surface, the size of near-surface turbulent eddies is observed to be proportional to their height above the surface. Flux–gradient relations for momentum and scalars in this case have been understood in the framework of the Monin–Obukhov hypothesis, the standard micrometeorological model (Panofsky and Dutton, 1984). Because of their simplicity, K-theory models are often applied to canopies. According to this hypothesis, profiles of scalars and horizontal wind in the layer just above a flat surface are approximately logarithmic with height. The profile of any quantity X is written $X \approx M(\ln(z/z_o) + \Psi(z/L))$, where the factor M depends directly on the surface flux of X and Ψ is a function to correct for profile changes related to buoyant air motions (known as thermals) near the ground. The function $\Psi(z/L)$ depends on the surface heat and momentum fluxes (through the Monin–Obukhov length L). The roughness length z_o is related to the size of obstacles near the surface (Panofsky and Dutton, 1984; Monteith and Unsworth, 1990). To accommodate closed canopies, the vertical coordinate is simply shifted by displacement height d, the hypothetical mean level of momentum absorption in the canopy. The displacement height d and roughness length z_o, though quite useful theoretically, are not known precisely. They are often estimated using curve fits to mean profiles on logarithmic plots, an inherently inaccurate technique. Estimates of $d \approx 0.6-0.7h$ have been made for temperate deciduous forests, increasing to $d \approx 0.9h$ in denser tropical forests (Shuttleworth, 1989).

The standard description also applies to gradients of quantities above canopies, but only above an intermediate "roughness sublayer," up to three to four canopy heights (Raupach *et al.*, 1991). In the roughness sublayer, fluxes are constant with height but the flux–gradient relations and turbulence properties do not quite follow the standard hypothesis, even when one allows for the displacement height. That the flux–profile relations Ψ are not the same just above rough forests as they are over flat surfaces has been referred to as the forest "anomaly" (Raupach, 1979). Fazu and Schwerdtfeger (1989) showed that exchange processes approach those of a plane surface layer two to four canopy heights above savannah scrubland in Australia. Fitzjarrald and Moore (1994) found that turbulence properties appropriate to a plane surface layer occurred at 4.6 canopy heights above an open lichen spruce woodland canopy in the subarctic. The forest anomaly appears to be related to the fact that eddies accomplishing the flux (see the following) are much larger than their height above the canopy, in contrast to turbulent flows over flat surfaces.

Momentum transport to a canopy depends on the roughness of the surface. The variation of z_o with tree density and canopy height is one way to distinguish a collection of trees from a forest. When canopies are sparse, one must consider each tree as a "roughness element." Laboratory and field studies indicate that the scaled roughness length z_o/h (where h is the height of the roughness elements) initially increases with increasing roughness element density λ (total roughness frontal area per element/ground area per element; Raupach *et al.*, 1991). As the density of trees in a canopy increases beyond $\lambda \approx 0.2$, each tree is in the "momentum shadow" of its neighbor, and z_o/h begins to decrease with λ. At this point, the canopy physically becomes a "forest," not simply a collection of trees.

Understanding physical transport mechanisms within the canopy strains conventional micrometeorological hypotheses. Heat transport in forest canopies sometimes moves "up the gradient," opposite to naive expectation (e.g., Denmead and Bradley, 1985). To students of turbulent transport, this phenomenon was not as great a surprise as it apparently was to forest canopy researchers. Relating flux to the local gradient of a quantity only makes sense if the characteristic size of the mixing element (the turbulent eddy) is small relative to changes in the mean gradient in question. Eddies that mix down into the canopy are largely determined by phenomena above the canopy. Intense but infrequent events may accomplish much of the mixing. During relatively long intervals between large events, for example, the mean temperature structure can recover. When the turbulent mixing occurs in sporadic events, the mean profile may have little to do with the turbulent transports. A large mixing event that would encounter a temperature inversion such as in Fig. 1 could lead to upward fluxes inside the canopy, a direction opposite to that predicted by the K-theory.

Raupach (1989) demonstrated that "up-gradient" mixing of scalars in plant canopies can be understood using elements of standard Lagrangian diffusion analysis. He treated the canopy as a collection of "chimneys" emitting the substance in question. He showed how linear K-theory is adequate for diffusion far from the source, but upward transport "against" the gradient is a property of near-field diffusion. One limitation is that some properties of the turbulent wind field [$U(z)$, $\sigma_w(z)$, among others] must be postulated *a priori*. Though Raupach applied the Lagrangian analysis beyond its formal range of validity, this technique explains many features of the forest anomaly.

Because of a widespread desire to believe in the simple diffusion analogy, or for reasons of computational efficiency, efforts to reinvigorate the K-theories to describe canopy exchange continue. Exchanges from an entire canopy to the atmosphere are sometimes idealized as being similar to a single "Big Leaf," in which an entire canopy is idealized to behave as does a single leaf. This is a conceptual resistance analogy incorporating the net effects of stomatal control (r_s), molecular diffusion very near the leaf (r_b), and turbulent transport (r_a) as resistors in series (Monteith and Unsworth, 1990). In this concept, stomata with saturated air at leaf temperature are linked via r_s to a molecular diffusive sublayer whose properties are described by r_b. It has had such wide appeal that observations of fluxes of water vapor and other trace gases from canopies have been presented by specifying the values of these resistors (F. M. Kelliher, R. Luning, and E.-D. Schulze, unpublished data, 1993). In diagnostic work, r_a can be directly determined from momentum flux observations. The resistance to molecular diffusion within millimeters of the leaf, r_b, must be found assuming that the sum of all the boundary layers very near leaves behaves as would a molecular boundary layer above a single large flat plate (Hicks *et al.*, 1987). The accuracy of the commonly used formulas is under question. The canopy resistance (r_s), which can be found as a residual after r_b and r_a are estimated, also carries their errors of estimation. A related approach for finding r_s is to consider relations that partition net radiation (Monteith and Unsworth, 1990). Embedded in this formalism is the idea that all leaves contribute to the canopy exchange process in a linear manner. Major difficulties in using the Big Leaf approach involve how to specify "surface," atmospheric reference temperatures, and the resistances themselves. Much of what is unknown in forest–atmosphere interaction is hidden in the resistors. There are also some practical limitations, for example, evaporation is only a weak function of the resistance (Raupach, 1990). This means that it can be extremely difficult to determine the canopy resistance r_c accurately from observations.

Profiles of temperature, wind speed, and other scalars within canopies have been simulated for transport models using networks of resistors to account for multiple layers (e.g., Baldocchi, *et al.*, 1987). Resistors are thought to account for stomatal control and turbulent exchange at many levels in a canopy. Such models have been criticized (Raupach, 1990) because it may not be possible to obtain sufficiently detailed observations to test model assumptions even at sophisticated field sites. The situation becomes more tenuous if the measurements must be extrapolated to represent large grid areas typically used in numerical models. If only bulk heat and moisture flux estimates for large areas are required, single Big Leaf models are just as effective as the multiple layer approach (Raupach and Finnigan, 1988; Raupach, 1990; McNaughton and Jarvis, 1991). Moore *et al.* (1993) discussed observational limitations to commonly used regional extrapolation techniques. To understand canopy microclimate, however, multiple layers in some form are essential.

IV. Estimating Turbulent Fluxes

We present an elementary review of flux measurement techniques. Refer to Lenschow (1993) for a more detailed technical discussion. In the laboratory or over short vegetation, one can estimate some surface scalar fluxes by putting a chamber on the ground, enclosing the vegetation (Hutchinson and Livingston, 1993), and observing concentration changes with time. Assessing sample representativity is a difficult task in chamber work (Denmead and Raupach, 1993). Except for studies of forest floor emissions, chambers are not very practical for forest work. The chamber would have to be enormous, and it would be difficult to maintain the canopy microclimate close to its natural state. An ambitious effort to measure rain forest metabolism and evapotranspiration was done in Puerto Rico during the 1960s (Odum and Jordan, 1970). A giant cylinder formed by plastic sheet enclosed a patch of forest 21 m high and 16 m wide. Transpiration and carbon uptake were estimated, but the observations were confined to the lower canopy. There was no clear way to assess how the altered air flow (from the plastic sheet) changed the results. Woodwell and Dykeman (1966) inferred forest respiration rates by observing nocturnal CO_2 buildup inside the canopy, effectively using the canopy as a large (albeit "leaky") chamber. Fitzjarrald and Moore (1990) pointed out that nocturnal turbulent events lead to CO_2 emissions from the top of this box that should also be considered (Fig. 5). Fan *et al.* (1990) studied CO_2 exchange in the Amazon rain forest and found that the buildup of CO_2 within the canopy and the flux through the forest top must be measured to understand net ecosystem uptake and respiration rate. At Harvard Forest, S. C. Wofsy (personal communication) reported that up to 80% of the reported respiration rate (Wofsy *et al.*, 1993) can be accounted for through CO_2 buildup in the canopy. Forest canopy activity has also been monitored by observing sap flow rates (Schulze *et al.*, 1985). Average estimates of evapotranspiration using sap flow observations compared well with those made using the eddy cor-

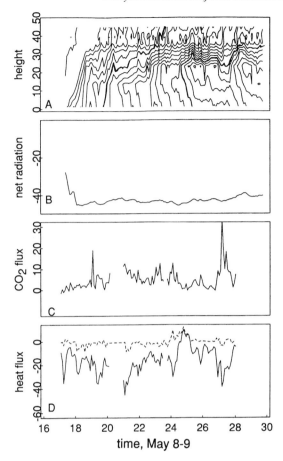

Figure 5 Structure and time series in the Amazon rain forest canopy at night. (A) Time–height section of Θ_v. For hours after midnight, 26 = 03 on the following day. The 301 K and 303 K isotherms are bold. (B) Net radiation (W m^{-2}). (C) CO_2 flux (ppm m s^{-1}). (D) Sensible heat flux at 39 m ($z \approx h$, solid) and at 23 m ($z \approx 0.5h$, dashed). From Fitzjarrald and Moore (1990), copyright by the American Geophysical Union.

relation technique; more than half of the sap flux in an experimental stand of 14 trees occurred in only three emergent trees (Köstner *et al.*, 1992).

Evaporation, sensible heat flux, CO_2 flux, and other scalar fluxes can be estimated using nonintrusive meteorological techniques. Three main approaches are used: (a) inference of fluxes from profile measurements; (b) for sensible heat and evaporation, indirect methods that rely on partitioning net radiation into Q_H and Q_E; and (c) direct flux estimation using the eddy correlation method.

Indirect measures of canopy–atmosphere exchange have been used for some time. These methods have relied on the basic energy balance relation

$$Q_* = Q_H + Q_E + Q_G + \text{canopy heat storage} \qquad (1)$$

in which Q_* is the net radiation, Q_H is the turbulent sensible heat flux, Q_E is the latent heat flux, and Q_G is the ground heat flux. In addition to (1), assumptions about the relationship of the fluxes to the gradients can be used to infer the fluxes from the more easily made gradient measurements, or if transfer coefficients (i.e., resistances) can be estimated, the Penman–Monteith relation (Monteith and Unsworth, 1990) can be used to estimate the latent and sensible heat fluxes. The Penman–Monteith technique was developed for use when limited data are available; it partitions the "available energy" $(Q_* - Q_G)$ into Q_H and Q_E and is referred to as a residual method. It requires the following assumptions: (a) linear flux–gradient relations are valid; exchange coefficients are assumed to be known, usually expressed as resistances; (b) errors of any type in estimation of the available energy are negligible; and (c) horizontal advection of heat or moisture is either known or negligible. Shuttleworth (1989) noted that tropical forests and midlatitude forests were quite similar from the point of view of displacement heights and exchange resistances. Residual methods will continue to be of great importance to climatologists, atmospheric modelers, and micrometeorologists. To understand the mixing phenomena themselves, however, one must consider more detailed observations.

Direct measures of the exchange of heat, moisture, momentum, and trace gases such as CO_2 are available through the eddy correlation method, in which very rapid (typically 10 Hz) measurements of the wind, temperature, humidity, and other scalars are made. Until recently most studies used data taken only for relatively short periods of time (an hour up to a few selected days). A few studies have extended these periods to years (Kaimal and Gaynor, 1983, for physical quantities; Wofsy *et al.*, 1993, for CO_2). In the eddy correlation method, vertical diffusion in turbulent air occurs when vertical motions are correlated with concentration or energy perturbations. The product of fluctuations of a quantity c' with the vertical component of the wind w', when averaged over a half hour or so, yields the turbulent flux of the quantity. One does not try to describe individual motions, but is content with a statistical result. For example, when upward-moving air is warmer on average than downward-moving air, vertical turbulent sensible heat flux results. It often surprises newcomers to the technique that the correlation coefficient between w and T, for example, r_{wT}, rarely exceeds 0.5 in turbulent air but is highly significant both in the statistical sense and in terms of its importance to the energy budget. Net flux of a quantity is found by calculating the covariance between vertical velocity and concentration perturbations, $F_x = \overline{w'X'} = r_{wX}\,\sigma_w\sigma_X$, where the overbar represents a time average, σ is the standard deviation, and r is a correlation coefficient. If one averages long enough, covariance statistics should not depend on the observation point if the canopy top is approximately uniform horizontally.

How long one must average data to get a good flux estimate and which area of the canopy the flux estimate represents are not fixed quantities. Each depends on the character of the turbulent environment in which the measurements are made. Detailed specification of the time required to achieve a good eddy flux estimate depends on the characteristic time scale of the turbulence (Tennekes and Lumley, 1972). This time scale finds the length of time one must sample to observe a single turbulent mixing "event." For the turbulent flux to be found accurately one must have an adequate statistical sample of events. This is achieved by averaging 10–20 of these turbulent mixing events. In a typical surface layer, it takes 20–30 minutes to achieve a good eddy flux estimate. In most cases, this averaging interval is short enough that diurnal variations can be adequately resolved. Eddy flux estimates obtained for shorter intervals are statistically uncertain and of limited utility.

The surface area of canopy represented by an eddy flux, the "footprint" of the observation, varies with time of day and wind direction. Estimates of this footprint have only been done with the aid of transport models. These models suffer from the same limitations of the Lagrangian diffusion technique. One must specify the mean and turbulent wind field to obtain the footprint estimate. To date, a footprint estimate that takes into account the effect of the roughness sublayer and possibility of diffusion out of the canopy has not been done, though the result is expected to be broadly similar to that obtained for plane surfaces. For example, a tower that extends 9 m above a plane surface with roughness length 0.06 m and displacement height of 0.3 m can capture 80% of the flux in an upwind sector 1 km upwind (Leclerc and Thurtell, 1990). Model results indicate that the footprint radius becomes much larger at night, as large as 6 km. We question, however, whether the model assumptions are still valid in stable conditions, when mixing may occur in strong but very infrequent bursts. The large size predicted for the footprints is good news because a flux estimate serves as an average over a relatively large area of canopy, but it places a restriction on the amount of horizontal canopy heterogeneity that the observer can tolerate. To do this kind of work over forests, one must seek an extensive area that is approximately homogeneous. In spite of the limitations, eddy correlation flux estimation is still the preferred technique.

Micrometeorological instrumentation for flux and gradient measurement is discussed by Wyngaard (1981) and Lenschow (1993). In addition to the familiar cup anemometers and aspirated psychrometers, one must measure three wind components at high frequency to find the eddy flux.

The most common techniques used in forest turbulence research are hot-film anemometers (Miller *et al.*, 1989) and sonic anemometers. Kaimal and Gaynor (1991) and Kaimal *et al.* (1990) discussed many of the practical problems associated with the use of sonic anemometers. Instruments capable of observing at 3–10 Hz are common in tower-based eddy correlation work. If one is interested only in the flux, it is not necessary to acquire data at such high frequency in the calculation. In many situations, grab-samples at 1 Hz suffice (Fitzjarrald and Moore, 1992), but the instrument must have a sufficiently short response time to sample very small eddies. Fast-response wind and concentration sensors are much more robust than they were only a few years ago. The large data-processing tasks demanded by the eddy correlation technique can easily be performed in the field with modern computers. Shuttleworth (1989) described eddy correlation observations in the Amazon over periods of several months. Our data acquisition of 10-Hz sonic anemometer data at Harvard Forest has been operating with few breaks since late 1991.

Not every forest canopy is appropriate for these types of measurements. Horizontal inhomogeneities can be important, as every observation has its own footprint. One way to assess how well one makes turbulent observations is to see how well the the energy budget is balanced locally. To do this, one must combine radiation measurements with turbulent flux estimates. The field of view of a typical net radiometer, for example, can be an order of magnitude smaller than the footprint for the turbulent fluxes. This difference in areas of reference explains why the energy budget has not been closed observationally to a very high degree over forests (Fitzjarrald and Moore, 1994). When all components of the energy balance just above the canopy have been estimated independently, there has regularly been a residual amounting to 5–15% or more of available energy (net radiation minus soil heat flux), due to either the different areas of representativity or lack of site homogeneity. That footprints for different types of instruments can be widely different adds to the desire for homogeneous surfaces.

Baldocchi *et al.* (1991) reviewed studies using meteorological techniques to estimate heat and water vapor fluxes. Verma *et al.* (1986) reported on heat, water, and CO_2 fluxes observed for several days above a midlatitude deciduous forest. Bulk surface (or stomatal) resistances found for a number of experiments are cited by Garratt (1992). Observations of carbon uptake estimated through eddy correlation measurements of CO_2 flux for two years at Harvard Forest are presented by Wofsy *et al.* (1993).

Eddy correlation estimates of turbulent heat and moisture flux inside the canopy have been made in several forests. The theoretical justification for using the technique in such an environment has not yet been presented convincingly. This is because one cannot be certain that the measurements represent undisturbed air flows near the forest floor. Though one would never make turbulence measurements in a flat surface layer just behind an obstacle, one can hardly object to obstacles perturbing the flow inside the canopy. However, results presented by Baldocchi and Meyers (1991) and Lee and Black (1992) indicate that plausible results can be obtained. Our subcanopy observations of heat flux at Harvard Forest (Fig. 2) are the longest subcanopy records yet obtained using the eddy correlation method. Continuing work there involves similar long-term observation of CO_2 and water vapor fluxes.

V. Turbulent Eddies in the Canopy Environment

We have discussed some aspects of the mean behavior of turbulent fluxes of sensible and latent heat and radiation absorption in canopies. Canopy structure modifies the physical fluxes. The flux divergences across the canopy determine its microclimate. Canopy researchers may also be concerned with the turbulent eddies themselves. Some processes of interest depend on extreme rather than mean statistics, for example, transport of spores or insects and the ventilation of trace gases.

What do the eddies that accomplish fluxes look like? Turbulent eddies within and just above the canopy are complex, three-dimensional structures. It has long been recognized that turbulence is organized, with relatively large amounts of transport occurring in short periods of intense activity. During these periods, there appear periods of organized flow, called coherent structures. In the day, they have been identified by sharp ramplike temperature signals at the canopy top (Antonia *et al.*, 1979). At times, large coherent structures above deciduous forests penetrate deep into the canopy (Gao *et al.*, 1989; Lu and Fitzjarrald, 1994). A cross section of a composite coherent structure (Fig. 6) illustrates the phase relationship among w, T, and u above and below the Harvard Forest canopy. When eddies pass overhead, vertical motions occur nearly simultaneously at different levels in the canopy. Ramps have been observed over different types of canopy (e.g., Finnigan, 1979; Bergström and Högström, 1989). Laboratory and field observations indicate that the characteristic horizontal length between structures is from $6-10h$ or 30 $(h-d)$, where h is canopy height and d is displacement height (Hussain, 1983; Raupach *et al.*, 1989, 1991; Paw U *et al.*, 1992). Baldocchi and Meyers (1991) presented a theoretical analysis to estimate the response time of subcanopy evaporation following a sweep of relatively dry air into the canopy airspace.

Where do the "eddies" come from? The rough surface of the forest canopy is partly responsible for the intensity and spacing of the eddies. Antonia *et al.* (1979) argued that coherent structures were determined by wind shear ($\Delta U/\Delta z$) rather than by thermal effects. An appealing explanation

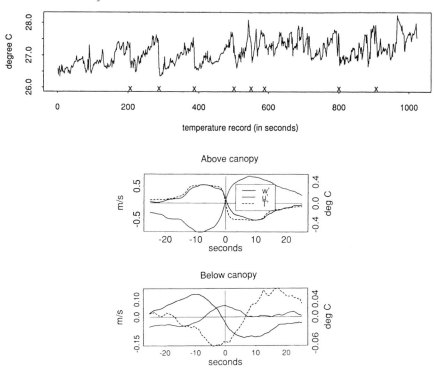

Figure 6   Coherent structures observed just above and inside the deciduous forest canopy at Harvard Forest. Top: time series of temperature illustrating ramps that occur during daytime convective conditions (from 10:14 to 10:34 A.M. on August 17, 1991). The following composites of coherent structures during summer at Harvard Forest illustrate vertical and horizontal velocity perturbations (w' and u') and temperature (T') during summer conditions. Middle: structure seen at $z \approx 1.5 \ h$. Bottom: structure seen at $z \approx 0.5 \ h$. From Lu and Fitzjarrald (1994). Reprinted by permission of Kluwer Academic Publishers.

for the origin of the coherent structures championed by Raupach *et al.* (1989) is that the structures result from shear instabilities in the layer just above the forest; the fluid can only stretch so much. This is an example of a "pure" fluid dynamics problem applied to the canopy exchange problem. In traditional fluid dynamics, a flow whose profile exhibits a change in curvature (like that common inside and just above canopies) is subject to a type of instability (Lamb, 1932). In its early stages, the motion connected to the instability resembles the coherent structures observed over canopies in many ways. The argument remains largely heuristic, because the results of the classical perturbation analysis would not be expected to hold in detail when an eddy reaches finite amplitude. Coherent structures tend to appear in groups, perhaps an indication of an influence of larger eddies on the wind shear in the layer just above the canopy (Lu and Fitzjarrald, 1994). The frequency of arrival of coherent structures may also reflect changes in the mean wind profile brought on by changes in stability or roughness that can occur over diurnal or annual cycles.

As turbulent motions penetrate into the canopy, higher-frequency components are removed. Energy shed in small vortices from leaves and branches is preferentially dissipated. This leaves the impression that the canopy is a "low-pass" filter. The high frequencies are removed by energy expended in moving branches and leaves. Comparing the spectrum of vertical velocity (σ_w^2 split into frequency categories) above and within the rain forest canopy (Fig. 7) illustrates this quenching effect on higher-frequency fluctuations. Below $z/h \approx 0.2$, Shaw and Zhang (1992) showed that the direct influence of air motion above the canopy was small. There, horizontal motions can be induced by pressure gradients associated with traveling coherent structures above the canopy. The motion deep within dense canopies is largely induced by the movement of eddies above the canopy. Lu and Fitzjarrald (1994) showed that in the limit of very thick canopies, little momentum flux is expected from such motions.

Eddies traveling above the canopy can induce other kinds of motions inside the canopy. Fitzjarrald and Moore (1990) found resonant oscillations

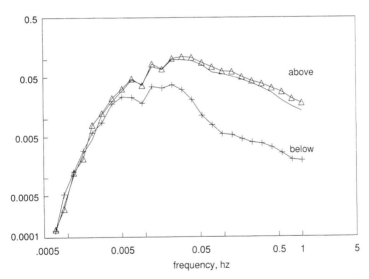

Figure 7 Power spectra of the vertical velocity observed in and above an Amazon rain forest canopy approximately 40 m tall at 45 m (solid line), 30 m (triangles), and 23 m (crosses). Abscissa is $f\,S_w(f)$ (m²s⁻²). From Fitzjarrald *et al.* (1990), copyright by the American Geophysical Union.

near the local buoyancy frequency determined by the density stratification in the air column between trees in the Amazon rain forest. Frequency-modulated w and T oscillations can take place when the wind speed above the canopy exceeds a certain threshold. An example of the phenomenon for the rain forest canopy illustrated in Fig. 1 is given in Fig. 8. Traveling eddies above the canopy are hypothesized to lead to pressure gradient oscillations that achieve resonance with the natural frequency of the stable interstitial air in the canopy. In this case, CO_2, and presumably other trace gases, were emitted to the atmosphere (Fig. 5) when the wave illustrated in Fig. 8 broke. We have also noted the presence of such oscillations within the canopy at Harvard Forest and just above the lichen woodland in northern Quebec. Other researchers have observed oscillations during the day. When there is significant in-canopy atmospheric stability, oscillations can develop even during situations that appear likely to remain quiescent.

Nocturnal mixing of warm air into the canopy may dry leaves. High relative humidity or liquid water on leaves are conditions favorable for the transport or germination of certain fungal spores at night because nocturnal exchanges have little importance for the daily heat or water balances,

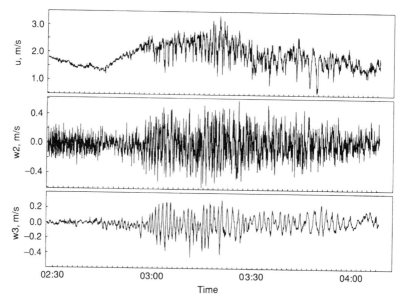

Figure 8 Nocturnal oscillatory motion observed inside and just above a 40-m-tall Amazon rain forest canopy in the same period as illustrated in Fig. 5. Top: horizontal wind speed at 45 m. Middle: vertical wind speed at 39 m. Bottom: vertical wind speed at 23 m. From Fitzjarrald and Moore (1990), copyright by the American Geophysical Union.

micrometeorologists have paid little attention to these transports. Because of their potential biological significance, they deserve closer attention from canopy ecologists.

The tropical rain forest canopy can be decoupled from the atmosphere above. Some evidence suggests that ventilation events related to strong cloud downdraft gust fronts, which are much more intense than the coherent structures just considered, may exchange air in the entire rain forest canopy. Fitzjarrald *et al.* (1990) described the influence of a cloud downdraft gust front that effectively exchanged the air within the Amazon rain forest canopy. Radar echoes indicated that a small cloud system passed over the observation site. The arrival of the gust and subsequent vertical motions within the canopy (Fig. 9) led to intense mixing in the normally quiescent rain forest canopy. This gust generated a short period of upward CO_2 flux, verifying that air from the deep canopy had been exchanged. Only a few such mixing events could be responsible for coupling the usually decoupled rain forest floor effectively with the atmosphere above. More work is needed in this area, as this type of extreme event may be of interest to canopy biologists.

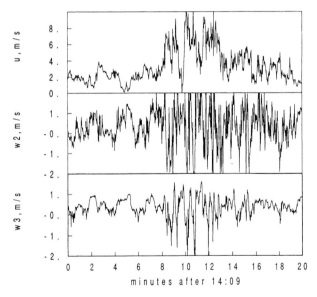

Figure 9   Canopy air motions in the Amazon rain forest canopy respond to the arrival of a cloud downdraft. Top: horizontal wind speed at 45 m. Middle: vertical wind speed at 39 m. Bottom: vertical wind speed at 23 m. From Fitzjarrald *et al.* (1990), copyright by the American Geophysical Union.

VI. Relation of Turbulent Motions to Transport of Biologically Important Phenomena

Forest canopies present special problems with respect to aerial transport of organisms for two reasons: (a) because most horizontal momentum is absorbed in the upper parts of the canopy, mean wind speeds are low; and (b) thermally stable conditions are caused by absorption of radiation in the upper canopy. Many forest fungi and insects that are dependent on aerial transport for some part of their life cycle are adapted to this poor transport environment. Studies of fungal spore populations in forests have documented the prevalence of small-spored (10 μm or less in diameter) fungi (DeGroot, 1968) in the subcanopy environment, where their low sedimentation speeds (<1 cm s^{-1}) allow them to remain suspended in low-speed air currents longer. Just as spores transported at night cannot be adequately described by simple eddy diffusion (K-theory) models (VanArsdel, 1967), so suspended particles in the subcanopy environment present the same difficulties.

Because many spores in the subcanopy environment are small, the efficiency with which they are impacted on surfaces tends to be low. The collection efficiency of spheres on cylinders is a function of the Stokes number: $S = v_s U/g d$, where v_s is the sedimentation velocity, U is the wind speed, g is the acceleration due to gravity, and d is the diameter of the cylinder. The collection efficiency increases nonlinearly with increasing Stokes number (Chamberlain, 1975), suggesting that intermittent gusts of wind that may be several times the mean wind speed will have a large effect on the impaction of small spores on surfaces.

Aylor (1990) and Aylor *et al.* (1981) examined the importance of the intermittency of turbulence within a canopy to the transport and deposition of larger (> 20 μm diameter) spores, for which inertial impaction is in general more important. The extreme events that make up less than 10% of the time may contain four or more times the kinetic energy than the average, resulting in a temporary decrease in the thickness of the laminar boundary layer next to the leaf or other surface. Organisms that are passively released into the air (e.g., some spores and many insect larvae suspended on silk) and dispersed often have a threshold speed above which they will become detached from the surface on which they reside. Such organisms will become airborne in gusts of wind even when the mean wind speed would not be sufficient to detach them. These events liberate spores and insects from their surface, transport them to a new location, and may result in more efficient inertial impaction on a new surface.

Another illustration of the importance of extreme events in the dispersion of biologically important materials is also provided by the interaction of insects with pheromone odor plumes (Murlis *et al.*, 1992). Statistically extreme events in the concentration distribution of pheromones play a disproportionate role in the orientation of insects to the source of the plume. This makes standard Gaussian plume modeling (common in air pollution studies) of limited value.

Aerial dispersal of spores, pollen, and larvae has been treated from the point of view of standard eddy diffusion models (Aylor, 1978; Fosberg and Peterson, 1986). This approach applies best to dispersal within the standard plane surface layer described earlier. It is widely recognized that such models do not apply to transport near the source when the source is a canopy (Raupach, 1989) owing to the violation of the assumption that the eddies are small relative to the plume size. How do materials produced at the soil surface or in the understory become transported out of the canopy and into the air above? One explanation lies in the fact that the coherent structures discussed earlier are accomplishing the transport. Canopy air motions consist of sweep (downward motions) followed by upward motion in the ejection phase. These motions include ejection phases (periods of slow updraft) perhaps as much as 0.5 m s^{-1} for 10 to 20 seconds (Gao *et al.*, 1989). Measurements of many coherent structures at Harvard Forest (Fig. 7) indicate that the mean updraft velocities are only 0.15 m s^{-1} for 10 seconds at $z/h = 0.5$. The intensity of the velocity changes as the coherent structure passes decreases with depth in the canopy.

The subcanopy environment within a few meters of the soil may experience direct mixing with the air above the canopy only infrequently. However, the transient pressure gradient changes that occur as coherent structures interact with the canopy (Shaw and Zhang, 1992; Lu and Fitzjarrald, 1994) can cause slow movement of air, suspended small organisms, or odors in unexpected ways. Reversal of plume direction and relatively slow dispersion are due to the stable conditions created by the canopy (Edmonds and Driver, 1974). Such conditions make the interpretation of experimental dispersal data difficult. As a result, studies of subcanopy dispersion are relatively few.

VII. Summary and Conclusions: Prospects for the Future

We frame our conclusions in terms of a proposed integrated canopy research project. The ideal canopy experiment of the future would integrate measurements of the physical "species" represented by heat, moisture, momentum, CO_2, and perhaps other trace gases, with those of the biological species of interest to ecologists. Because the mean condition changes in time only if there is spatial divergence in the flux, flux measurement sampling at a number of locations vertically and horizontally would be used. The mean conditions of temperature, humidity, wind speed, and components of the net radiation through the canopy would be measured. Long time series of automated measurements reveal changes on the wide variety of time scales that are significant to forest biology. The measurements should be made continuously to capture the most fundamental time scales of interest. These include phenomena lasting only a few seconds, changes manifest during the diurnal cycle, and seasonal variation.

Such a suite of physical measurements has been made over the long term in a few locations such as at Harvard Forest. Micrometeorologists and canopy researchers need to stretch the current limits of conventional measurements. Our hypothetical experiment could include measurements of forest sounds over a range of frequencies. Many biological species and their activities could be detected by this technique, and other phenomena or canopy attributes could also be studied. Because the sound of the wind in the trees is produced by the shedding of turbulent eddies from leaves and branches, the sound of the forest directly relates canopy structure and turbulence structure. Flow visualization techniques using neutrally buoyant soap bubbles (Niklas, 1985; Moore, 1987) and video recording would help elucidate the slow, complex movements in the subcanopy space. Such visualizations could be examined with some of the image analysis techniques now available.

Transport of spores and insects within and beneath canopies has not been adequately studied. The usual sampling techniques for such organisms have poor time resolution or are very labor-intensive, so that relating such samples to changes in the transport of physical quantities is difficult. Automated detection of small particles or insects (e.g., a vertical profile of bug-zappers) might reveal diurnal fluctuations in these populations. Direct measures of the flux of organisms or propagules await the development of instruments that can resolve concentration changes on short time scales.

Intermittent nocturnal mixing and its impact on leaf wetness and available free water on leaves and other surfaces should also be studied. Because so many microorganisms are dependent on free water for processes such as spore release (Meredith, 1973), these short-term events, which are of little importance to the energy balance, could be quite important biologically. Detailed measurements of the water vapor flux in conjunction with leaf wetness measurements are needed. Long-running observations must be made to track seasonal variation in the canopy microclimate, an approach that the forest meteorological community has been slow to recognize. Even now (spring of 1994), publications in forest meteorology based on study of only a few hours' data appear. It is clear that increased presentation of extreme event statistics in the future will be of great value to canopy researchers.

Acknowledgments

Portions of this work and observations at Harvard Forest by the authors were supported by the U.S. Department of Energy's (DOE) National Institute for Global Environmental Change (NIGEC) through the NIGEC Northeast Regional Center at Harvard University (DOE Cooperative Agreement No. DE-FC03-90ER61010, Subcontract 901214 from Harvard University to the Research Foundation of the State University of New York).

References

Antonia, R. A., Chambers, A. J., Friehe, C. A., and van Atta, C. W. (1979). Temperature ramps in the atmospheric surface layer. *J. Atmos. Sci.* 36, 99–108.

Aylor, D. E. (1978). Dispersal in time and space: Aerial pathogens. *In* "Plant Disease: An Advanced Treatise" (J. G. Horsfall and E. B. Cowling, eds.), Vol. 2, pp. 159–180. Academic Press, New York.

Aylor, D. E. . (1990). The role of intermittent wind in the dispersal of fungal pathogens. *Annu. Rev. Phytopathol.* 28, 73–92.

Aylor, D. E., McCartney, H. A., and Bainbridge, A. (1981). Deposition of particles liberated in gusts of wind. *J. Appl. Meteorol.* 20, 1212–1221.

Baldocchi, D. D., and Meyers, T. P. (1991). Trace gas exchange above the floor of a deciduous forest. 1. Evaporation and CO_2 efflux. *J. Geophys. Res.* 96, (D4), 7271–7285.

Baldocchi, D. D., Hicks B. B., and Camara, P. (1987). A canopy stomatal resistance model for gaseous deposition to vegetated surfaces. *Atmos. Environ.* 21, 91–101.

Baldocchi, D. D., Luxmore, R. J., and Hatfield, J. L. (1991). Discerning the forest from the trees: An essay on scaling canopy stomatal conductance. *Agric. For. Meteorol.* 54, 197–226.

Bergström, H. and Hogström, U. (1989). Turbulent exchange above a pine forest. Part II. Organized structures. *Boundary-Layer Meteorol.* 49, 231–263.

Black, T. A., and Kelliher, F. M. (1989). Processes controlling understory evapotranspiration. *Philos. Trans. R. Soc. London, Ser. B* 324, 207–228.

Chamberlain, A. C. (1975). The movement of particles in plant communities. *In* "Vegetation and the Atmosphere" (J. L. Monteith, ed.), Vol. 1, pp. 155–203. Academic Press, London.

Chen, J. M., Black, T.A., and Adams, R.S. (1991). Evaluation of hemispherical photography for determining plan area index and geometry of a forest stand. *Agric. For. Meteorol.* 56, 129–143.

Cionco, R. M. (1972). A wind–profile index for canopy flow. *Boundary–Layer Meteorol.* 3, 255–263.

DeGroot, R. C. (1968). Diurnal cycles of airborne spores produced by forest fungi. *Phytopathology* 58: 1223–1229.

Denmead, O. T., and Bradley, E. F. (1985). Flux–gradient relationships in a forest canopy. *In* "The Forest: Atmosphere Interaction" (B. H. Hutchison and B. B. Hicks, eds.), pp 421–442. Reidel Publ., Dordrecht, The Netherlands.

Denmead, O. T., and Raupach, M. R. (1993). Methods for measuring atmospheric gas transport in agricultural and forest systems. *ASA Spec. Publ.* 55, 1–206.

Edmonds, R. L., and Driver, C.H. (1974). Dispersion and deposition of spores of Fomes annosus and fluorescent particles. *Phytopathology* 64, 1313–1321.

Fan S.-M., Wofsy, S. C., Bakwin, P. S., Jacob, D. J., and Fitzjarrald, D. R. (1990). Atmosphere–biosphere exchange of CO_2 and O_3 in the central Amazon forest. *J. Geophys. Res.* 95, 16851–16864.

Fazu, C., and Schwerdtfeger, P. (1989). Flux–gradient relationships for momentum and heat over a rough natural surface. *Q. J. R. Meteorol. Soc.* 115, 335–352.

Finnigan, J. J. (1979) Turbulence in waving wheat. II. Structure of momentum transfer. *Boundary-Layer Meteorol.* 16, 213–236.

Fitzjarrald, D. R., and Moore, K. E. (1990). Mechanisms of nocturnal exchange between the rain forest and the atmosphere. *J. Geophys. Res.* 95(D10), 16839–16850.

Fitzjarrald, D. R., and Moore, K. E. (1992). Turbulent transport over tundra. *J. Geophys. Res.* 97(D15), 16717–16729.

Fitzjarrald, D. R., and Moore, K. E. (1994). Growing season boundary layer climate and surface exchanges in a subarctic lichen woodland. *J. Geophys. Res.* 99(D1), 1899–1917.

Fitzjarrald, D., Stormwind, B., Fisch, G., and Cabral, O. (1988). Turbulent transport observed just above the Amazon forest. *J. Geophys. Res.* 93(D2), 1551–1563.

Fitzjarrald, D. R., Moore, K. E., Cabral, O. M. R., Scolar, J., and de Abreu Sá, L. D. (1990). Daytime turbulent exchange between the Amazon forest and the atmosphere. *J. Geophys. Res.* 95(D10), 16825–16838.

Fosberg, M. A., and Peterson, M. (1986). Modeling airborne transport of gypsy moth (Lepidoptera: Lymantriidae) larvae. *Agric. For. Meteorol.* 38, 1–8.

Fox, D. G. (1985). Forestry, *In* "Handbook of Applied Meteorology" (D. D. Houghton, ed.), Wiley, New York.

Gao, W., Shaw, R. H., and Paw U, K. T. (1989). Observation of organized structure in turbulent flow within and above a forest canopy. *Boundary-Layer Meteorol.* 47, 349–377.

Garratt, J. R. (1992). "The Atmospheric Boundary Layer." Cambridge Univ. Press, New York.

Hicks, B. B., Baldocchi, D. D., Meyers, T. P., Hosker, R. P., Jr., and Matt, D. R. (1987). A preliminary multiple resistance routine for deriving dry deposition velocities from measured quantities. *Water, Air, Soil Pollut.* 35, 311–330.

Horst, T. W., and Weil, J. D. (1992). Footprint estimation for scalar flux measurements in the atmospheric surface layer. *Boundary-Layer Meteorol.* 59, 279–296.

Huber, L., and Gillespie, T. J. (1992). Modeling leaf wetness in relation to plant-disease epidemiology. *Annu. Rev. Phytopathol.* 30, 553–577.

Hussain, A. K. (1983). Coherent structures—Reality and myth. *Phys. Fluids* 26, 2816–2850.

Hutchinson, B. A., and Hicks, B. B. (1985). *The Forest–Atmosphere Interaction*. Reidel Publ., Hingham, MA.

Hutchinson, G. L., and Livingston, G. P. (1993). Use of chamber systems to measure trace gas fluxes. *ASA Spec. Publ.* 55, 1–206.

Jarvis, P. G., and McNaughton, K. G. (1986). Stomatal control of transpiration: Scaling up from leaf to region. *Adv. Ecol. Res.* 15, 1–48.

Kaimal, J. C., and Gaynor, J. E. (1983). The Boulder Atmospheric Observatory. *J. Appl. Meteorol.* 5, 863–880.

Kaimal, J. C., and Gaynor, J. E. (1991) Another look at sonic thermometry. *Boundary-Layer Meteorol.* 56, 401–410.

Kaimal, J. C., Gaynor, J. E., Zimmerman, H. A., and Zimmerman, G. A. (1990). Minimizing flow distortion errors in a sonic anemometer, *Boundary-Layer Meteorol.* 53, 103–115.

Köstner, B. M. M., Schulze, E.-D., Kelliher, F. M., Hollinger, D. Y., Byers, J. N., Hunt, J. E., McSeveny, T. M., Meserth, R., and Weir, P. L. (1992). Transpiration and canopy conductance in a pristine broad-leaved forest of *Nothogagus*: An analysis of xylem sap flow and eddy correlation measurements. *Oecologia* 91, 350–359.

Lamb, H. (1932). "Hydrodynamics," Dover, New York.

Leclerc, M. Y., and Thurtell, G. W. (1990). Footprint prediction of scalar fluxes using a Markovian analysis. *Boundary-Layer Meteorol.* 52, 247–258.

Lee, X., and Black, T. A. (1992). Atmospheric turbulence within and above a douglas-fir stand. Part II. Eddy fluxes of sensible heat and water vapor. *Boundary-Layer Meteorol.* 64, 369–389.

Lenschow, D. H. (1993). Micrometeorological techniques for measuring biosphere–atmosphere trace gas exchange. *In* "Methods in Ecology: Trace Gases" (P. Matson and R. Harriss, eds.).

Lu, C.-H., and Fitzjarrald, D. R. (1994). Seasonal and diurnal variation of coherent structures over a deciduous forest. *Boundary-Layer Meteorol.* 69, 43–69.

McNaughton, K. G., and Jarvis, P. G. (1991). Effects of spatial scale on stomatal control of transpiration. *Agric. For. Meteorol.* 54, 279–301.

Meredith, D. S. (1973). Significance of spore release and dispersal mechanisms in plant disease epidemiology. *Annu. Rev. Phytopathol.* 11, 313–342.

Miller, D. R., Lin, J. D., Wang, Y. S., and Thistle, H. W. (1989). A triple hot-film and wind octant combination probe for turbulent air flow measurements in and near plant canopies. *Agric. For. Meteorol.* 44, 353–368.

Monteith, J. L., and Unsworth, M. H. (1990). "Principles of Environmental Physics." Edward Arnold, London.

Moore, K. E. (1987). Nocturnal spore dispersal in young plantations: A micrometeorological examination of the night-time surface layer. Ph.D. Dissertation, State University of New York at Albany.

Moore, K. E., Fitzjarrald, D. R., and Ritter, J. A. (1993). How well can regional fluxes be derived from smaller-scale estimates? *J. Geophys. Res.* 98(D4), 7187–7198.

Murlis, J., Elkinton, J. S., and Carde, R. T. (1992). Odor plumes and how insects use them. *Annu Rev. Entomol.* 37, 505–532.

Niklas, K. J. (1985). Wind pollination—A study in controlled chaos. *Am. Sci.* 73, 462–470.

Odum, H. T., and Jordan, C. F. (1970). Metabolism and evapotranspiration of the lower forest in a giant plastic cylinder. *In* "A Tropical Rain Forest. A Study of Irradiation and Ecology at El Verde, Puerto Rico" (H. T. Odum and R. F. Pigeon, eds.). Div. Tech. Inf., U. S. Atomic Energy Commission, Oak Ridge, TN.

Panofsky, H. A., and Dutton, J. A. (1984). "Atmospheric Turbulence." Wiley, New York.

Parker, G. G., O'Neill, J. P., and Higman, D. (1989). Vertical profile and canopy organization in a mixed deciduous forest. *Vegetatio* 89, 1–12.

Paw U, K. T., Brunet, Y., Collineau, S., Shaw, R. H., Maitani, T., Qiu, J., and Hipps, L. (1992). On coherent structures in turbulence above and within agricultural plant canopies. *Agric. For. Meteorol.* 61, 55–68.

Raupach, M. R. (1979). Anomalies in flux–gradient relationships over forest. *Boundary-Layer Meteorol.* 16, 467–486.

Raupach, M. R. (1989). Stand overstory processes. *Philos. Trans. R. Soc. London, Ser B* 324, 175–185.

Raupach, M. R. (1990). Vegetation–atmosphere interaction in homogeneous and heterogeneous terrain: Some implications of mixed-layer dynamics. *Vegetatio* 91, 105–120.

Raupach, M. R., and Finnigan, J. J. (1988). Single layer models of evaporation from plant canopies are incorrect but useful, whereas multilayer models are correct but useless: Discuss. *Aust. J. Plant Physiol.* 15, 705–716.

Raupach, M. R., Finnigan, J. J., and Brunet, Y. (1989). Coherent eddies in vegetation canopies. *Proc. Australas. Conf. Heat Mass Transfer, 4th,* Christchurch, NZ, 1989.

Raupach, M. R., Antonia, R. A. and Rajagopalan, S. (1991). Rough-wall turbulent boundary layers. *Appl. Mech. Rev.* 44, 1–25.

Schultze, E.-D., Küppers, M., Matyssek, R., Penka M., Zimmermann, R. Vasicek, F., Gries, W., and Kucera, J., (1985). Canopy transpiration and water fluxes in the xylem of the trunk of *Larix* and *Picea* trees—A comparison of xylem flow, propometer and cuvette measurements. *Oecologia* 66, 475–483.

Shaw, R. H., and Zhang, X. J. (1992). Evidence of pressure-forced turbulent flow in a forest. *Boundary-Layer Meteorol.* 58, 273–288.

Shaw, R. H., den Hartog, G., and Neuman, H. H. (1988). Influence of foliar density and thermal stability on profiles of Reynolds stress and turbulence intensity in a deciduous forest. *Boundary-Layer Meteorol.* 45, 391–409.

Shuttleworth, W. J. (1989). Micrometeorology of temperate and tropical forest. *Philos. Trans. R. Soc. London, Ser. B* 324, 207–228.

Stull, R. B. (1988). "An Introduction to Boundary Layer Meteorology." Kluwer Academic Publishers, Dordrecht, The Netherlands.

Tennekes, H., and Lumley, J. L. (1972). "A First Course in Turbulence." MIT Press, Cambridge, MA.

Van Arsdel, E. P. (1967). The nocturnal diffusion and transport of spores. *Phytopathology* 57, 1221–1229.

Verma, S. B., Baldocchi, D. D., Anderson, D. E., Matt, D. R., and Clement, R. J. (1986). Eddy fluxes of CO_2, water vapor, and sensible heat over a deciduous forest. *Boundary-Layer Meteorol.* 36, 71–91.

Wofsy, S. C., Goulden, M. L., Munger, J. W., Fan, S.-M., Bakwin, P. S., Daube, B. C., Bassow, S. L., and Bazzaz, F. A. (1993). Net exchange of CO_2 in midlatitude forests. *Science* 260, 1314–1317.

Woodwell, G. M., and Dykeman, W. R. (1966). Respiration of a forest measured by CO_2 accumulation during temperature inversions. *Science* 154, 1031–1034.

Wyngaard, J. C. (1981). Cup, propeller, vane, and sonic anemometers in turbulence research, *Annu. Rev. Fluid Mech.* 13, 399–424.

4

Structure and Microclimate of Forest Canopies

Geoffrey G. Parker

I. Introduction

The canopy is both a unique subsystem of the forest and the site of fundamental interactions between vegetation and the physical environment. Other authors in this volume view the canopy as a habitat and discuss the distribution, abundance, and ecological relations of its associated species. In this chapter, I explore canopy–atmosphere interactions for common environmental parameters and the influence of canopy organization on forest environment.

I summarize the general structural features and microclimates peculiar to forest canopies, compare how some forests differ in these characteristics, and explore how structure affects microclimate. I do not discuss the processes or rates of exchange of matter and energy between the forest and the atmosphere (see Chapter 3), nor consider the issue of scaling observations from leaf to canopy or from canopy to region. My principal focus is on the structure and environment of closed, continuous forests. More detail on forest influence on environment is in Kittredge (1948), Geiger (1965), and Lee (1983). The environments and atmospheric interactions of forests are considered in several volumes on physiological plant ecology (Gates, 1980; Landsberg, 1986), micrometeorology or microclimatology (Rosenberg, 1974; Monteith, 1975, 1976; Jones, 1983; Oke, 1987; Arya, 1988), and specialized multiauthored works (Hutchison and Hicks, 1985; Pearcy *et al.*, 1989; Russell *et al.*, 1989).

Copyright © 1995 by Academic Press, Inc.
All rights of reproduction in any form reserved.

I will advance several themes. First, forest canopy structure, though broadly understood, has not been clearly defined and is rarely represented in a manner allowing cross-system comparisons. Second, measurements of microclimate have concentrated on mean values of selected environmental characteristics, usually at few locations and short time scales. Spatial variation is rarely assessed and the capacity to predict ecologically meaningful variation is limited. Finally, studies of the relation between canopy structure and function, though increasing in number, are still uncommon. Theories of several canopy–atmosphere interactions (e.g., interception of radiation, momentum, and precipitation) are well developed; data relevant to these interactions are less abundant.

I will illustrate aspects of the structure and function of canopies with examples from several studies in temperate deciduous forests. Reference is made to work in tall, mixed-species forest on the U.S. coastal plain dominated by *Liriodendron tulipifera* (Parker *et al.,* 1989) and in mixed-oak forests in the southern Appalachian Mountains (Parker *et al.,* 1993).

II. Structure of Forest Canopies

A. Definitions

The canopy is the combination of all leaves, twigs, and small branches in a stand of vegetation; it is the aggregate of all the crowns. The canopy is a region as well as a collection of objects (Carroll, 1980). "Canopy structure" is the organization in space and time, including the position, extent, quantity, type and connectivity, of the aboveground components of vegetation (e.g., Maser, 1989; Norman and Campbell, 1989; Nobel *et al.,* 1993). It is often useful to consider the open spaces between canopy elements and the atmosphere contained within and between crowns as part of the canopy.

This definition of canopy departs from other concepts that are restricted to the uppermost covering of vegetation, that is, the "roof" of the forest, the layer above the living limbs of the larger stems (e.g., Norse, 1990). This distinction is useful because in many stands, structural elements are distributed throughout the height of the forest; the covering layer is not always clearly definable. Similarly, the forest environment changes continuously from top to bottom; it is not clear how the gradient should be objectively subdivided. Finally, the dynamic nature of both canopy structure and environment makes the discrimination of boundaries difficult.

Separate study objectives require different operational representations of canopy structure, even for the same forest. For an investigation focused on organisms inhabiting woody surfaces (e.g., epiphytes), the canopy may be conceived as a network of connecting limbs (Nychka and Nadkarni, 1994). The canopy also could be conceptualized as a community of leaves and studied demographically (Harper, 1989). To consider the patterns of environment and whole-canopy exchange of matter and energy, I will consider the canopy as a three-dimensional porous medium, having both passive and active surfaces.

Many terms have been used interchangeably in reference to canopy structure, but they emphasize different aspects. *Physiognomy* focuses on the shapes of individual crowns. *Architecture* describes the growth patterns and resultant forms of stems (e.g., Hallé *et al.*, 1978; Oldeman, 1990). *Organization* has implied the statistical distribution of canopy components (or important characteristics) in space or time (e.g., Hollinger, 1989). Canopy *texture* refers to the sizes of the crown units composing the overstory that are apparent from above the stand.

B. The Units of Canopy Structure

The proximate units of canopy structure are the crowns of trees; the ultimate units are its leaves and twigs (Evans, 1972). However, as a hectare of closed canopy forest may have millions of leaves and many kilometers of twigs, most studies deal with these units statistically or focus on other levels of organization. Many scales of organization are evident (e.g., Kruijit, 1989): foliage may be clumped (often along branches and branch tips) and/or arranged in clusters, branch systems, and crownlets (Kira *et al.,* 1969; Hallé *et al.,* 1978; Bourgeron, 1983; Whitmore, 1984). Crowns are themselves sometimes grouped (Oldeman, 1990).

Canopy structure can be characterized at several levels of detail. It is most commonly summarized by a characteristic dimension or descriptor, for example, maximum tree height (h_{max}) or the mean height of the dominant trees (h), number or biomass density of the elements (stems ha^{-1} or Mg ha^{-1}), canopy cover (the fraction of sky not obscured by vegetation), or the leaf area index (LAI, m^2 m^{-2}), the ratio of the total one-sided leaf area to the projected ground area. Branch area index (BAI) is the area index of nonleafy material, and the total plant area index (PAI) integrates the two (i.e., PAI = LAI + BAI). Another common descriptor is the mean leaf area per unit volume (leaf area density, LAD, m^2 m^{-3}). Less common are one-dimensional summaries of spatially averaged vertical conditions, such as the height distribution of leaf area [$L(z)$, the foliage–height distribution, $\Sigma L(z) = $ LAI], the total foliage or branch area, and the leaf angle distribution. Rarely is it possible to specify the three-dimensional organization of canopy elements [i.e., $L(x,y,z)$]. Whatever characteristic is emphasized, canopy structure is usually described by mean conditions, without an assessment of variation.

Forests tend to have higher LAI and lower LAD than other community types in a given environment (Tadaki, 1977). LAD is inversely related to stand height. LAD in herbaceous stands ranges from 2 to 4 m^2 m^{-3}, whereas

Table I   Ranges of Leaf Biomass, Left Area Index (LAI), and Mean Leaf Area Density (LAD) Observed in Different Forest Types[a]

Forest type	Leaf biomass (Mg ha^{-1})	LAI (ha ha^{-1})	LAD (m^2 m^{-3})
Deciduous broadleaf	2–3	4–6	0.1–0.3
Evergreen broadleaf	7–11	7–12	0.2–0.5
Deciduous conifer	2–3	5–7	0.1–0.4
Pinus	5–6	7–12	0.2–0.5
Evergreen conifer	9–15	15–20	0.3–0.7

[a]Extended from Tadaki (1977).

the values in forests are typically much lower, 0.2–0.4 m^2 m^{-3} (Kira *et al.*, 1969; Monsi *et al.*, 1973; Parker *et al.*, 1989); LAD in individual foliage layers can exceed these ranges (Table 1).

The surfaces of the canopy are the most important feature for atmospheric interactions. The height distribution of surface area and biomass for leaves and stems has been measured in a midsuccessional, mixed-species deciduous forest (Fig. 1). Biomass of stem tissues, which decreases steeply with height in the forest, far exceeds that of leaves. However, leaf area domi-

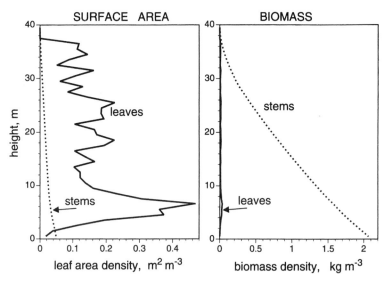

Figure 1   Mean vertical structure of canopy surfaces (left) and biomass (right) in a mixed-species deciduous forest on the mid-Atlantic coastal plain, United States. (From Parker *et al.*, 1989. Reprinted by permission of Kluwer Academic Publishers.)

nates the total aboveground surface area at all canopy levels. This multimodal vertical distribution of leaf area is common in mixed-species stands. The leaf area/total area ratio declines with increasing stem size (Whittaker and Woodwell, 1968). Thus, in older stands a larger proportion of total aboveground surface area is in stems and bark than in younger ones (i.e., LAI/PAI declines).

C. Vertical Organization

Canopy elements (tree height, species, or foliage) may be nonuniformly distributed with height (Smith, 1973; Richards, 1983; Oliver and Larson, 1990). Such patterns derive from species differences in growth form and shade tolerance, and on the stand developmental stage (Wierman and Oliver, 1979; Guldin and Lorimer, 1985; Terborgh, 1985). The vertical divergence of species height during forest development is called *differentiation* (Bicknell, 1982; Oliver and Larson, 1990). Vertical sorting of species leaf area has been demonstrated in mixed-species stands (Fig. 2).

In some stands, canopy leaves and branches are organized into distinct, nonoverlapping layers (*strata*). Such stratification is most easily recognized in single-species forests (especially managed stands) or in single-cohort stands, particularly at early stages of development. Early work in tropical vegetation suggested that stratification was a conspicuous characteristic of rain forests (e.g., Davis and Richards, 1933; Newman, 1954). Because vertical

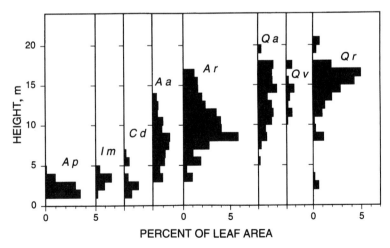

Figure 2 Foliage–height profile for a mixed-species forest in the southern Appalachian Mountains, United States, giving the percentage of leaf area for the major species. Species are coded Ap (*Acer pennsylvanicum*), Im (*Ilex montana*), Cd (*Castanea dentata*), Aa (*Amelanchier arborea*), Ar (*Acer rubrum*), Qa (*Quercus alba*), Qv (*Quercus velutina*), and Qr (*Quercus rubra*). (Reproduced from Parker *et al.*, 1993, with permission.)

organization was rarely quantified, the identification of strata tended to be subjective (Richards, 1983), and distinct vertical layering has not been universally recognized (Hallé *et al.*, 1978; UNESCO/UNEP/FAO, 1978; Whitmore, 1984).

The question of stratification was a contentious issue (Bourgeron, 1983; Brünig, 1983; Longman and Jenik, 1987) partly because at least three distinct characteristics were confused in reference to vertical layering: tree height, species, and biomass (Smith, 1973). For silviculture and forest dynamics, the distribution of crown heights was the important characteristic (Oliver and Larson, 1990). Forest community studies referred to the layering of species within the forest (e.g., Heinsdijk, 1957; Ogawa *et al.*, 1965). The literature on forest microclimate and physiology has concentrated on the vertical organization of biomass and surface area (e.g., Nobel *et al.*, 1993).

Several categories of canopy levels (stories) are recognized. The *overstory* comprises the crowns that are fully (dominant) or largely (codominant) illuminated from above. The *understory* (or *subcanopy*) includes the woody plants in the lowest shady layers. The *midcanopy* is a transitional region between understory and overstory and has crowns that are partly (intermediate) illuminated or overtopped (suppressed). The *ground layer* includes the seedlings of woody plants and other, herbaceous vegetation just above the *forest floor*, which includes the litter. In some stands, an irregular zone of extremely tall crowns (*emergents*) rises above the main canopy. The *outer canopy* is the canopy surface immediately adjacent to the atmosphere; it has variable height and irregular shape.

Richards (1952) distinguished and named distinct layers in tropical rain forests: the more or less continuous canopy of larger trees comprising the main canopy is the B layer. In the A layer are the occasional emergents rising above the B layer. The C and D layers, usually less continuous than the B, are also defined for midcanopy and understory. The ground vegetation is called the E layer. Similar schemes are employed in other forest types, particularly in silvicultural applications (Smith, 1962; Oliver and Larson, 1990).

Height is the most tractable axis for representing forest structure; many models of canopy–atmosphere interactions are one dimensional. Vertical variation in structure, environment, or flux is conventionally represented in graphs with height on the vertical axis. Height is only a crude proxy for environment characteristics in forests; for photosynthetically active radiation (PAR), for example, absolute height is less important than proximity to the outer canopy. Canopy position is usually measured from the ground surface (z, height) but is often scaled to the canopy height (z/h). Others reference canopy position to the top of the forest [depth, $h - z$, or scaled depth $(h - z)/h$].

In many stands, particularly in crops and plantations, the foliage height distribution is unimodal and elevated; the leaf area below $0.25h$ is ignored in some studies (Munn, 1966). However, the vertical distribution of leaf area in mixed-species or multicohort stands is rarely this simple and can have one or more peaks (Monsi *et al.,* 1973; Ross, 1975; Rauner, 1976; Aber, 1979; Franklin *et al.,* 1981; Hedman and Binkley, 1988; Parker *et al.,* 1989). In some older forests stands, $L(z)$ may be nearly uniform (Aber, 1979).

Leaf size, thickness, shape, and tissue chemistry differ between forest understory and overstory (e.g., Boardman, 1977; Kramer and Kozlowski, 1979). The display of foliage is also dependent on canopy position. Leaves tend to be held more vertically in the upper canopy than in the understory (Ford and Newbould, 1971; Hutchison *et al.,* 1986; Hollinger, 1989); leaf azimuth tends not to be preferentially oriented. De Wit (1965) described distributions of leaf inclination angles, including *planophile* (preferentially horizontal), *erectophile* (mostly vertical), *plagiophile* (mostly oblique), and *extremophile* (both horizontal and vertical).

D. Horizontal Variation

Forests are not spatially uniform, but are horizontally heterogeneous at various scales (e.g., Kira *et al.,* 1969; Fritschen, 1985; Kruijit, 1989). The various sizes of features of canopy organization are mirrored in corresponding distributions of foliage-free spaces. Light gaps are holes in the canopy extending to the forest floor that permit the penetration of unscattered light (Runkle, 1985; Canham *et al.,* 1990). Gaps of various sizes originate from a variety of causes; their numbers are inversely related to size (e.g., Brokaw, 1985; Sanford *et al.,* 1986). Intercrown spacing and horizontal variation in leaf density increase greatly with height as the number of crowns declines (Ishizuka, 1984; Hubbell and Foster, 1986).

Categories of local organization are nonetheless readily recognized in forests. There are small clearings; open areas with little understory and high overstory ("cathedral" type of organization); places with extremely dense understories only; and other areas thick with leaves at all levels. As many as four tree crowns may overlap at any one position. In older forests, such variations may reflect the mosaic of patch ages (Whitmore, 1984; Oldeman, 1990). Crowns of adjacent trees may overlap like roof shingles (*imbrication*) on hillsides; older forest edges may present a wall of foliage ("river edge effect," Hallé *et al.,* 1978).

J. H. Connell, M. D. Lowman, and I. R. Noble (unpublished data) suggested that variation in canopy vertical structure can be classified with regard to the presence or absence of canopy stories. A three-tiered stand could have as many as eight distinct structural classes (Fig. 3). The classical light gap, vegetation-free at all levels (Brokaw, 1982), is a relatively rare form of several vertical structures that occur in closed-canopy stands. This

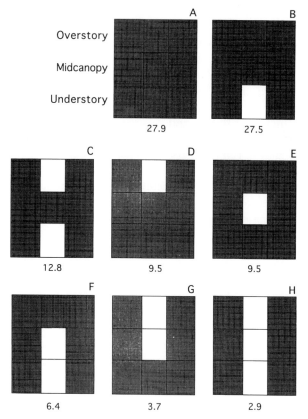

Figure 3   Forest layer spectrum for the stand in Fig. 2 giving the percentage of the stand in each of eight categories of vertical structure characterized by the presence or absence of overstory, midcanopy, or understory (following the approach of J. H. Connell, M. D. Lowman, and I. R. Noble, unpublished data). Unshaded segments indicate absent strata. Numbers give the percentage of the stand in each structural class. For example, panel H indicates a location without foliage at any level (the classical gap), which represents 2.9% of the forest.

approach, however, does not yield information on the spatial scale of individual open spaces.

In some forests, individual subcrowns and crowns are clearly separated, with intervening vegetation-free borders (Kira *et al.,* 1969). This "crown shyness" (Jacobs, 1955; Ng, 1977) is most common in single-species and single-cohort stands (especially plantations) and in stands on windy sites. It is probably maintained by wind-induced abrasion between adjacent crowns (Putz *et al.,* 1984).

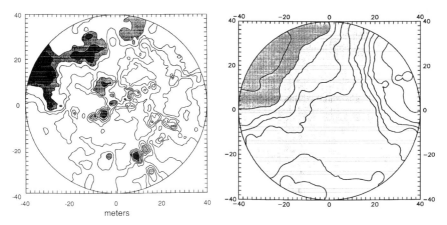

Figure 4 Topography of the outer canopy (left) and of the ground beneath the forest (right) in a young dry forest near Panama City, Panama. The contour intervals are 1 m for the ground surface and 5 m for the canopy. (From G. Parker, unpublished data.)

E. The Outer Canopy

The outer canopy is the layer of leaves and branches at the atmospheric interface; its undulating surface can slope laterally and may intersect the ground (e.g., in forest openings). The complex topography of this region, with features such as steep walls, canyons, and broad ridges (Parker, 1993; and Fig. 4), is generally not apparent from the ground. The surface area of the outer canopy is usually far greater than (e.g., more than twice) the area of ground below (Ford, 1976; Miller and Lin, 1985; Parker *et al.*, 1992). The surface is often punctuated with gaps, but can be very smooth in forests that experience recurring strong winds.

F. Quantification of Canopy Structure

The difficulties of canopy access impose severe limitations on the quantification of canopy structure (Denison *et al.*, 1972; Moffett and Lowman, Chapter 1). Direct sampling within the intact canopy space is rarely possible in forests (but see Ford and Newbould, 1971; Beadle *et al.*, 1982; Hutchison *et al.*, 1986). Consequently, most structural descriptions are simplifications, often one-dimensional representations of a spatially averaged characteristic, such as $L(z)$.

The appropriate characteristic of canopy structure ultimately depends on the phenomenon studied. The height distribution of leaf surface area, optical properties, and inclination angles are purported to be sufficient to describe critical features of radiation absorption (Campbell and Norman,

1989; Norman and Campbell, 1989). However, information on leaf angle is not essential for understanding wind velocities or momentum absorption.

Early characterizations of canopy structure were caricatures of the crowns of larger trees in a representative strip of forest, usually including both side and top views (e.g., Richards, 1952; Holdridge *et al.*, 1971). These *profile diagrams* are useful descriptions, particularly for illustrating structural aspects of stand classification (e.g., Kuiper, 1988; Oldeman, 1990), variation along environmental gradients (e.g., Beard, 1944), or changes during succession (e.g., Uhl and Jordan, 1984). Others employed diagrams of vertical structure, idealized by life form and habit (e.g., Dansereau, 1951) or by patterns of growth (Hallé *et al.*, 1978). However, such profiles tend to reflect the peculiarities of the chosen plot and provide little quantification of vertical organization of the whole stand.

Whole-canopy $L(z)$ may be estimated by assembling structural measurements of individual crowns made from observations taken from the ground (e.g., Kruijit, 1989) or from harvested stems (Beadle *et al.*, 1982; Massman, 1982). Foliage–height distributions of individual crowns have been summarized with various distributions: (a) triangular (Kinerson and Fritschen, 1971; Shaw and Pereira, 1982); (b) Gaussian (Stephens, 1969; Jarvis *et al.*, 1976); (c) Weibull (Yang *et al.*, 1993); and (d) beta (Massman, 1982). The assembly of canopy structure from crown measurements works best for crowns with "well-behaved" shapes (e.g., conifers in plantations). Mixed-species stands, which have diverse growth forms, ages, and heights, are difficult to quantify structurally (Campbell and Norman, 1989).

Although $L(z)$ can be sampled directly in grass, crop, or shrub canopies (e.g., Warren-Wilson, 1965), it is rarely possible to obtain *in situ* measurements in forests (but see Miller and Lin, 1985). The method of optical point-quadrats (MacArthur and Horn, 1969) can yield the relative $L(z)$ in forests (e.g., Aber, 1979; Hedman and Binkley, 1988; Parker *et al.*, 1989) but the method is time-consuming. Hemispheric (fish-eye) photography can assess the potential light environment at a point (Chazdon and Field, 1987; Becker *et al.*, 1989; Rich, 1990; Smith *et al.*, 1992) or, in some cases, estimate the LAI (Neumann *et al.*, 1989; Chen *et al.*, 1991; Martens *et al.*, 1993).

Forest structural attributes have been inferred from the effects of the canopy on measurements of environmental variables. These *inversion* techniques (e.g., Welles, 1990), depend on a robust relationship between structure and behavior. The expression describing behavior as a function of structure can sometimes be inverted and solved, yielding aspects of structure as a function of performance. Such methods are extensively developed for radiation-specific attributes (Norman, 1979, 1982; Lang, 1987; Perry *et al.*, 1988; Pierce and Running, 1988). Several devices are now available to facilitate the estimation of LAI from in-canopy light measurements (Welles, 1990; Martens *et al.*, 1993).

Kinerson and Fritschen (1971) inferred some aspects of canopy vertical structure $[L(z)]$ in a Douglas-fir forest from its effect on mean wind profiles. The difference between the predicted logarithmic wind profile and the measured wind profile had a shape similar to the downward cumulative $L(z)$. Wind profiles are not routinely employed to estimate canopy structure.

Some features of canopy structure can be estimated from the spectral quality of canopy light. Jordan (1969) used the ratio of red and far-red light transmitted through a tropical forest canopy to estimate its LAI. Remotely sensed reflectance can provide estimates of regional canopy structure (e.g., Leckie, 1990; Gholz *et al.*, 1991). Several measures calculated from combining different reflectance bands relate to the amount of green canopy biomass, such as the normalized difference vegetation index (NDVI), or "greenness," index:

$$\text{NDVI} = \frac{R_{nir} - R_{red}}{R_{nir} + R_{red}}$$

where R_{nir} and R_{red} are the canopy reflectances in the near-infrared and red wavelength bands, respectively. This index takes advantage of the strong reflectance in the near-infrared but weak reflectance in the red wavelengths of green canopies (Smith *et al.*, 1991). Other remote-sensing techniques use a similar approach, with sensors of different wavebands on several airborne or satellite platforms. Though certain interferences (correction for atmospheric absorption, phenological differences in canopy reflectance, reflectance of ground beneath the canopy) must be considered, such methods can be calibrated to yield reasonable predictions of LAI or leaf biomass (e.g., Spanner *et al.*, 1990). Other techniques can yield information beyond the amount of foliage biomass. For example, a profiling airborne laser can simultaneously sense the elevation of both the ground and canopy, yielding the contour of canopy height over a long transect (Krabill *et al.*, 1984; Leckie, 1990).

Paralleling the advances in access methodology (Moffett and Lowman, Chapter 1), the means for quantifying and visualizing canopy structure are growing in diversity and sophistication. The canopy surface has been described as a fractal object (Zeide and Pfeifer, 1991) and as a sum of numerous spatial wavelets (Bradshaw and Spies, 1992). Canopies have also been examined with two-dimensional spectral analysis (Ford, 1976) and three-dimensional tomography (Vanderbilt, 1985).

G. Temporal Changes

Canopy structure changes seasonally in all forests but is most dramatic in completely deciduous stands (e.g., Parker *et al.*, 1989; Fig. 5). Even in evergreen forests, the quantity of leaf area varies over the year (Ford and New-

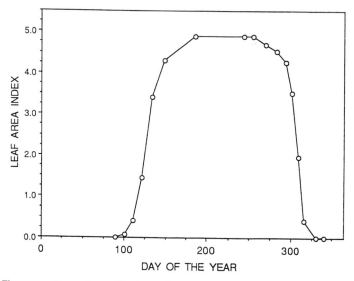

Figure 5   Change in total LAI over the course of a year for the stand in Fig. 1.

bould, 1971; Kinerson *et al.,* 1974; Gholz *et al.,* 1991; Hollinger *et al.,* 1994).
More substantial changes occur on a successional time scale (Fig. 6); the
total amount of leaf area stabilizes early in stand development, but its verti-

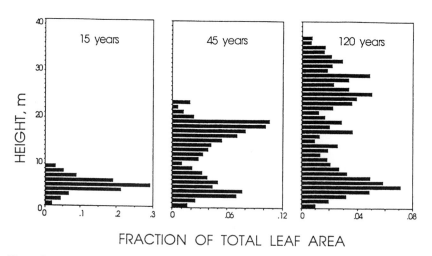

Figure 6   Examples of vertical canopy structures in different-aged stands of the tulip–poplar
association in the mid-Atlantic coastal plain, United States. (From G. Parker, unpublished
data.)

cal and horizontal distributions alter more slowly (Tadaki, 1977; Aber, 1979; Covington and Aber, 1980; Waring and Schlesinger, 1985; Oliver and Larson, 1990). Growing crowns ultimately come into contact with each other; when this happens throughout the stand, the canopy is said to be closed (the height of this contact is called the closure height). Subsequent differences in crown elongation cause height differentiation. The appearance of understory and shade-tolerant species initiates another layer of leaves. As overstory trees begin to die, stand leaf area declines slightly. Older stands may have crowns throughout vertical canopy space, but exhibit much spatial variation in canopy structure.

III. Microclimate

A. General Considerations

Canopy microclimate is ultimately determined by the stand macroclimate; the rhythms of change above and within the forest are set by the cycles of annual and diurnal heating and by the movements of air masses and clouds. Additionally, very short-term events associated with large penetrating eddies recurring on the time scale of less than a minute are important in exchange processes and the ventilation of lower canopy layers. Because most observations of forest environment are taken from measurements at a single location with averaging times of 5–30 minutes, important high-frequency events may be missed.

Some environmental variables (e.g., wind) are influenced by broad-scale canopy features whereas others (e.g., irradiance of visible light) depend more on the local arrangement of elements. The amount of *throughfall* (precipitation reaching the forest floor) depends on the pathways of droplet percolation. The amount of beam radiation received at a point depends on the obstructions along the path of the sun over time. Wind speed and direction depend on the disposition of obstacles upwind and can be very complex in the vicinity of isolated crowns, *windbreaks* (crowns in rows), or vegetation boundaries (McNaughton, 1989).

Several distinct aerodynamic regimes of the surface boundary layer are recognized above and within the canopy. In the *inertial sublayer* (up to several canopy heights above the stand), wind profiles tend to be semilogarithmic, the mean flow is largely horizontal, and turbulent transport of momentum is negligible (i.e., turbulence production is balanced by dissipation). In the *roughness sublayer* (close to and within the forest), profiles are complex and flows are three dimensional, and because the production and dissipation of turbulence are not balanced, net transport may be upward or downward. Additional microclimatic strata may be defined within the canopy based on the direction of the heat and momentum fluxes (e.g., Raupach and Thom, 1981).

B. Precipitation

Precipitation is intercepted, retained, and redistributed by the canopy. Water ultimately evaporates from the canopy (*interception*) or drips through (throughfall) or runs down the stems (*stemflow*) to the forest floor (e.g., McCune and Boyce, 1992). The chemical composition of precipitation may also be altered dramatically by the canopy. Parker (1983) and Coxson and Nadkarni (Chapter 20) discuss the acquisition and cycling of solutes by canopies.

In general, between 10 and 30% of incident precipitation is intercepted and evaporated from the canopy. From 1 to 3 mm of water can be retained in the canopy at a given time (*interception capacity*), with greater amounts held in coniferous than in broad-leaved stands. Of the water eventually reaching the forest floor, up to 85% is throughfall and 0–30% is stemflow. Many crown and canopy characteristics affect the retention and redistribution of the amount of precipitation, including species, leaf shape, leaf texture, stem branching, bark roughness, canopy height, and canopy closure (e.g., Kittredge, 1948; Doley, 1981). Models of the precipitation interception by canopies are well developed (e.g., Rutter *et al.*, 1971, 1975; Gash, 1979).

C. Radiation

Radiation absorption in the canopy is dependent on the distribution of leaves and leaf optical properties in the path of direct, diffuse, and scattered light. In general, forest leaves absorb most ($\geq 80\%$) of the incident short-wave radiation (< 700 nm); they transmit and reflect the remainder (e.g., Gates, 1980). More than half the radiation of the longer wavelengths (> 700 nm) penetrates canopy leaves; the remainder is largely reflected. The absorption and transmission spectra of leaves depend on species, leaf surface, age, and the angle of light (Gates, 1980).

Endler (1993) distinguished five types of light environment within forests, each with a characteristic color. Shade light in forests is depleted in the blue and red wavelengths relative to incident radiation; green wavelengths are less strongly absorbed (Federer and Tanner, 1966; Holmes, 1981). Understory light in coniferous forests appears to be slightly "bluer" than under broad-leaved hardwoods (Morgan and Smith, 1981). Most dramatic is the relative depletion in the far-red wavelengths; the red:far-red ratio is much narrower in the understory than outside the forest (Smith, 1982; Lee, 1987). The light quality of sunflecks resembles that of direct beam light (less blue and more reddish than shade light). Ultraviolet (UV) light is strongly absorbed by leaves except in open or disturbed canopies, where diffuse light richer in UV can penetrate (Brown *et al.*, 1994).

Because their surface roughness reduces backscattering to the atmosphere, forest canopies are generally more efficient absorbers of shortwave radiation than are other forms of vegetation (Shuttleworth, 1989). The reflection coefficient (*albedo*) of temperate forests ranges from 0.08 to 0.13 in conifers (Jarvis *et al.*, 1976), 0.10 to 0.12 in growing-season hardwoods, and ≈0.13 for tropical rain forest (Shuttleworth, 1989). Stanhill (1970) found that albedo decreased with vegetation height, from about 0.25 when less than 1 m high to 0.11 for stands above 20 m. This effect may reflect the dependence of roughness on height. Albedo also increases at higher solar zenith angles.

Light transmission deceases rapidly below the level of canopy closure. Average light levels decline with depth thereafter, but there is substantial spatial variability in canopy light environments, particularly in the overstory (Fig. 7). However, transmittance at any level is closely related to the total leaf area from the top of the forest. Monsi and Saeki (1953) proposed a general description of light in canopies, following the Beer–Lambert law for transmission in turbid media:

$$\frac{I}{I_0} = e^{-kL}$$

where I and I_0 are the illuminance at level z and the top of the canopy, L is the total leaf area downward through level z, and k is the light extinction

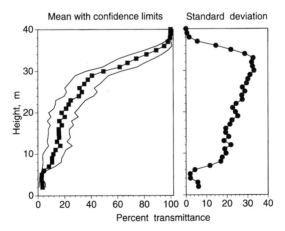

Figure 7 Vertical profile of the transmittance of photosynthetically active radiation (PAR) averaged over profiles at 26 different locations measured from a helium balloon at midday in the forest of Fig. 1. (From G. Parker, unpublished data.)

coefficient. A considerable theory of optics in vegetation canopies has developed with many modifications and extensions to this expression (e.g., Ross, 1975; Norman, 1979). The distribution of leaf angles has a very strong effect on light penetration (e.g., de Wit, 1965): theoretically, the coefficient of extinction, k, ranges from 0.5 for canopies of randomly distributed leaves with spherical orientation to 1.0 in planophile canopies.

Although this expression was originally intended only for monochromatic, direct-beam light interacting with a homogeneous medium of randomly oriented elements, it has been used widely for empirical estimates of canopy leaf area from measurements of light transmittance. The relationship must generally be calibrated for a given stand (e.g., Chason *et al.*, 1991; Brown and Parker, 1994). It also satisfactorily represents the attenuation of global (direct plus diffuse) radiation in canopies (Fig. 8). In some complex forests (e.g., woodlands or canopies with emergent crowns), however, the relation is not so smooth (e.g., Yoda, 1978).

The annual changes in light transmittance in a completely deciduous forest illustrate the influence of foliage on the distribution of PAR in time and height (Fig. 9). Whereas 30–40% of incident light reaches the forest floor in the leafless season, only 1–2% penetrates when the canopy is complete.

Some suggest the one-dimensional interception of radiation can be adequately characterized with information on $L(z)$, leaf angle distributions,

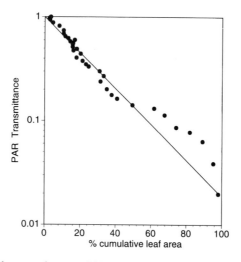

Figure 8   Relation between the mean PAR transmittance and the cumulative downward fraction of leaf area by 1-m height intervals in the stand of Fig. 1. The slope of the relation between transmittance and cumulative leaf area index is the empirical extinction coefficient, k (=0.63 here, for LAI=6).

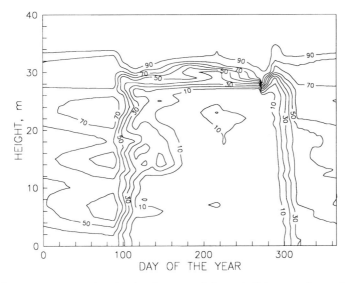

Figure 9   Height–time section showing the seasonal change in light transmittance in the forest of Fig. 1.

leaf optical properties, and leaf clumping (Norman 1979, 1982; Jarvis and Leverenz, 1983). However, such measurements are available for few stands.

D. Wind Speed

Wind is rapidly decelerated in the layer just above the forest (Fig. 10). The velocity profile in this layer is commonly described as though the wind were reacting to a rough surface displaced above the ground by a distance d (the zero plane displacement). This is equivalent to the mean height of momentum absorption (Thom, 1971), with a gradient controlled by the roughness of the surface (described by the roughness length for momentum, z_o) (e.g., Raupach and Thom, 1981). The mean wind velocity at height z is given by

$$u(z) \;=\; \frac{u_*}{k} \ln \left(\frac{z-d}{z_0} \right)$$

where u_* is the friction velocity and k is von Karman's constant (0.41). This expression applies only under neutral conditions and must be corrected when the air is stable or unstable.

Both the roughness and displacement height of canopies depend on the amount and distribution of canopy material, and also on wind speed itself (Brünig, 1970; Thom, 1971). Estimates are often based on the canopy

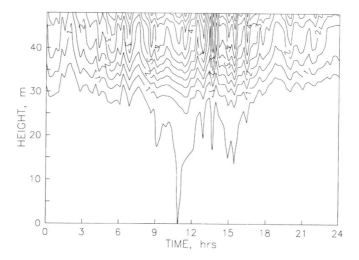

Figure 10   Height–time section of diurnal wind velocity near the end of the growing season (September 26) in the forest of Fig. 1. Velocity contours are 0.5 m s^{-1}. Note the incursion of stronger winds deep in the canopy during the day.

height; z_o/h is typically 0.1 (0.08—0.12) and d/h ranges from 0.6 to 0.7 (Monteith and Unsworth, 1990). In forests, the displacement height tends to be higher in the canopy than in agricultural crops, and the roughness length tends to be lower in forests than in crops. A simulation experiment (Shaw and Pereira, 1982) suggested that d increases with both plant density and the mean height of the canopy area, but that z_o rises and then declines as plant density increases.

The drag coefficient of a deciduous canopy is not strongly affected by the presence of leaves (Dolman, 1986); apparently boles, branches, and twigs are nearly as efficient drag elements as leaves. However, the canopy depth to which momentum penetrates is affected by the presence of leaves (Shaw *et al.*, 1988).

Within the canopy layers, local heterogeneity in structure complicates the wind field and hinders application of the ideal aerodynamic approach. Empirical relationships using power or exponential relations can describe wind profiles inside the forest (e.g., Landsberg and James, 1971; Thom, 1971). Cionco (1965) described the mean wind velocity within the canopy as

$$u(z) = u_h e^{a(z/h-1)}$$

where u_h is the mean wind velocity at the top of the canopy and a is an attenuation coefficient for momentum (the wind profile index). The a parameter ranges between 0.3 and 3.0 and increases with canopy density and flexibility (Cionco, 1972). Unfortunately, this description of canopy wind regime applies poorly in the lowest levels of the canopy (<0.2 h).

In some stands, wind speeds do not decline monotonically with depth but exhibit a secondary maximum in the lowermost canopy levels, often at about 0.1h (Allen, 1968; Bergen, 1971; Oliver, 1975; Jarvis *et al.*, 1976; Shaw, 1977). This has been observed either in stands of simple structure that lack vegetation layers at the bottom (the "trunk space") or at forest boundaries, where winds may blow through for distances equal to several canopy heights (McNaughton, 1989).

The canopy acts as a filter of high-frequency gusts, arresting small-scale fluctuations but permitting the penetration of large eddies (Baldocchi and Meyers, 1988; Fitzjarrald *et al.*, 1990). The depth to which eddies penetrate depends on canopy density, the strength of the eddy, and, especially, the stability of the canopy air column (Denmead and Bradley, 1985; Shaw *et al.*, 1988). Sigmon *et al.* (1983) showed that the presence of leaves affected differences in barometric pressure fluctuations in a deciduous forest.

Much of the total transport in forests occurs during a small fraction of the time. Periods of relative quiescence are punctuated with gusts that penetrate deeply into the canopy. With high-frequency sensors, rapid vertical motions and temperature deviations (*organized* or *coherent structures*) may be observed almost simultaneously at several canopy levels (Crowther and Hutchings, 1985; Gao *et al.*, 1989; Fitzjarrald and Moore, Chapter 3). During such events, slow temperature increases (or declines) are followed by abrupt declines (or increases). Several kinds of motions have been identified: an *ejection*, or an upward motion of relatively slow air from near the ground, and a *sweep*, or a downward motion of relatively fast air from above (Raupach and Thom, 1981; Baldocchi and Meyers, 1988; Bergström and Högström, 1989). Ejections are more important for momentum transport within the canopy, but sweeps are increasingly critical above the canopy (Gao *et al.*, 1989). The eddies responsible for organized structures have vertical scales on the order of the height of the forest. They may last tens of seconds, with an interval between eddies on the order of 30–100 seconds. The low frequency of most profiled observations (5–30 minutes) are insufficient to detect such events (Denmead and Bradley, 1985; Finnigan, 1985).

E. Temperature and Humidity

During midday, turbulence is effective in promoting transport in the canopy. Consequently, temperature and humidity gradients tend to be weak (Shaw and Pereira, 1982) (Figs. 11 and 12). The absolute humidity can remain rather constant at a forest level over a period of days. Because of this, temperature and relative humidity vary inversely. The *saturation deficit* (the drying power of the air) also changes throughout the forest, from

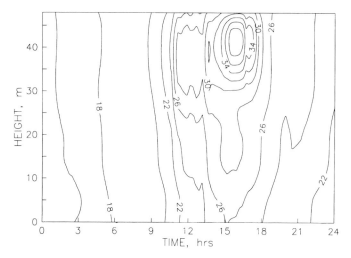

Figure 11 Height–time section of diurnal air temperature on the same day as Fig. 10. Temperature contours are 2 degrees. Note the hot spot in the upper canopy in the midafternoon and the weak gradients at night.

highly unsaturated during the day to near saturation at some levels at night (e.g., Cachan, 1963; Lemon *et al.*, 1970; Aoki *et al.*, 1978; Elias *et al.*, 1989). This often stimulates dew formation. Measurements of the temperature–

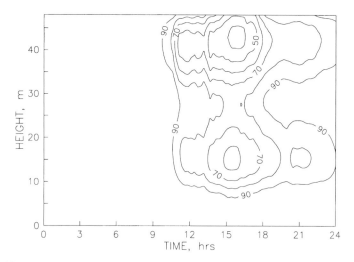

Figure 12 Height–time section of diurnal relative humidity (RH) on the same day as Fig. 10. Contours are 10% RH units. Humidities were near saturation (>95%) in the rest of the day outside the contours shown. Note the decline in RH in the midafternoon.

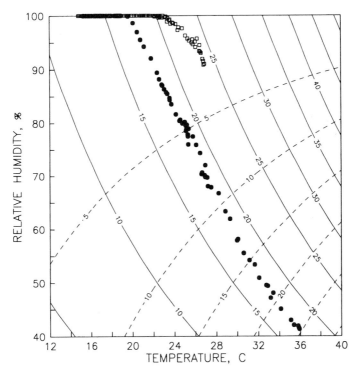

Figure 13 Temperature and humidity covariation just above the forest (solid circles) and at 14 m (open squares) height on the same day as Fig. 10, plotted as loci of successive points 15 minutes apart. Also shown are isolines of water vapor content (solid lines, g m^{-3}) and vapor deficit (dashed lines, g m^{-3}).

humidity relation at two levels in a deciduous canopy over the course of a day show the relative constancy of water vapor content despite wide variation in temperature, relative humidity, and vapor deficit (Fig. 13).

Most incident short-wavelength radiation is absorbed in the outer canopy, the *active layer* for heat exchange. This layer reradiates heat more readily than lower levels in the forest and cools rapidly on clear, still nights. The cooling at the top can establish a stable inversion, which effectively divides the canopy into two thermal zones. When heating resumes in the morning, the stability is broken up. This stratification is weaker than that observed in the temperature cycle of monomictic lakes. It can be punctuated by ventilating events, even at night (Raupach, 1989; Fitzjarrald *et al.*, 1990; Fitzjarrald and Moore, 1990).

Profiles of temperature and moisture have been simulated (e.g., Waggoner, 1975; Norman, 1979; Meyers and Paw U, 1987), but the estimation depends strongly on leaf stomatal behavior and is difficult to achieve. When present, forest understories can make a large contribution to the stand evapotranspiration (Black and Kelliher, 1989; Kelliher *et al.*, 1990; Baldocchi and Meyers, 1991).

F. Trace Gases and Particles

Trace gas and particle concentrations often have relatively weak gradients within the canopy, but mean concentrations are typically lower in the understory than in the overstory (but see Lovett and Lindberg, 1992). For models of particle, gas, and vapor gradients in canopies, see Wiman and Agren (1985), Wiman *et al.* (1985), and Lovett and Lindberg (1992).

For CO_2, however, the active layer of the canopy is an enormous sink, and CO_2 concentrations are often slightly depressed in the overstory during the daytime (Lemon *et al.*, 1970; Saeki, 1973; Allen and Lemon, 1976; Aoki *et al.*, 1978; Elias *et al.*, 1989). The layer near the ground is a source of CO_2, arising from root and soil respiration and decomposition. A pronounced CO_2 maximum often develops in the understory late at night, especially under stable conditions (Odum *et al.*, 1970; Landsberg, 1986; Kira and Yoda, 1989; Shuttleworth, 1989; Bazzaz and Williams, 1991) (Fig. 14). The rate of increase in understory carbon dioxide concentrations during nighttime inversions was employed by Woodwell and Dykeman (1966) to estimate whole-forest respiration.

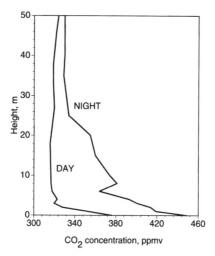

Figure 14   Daytime and nighttime profiles of CO_2 concentration (parts per million) in the forest of Fig. 1.

G. The Complex Vertical Gradient

The environment just above and within the canopy is a transition region between the free atmosphere and conditions near the forest floor. The mean values of many quantities change continuously with height, though often in a complicated and nonmonotonic manner. Because separate atmospheric parameters have different sinks and sources, their profiles are not identical. For example, leaf stomates are sources of water vapor but net sinks for CO_2. The active layer for radiation exchange is not necessarily coincident with that for momentum.

The range of variation in environment also differs between the upper and lower canopy layers (e.g., Fig. 13). The top has a marked diurnal fluctuation in almost every characteristic; variation is progressively reduced with depth in the canopy (Richards, 1952; Whitmore, 1984; Longman and Jenik, 1987). Relative to the outer canopy, the understory is reliably moist, dark, and still.

Also within the canopy is a gradient of connectivity: the outer canopy is clearly more closely coupled with the free atmosphere than is the understory. Cionco (1983, 1985) described a measure of the coupling of air flow in the canopy to that in the surface layer above (the *coupling ratio parameter*, R_c), defined as the ratio of mean wind speed within the canopy to that above it [$R_c = u(0.25h)/u(1.4h)$].

The connectivity between within-canopy layers and the forest exterior depends on the degree to which the large-scale environment is perceived at the level of individual canopy elements. For some environmental variables (e.g., the humidity at the leaf surface), the coupling is mediated by biological processes (largely stomatal control). Jarvis and McNaughton (1986) proposed an index, Ω, of the connection between leaf surface and free airstream values of water vapor deficits, for the case of transpiration. Omega reflects the decoupling of the two environments and ranges between 1 (completely uncoupled) and 0 (extremely strong coupling).

H. Transport in Canopies

Vertical profiles of important environmental parameters are related to the disposition of canopy material. The turbulent exchange of material between forest and atmosphere is in turn related to the stability of the atmosphere, the canopy structure, and the biological controls at canopy surfaces. The broad-scale atmosphere, canopy structure, and biological responses together influence and are influenced by canopy environments and transport processes.

The inference of canopy environment from canopy structure and of atmospheric exchanges from structure and environment is difficult. For example, because of the large scale of the eddies transporting material between canopy and atmosphere, neither the direction nor the magnitude of fluxes can be inferred from the curvature of profiles within the canopy (e.g., Denmead and Bradley, 1985; Paw U and Meyers, 1989). Fitzjarrald and Moore (Chapter 3) discuss why environmental gradients in forests are not reliable for predicting fluxes, as was once believed.

IV. Canopy Structure of Different Forest Types

There is enormous diversity within latitudes in canopy structure. Most temperate-zone work has focused on younger stands, often conifer plantations, and evergreen lowland rain forests have preoccupied most tropical researchers. Although few suitable measurement sets exist to compare forests in different places, some aspects of the archetypal tropical and temperate forest stands can be compared.

Tropical rain forests have greater annual leaf production (Medina and Klinge, 1983), standing canopy biomass (O'Neill and DeAngelis, 1981), and a higher leaf area index (Tadaki, 1977; DeAngelis *et al.,* 1981; Anderson, 1981). Foliage–height profiles $[L(z)]$ are rarely reported for tropical forest (see Kira and Yoda, 1989), but they are commonly presumed to be relatively uniform (e.g., Shuttleworth, 1989). Temperate canopies are often pictured as unimodal and elevated.

Rain forests tend to have lower relative illuminance of visible light at the forest floor (often <1%, Grubb and Whitmore, 1967; Leigh, 1975; Chazdon and Fetcher, 1984), whereas temperate canopies tend to permit higher transmittance (hardwoods in leaf: 1–3%, pines: >5%, hemlocks: <5%). Shuttleworth (1989), however, claimed that many of the bulk characteristics of temperate and tropical canopies (albedo, aerodynamic roughness, illuminance, interception capacity, and the behavior of stomatal conductance) may be roughly similar, given the reliability of the estimates. He suggested that more important distinctions between forests of different latitudes arise in the interaction of structure and local microclimate. Although structures may not differ significantly, the manner by which structure affects energy partitioning, transport, and microclimate may be quite distinct.

V. Summary, Conclusions, and Recommendations

The canopy is the primary site of interaction between the biosphere and atmosphere. The amount and spatial organization of aboveground plant parts influence both the atmospheric environment within canopies and the exchange of material and energy with the lower atmosphere. In the forest, atmospheric characteristics are strongly modified by canopy structure in two general ways. First, canopy surfaces act as passive drag elements and exchange surfaces for the absorption of wind energy, the dissipation of turbulence, and the exchange of radiation. Second, canopy surfaces actively participate in exchanges of biologically important compounds, such as CO_2 and water vapor.

Forest canopies are distinct from other forms of vegetation because they are "dense, extensive, tall and perennial" (Shuttleworth, 1989). They have more biomass, greater surface area, and lower average leaf area density. The structural complexity of forests makes them aerodynamically rougher than other forms of vegetation (except possibly woodlands). This increases the effectiveness of daytime turbulent mixing, which in turn may reduce environmental gradients within forests.

Forest structure is generally recognized as consisting of the amount and distribution in space of leaves, stems, twigs, and branches, but there are nearly as many distinct measures of canopy structure as there are canopy research programs. Because of the variation in attributes considered, comparisons are restricted to the most commonly measured attributes. Environmental variables are usually measured at only a few locations; spatial variation is rarely assessed (e.g., Kinerson and Fritschen, 1971; Hutchison and Baldocchi, 1989). Representativeness of long-term measurements is unknown and processes that react to structural heterogeneity, such as local convection induced by differences in heat penetration, cannot be assessed.

The bulk of observations of canopy environment are average descriptions of processes and average descriptions of motivating structures. Quantification of structure has been inadequate, even for studies directly focused on forest–atmosphere interactions (Fritschen, 1985). The lack of detailed information limits our capacity to generalize about the importance of canopy structure and climate in controlling canopy environment.

The theoretical basis of canopy–environment interactions is better developed than the observations to test or validate them. Most current models deal with ideal cases and yield predictions about mean conditions. There is a need to extend these predictions to the understanding of particular environments. More empirical studies could be focused on transitional environments, transient regions such as within forest clearings (gaps), and environments that exhibit wide variation, such as the outer forest canopy. The success in understanding mean conditions should be extended to appreciating the variation that organisms undoubtedly perceive.

Much canopy micrometeorology has focused on stands of simple structure, particularly single-species, single-aged forests or crops with elevated, unimodal canopies. Studies of such situations have propelled the development of several useful descriptive and predictive models. However, many stands are not so simple. There is a need to focus attention on the "non-ideal" canopies—mixed-species forests, multiple-cohort stands, and those with partial or complete deciduous seasons.

Acknowledgments

This work is dedicated to the memory of Alan P. Smith, canopy pioneer. I thank Martin Brown, Joe Connell, Dave Fitzjarrald, David Ford, Jerry Franklin, Mike Goltz, Gary Lovett, Kathleen Moore, Nalini Nadkarni, Catherine Parker, and Peter Stone for their data, discussions, and support.

References

Aber, J. D. (1979). Foliage–height profiles and succession in northern hardwood forest. *Ecology* 60, 18–23.

Allen, L. H., Jr. (1968). Turbulence and wind speed spectra within a Japanese larch plantation. *J. Appl. Meteorol.* 7, 73–78.

Allen, L. H., Jr., and Lemon, E. R. (1976). Carbon dioxide exchange and turbulence in a Costa Rican tropical rain forest. *In* "Vegetation and the Atmosphere" (J. L. Monteith, ed.), Vol. 2, pp. 265–308. Academic Press, London.

Anderson, M. C. (1981). The geometry of leaf distributions in some south-eastern Australian forests. *Agric. Meteorol.* 25, 195–205.

Aoki, M., Yabuki, K., and Koyama, H. (1978). Micrometeorology of Pasoh forest. *Malay. Nat. J.* 30, 149–159.

Arya, S. P. (1988). "Micrometeorology." Academic Press, San Diego.

Baldocchi, D. D., and Meyers, T. P. (1988). Turbulence structure in a deciduous forest. *Boundary-Layer Meteorol.* 43, 345–364.

Baldocchi, D. D., and Meyers, T. P. (1991). Trace gas exchange above the floor of a deciduous forest. 1. Evaporation and CO_2 flux. *J. Geophys. Res.* 96(D4), 7271–7285.

Bazzaz, F. A., and Williams, W. E. (1991). Atmospheric CO_2 concentrations within a mixed forest: Implications for seedling growth. *Ecology* 72, 12–16.

Beadle, C. L., Talbot, H., and Jarvis, P. G. (1982). Canopy structure and leaf area index in a mature Scots pine forest. *Forestry* 55, 105–123.

Beard, J. S. (1944). Climax vegetation in tropical America. *Ecology* 25, 127–158.

Becker, P., Erhart, D. W., and Smith, A. P. (1989). Analyses of forest light environments. Part I. Computerized estimation of solar radiation from hemispherical canopy photographs. *Agric. For. Meteorol.* 44, 217–223.

Bergen, J. D. (1971). Vertical profiles of windspeed in a pine stand. *For. Sci.* 17, 314–321.

Bergström, H., and Högström, U. (1989). Turbulent exchange above a pine forest. II. Organized structures. *Boundary-Layer Meteorol.* 49, 231–263.

Bicknell, S. H. (1982). Development of canopy stratification during early succession in northern hardwoods. *For. Ecol. Manage.* 4, 41–51.

Black, T. A., and Kelliher, F. M. (1989). Processes controlling understory evapotranspiration. *Philos. Trans. R. Soc. London, Ser. B* 324, 207–231.

Boardman, N. K. (1977). Comparative photosynthesis of sun and shade plants. *Annu. Rev. Plant Physiol.* 28, 355–377.

Bourgeron, P. S. (1983). Spatial aspects of vegetation structure. *In* "Tropical Rain Forest Ecosystems: Structure and Function" (F. B. Golley, ed.), pp. 29–47. Elsevier, Amsterdam.

Bradshaw, G. A., and Spies, T. A. (1992). Characterizing gap structure in forests using wavelet analysis. *J. Ecol.* 80, 205–215.

Brokaw, N. V. L. (1982). The definition of treefall gap and its effects on measures of forest dynamics. *Biotropica* 14, 158–160.

Brokaw, N. V. L. (1985). Treefalls, regrowth, and community structure in tropical forests. *In* "The Ecology of Natural Disturbances and Patch Dynamics" (S. T. A. Pickett and P. S. White, eds.), pp. 53–69. Academic Press, Orlando, FL.

Brown, M. J., and Parker, G. G. (1994). Canopy light transmittance in a chronosequence of mixed-species deciduous forests. *Can. J. For. Res.* 24, 1694–1703.

Brown, M. J., Parker, G. G., and Posner, N. (1994). A survey of ultraviolet-B radiation in forests. *J. Ecol.* 82, 843–854.

Brünig, E. F. (1970). Stand structure, physiognomy and environmental factors in some lowland forests in Sarawak. *Trop. Ecol.* 11, 26–43.

Brünig, E. F. (1983). Vegetation structure and growth. *In* "Tropical Rain Forest Ecosystems: Structure and Function" (F. B. Golley, ed.), pp. 49–75. Elsevier, Amsterdam.

Cachan, P. (1963). Signification écologique des variations microclimatiques verticales dans la forêt sempervirente de basse Côte d'Ivoire. *Ann. Fac. Sci. Dakar* 8, 89–155.

Campbell, G. S., and Norman, J. M. (1989). The description and measurement of plant canopy structure. *In* "Plant Canopies: Their Growth, Form and Function" (G. Russell, B. Marshall, and P. Jarvis, eds.), pp. 1–19. Cambridge Univ. Press, Cambridge, UK.

Canham, C. D., Denslow, J. S., Platt, W. J., Runkle, J. R., Spies, T. A., and White, P. S. (1990). Light regimes beneath closed canopies and tree-fall gaps in temperate and tropical forests. *Can. J. For. Res.* 20, 620–631.

Carroll, G. L. (1980). Forest canopies: Complex and independent subsystems. *In* "Forests: Fresh Perspectives from Ecosystem Analysis" (R. H. Waring, ed.), pp. 87–107. Oregon State Univ. Press, Corvallis.

Chason, J. W., Baldocchi, B. B., and Huston, M. A. (1991). A comparison of direct and indirect methods for estimating canopy leaf area. *Agric. For. Meteorol.* 57, 107–128.

Chazdon, R. L., and Fetcher, N. (1984). Photosynthetic light environments in a lowland tropical rainforest in Costa Rica. *J. Ecol.* 72, 553–564.

Chazdon, R. L., and Field, C. B. (1987). Photographic estimation of photosynthetically active radiation: Evaluation of a computerized technique. *Oecologia* 73, 525–533.

Chen, J. M., Black, T. A., and Adams, R. S. (1991). Evaluation of hemispherical photography for determining plant area index and geometry of a forest stand. *Agric. For. Meteorol.* 56, 129–143.

Cionco, R. M. (1965). A mathematical model for air flow in a vegetation canopy. *J. Appl. Meteorol.* 4, 517–522.

Cionco, R. M. (1972). A wind profile index for canopy flow. *Boundary-Layer Meteorol.* 3, 255–263.

Cionco, R. M. (1983). On the coupling of canopy flow to ambient flow for a variety of vegetation types and densities. *Boundary-Layer Meteorol.* 26, 325–335.

Cionco, R. M. (1985). Modeling windfields and surface layer wind profiles over complex terrain and within vegetation canopies. *In* "The Forest–Atmosphere Interaction" (B. A. Hutchison and B. B. Hicks, eds.), pp. 501–520. Reidel Publ., Dordrecht, The Netherlands.

Covington, W. W., and Aber, J. D. (1980). Leaf production during secondary succession in northern hardwoods. *Ecology* 61, 200–204.

Crowther, J. M., and Hutchings, N. J. (1985). Correlated windspeeds in a spruce forest. *In* "The Forest–Atmosphere Interaction" (B. A. Hutchison and B. B. Hicks, eds.), pp. 543–561. Reidel Publ., Dordrecht, The Netherlands.

Dansereau, P. (1951). Description and recording of vegetation upon a structural basis. *Ecology* 32, 172–229.

Davis, T. A. W., and Richards, P. W. (1933). The vegetation of Moraballi Creek, British Guiana: An ecological study of a limited area of tropical rain forest. *J. Ecol.* 21, 350–384.

DeAngelis, D. L., Gardner, R. H., and Shugart, H. H. (1981). Productivity of forest ecosystems studied during the IBP: The Woodlands data set. *In* "Dynamic Properties of Forest Ecosystems" (D. E. Reichle, ed.), pp. 567–672. Cambridge Univ. Press, Cambridge, UK.

Denison, W. C., Tracey, D. M., Rhoades, F. M., and Sherwood, M. (1972). Direct, non-destructive measurements of biomass and structure in living, old-growth Douglas-fir. *In* "Research on Coniferous Forest Ecosystems" (J. F. Franklin, L. J. Dempster, and R. H. Waring, eds.), pp. 147–158. Pacific Northwest and Range Experiment Station, Portland, OR.

Denmead, O. T., and Bradley, E. F. (1985). Flux–gradient relationships in a forest canopy. *In* "The Forest–Atmosphere Interaction" (B. A. Hutchison and B. B. Hicks, eds.), pp. 421–442. Reidel Publ., Dordrecht, The Netherlands.

de Wit, C. T. (1965). "Photosynthesis of Leaf Canopies." Agric. Res. Rep. No. 663. Cent. Agric. Publ. Doc., Wageningen.

Doley, D. (1981). Tropical and subtropical forests and woodlands. *In* "Water Deficits and Plant Growth" (T. T. Kozlowski, ed.), pp. 209–307. Academic Press, New York.

Dolman, A. J. (1986). Estimates of surface roughness length and zero plane displacement for a foliated and non-foliated oak canopy. *Agric. For. Meteorol.* 36, 241–248.

Eliáš, P., Kratochvílová, I., Janouš, D., Marek, M., and Masarovičová, E. (1989). Stand microclimate and physiological activity of tree leaves in an oak–hornbeam forest. I. Stand microclimate. *Trees* 4, 227–253.

Endler, J. A. (1993). The color of light in forests and its implications. *Ecol. Monogr.* 63, 1–27.

Evans, G. C. (1972). "The Quantitative Analysis of Plant Growth." Blackwell, London.

Federer, C. A., and Tanner, C. B. (1966). Spectral distribution of light in the forest. *Ecology* 47, 555–560.

Finnigan, J. J. (1985). Turbulent exchange in flexible plant canopies. *In* "The Forest–Atmosphere Interaction" (B. A. Hutchison and B. B. Hicks, eds.), pp. 443–480. Reidel Publ., Dordrecht, The Netherlands.

Fitzjarrald, D. R., and Moore, K. E. (1990). Mechanisms of nocturnal exchange between the rain forest and atmosphere. *J. Geophys. Res.* 95(D10), 16839–16850.

Fitzjarrald, D. R., Moore, K. E., Cabral, O. M. R., Scolar, J., Manzi, A. O., and de Abreu Sa, L. D. (1990). Daytime turbulent exchange between the Amazon forest and the atmosphere. *J. Geophys. Res.* 95(D10), 16825–16838.

Ford, E. D. (1976). The canopy of a Scots pine forest: Description of a surface of complex roughness. *Agric. Meteorol.* 17, 9–32.

Ford, E. D., and Newbould, P. J. (1971). The leaf canopy of a coppiced deciduous woodland. I. Development and structure. *J. Ecol.* 59, 843–862.

Franklin, J. F., Cromack, K., Jr., Denison, W., McKee, A., Maser, C., Sedell, J., Swanson, F., and Juday, J. (1981). Ecological characteristics of old-growth Douglas fir forests. *USDA For. Serv. Gen. Tech. Rep.* PNW-118.

Fritschen, L. J. (1985). Characterization of boundary conditions affecting forest environmental phenomena. *In* "The Forest–Atmosphere Interaction" (B. A. Hutchison and B. B. Hicks, eds.), pp. 3–23. Reidel Publ., Dordrecht, The Netherlands.

Gao, W., Shaw, R. H., and Paw U, K. T. (1989). Observation of organized structure in turbulent flow within and above a forest canopy. *Boundary-Layer Meteorol.* 47, 349–377.

Gash, J. H. C. (1979). An analytical model of rainfall interception by forests. *Q. J. R. Meteorol. Soc.* 105, 43–55.

Gates, D. M. (1980). "Biophysical Plant Ecology." Springer-Verlag, New York.

Geiger, R. (1965). "The Climate near the Ground." Harvard Univ. Press, Cambridge, MA.

Gholz, H. L., Cropper, W. P., Jr., McKelvey, K., Ewel, K. C., Teskey, R. O., and Curran, P. J.

(1991). Dynamics of canopy structure and light interception in *Pinus elliotii* stands in northern Florida. *Ecol. Monogr.* 61, 33–51.

Grubb, P. J., and Whitmore, T. C. (1967). A comparison of montane and lowland forest in Ecuador. III. The light reaching the ground vegetation. *J. Ecol.* 47, 33–57.

Guldin, J. M., and Lorimer, C. G. (1985). Crown differentiation in even-aged northern hardwood forests of the Great Lakes region, U.S.A. *For. Ecol. Manage.* 10, 65–86.

Hallé, F., Oldeman, R. A. A., and Tomlinson, P. B. (1978). "Tropical Trees and Forests: An Architectural Analysis." Springer-Verlag, Berlin.

Harper, J. L. (1989). Canopies as populations. *In* "Plant Canopies: Their Growth, Form and Function" (G. Russell, B. Marshall, and P. Jarvis, eds.), pp. 105–128. Cambridge Univ. Press, Cambridge, UK.

Hedman, C. W., and Binkley, D. (1988). Canopy profiles of some Piedmont hardwood forests. *Can. J. For. Res.* 18, 1090–1093.

Heinsdijk, D. (1957). The upper story of tropical forests. *Trop. Woods* 107, 66–83; 108, 31–45.

Holdridge, L. R., Grenke, W. C., Hatheway, W. H., Liang, T., and Tosi, J. A., Jr. (1971). "Forest Environments in Tropical Life Zones: A Pilot Study." Pergamon, Oxford.

Hollinger, D. Y. (1989). Canopy organization and foliage photosynthetic capacity in a broadleaved evergreen montane forest. *Funct. Ecol.* 3, 53–62.

Hollinger, D. Y., Kelliher, F. M., Byers, J. N., Hunt, J. E., McSeveny, T. M., and Weir, P. L. (1994). Carbon dioxide exchange between an undisturbed old-growth temperate forest and the atmosphere. *Ecology* 75, 134–150.

Holmes, M. G. (1981). Spectral distribution of radiation in plant communities. *In* "Plants and the Daylight Spectrum" (H. Smith, ed.), pp. 147–158. Academic Press, London.

Hubbell, S. P., and Foster, R. B. (1986). Canopy gaps and the dynamics of a neotropical forest. *In* "Plant Ecology" (M. J. Crawley, ed.), pp. 77–96. Blackwell, Boston.

Hutchison, B. A., and Baldocchi, D. D. (1989). Forest meteorology. *In* "Analysis of Biogeochemical Cycling Processes in Walker Branch Watershed" (D. W. Johnson and R. I. Van Hook, eds.), pp. 21–95. Springer-Verlag, New York.

Hutchison, B. A., and Hicks, B. B., eds. (1985). "The Forest–Atmosphere Interaction." Reidel Publ., Dordrecht, The Netherlands.

Hutchison, B. A., Matt, D. R., McMillen, R. T., Gross, L. J., Tajchman, S. J., and Norman, J. R. (1986). The architecture of a deciduous forest canopy in eastern Tennessee, U.S.A. *J. Ecol.* 74, 635–646.

Ishizuka, M. (1984). Spatial pattern of trees and their crowns in natural mixed forests. *Jpn. J. Ecol.* 34, 421–430.

Jacobs, M. R. (1955). "Growth Habits of the Eucalypts." Forestry and Timber Bureau, Commonwealth Government Printer, Canberra.

Jarvis, P. G., and Leverenz, J. W. (1983). Productivity of temperate, deciduous and evergreen forests. *In* "Physiological Plant Ecology. IV. Ecosystem Processes: Mineral Cycling, Productivity and Man's Influences" (O. L. Lange, P. S. Nobel, C. B. Osmond, and H. Ziegler, eds.), pp. 233–280. Springer-Verlag, Berlin.

Jarvis, P. G., and McNaughton, K. G. (1986). Stomatal control of transpiration: Scaling up from leaf to region. *Adv. Ecol. Res.* 15, 1–49.

Jarvis, P. G., James, G. B., and Landsberg, J. J. (1976). Coniferous forests. *In* "Vegetation and the Atmosphere" (J. L. Monteith, ed.), Vol. 2, pp. 171–240. Academic Press, London.

Jones, H. G. (1983). "Plants and Microclimate." Cambridge Univ. Press, Cambridge, UK.

Jordan, C. F. (1969). Derivation of leaf area index from quality of light on the forest floor. *Ecology* 50, 663–666.

Kelliher, F. M., Whitehead, D., McAneney, K. J., and Judd, M. J. (1990). Partitioning evapotranspiration into tree and understory components in two young *Pinus radiata* D. Don stands. *Agric. For. Meteorol.* 50, 211–227.

Kinerson, R. S., and Fritschen, L. J. (1971). Modeling a coniferous forest. *Agric. Meteorol.* 8, 439–445.

Kinerson, R. S., Higgenbotham, K. O., and Chapman, R. C. (1974). The dynamics of foliage distribution within a forest canopy. *J. Ecol.* 11, 347–353.

Kira, T., and Yoda, K. (1989). Vertical stratification in microclimate. *In* "Tropical Rain Forest Ecosystems" (H. Lieth and M. J. A. Werger, eds.), pp. 55–71. Elsevier, Amsterdam.

Kira, T., Shinozaki, K., and Hozumi, K. (1969). Structure of forest canopies as related to their primary productivity. *Plant Cell Physiol.* 10, 129–142.

Kittredge, J. (1948). "Forest Influences." McGraw-Hill, New York.

Krabill, W. B., Collins, J. G., Link, L. E., Swift, R. N., and Butler, M. L. (1984). Airborne laser topographic mapping results. *Photogramm. Eng. Remote Sens.* 50, 685–694.

Kramer, P. J., and Kozlowski, T. T. (1979). "Physiology of Woody Plants." Academic Press, Orlando, FL.

Kruijit, B. (1989). Estimating canopy structure of an oak forest at several scales. *Forestry* 62, 269–284.

Kuiper, L. C. (1988). "The Structure of Natural Douglas-fir Forests in Western Washington and Western Oregon," Paper 88-5. Agric. Univ. Wageningen, Wageningen.

Landsberg, J. J. (1986). "Physiological Ecology of Forest Production." Academic Press, London.

Landsberg, J. J., and James, G. B. (1971). Wind studies in plant canopies: Studies on an analytical model. *J. Appl. Ecol.* 8, 729–741.

Lang, A. R. G. (1987). Simplified estimate of leaf area index from transmittance of the sun's beam. *Agric. For. Meteorol.* 41, 179–186.

Leckie, D. G. (1990). Advances in remote sensing technologies for forest surveys and management. *Can. J. For. Res.* 20, 464–483.

Lee, D. W. (1987). The spectral distribution of radiation in two neotropical rainforests. *Biotropica* 19, 161–166.

Lee, R. (1983). "Forest Microclimatology." Columbia Univ. Press, New York.

Leigh, E. G., Jr. (1975). Structure and climate in tropical rain forest. *Annu. Rev. Ecol. Syst.* 6, 67–86.

Lemon, E., Allen, L. H., Jr., and Müller, L. (1970). Carbon dioxide exchange of a tropical rain forest. Part II. *BioScience* 20, 1054–1059.

Longman, K. A., and Jenik, J. (1987). "Tropical Forest and its Environment." Longman, London.

Lovett, G. M., and Lindberg, S. E. (1992). Concentration and deposition of particles and vapors in a vertical profile through a forest canopy. *Atmos. Environ.* 26A, 1469–1476.

MacArthur, R. H., and Horn, H. S. (1969). Foliage profiles by vertical measurements. *Ecology* 50, 802–804.

Martens, S. N., Ustin, S. L., and Rousseau, R. A. (1993). Estimation of the canopy leaf area index by gap fraction analysis. *For. Ecol. Manage.* 61, 91–108.

Maser, C. (1989). "Forest Primeval: The Natural History of an Ancient Forest." Sierra Club Books, San Francisco.

Massman, W. J. (1982). Foliage distribution in old-growth coniferous tree canopies. *Can. J. For. Res.* 12, 10–17.

McCune, B. C., and Boyce, R. L. (1992). Precipitation and the transfer of water, nutrients and pollutants in tree canopies. *Trends Ecol. Evol.* 7, 4–7

McNaughton, K. G. (1989). Micrometeorology of shelter belts and forest edges. *Philos. Trans. R. Soc. London, Ser. B* 324, 351–368.

Medina, E., and Klinge, H. (1983). Productivity of tropical forests and tropical woodlands. *In* "Physiological Plant Ecology. IV. Ecosystem Processes: Mineral Cycling, Productivity and Man's Influences" (O. L. Lange, P. S. Nobel, C. B. Osmond, and H. Ziegler, eds.), pp. 281–303. Springer-Verlag, Berlin.

Meyers, T. P., and Paw U, K. T. (1987). Modelling the plant canopy microenvironment with higher-order closure principles. *Agric. For. Meteorol.* 41, 143–163.

Miller, D. R., and Lin, J. D. (1985). Canopy architecture of a red maple edge stand measured by a point drop method. *In* "The Forest–Atmosphere Interaction" (B. A. Hutchison and B. B. Hicks, eds.), pp. 59–70. Reidel Publ., Dordrecht, The Netherlands.

Monsi, M., and Saeki, T. (1953). Über den Lichtfaktor in den Pflanzegesellschaften und seine Bedeutung für die Stoffproduktion. *Jpn. J. Bot.* 14, 22–52.

Monsi, M., Uchijima, Z., and Oikawa, T. (1973). Structure of foliage canopies and photosynthesis. *Annu. Rev. Ecol. Syst.* 4, 301–327.

Monteith, J. L., ed. (1975). "Vegetation and the Atmosphere," Vol. 1. Academic Press, London.

Monteith, J. L., ed. (1976). "Vegetation and the Atmosphere," Vol. 2. Academic Press, London.

Monteith, J. L., and Unsworth, M. H. (1990). "Principles of Environmental Physics." Edward Arnold, London.

Morgan, D. C., and Smith, H. (1981). Non-photosynthetic responses to light quality. *In* "Physiological Plant Ecology. I. Responses to the Physical Environment" (O. L. Lange, P. S. Nobel, C. B. Osmond, and H. Ziegler, eds.), pp. 109–134. Springer-Verlag, Berlin.

Munn, R. E. (1966). "Descriptive Micrometeorology." Academic Press, New York.

Neumann, H. H., den Hartog, G., and Shaw, R. H. (1989). Leaf area measurements based on hemispheric photographs and leaf-litter collection in a deciduous forest during autumn leaf-fall. *Agric. For. Meteorol.* 45, 325–345.

Newman, I. (1954). Locating strata in tropical rain forest. *J. Ecol.* 42, 218–219.

Ng, F. S. P. (1977). Shyness in trees. *Nat. Malays.* 2, 34–37.

Nobel, P. S., Forseth, I. N., and Long, S. P. (1993). Canopy structure and light interception. *In* "Photosynthesis and Production in a Changing Environment: A Field and Laboratory Manual" (D. O. Hall, J. M. O. Scurlock, H. R. Bolhàr-Nordenkampf, R. C. Leegood, and S. P. Long, eds.), pp. 79–90. Chapman & Hall, London.

Norman, J. M. (1979). Modelling the complete crop canopy. *In* "Modification of the Aerial Environment of Crops" (B. Barfield and J. Gerber, eds.), pp. 249–277. Am. Soc. Agric. Eng., St. Joseph, MO.

Norman, J. M. (1982). Simulation of microclimates. *In* "Biometeorology in Integrated Pest Management" (J. L. Hatfield and I. J. Thomason, eds.), pp. 65–99. Academic Press, New York.

Norman, J. M., and Campbell, G. S. (1989). Canopy structure. *In* "Plant Physiological Ecology: Field Methods and Instrumentation" (R. W. Pearcy, J. R. Ehleringer, H. A. Mooney, and P. W. Rundel, eds.), pp. 301–325. Chapman & Hall, London.

Norse, E. A. (1990). "Ancient Forests of the Pacific Northwest." Island Press, Washington, DC.

Nychka, D., and Nadkarni, N. M. (1994). Spatial analysis of points on tree structures: The distribution of epiphytes on tropical trees. *Biometrics* (in press).

Odum, H. T., Drewry, G., and Kline, J. R. (1970). Climate at El Verde. *In* "A Tropical Rain Forest: A Study of Irradiation and Ecology at El Verde, Puerto Rico" (H. T. Odum and R. F. Pigeon, eds.), pp. B347–B418. U.S. Atomic Energy Commission, Oak Ridge, TN.

Ogawa, H., Yoda, K., Kira, T., Ogino, K., Shidei, T., Ratanawongse, D., and Apasutaya, C. (1965). Comparative ecological study on three main types of forest vegetation in Thailand. I. Structure and floristic composition. *Nat. Life Southeast Asia* 4, 13–49.

Oke, T. R. (1987). "Boundary Layer Climates." Methuen, London.

Oldeman, R. A. A. (1990). "Forests: Elements of Silvology." Springer-Verlag, Berlin.

Oliver, C. D., and Larson, B. C. (1990). "Forest Stand Dynamics." McGraw-Hill, New York.

Oliver, H. R. (1975). Wind speeds within the trunk space of a pine forest. *Q. J. R. Meteorol. Soc.* 101, 168–169.

O'Neill, R. V., and DeAngelis, D. L. (1981). Comparative productivity and biomass relations of forest ecosystems. *In* "Dynamic Properties of Forest Ecosystems" (D. E. Reichle, ed.), pp. 411–449. Cambridge Univ. Press, Cambridge, UK.

Parker, G. G. (1983). Throughfall and stemflow in the forest nutrient cycle. *Adv. Ecol. Res.* 13, 57–133.

Parker, G. G. (1993). Structure and dynamics of the outer canopy of a Panamanian dry forest. *Selbyana* 14, 5.

Parker, G. G., O'Neill, J. P., and Higman, D. (1989). Vertical profile and canopy organization in a mixed deciduous forest. *Vegetatio* 89, 1–12.

Parker, G. G., Smith, A. P., and Hogan, K. P. (1992). Access to the upper forest canopy with a large tower crane. *BioScience* 42, 664–670.

Parker, G. G., Hill, S. M., and Kuehnel, L. A. (1993). Decline of understory American chestnut (*Castanea dentata* (Marsh.) Borkh.) in a southern Appalachian forest. *Can. J. For. Res.* 23, 259–266.

Paw U, K. T., and Meyers, T. P. (1989). Investigations with a higher-order canopy turbulence model into mean source–sink levels and bulk canopy resistances. *Agric. For. Meteorol.* 47, 259–271.

Pearcy, R. M., Ehleringer, J. R., Mooney, H. A., and Rundel, P. W. eds. (1989). "Plant Physiological Ecology: Field Methods and Instrumentation." Chapman and Hall, London, UK.

Perry, S. G., Fraser, A. B., Thomson, D. W., and Norman, J. M. (1988). Indirect sensing of plant canopy structure with simple radiation measurements. *Agric. For. Meteorol.* 42, 255–278.

Pierce, L. L., and Running, S. W. (1988). Rapid estimation of coniferous forest leaf area index using a portable integrating radiometer. *Ecology* 69, 1762–1767.

Putz, F. E., Parker, G. G., and Archibald, R. M. (1984). Mechanical abrasion and intercrown spacing. *Am. Midl. Nat.* 112, 24–28.

Rauner, J. L. (1976). Deciduous forests. *In* "Vegetation and the Atmosphere" (J. L. Monteith, ed.), Vol. 2, pp. 241–264. Academic Press, London.

Raupach, M. R. (1989). Stand overstory processes. *Philos. Trans. R. Soc. London, Ser. B* 324, 175–190.

Raupach, M. R., and Thom, A. S. (1981). Turbulence in and above plant canopies. *Annu. Rev. Fluid Mech.* 13, 97–129.

Rich, P. M. (1990). Characterizing plant canopies with hemispherical photographs. *Remote Sens. Environ.* 5, 13–29.

Richards, P. W. (1952). "The Tropical Rain Forest." Cambridge Univ. Press, Cambridge, UK (with corrections in 1975).

Richards, P. W. (1983). The three dimensional structure of tropical rain forest. *In* "Tropical Rain Forest: Ecology and Management" (S. L. Sutton, T. C. Whitmore, and A. C. Chadwick, eds.), pp. 3–10. Blackwell, Oxford.

Rosenberg, N. J. (1974). "Microclimate: The Biological Environment." Wiley, New York.

Ross, J. (1975). Radiative transfer in plant communities. *In* "Vegetation and the Atmosphere," (J. L. Monteith, ed.), Vol. 1, pp. 13–52. Academic Press, London.

Runkle, J. R. (1985). Disturbance regimes in temperate forest. *In* "The Ecology of Natural Disturbances and Patch Dynamics" (S. T. A. Pickett and P. S. White, eds.), pp. 17–33. Academic Press, Orlando, FL.

Russell, G., Marshall, B., and Jarvis, P., eds. (1989). "Plant Canopies: Their Growth, Form and Function." Cambridge Univ. Press, Cambridge, UK.

Rutter, A. J., Hershaw, K. A., Robins, P. C., and Morton, A. J. (1971). A predictive model of rainfall interception in forests. I. Derivation of the model and comparison from observations in Corsican pine. *Agric. Meteorol.* 9, 367–384.

Rutter, A. J., Morton, A. J., and Robins, P. C. (1975). A predictive model of rainfall interception in forests. II. Generalizations of the model and comparison with observations in some coniferous and hardwood stands. *J. Appl. Ecol.* 12, 367–380.

Saeki, T. (1973). Distribution of radiant energy and CO_2 in terrestrial plant communities. *In* "Photosynthesis and Productivity in Different Environments" (J. P. Cooper, ed.), pp. 297–327. Cambridge Univ. Press, Cambridge, UK.

Sanford, R. L., Jr., Braker, H. E., and Hartshorn, G. S. (1986). Canopy openings in a primary neotropical lowland forest. *J. Trop. Ecol.* 2, 277–282.

Shaw, R. H. (1977). Secondary wind speed maxima inside plant communities. *J. Appl. Meteorol.* 16, 514–521.

Shaw, R. H., and Pereira, A. R. (1982). Aerodynamic roughness of a plant canopy: A numerical experiment. *Agric. Meteorol.* 26, 51–56.

Shaw, R. H., den Hartog, G., and Neumann, H. H. (1988). Influence of foliar density and thermal stability on profiles of Reynolds stress and turbulence intensity in a deciduous forest. *Boundary-Layer Meteorol.* 45, 391–409.

Shuttleworth, W. J. (1989). Micrometeorology of temperate and tropical forests. *Philos. Trans. R. Soc. London, Ser. B* 324, 299–334.

Sigmon, J. T., Knoerr, K. R., and Shaughnessy, E. J. (1983). Microscale pressure fluctuations in a mature deciduous forest. *Boundary-Layer Meteorol.* 27, 345–358.

Smith, A. P. (1973). Stratification of temperate and tropical forest. *Am. Nat.* 107, 671–683.

Smith, A. P., Hogan, P., and Idol, J. R. (1992). Spatial and temporal patterns of light and canopy structure in a lowland tropical moist forest. *Biotropica* 24, 503–511.

Smith, D. M. (1962). "The Practice of Silviculture." Wiley, New York.

Smith, H. (1982). Light quality, photoreception and plant strategy. *Annu. Rev. Plant Physiol.* 33, 481–518.

Smith, N. J., Borstad, G. A., Hill, D. A., and Kerr, R. C. (1991). Using high-resolution airborne spectral data to estimate forest leaf area and stand structure. *Can. J. For. Res.* 21, 1127–1132.

Spanner, M. A., Pierce, L. L., Peterson, D. L., and Running, S. W. (1990). Remote sensing of temperate coniferous forest leaf area index—The influence of canopy closure, understory vegetation and background reflectance. *Int. J. Remote Sens.* 11, 95–111.

Stanhill, G. (1970). Some results of helicopter measurements of albedo. *Sol. Energy* 13, 59–66.

Stephens, G. R. (1969). Productivity of red pine. 1. Foliage distribution in tree crown and stand canopy. *Agric. Meteorol.* 6, 275–282.

Tadaki, Y. (1977). Some discussions on the leaf biomass of forest stands and trees. *Tokyo For. Exp. Stn. Bull.* 184, 135–161.

Terborgh, J. (1985). The vertical component of plant species diversity in temperate and tropical forests. *Am. Nat.* 126, 760–776.

Thom, S. (1971). Momentum absorption by vegetation. *Q. J. R. Meteorol. Soc.* 97, 414–428.

Uhl, C., and Jordan, C. F. (1984). Vegetation and nutrient dynamics during the first five years of succession following forest cutting and burning in the upper Rio Negro region of Amazonia. *Ecology* 65, 1476–1490.

UNESCO/UNEP/FAO (1978). "Tropical Forest Ecosystems." UNESCO-UNEP, Paris.

Vanderbilt, V. C. (1985). Measuring plant canopy structure. *Remote Sens. Environ.* 18, 281–294.

Waggoner, P. E. (1975). Micrometeorological models. *In* "Vegetation and the Atmosphere" (J. L. Monteith, ed.), Vol. 1, pp. 205–228. Academic Press, London.

Waring, R. H., and Schlesinger, W. H. (1985). "Forest Ecosystems: Concepts and Management." Academic Press, Orlando, FL.

Warren-Wilson, J. (1965). Stand structure and light penetration. 1. Analysis by point quadrats. *J. Appl. Ecol.* 2, 383–390.

Welles, J. M. (1990). Some indirect methods for evaluating canopy structure. *Remote Sens. Environ.* 5, 31–43.

Whitmore, T. C. (1984). "Tropical Rain Forests of the Far East." Oxford Univ. Press, Oxford.

Whittaker, R. H., and Woodwell, G. M. (1968). Surface area relations of woody plants and forest communities. *Am. J. Bot.* 54, 931–939.

Wierman, C. A., and Oliver, C. D. (1979). Crown stratification by species in even-aged mixed stands of Douglas-fir—Western hemlock. *Can. J. For. Res.* 9, 1–9.

Wiman, B. L. B., and Agren, G. I. (1985). Aerosol depletion and deposition in forests—A model analysis. *Atmos. Environ.* 19, 335–347.

Wiman, B. L. B., Agren, G. I., and Lannefors, H. O. (1985). Aerosol concentration profiles within a mature coniferous forest—Model versus field results. *Atmos. Environ.* 19, 363–367.

Woodwell, G. M., and Dykeman, W. R. (1966). Respiration of a forest measured by CO_2 accumulation during temperature inversions. *Science* 154, 1031–1034.

Yang, X., Miller, D. R., and Montgomery, M. E. (1993). Vertical distribution of canopy foliage and biologically active radiation in a defoliated/refoliated hardwood forest. *Agric. For. Meteorol.* 67, 129–146.

Yoda, K. (1978). The three-dimensional distribution of light intensity in a tropical rain forest in West Malaysia. *Jpn. J. Ecol.* 24, 247–254.

Zeide, B., and Pfeifer, P. (1991). A method for estimating the fractal dimension of tree crowns. *For. Sci.* 37, 1253–1265.

II

Organisms in Tree Canopies

5

Measuring Arthropod Biodiversity in the Tropical Forest Canopy

Terry L. Erwin

*These 19 trees . . . produced 955 species of beetles,
excluding weevils.*
—*T. L. Erwin (1982)*

A total of 41,844 individuals were sorted and identified.
—*V. C. Moran and T. R. E. Southwood (1982)*

*51,600 canopy arthropods were collected . . .
[and] 759 species were recognized.*
—*Y. Basset (1991a)*

23,874 arthropods were collected . . . represent[ing] 3059 species.
—*N. E. Stork (1991)*

We obtained a total of 4840 beetles representing 633 species.
—*A. Allison et al. (1993)*

I. Introduction

Tropical arboricolous arthropods were observed in the early 1800s in the "great forests near the equator in South America" and were later described by Henry Walter Bates (1884). Despite Bates' keen observations, more than a century passed before insects and their relatives were routinely observed in the forest canopy or collected directly from it (Collyer, 1951). William Beebe and collaborators (1917) also recognized that the canopy held marvels of biological interactions, but "gravitation and tree-trunks swarming

Copyright © 1995 by Academic Press, Inc.
All rights of reproduction in any form reserved.

with terrible ants" kept him at bay. Frank Chapman, the canopy pioneer (of sorts), had taken to viewing the treetops from his "Tropical Air Castle" in Panama in the 1920s, but his interest was vertebrate-oriented, his perch was a tower (Chapman, 1929; p. 187), and his arthropod observations were casual.

Prior to the early 1950s, disparate records and facts about arthropod species of the canopy's crown rim, inner canopy branches, boles, and epiphytes came from felled trees or trapping techniques [e.g., ultraviolet (UV) and mercury vapor lights] that dislodged or drew individuals from their natural (normal) microhabitats. No techniques were available to examine arthropod stratification nor to look for upper crown or crown-rim specialists. Collyer (1951) seems to have been the first to use chemical "knockdown" to measure arthropod abundance in fruit trees. His study was published in a research station bulletin directed to applied researchers, and it received little attention. Southwood's (1960, 1961) scientific interest in arboreal insects began when he assembled data from Hawaii, Russia, Sweden, Cyprus, and Great Britain. Unlike Collyer, Southwood's interest was directed toward basic research. However, his early studies were founded on faunal and floral lists rather than on field samples.

After Collyer, Martin (1966) and Gagné and Martin (1968) were the first to collect canopy arthropods *in situ*, first in temperate-zone pine plantations of Canada and later in acacia forests of Hawaii (Gagné, 1979). Their studies, like Collyer's, were economically oriented. Southwood's earlier interest in trees as islands was echoed in Central America by Janzen (1968, 1973) and by Opler (1974) in North America, Taksdal (1965) in Norway, and Evans (1966) in Australia and New Zealand, and was later reviewed by Southwood and Kennedy (1983). None of these other authors was working in the canopy itself. My field work with Lubin and Montgomery in Panama in 1974 (Erwin and Scott, 1980; Erwin, 1982), with Adis in 1979 in Brazil (Erwin, 1983a) and in Peru (Erwin, 1983a, b, 1991b; Farrell and Erwin, 1988) began the first sustained basic research program concerned solely with arthropod samples from the canopy itself. In the late 1970s and early 1980s, Southwood and his colleagues and students turned to chemical knockdown to study arboreal temperate faunas in England and South Africa (Southwood *et al.*, 1982a, b; Moran and Southwood, 1982).

In 1979, Southwood *et al.* sampled using a hydraulically applied pyrethrum spray of visible liquid droplets (following Martin, 1966), rather than an insecticidal "fogging" technique (Roberts, 1973), where the pyrethrum is nearly gaseous (droplets of less than 5 μm). This group thus established a sampling regime in the Eastern Hemisphere parallel to, but not directly comparable with, that under way in Neotropical forests (Erwin and Scott, 1980; Adis *et al.*, 1984; Erwin, 1983a, b). A third system, "smoking," was established in Japanese tree plantations (Hijii, 1983), where a canister or two of insecticide (Varsan P Jet No. 1) was released into a tree that was enclosed by a tent. A historical account of canopy studies summarized various "true" fogging and smoking applications (Erwin, 1989a). Since the late 1980s, a plethora of short-term and sustained studies of canopy arthropods have been implemented in tropical and temperate zones using a variety of *in situ* branch-cutting, trapping, and/or "fogging" techniques (e.g., Teulon and Penman, 1987).

Despite over four decades of interest, both remote and *in situ* canopy and subcanopy arthropod studies resulting in samples of thousands of species and tens of thousands of specimens have produced a relatively small body of work on the "little things" (Wilson, 1987) that inhabit this latest frontier of hyperdiverse taxa to be studied. Reasons (and excuses) abound, but aside from canopy access, a fundamental factor is the enormous task of recording and extracting data from such diverse and abundant samples, particularly the rich tropical ones, in a short period of time (Hevel, 1988; Erwin, 1989b; Steiner, 1989; Ashe, 1991; Erwin and Pogue, 1992). The task is further compounded by the paucity of taxonomists with time, funds, or interest to tackle the truly horrendous problem of identifying all of these arthropods.

Sampling design for canopy and subcanopy arthropod diversity outpaced processing methods by a factor of at least four. It took roughly five years to turn Martin/Gagné/Southwood spraying and Roberts fogging techniques into reliable tools for acquiring comparable samples, but it took some 20+ additional years to design and implement a processing and analysis system to handle the specimens and data on a scale equal to the task (Erwin, 1991a, 1992a, unpublished data; Erwin and Pogue, 1992). Studies initiated before the establishment of this bulk processing and analysis system often were restricted to a few tree or arthropod taxa, single tree species, single microhabitats in trees, or trees with small faunas (e.g., Erwin and Scott, 1980; Erwin, 1982; Majer and Recher, 1988; Watanabe and Ruaysoongnern, 1989; Nadkarni and Longino, 1990; Stork, 1991; Tobin, 1991; Allison *et al.*, 1993; Basset, 1993a), or made identifications at the order or family-group level (Adis *et al.*, 1984; E. Guilbert *et al.*, unpublished data; Kitching *et al.*, 1993). Hence, only tantalizing tidbits of information have emerged from tropical and subtropical studies and virtually no global comparative analyses have appeared. The temperate zone is not much better off in this respect (e.g., Futuyma and Gould, 1979; Ohmart and Voigt, 1981; Southwood *et al.*, 1982a,b; Ohmart *et al.*, 1983; Majer *et al.*, 1990; Recher *et al.*, 1991; N. Stork and Hammond, unpublished data).

The purpose of this chapter is to review what has been accomplished with arthropod diversity sampling techniques in tropical forest studies, and in temperate studies that are associated with the tropical fauna, thereby providing an historic context for this expanding field of biological investigation. I also explain and elaborate current inventory techniques and processing systems. Last, I explore future options in our studies of arboricolous arthropod diversity. A rich applied entomology literature on arthropods that infest crowns of timber stands exists, but is not considered in this chapter, except where such studies played a role in developing new sampling techniques.

II. Methods and Materials

A. Canopy and Subcanopy Arthropod Diversity Projects and Sampling Methods

Hand-extracting, branch-clipping and bagging, trapping, smoking, and hydraulic spraying of trees either target certain species or groups of species or are limited in some other fashion in providing a complete sample of arthropod diversity (Erwin, 1989a). True fogging techniques, that is, dynamic dispersal of insecticidal droplets of 5 μm in diameter, when used correctly, are superior to these other methods, as they acquire nearly complete samples of substrate surface arthropod diversity, especially in tropical forests (Erwin, 1983a,b, 1989a). Erwin (1989a) and Basset (1991a) used rope access techniques (adapted from Perry, 1978; Lowman, 1984; Nadkarni, 1984) to fog or gas from within the canopy. N. Stork and Hammond (personal communication) used a mechanical arm mounted on a truck ("cherry picker") to elevate a person and fogging machine to canopy level in a park in London. Basset (1988, 1990) developed techniques for gassing (CO_2) and Malaise-window trapping for canopy arthropods, and Majer *et al.* (1990) used a "misting" method. Nadkarni and Longino (1990) and H. Frania (personal communication) independently developed techniques for extracting leaf litter and associated arthropods from the canopy and subcanopy. Erwin (1991b) extracted the arthropods from suspended leaf litter *in situ* by fogging. Several workers, such as R. Colwell and J. Longino in Costa Rica and Y. Basset in Papua New Guinea (personal communication) are currently testing a comparative series of sampling methods in the canopy and subcanopy. Several entomologists used a "canopy raft" provided by F. Hallé and his team in French Guyana and in Cameroon (Hallé and Blanc, 1990). This raft was placed on the canopy with a large dirigible and was used as a stationary observation platform. The dirigible was also used in Cameroon to move a small sledge over the canopy while M. Lowman and M. Moffett (personal communication) made aerial sweep samples from the upper crown rim. Basset *et al.* (1992) used light-traps on the platform and took branch clipping samples from both the platform and sledge. Various workers have used canopy-suspended light traps or sticky traps (e.g., Wolda,

1978; Lowman, 1982; Broadhead and Wolda, 1985). The construction crane of the Smithsonian Tropical Research Institute, with a small, suspended, and movable gondola (Parker *et al.*, 1992), Hallé's canopy raft (Hallé and Blanc, 1990), and canopy walkways have not yet been used for making large-scale inventories, although they have been used for ecological studies involving arthropods (Wint, 1983; Lowman, 1985; Basset and Arthington, 1992).

In the following, I summarize the history of sustained field projects that have resulted in, or promise, publications on arboricolous insects. Other excellent one-time studies are mentioned elsewhere in this chapter.

1. A. Allison and Company: Sampling Years, 1987 to Present   A. Allison and S. Miller of the Bishop Museum, Honolulu, began a study of New Guinean fagaceous trees in 1987 and enlisted the taxonomic expertise of G. A. Samuelson. Recently, Y. Basset has collaborated on the project. Their goals are to sample southern-oak tree (*Castanopsis*) canopies on an altitudinal transect. They focus on invertebrate herbivores at three different altitudes to develop a canopy arthropod data base. They used fogging techniques adapted from Erwin (1983a,b, 1989a) and Basset's techniques from Australia (see following description). They have a large fauna, so specimen processing has slowed the analysis, as it has in most other studies (Allison *et al.*, 1993). Taxonomists at the Bishop Museum, particularly Samuelson, processed, sorted, and identified the bulk of the specimens.

2. Y. Basset and Company: Sampling Years, 1985–1989   Yves Basset, with the help of his taxonomist colleagues (some 40 of them, 11 outside of Australia), has produced an admirable body of work on the crown of the subtropical tree species *Argyrodendron actinophyllum* Edlin (Sterculiaceae) near Brisbane, Australia. Basset worked with this tree species over a period of four and one-half years as his Ph.D. project (1993a). He documented trapping techniques and all arthropods in terms of abundance, trophic structure of the community, species richness, body length, spatial and seasonal distributions, diel activity, and characteristics of the tree species itself (Basset, 1988, 1990, 1991a–e, 1992a–d, 1993a,b; Basset and Arthington, 1992; Basset and Kitching, 1991; Basset and Springate, 1992). Basset collected over 50,000 individual arthropods distributed across 760 species during this study, 660 of which were identified to at least the generic level. Mt. Glorious, near Brisbane, where the study was carried out, is subtropical (Basset, 1991a), which perhaps explains the rather small number of species in this tree. The field work was continuous and extensive, using gassing with CO_2 (termed "restricted canopy fogging"), hand collecting, beating, and arboreal Malaise-window trapping (termed "composite interception trap"), which constituted a multifold methodology blitz on the canopy fauna of the target tree. One exciting aspect of this study is that sampling was continuous over a long period, so diel and seasonal cycles could be quantified (e.g., Basset, 1991c,d; Basset and Springate, 1992). Remaining in the field for long periods of time is not a luxury most researchers have. Basset (1991a, Appendix II) enlisted taxonomists from Australian and overseas institutions to provide identifications, but in some cases did the sorting of taxa himself.

3. T. L. Erwin and Company: Sampling Years, 1974 to Present   My studies have included Panama (1974), Brazil (1979), Peru (1981 to present at four sites), Bolivia (1987), and Ecuador (1994 at three sites). I began with a study of beetles from the canopy of a single tree species, *Luehea seemannii* Triana and Planch (Tiliaceae), and then examined specific types of forests. I am now studying microhabitats within different local forest types across geographic landscapes to look at species turnover in time and space, as well as guild member replacement and microhabitat fidelity of beetle species. I have reported on trophic groups, species richness, densities, biomass, abundances, and species turnover (Erwin, 1982, 1983a,b, 1988, 1989a, 1991b; Erwin and Scott, 1980), and have compared herbivore/predator distribution across forest types (Farrell and Erwin, 1988). Much of my time in the early 1980s was devoted to the development of fogging techniques and design of permanent plots where studies could be repeated and tested (Erwin, 1991c; Erwin and Kabel, 1991). More recently, I have developed bulk specimen preparation and data-gathering techniques, and computer programming to handle the very large data sets from the canopy and subcanopy trees and specimens (T. Erwin, unpublished data). I have done all the taxonomic sorting of the beetles, my target taxon in these studies. J. Lawrence, A. Newton, and M. Thayer assisted me in setting up a synoptic family collection of beetles in the early 1980s. Various resident and visiting coleopterists who have used specimens from this collection during the past ten years have corrected my errors. The ANTSE function (an arthropod sorting center) of the Department of Entomology at the Smithsonian Institution (Erwin, 1992a) and my assistant Michael G. Pogue are responsible for sorting and distribution of all other taxa.

4. R. L. Kitching and Company: Sampling Years, 1987 to Present   R. L. Kitching joined M. Lowman (see the following) in 1987 to add a fogging component to her studies. They developed an ambitious program to compare canopy arthropod faunas from temperate New South Wales to northern tropical Queensland (Kitching *et al.,* 1993). Beginning with methodology and an ordinal overview in their first paper, the team plans to offer "detailed results on particular orders and/or forest types" (Kitching *et al.,* 1993). Taxonomists at Australian museums have aided in identifying specimens.

5. M. Lowman and Company: Sampling Years, 1979 to Present M. Lowman, who is interested in herbivory, carried out ecological studies in Australia involving access to the canopy in the early 1980s. In her early work, she used branch-clipping techniques, light traps, sweep nets, beating trays, and *in situ* measures and observations. In 1987, R. L. Kitching joined her canopy group and their studies expanded to include fogging. Lowman is interested in insect herbivore impact in Australian rain forests and has approached this from leaf-growth dynamics. She focused on spatial and temporal variability in defoliation in specific tree groups. Beginning in 1990, she has undertaken canopy insect studies in temperate deciduous forests, using a permanent platform and walkway for access. With undergraduate students, she has been comparing insect abundance between upper and lower canopy strata using Malaise-trapping in addition to the array of techniques employed in Australia.

6. Sir T. R. E. Southwood and Company: Sampling Year, 1979 T. R. E. Southwood and his team are the only researchers to have sampled amphitropically in a truly comparative mode. They gathered their data from tree species that were either native or introduced into England and South Africa (Southwood and Kennedy, 1983). Where possible, they sampled from the same species of trees in both places and recorded guild composition (Moran and Southwood, 1982), species richness, abundance, and biomass (Southwood *et al.*, 1982a) to examine niche breath and the accumulation of species in new habitats (i.e., introduced trees). Kennedy and Southwood (1984) examined species richness in relation to area (sizes of trees and their abundance on the landscape) as a follow-up to Southwood's original paper (1960). Taxonomic identification at the morphospecies level was made by Southwood, his team members, and numerous taxonomists in Britain and South Africa. Southwood *et al.* (1979) used scaffolding to gain access to the canopy at about 6 m off the ground, where they applied a vacuum suction device to sample insects.

7. N. Stork and Company: Sampling Years, 1982–1985 N. Stork and his colleagues from The Natural History Museum, London, undertook studies in Brunei and Sulawesi, the latter expedition lasting a year and involving several teams of investigators. Stork examined trophic structure of the community across tree and forest types, species richness, seasonality, density, and abundance in relation to body size, and altitudinal comparisons of all of these (Stork, 1987a,b, 1991; Stork and Brendell, 1990; Morse *et al.*, 1988). Noyes (1989) used Stork's fogged Hymenoptera from Sulawesi for a comparative study of five collecting techniques. N. Stork and Hammond (personal communication) used their experience in Southeast Asia to study the arthropods of oak trees in a London park. Stork enlisted Hammond and other taxonomists at The Natural History Museum to sort, prepare, and identify the specimens.

III. Results

The goal of all studies has been to characterize the canopy fauna in terms of numbers of taxa and their abundance, biomass, and distribution through time. Some have discussed taxa as they are assembled in feeding guilds. Others have measured the environment, for example, rates of herbivory through leaf examination and monitoring. Seasonality, diel activity, and spatial (particularly altitudinal) turnover have received some attention. Faunal stratification and both micro- and macrogeographic species turnover have been recently studied.

Overall, an average of 4.64 years elapsed between the initiation of sampling and the first publication with arthropod data (Fig. 1). An average of 5750 specimens was processed per year. Because some studies extracted data at the ordinal or family level, it is not possible to summarize processing at the species level for all studies.

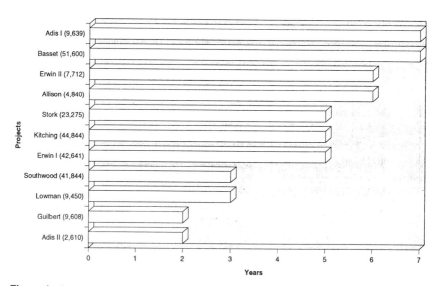

Figure 1   Lapse time between initial sampling of project and date of first publication with data resulting from that sampling. Total number of individual arthropods collected and studied for each project is given in parentheses. Adis II is an "ant" project with Dr. Ana Harada (personal communication). Erwin I is Panama; Erwin II is Tambopata, Peru. See project descriptions in text for others.

Methods of canopy sampling have been so various that comparing results of different investigators is almost useless except for making very broad generalizations (Guilbert *et al.,* 1994). The studies of each individual team, however, provide results that can form a focus for global cooperation and future conformity in sampling and analysis techniques.

A. What We Really Know

Very little is known about the canopy and subcanopy arthropod faunas in the forests around the world in the tropics, subtropics, and temperate zones (Erwin, 1992b). Sustained or single-event field studies using sampling techniques that result in huge collections (fogging, smoking, gassing) have taken place at only 20 sites around the world (Fig. 2) from a total of about 500 individual trees representing less than 100 tree species (less than 0.002% of possible tree species on the planet). Many of the tree species (e.g., in Manaus, the Iquitos area, and Sulawesi) either have not been identified or their names are unreported.

Quite possibly, crown faunas of England and other parts of Europe are the best known because of the long-held interest in biotope analysis, which emphasizes the study of whole faunas and floras rather than the theory of community ecology. Another reason is the far smaller, better-known arthropod faunas of Great Britain and the rest of Europe and the long tradition

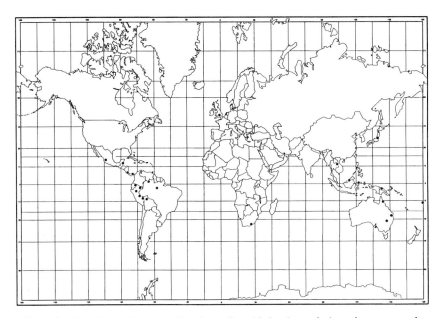

Figure 2   Sites (•) at which megadiversity studies with fogging techniques have occurred.

of taxonomy. Tropical studies, on the other hand, have only recently begun and include a huge fauna of arthropods in a diverse array of vegetation types.

IV. Discussion

A. Strategies of Recent Studies

1. Analyses   Studies of canopy and subcanopy insects have included many spatial scales: investigations of a single tree or tree species (Erwin and Scott, 1980; Erwin, 1982; Basset, 1993a), multiple tree species and transects (Lowman, 1982; Moran and Southwood, 1982; Erwin, 1983a,b; Adis *et al.,* 1984; Stork, 1987b; Stork and Brendell, 1990), latitudinal comparisons of forest types and tree species (Kitching *et al.,* 1993), and microhabitat comparisons in canopy, subcanopy, understory, and leaf litter (T. Erwin, unpublished data). These studies have used single sampling techniques such as fogging, or a variety of combined trapping and gassing methods. New studies focus on testing methods as the first priority (R. Colwell and J. Longino, personal communication) or an integral part of the overall study (Y. Basset, personal communication). Others gather data about the fauna as the first priority, having selected one type of sampling device.

Analyses have also varied considerably, ranging from simple descriptive statistics (sums, means, standard deviations, and application of some indexing procedures) to multivariant techniques, for example, principal component analysis, detrended correspondence analysis, and path analysis (Basset, 1991b, 1992a; E. Guilbert, personal communication). Canopy data have been used to examine relationships between body length and population parameters (Morse *et al.,* 1985), or to test the theory of fractal dimensions in plants (Lawton, 1986). Very few of the early canopy sampling projects began with asking questions beyond "what is up there?" Most did not use canopy sampling techniques to addressing ecological theory. Those beginning with such questions did not set rigorous sampling procedures that could be repeated. One reason for this general and unfocused approach was because canopy arthropod observations were virtually nonexistent until 20 years ago; there was very little on which to formulate hypotheses except as an extension of research at ground level. Canopy access was not feasible until recently, so statistically sound sampling regimes were not designed. Another reason is that taxonomists, not theory-based ecologists, pioneered this field, thus the questions "what is up there?" and "how, when, and where are they distributed?" were (and are) valid ones.

B. Defining the Systematist's Role

1. Traditions   Traditional ways are often plodding ways. In the study of insects and their relatives, tradition has resulted in less than adequate taxonomic and biogeographic knowledge and poor prospects for learning enough about insect biodiversity to plan for its conservation. The arguments against tradition have mainly been focused on response time from entomologists to get their samples ready for analysis, which has been a problem in canopy studies (Fig. 1). Tradition requires that specimens be prepared and named; the former takes days for a few hundred specimens depending on process (pinning, pointing, double-pointing, spreading), whereas the latter may require a trip to a European museum to study type specimens in order to confirm or discover a name. It may be years before large faunal-representative samples from a single site are ready for analysis if traditions are followed! The reason for Latin names is to track literature and improve consistency in communication. With perhaps 80% or more of tropical species undescribed, however, there is little information to track if recent estimates of the total numbers of species are near correct. Most insect species that are described from the tropics are known from one or a few specimens collected at a single site in the eighteenth and nineteenth centuries and described in two or three lines of text. This information is far too meager to wait for years for research to track. In general, canopy studies of the teams described here have relied on "morphospecies" as the OTU (operational taxonomic unit) on which counts and measures are made, thus avoiding the traditional delay. However, in these studies, the use of "morphospecies" as a substitution for traditions has not ensured consistency of identifications nor in all cases resulted in retrievable vouchers for the data sets, a problem that has been addressed before (Erwin, 1991a).

2. Interim Taxonomy   To avoid excessive delays between field work and data analysis, I developed a system of an interim taxonomy and synoptic processing (Erwin, 1991a, unpublished data; Erwin and Pogue, 1992). This process eliminates 95% of the preparation time required by traditional approaches and all of the time devoted to applying Latin names, while making all the specimens immediately available for study and each specimen and species trackable. It guarantees that a "type" exists and that species are vouchered for easy future checking. Further, the system provides readily available specimens for taxonomists to use for systematics and biogeography, which are the underpinnings of the entire endeavor of biology (May, 1990).

3. Expert Systems   Investigators cannot identify the very large samples usually obtained by fogging tree crowns and subcanopy microhabitats quickly enough by using printed keys to make associated data available for immediate purposes such as conservation decisions. Computer-based identification systems such as that pioneered by F. C. Thompson of the Systematic Entomology Laboratory, USDA (and now available from others in a few forms of software), will be needed to keep pace with the increasing flood of canopy and subcanopy arthropod species needing to be named. Once a synoptic collection of species representatives increases to more than one drawer of specimens, efficiency in matching newly processed specimens considerably decreases. Use of an interactive expert system with large amounts of disk space (or other quick-access hardware) for character storage and retrieval will speed the process of "match-identification."

C. Future Options

1. Single Investigators The most common mode of investigation is that of single researchers. A few senior scientists with their students or assistants (usually on shoestring budgets) were the first to breach the forest-top and subcanopy frontier in disparate parts of the world. However, these studies were not connected in any way, nor were methods comparable.

2. Team Investigations Global teamwork, using comparable methods not now in existence, is the only sure way of quickly getting the large data sets needed for useful, comparable results. Agreeing on the right protocol will be difficult: every investigator believes his or her approach is the most appropriate. Finding sufficient funding seems impossible. Given sufficient funding, however, teams involving available experts, together with their students and assistants, could probably be assembled to determine, at least to an order of magnitude, the amount of diversity in canopy and subcanopy habitats. In the process, they will gather much information that is enormously valuable to taxonomists, biogeographers, and ecologists. Even better would be teams of all interested disciplines set up to design the protocols so that the logistically difficult-to-obtain information is available for study using several approaches across the disciplines.

3. Walkways, Rafts, and Crane-gondolas These expensive devices are limited in that they allow access only to adjacent canopy (walkways and crane-gondolas) or are temporarily stationary until they are moved (in the case of rafts, which are logistically time-consuming and costly). For the inventory work necessary to measure arthropod diversity across the multitude of tree species in most tropical forests, they are not useful. However, in the small area in which such mechanisms are established, they are useful for answering ecological questions about those taxa within reach and particularly for long-term studies involving the circumscribed tree and arthropod species (in the case of walkways and crane-gondolas).

4. Standardization Standardization is the most difficult problem of all for canopy workers to solve, as illustrated by two case histories: (1) In the early 1980s, N. Stork accompanied me to Peru to learn my fogging technique. Upon his return to London, he was successful in getting funding to take an expedition to Brunei (and later to Sulawesi). Unfortunately, he acquired a fogger in Europe with different specifications than mine and selected a different pyrethrum formulation made in England. Quality and quantity of the catch are influenced primarily by the manner and power of the insecticide dispersal and its knockdown potential. His regime could not be considered equal in a rigorous comparison, so our analyses are not comparable. (2) Allison *et al.* (1993) also based their methods on my early work in Peru for their studies in Papua New Guinea. However, despite acquiring a fogger exactly like mine, they loaded it with another insecticide that forms a different solution percentage in the carrier. Thus, their knockdown may be qualitatively different than mine or Stork's and nobody's results can be compared.

V. Conclusions

Terrestrial arthropods may constitute 70–97% of biodiversity on the planet, according to some recent estimates. Not only do these 6- and 8-legged forms, most of which fly, make up the great majority of all species lists, their interactions with each other and with their biotic and abiotic environment and their genetically variant populations also swamp those of "higher" organisms in quantity. If we are to understand anything about the world in which we live, canopy and subcanopy arthropod taxa cannot be ignored. No reasoning person would accept a painting of the French master impressionist Georges Seurat with only five or six points of paint, nor would anyone accept a star map with only the 20 brightest stars. Why should anyone do the same with our picture, or map, of life on earth? Technology is now available to get the job done. The biological and conservation communities should make our view of the planet more complete. Perhaps through this improved view, our conservation problems may be more readily solved. Both sampling and processing the samples are now less daunting than ever before. It is time to extract crucial data so that we will know the biodiversity with which we share the planet.

Over the past 30 years, significant advances in the study of canopy and subcanopy arthropods have occurred. Although debates still abound on how to compare and use the data, and what levels of scaling are appropriate for making regional or global generalizations, there is little argument that very large samples can be efficiently extracted and organized for study. Modern computer technology assists in this organization, allows rapid access to data, and performs analytical statistics. Acceptance of an interim taxonomy system markedly speeds the reporting process; these new sample preparation and storage techniques promote follow-up taxonomic studies and ensure that all studies are fully vouchered in museums to facilitate future investigations.

A. Recommendations

Arthropods, particularly insects, are not randomly distributed across the environment; they have vastly different evolutionary histories. Their traits are adapted to the microenvironments in which they are found most of the time. Adult insects that are observed out of their "niche" will soon return if they have the chance. Immature individuals are more closely tied to their own microenvironment. The degree to which the foregoing is true depends on the trophic group to which the species belongs, and also on the latitude and altitude. Generalist arthropods (in the broad sense) tend to occur in tropical, open, wet situations and in areas where there is little annual fluctuation in climate; specialists are often more common nearer the equator in forest canopies and at higher latitudes and altitudes (Erwin, 1985).

Neotropical forests can have as many as 473 species of trees (5 cm or more DBH) in one hectare (Valencia *et al.,* 1994), all of which support their special faunules, or microcommunities, of arthropods distributed within fractal universes (T. Erwin, unpublished data) in relation to microclimate and to the chemistry and architectural arrangement of the substrate. Some species and faunules may be host[1] specific. Designs for field sampling regimes should consider architecturally limited taxa. Previous fogging samples from all teams consisted of mixed, or potentially mixed, microhabitats within the crown, and therefore data analyses can only be descriptive of the general fauna in a tree.

What is needed is a workshop at which current and potential students of canopy arthropods come together to sort out methodology and develop appropriate sampling techniques. The aim of this workshop would be to generate the most complete, and readily comparable, sets of data in the shortest amount of time. Funding strategies can then be devised to accomplish both scientific and conservation objectives. Data should be collected to address basic questions and to meet the immediate needs of conservation planners. General inventory and biodiversity projects that have vertebrates and plants as the focus, which are commonly used in the conservation community, should add arthropods to the lists of data points. Only in this way can insects and their relatives be added to rapid assessment and survey strategies for conservation purposes and at the same time it would be possible to develop consistent data sets with which to address ecological questions.

VI. Summary

I summarize the history of canopy arthropod studies, particularly focusing on the fogging method of sampling. Major projects are described and

---

[1] Host specificity is often thought of as an herbivore that is restricted to plant species for its food; however, other restrictive "host" relations do exist, including parasites, hyperparasites, fungi feeders, and microhabitat architecture, which provide nutrients or substrates required for development or reproduction.

their general goals and results are compared. No projects are the same in their methods or goals; this has resulted in noncomparable data sets across sites studied by different teams. Recommendations to alleviate these problems are presented primarily by encouraging the development of coordinated efforts in a workshop attended by all interested teams and interested students of biodiversity. Specifically, sampling methods and a set of basic questions agreed to by all participating parties must be exactly the same at all sites to achieve global comparisons. We need to know what lives in the tropical forest canopy and subcanopy (taxonomy), why arthropods are there (evolution), and how the lineages and species are distributed in time and through microhabitats (ecology) and across geographic space (biogeography).

Acknowledgments

I thank Michael Pogue for his continued and diligent work in obtaining reference material and Grace Servat for her comments on early drafts of the manuscript. I especially thank Egbert Leigh, Jr., for his critical "nonanonymous" review, which substantially helped the final draft.

References

Adis, J., Lubin, Y. D., and Montgomery, G. G. (1984). Arthropods from the canopy of inundated and terra firme forests near Manaus, Brazil, with critical consideration of the pyrethrum-fogging technique. *Stud. Neotrop. Fauna Environ.* 19, 223–236.

Allison, A., Samuelson, G. A., and Miller, S. E. (1993). Patterns of beetle species diversity in New Guinea rain forest as revealed by canopy fogging: Preliminary findings. *Selbyana* 14, 16–20.

Ashe, J. S. (1991). A resource management report. *Insect Coll. News* 6, 15–16.

Basset, Y. (1988). A composite interception trap for sampling arthropods in tree canopies. *J. Aust. Entomol. Soc.* 27, 213–219.

Basset, Y. (1990). The arboreal fauna of the rainforest tree *Argyrodendron actinophyllum* as sampled with restricted fogging: Composition of the fauna. *Entomologist* 109, 173–183.

Basset, Y. (1991a). The taxonomic composition of the arthropod fauna associated with an Australian rainforest tree. *Aust. J. Zool.* 39, 171–190.

Basset, Y. (1991b). Influence of leaf traits on the spatial distribution of insect herbivores associated with an overstory rainforest tree. *Oecologia* 87, 388–393.

Basset, Y. (1991c). Leaf production of an overstory rainforest tree and its effects on the temporal distribution of associated insect herbivores. *Oecologia* 88, 211–219.

Basset, Y. (1991d). The seasonality of arboreal arthropods foraging within an Australian rainforest tree. *Ecol. Entomol.* 16, 265–278.

Basset, Y. (1991e). The spatial distribution of leaf damage, galls and mines within an Australian rainforest tree. *Biotropica* 23, 271–281.

Basset, Y. (1992a). Influence of leaf traits on the spatial distribution of arboreal arthropods within an overstory rainforest tree. *Ecol. Entomol.* 17, 8–16.

Basset, Y. (1992b). Activité et distribution spatiale de l'entomofaune mobile associée à un arbre des forêts humides d'Australie. *Vie Milieu* 42, 253–262.

Basset, Y. (1992c). Synecology and aggregation patterns of arboreal arthropods associated with an overstory rainforest tree in Australia. *J. Trop. Ecol.* 8, 317–327.

Basset, Y. (1992d). Host specificity of arboreal and free-living insect herbivores in rain forests. *Biol. J. Linn. Soc.* 47, 115–133.

Basset, Y. (1993a). Patterns in the organization of the arthropod community associated with an Australian rainforest tree: How distinct from elsewhere? *Selbyana* 14, 13–15.

Basset, Y. (1993b). Arthropod species-diversity and component communities in the rainforest canopy: Lessons from the study of an Australian tree. *Trop. Zool.* 1, 19–30.

Basset, Y., and Arthington, A. H. (1992). The arthropod community associated with an Australian rainforest tree: Abundance of component taxa, species richness and guild structure. *Aust. J. Ecol.* 17, 89–98.

Basset, Y., and Kitching, R. L. (1991). Species number, species abundance and body length of arboreal arthropods associated with an Australian rainforest tree. *Ecol. Entomol.* 16, 391–402.

Basset, Y., and Springate, N. D. (1992). Diel activity of arboreal arthropods associated with a rainforest tree. *J. Nat. Hist.* 26, 947–952.

Bates, H. W. (1884). Biologia Centrali-Americana. *Insecta, Carabidae, Cicindelidae, Suppl.* 1, 257–299.

Beebe, W., Hartley G. I., and Howes, P. G. (1917). "Tropical Wild Life in British Guiana." N.Y. Zool. Soc., New York.

Broadhead, E., and Wolda, H. (1985). The diversity of Psocoptera in two tropical forests in Panamá. *J. Anim. Ecol.* 54, 739–754.

Chapman, F. M. (1929). "My Tropical Air Castle: Nature Studies in Panama." Appleton, New York.

Collyer, E. (1951). A method for the estimation of insect populations of fruit trees. *Rep.—East Malling Res. Stn. (Maidstone, Engl.), 1949–1950,* pp. 148–151.

Erwin, T. L. (1982). Tropical forests: Their richness in Coleoptera and other arthropod species. *Coleopterists Bull.* 36, 74–75.

Erwin, T. L. (1983a). Tropical forest canopies, the last biotic frontier. *Bull. Entomol. Soc. Am.* 29, 14–19.

Erwin, T. L. (1983b). Beetles and other arthropods of the tropical forest canopies at Manaus, Brasil, sampled with insecticidal fogging techniques. *In* "Tropical Rain Forests: Ecology and Management" (S. L. Sutton, T. C. Whitmore, and A. C. Chadwick, eds.), pp. 59–75. Blackwell, Oxford.

Erwin, T. L. (1985). The taxon pulse: A general pattern of lineage radiation and extinction among carabid beetles. *In* "Taxonomy, Phylogeny, and Zoogeography of Beetles and Ants: A Volume Dedicated to the Memory of Philip Jackson Darlington, Jr., 1904–1983" (G. E. Ball, ed.), pp. 437–472. Dr. W. Junk Publ., The Hague, The Netherlands.

Erwin, T. L. (1988). The tropical forest canopy: The heart of biotic diversity. *In* "Biodiversity" (E. O. Wilson, ed.), pp. 123–129. National Academy Press, Washington, DC.

Erwin, T. L. (1989a). Canopy arthropod biodiversity: A chronology of sampling techniques and results. *Rev. Peru. Entomol.* 32, 71–77.

Erwin, T. L. (1989b). Sorting tropical forest canopy specimens. *Insect Coll. News* 2, 8.

Erwin, T. L. (1991a). The significance of diversity: New challenges for the entomologist. *In* "Entomology Serving Society: Emerging Technologies and Challenges" (S. B. Vinson and R. L. Metcalf, eds.), pp. 25–39. Entomological Society of America, Lanham, MD.

Erwin, T. L. (1991b). Natural history of the carabid beetles at the BIOLAT Rio Manu Biological Station, Pakitza, Perú. *Rev. Peru. Entomol.* 33, 1–85.

Erwin, T. L. (1991c). Establishing a tropical species co-occurrence database. Part 1. A plan for developing consistent biotic inventories in temperate and tropical habitats. *Mem. Mus. Hist. Nat.* No. 20, 1–16.

Erwin, T. L. (1992a). ANTSE—A network and tracking system for entomology. *Insect Coll. News* 7, 6–7.

Erwin, T. L. (1992b). A current vision of insect diversity. *In* "México Ante los Retos de la Biodiversidad (Mexico Confronts the Challenges of Biodiversity)" (J. Sarukhán and R. Dirzo, eds.), pp. 91–97. Comisión Nacional para el Conocimiento y Uso de la Biodiversidad, Mexico City.

Erwin, T. L., and Kabel, M. (1991). Establishing a tropical species co-occurrence database. Part 2. An automated system for mapping dominant vegetation. *Mem. Mus. Hist. Nat.* No. 20, 17–36.

Erwin, T. L., and Pogue, M. G. (1992). Specimen management: Prep time—prep costs. *Insect Coll. News* 8, 11.

Erwin, T. L., and Scott, J. C. (1980). Seasonal and size patterns, trophic structure, and richness of Coleoptera in the tropical arboreal ecosystem: The fauna of the tree *Luehea seemannii* Triana and Planch in the Canal Zone of Panama. *Coleopterists Bull.* 34, 305–322.

Evans, J. W. (1966). The leafhoppers and froghoppers of Australia and New Zealand. *Mem. Aust. Mus.* 12, 1–347.

Farrell, B. D., and Erwin, T. L. (1988). Leaf-beetles (Chrysomelidae) of a forest canopy in Amazonian Perú: Synoptic list of taxa, seasonality and host-affiliations. *In* "The Biology of the Chrysomelidae" (P. Jolivet, E. Petitpierre, and T. Hsiao, eds.), pp. 73–90. Dr. W. Junk Publ., The Hague, The Netherlands.

Futuyma, D. J., and Gould, F. (1979). Associations of insects and plants in a deciduous forest. *Ecol. Monogr.* 49, 33–50.

Gagné, W. C. (1979). Canopy-associated arthropods in *Acacia koa* and *Metrosideros* tree communities along an altitudinal transect on Hawaii Island. *Pac. Insects* 21, 56–82.

Gagné, W. C., and Martin, J. L. (1968). The insect ecology of red pine plantations in central Ontario. V. The Coccinellidae (Coleoptera). *Can. Entomol.* 100, 835–846.

Guilbert, E., Baylac, M., and Najt, J. (1994). Canopy arthropod diversity in a New Caledonian primary forest sampled by fogging. *Pan-Pac. Entomol.* (in press).

Hallé, F. and Blanc, P., eds. (1990). "Biologie d'une Canopée de Forêt Equatoriale. Rapport de Mission: Radeau des Cimes Octobre–Novembre 1989. Guyane Française." Xylochimie, Paris.

Hevel, G. F. (1988). Specimen preparation costs: Time and money. *Insect Coll. News* 1, 6–7.

Hijii, N. (1983). Arboreal arthropod fauna in a forest. I. Preliminary observations on seasonal fluctuations in density, biomass, and faunal composition in a *Chamaecyparis obtusa* plantation. *Jpn. J. Ecol.* 33, 435–444.

Janzen, D. H. (1968). Host plants as islands in evolutionary and contemporary time. *Am. Nat.* 102, 592–595.

Janzen, D. H. (1973). Host plants as islands. II. Competition in evolutionary and contemporary time. *Am. Nat.* 107, 786–790.

Kennedy, C. E. J., and Southwood, T. R. E. (1984). The number of species of insects associated with British trees: A re-analysis. *J. Anim. Ecol.* 53, 455–478.

Kitching, R. L., Bergelson, J. M., Lowman, M. D., McIntyre, S., and Carruthers, G. (1993). The biodiversity of arthropods from Australian rainforest canopies: General introduction, methods, sites, and ordinal results. *Aust. J. Ecol.* 18, 181–191.

Lawton, J. H. (1986). Surface availability and insect community structure: The effects of architecture and fractal dimension of plants. *In* "Insects and the Plant Surface" (B. E. Juniper and T. R. E. Southwood, eds.), pp. 317–331. Edward Arnold, London.

Lowman, M. D. (1982). Seasonal variation in insect abundance among three Australian rain forests, with particular reference to phytophagous types. *Aust. J. Ecol.* 7, 353–361.

Lowman, M. D. (1984). An assessment of techniques for measuring herbivory: Is rainforest defoliation more intense than we thought? *Biotropica* 16, 14–18.

Lowman, M. D. (1985). Temporal and spatial variability in insect grazing of the canopies of five Australian rainforest species. *Aust. J. Ecol.* 10, 7–24.

Majer, J. D., and Recher, H. F. (1988). Invertebrate communities in Western Australian eucalypts: A comparison of branch clipping and chemical knockdown procedures. *Aust. J. Ecol.* 13, 269–278.

Majer, J. D., Recher, H. F., Perriman, W. S., and Achuthan, N. (1990). Spatial variation of invertebrate abundance within the canopies of two Australian eucalypt forests. *Stud. Avian Biol.* 13, 65–72.

Martin, J. L. (1966). The insect ecology of red pine plantations in central Ontario. IV. The crown fauna. *Can. Entomol.* 98, 10–27.

May, R. M. (1990). How many species? *Philos. Trans. R. Soc. London, Ser. B* 330, 293–304.

Moran, V. C., and Southwood, T. R. E. (1982). The guild composition of arthropod communities in trees. *J. Anim. Ecol.* 51, 289–306.

Morse, D. R., Lawton, J. H., Dodson, M. M., and Williamson, M. H. (1985). Fractal dimension of vegetation and the distribution of arthropod body lengths. *Nature (London)* 314, 731–733.

Morse, D. R., Stork, N. E., and Lawton, J. H. (1988). Species number, species abundance and body length relationships of arboreal beetles in Bornean lowland rain forest trees. *Ecol. Entomol.* 13, 25–37.

Nadkarni, N. M. (1984). Epiphyte biomass and nutrient capital of a Neotropical elfin forest. *Biotropica* 16, 249–256.

Nadkarni, N. M., and Longino, J. T. (1990). Invertebrates in canopy and ground organic matter in a Neotropical montane forest, Costa Rica. *Biotrpica* 22, 286–289.

Noyes, J. (1989). A study of five methods of sampling Hymenoptera (Insecta) in a tropical rainforest, with special reference to the parasites. *J. Nat. Hist.* 23, 285–298.

Ohmart, C. P., and Voigt, W. G. (1981). Arthropod communities in the crowns of the natural and planted stands of *Pinus radiata* (Monterey Pine) in California. *Can. Entomol.* 113, 673–684.

Ohmart, C. P., Stewart, L. G., and Thomas, J. R. (1983). Phytophagous insect communities in the canopies of three *Eucalyptus* forest types in south-eastern Australia. *Aust. J. Ecol.* 8, 395–403.

Opler, P. A. (1974). Oaks as evolutionary islands for leaf-mining insects. *Am. Sci.* 62, 67–73.

Parker, G. G., Smith, A. P., and Hogan, K. P. (1992). Access to the upper canopy with a large tower crane. *BioScience* 42, 664–670.

Perry, D. R. (1978). A method of access into the crowns of emergent and canopy trees. *Biotropica* 10, 155–157.

Recher, H. F., Majer, J. D., and Ford, H. A. (1991). Temporal and spatial variation in the abundance of eucalypt canopy invertebrates: The response of forest birds. *Acta Congr. Int. Ornithol., 20th, 1990,* Vol. 3, pp. 1568–1575.

Roberts, H. R. (1973). Arboreal Orthoptera in the rain forest of Costa Rica collected with insecticide: A report on the grasshoppers (Acrididae) including new species. *Proc. Acad. Nat. Sci. Philadelphia* 125, 46–66.

Southwood, T. R. E. (1960). The abundance of the Hawaiian trees and the number of their associated insect species. *Proc. Hawaii. Entomol. Soc., 1959,* pp. 299–303.

Southwood, T. R. E. (1961). The number of species of insects associated with various trees. *J. Anim. Ecol.* 30, 1–8.

Southwood, T. R. E., and Kennedy, C. E. J. (1983). Trees as islands. *Oikos* 41, 359–371.

Southwood, T. R. E., Brown, V. K., and Reader, P. M. (1979). The relationships of plant and insect diversities in succession. *Biol. J. Linn. Soc.* 12, 327–348.

Southwood, T. R. E., Moran, V. C., and Kennedy, C. E. J. (1982a). The richness, abundance, and biomass of the arthropod communities on trees. *J. Anim. Ecol.* 51, 635–650.

Southwood, T. R. E., Moran, V. C., and Kennedy, C. E. J. (1982b). The assessment of arboreal insect fauna: Comparisons of knockdown sampling and faunal lists. *Ecol. Entomol.* 7, 331–340.

Steiner, W. E. (1989). New collections from Madagascar at NMNH with notes on processing alcohol collections. *Insect Coll. News* 2, 8–9.

Stork, N. E. (1987a). Guild structure of arthropods from Bornean rain forest trees. *Ecol. Entomol.* 12, 69–80.

Stork, N. E. (1987b). Arthropod faunal similarity of Bornean rain forest trees. *Ecol. Entomol.* 12, 219–226.

Stork, N. E. (1991). The composition of the arthropod fauna of Bornean lowland rain forest trees. *J. Trop. Ecol.* 7, 161–180.

Stork, N. E., and Brendell, M. J. D. (1990). Variation in the insect fauna of Sulawesi trees with season, altitude and forest type. *In* "Insects and the Rainforest of South East Asia (Wallacea)" (W. J. Knight and J. D. Halloway, eds.), pp. 173–190. Royal Entomological Society of London, London.

Taksdal, G. (1965). Hemiptera (Heteroptera) collected on ornamental trees and shrubs at the Agricultural College of Norway. *As. Nor. Entomol. Tidsskr.* 13, 5–10.

Teulon, D. A. J., and Penman, D. R. (1987). Vertical stratification of sticky board catches of leafhopper adults (Hemiptera: Cicadellidae) within apple orchards. *N. Z. Entomol.* 9, 100–103.

Tobin, J. E. (1991). A Neotropical rainforest canopy ant community: Some ecological considerations. *In* "Ant–Plant Interactions" (C. R. Huxley and D. F. Cutler, eds.), pp. 536–538. Oxford Univ. Press, Oxford.

Valencia, R., Balslev, H., and Paz y Miño C., G. (1994). High tree alpha-diversity in Amazonian Ecuador. *Biodiversity Conservation* (in press).

Watanabe, H., and Ruaysoongnern, S. (1989). Estimation of arboreal density in a dry evergreen forest in northeastern Thailand. *J. Trop. Ecol.* 5, 151–158.

Wilson, E. O. (1987). The little things that run the world, *Conserv. Biol.* 1, 344–346.

Wint, G. R. W. (1983). Leaf damage in tropical rain forest canopies. *In* "Tropical Rain Forests: Ecology and Management" (S. L. Sutton, T. C. Whitmore, and A. C. Chadwick, eds.), pp. 229–240. Blackwell, Oxford.

Wolda, H. (1978). Fluctuations in abundance of tropical insects. *Am. Nat.* 112, 1017–1045.

6

Ecology and Diversity of
Tropical Forest Canopy Ants

John E. Tobin

I. Introduction

In terms of their biomass and their multiple effects on other species, ants
are the dominant arthropod family in lowland tropical forest canopies. By
virtue of their catholic diet and their associations with animal, plant, and
fungal species, ants have a significant impact at all trophic levels. Canopy
ants display a wide variety of foraging strategies, nesting habits, and patterns
of colony organization. In this chapter, I present a broad overview of the
biology of forest canopy ants, with an emphasis on lowland rain forest species. I then summarize information on the abundance and diversity of ant
assemblages in forest canopies. Finally, I discuss practical issues raised by
the diversity of arboreal ants and other canopy arthropods in light of the
present state of taxonomy and current rates of tropical habitat transformation. I conclude that the urgency of the situation calls for a partial reorder-
ing of our priorities in taxonomic and ecological research.

Arboreal ants have drawn the attention of biologists for many years. However, the obvious difficulties involved in collecting data and specimens in
forest canopies have, until recently, precluded us from quantifying the diversity and patterns of assembly of arboreal ant species. The emerging picture is one of surprising abundance and diversity of ants. Recent studies
have shown that the dominant, aggressive ant species with large colonies
and extensive foraging territories (the presence of which is often painfully
apparent to canopy researchers) make up only a small part of total ant di-

Copyright © 1995 by Academic Press, Inc.
All rights of reproduction in any form reserved.

versity. A large majority of the ant species richness in lowland forest canopies is made up of inconspicuous species with small colonies and limited foraging territories that exploit the space and resources left by the dominants. This surprising diversity notwithstanding, the most notable feature of the ant assemblages has not been their diversity, which pales in comparison with that of other important canopy arthropod families, but their abundance. An extreme example is in western Amazonian canopies, where ants can be more abundant than all other arthropods combined (Erwin, 1989; Tobin, 1991). Even where they do not comprise an absolute majority of the arthropods, the family Formicidae is, with few exceptions, the largest single contributor to biomass in lowland rain forest canopy samples. Explaining the historical, behavioral, and physiological factors that have led to the overwhelming dominance of the ants in lowland canopy communities will be a leading undertaking in insect ecology in the future.

II. Biology of Canopy Ants

The distribution of arboreality in the family Formicidae suggests that the trait evolved and was secondarily lost numerous times during the evolutionary history of the family (cf. Baroni Urbani *et al.,* 1992). Though several ant subfamilies are predominantly arboreal (most notably the Pseudomyrmecinae), several others are strictly terrestrial. In a few subfamilies, however, arboreality appears to be distributed at random. In several large myrmicine genera (e.g., *Pheidole*), one species may be restricted to the canopy while a closely related congener is a ground dweller. A number of species are known to nest both in the ground and in canopy humus at the same site (Longino and Nadkarni, 1990). In a few cases, altitudinally wide-ranging species appear to nest either in the ground or in the canopy, depending on the altitude and microclimatic conditions of the site (J. T. Longino, personal communication).

Relatively little is documented on the morphological and behavioral specializations associated with arboreality in ants. Several of the behavioral adaptations associated with ant–plant symbioses and with nest-weaving have been discussed by Hölldobler and Wilson (1990). In his study of the ant fauna of New Guinea, Wilson (1959) tested the hypothesis that arboreality is associated with spinescence, and concluded that there is no definite correlation between the two. Although there may be other external morphological specializations associated with arboreality, the most distinctive difference between canopy and ground ant species is the difference in average size (J. E. Tobin, personal observation; S. P. Cover, personal communication). In tropical forest litter, there is a huge diversity of minute, often predatory ant species. Ants in this size class are poorly represented in the canopy, which may be a consequence of the available nesting and foraging sites. Small, hollow twigs are the most widely available nesting sites in leaf litter; in the trees, many species build their own nests or dwell inside large, hollow branches. Because smaller ants are more vulnerable to environmental stress, they are better adapted to the more protected environment of the forest floor (S. P. Cover, personal communication).

Given the wide temperature fluctuations and the lower humidity characteristic of forest canopies, one predicted adaptation in canopy ants is an increased ability to withstand desiccation. Another distinction between arboreal and terrestrial life is the geometric complexity of the canopy, which presumably makes orientation and homing more difficult than on the ground. Some ground-foraging ants find their way back to their colonies by "canopy orientation" (Hölldobler, 1980) or by reference to the pattern of polarized light in the sky (Wehner, 1987). For foragers that are already in the canopy, and for which the closest route to the colony is not a straight line along the two-dimensional surface of the ground, orientation and navigation may be complicated. Novel cognitive abilities and orientation strategies that may have evolved in response to the structural complexity of the canopy need investigation.

Although many or most ground-dwelling ant species are thought to be primarily predators and scavengers, canopy ants appear to exploit a wider range of food sources. In addition to hunting and gathering dead prey, arboreal ants collect floral and extrafloral nectar, plant and fruit saps, arthropod exudates, food bodies, and seeds (Tobin, 1994). Extrafloral nectar and food bodies have long been assumed to reduce herbivory by attracting ant visitors, but direct evidence for this has accumulated only recently (e.g., Bentley, 1977; Buckley, 1982; Beattie, 1985). The taxonomic distribution of food bodies and extrafloral nectaries have been reviewed by O'Dowd (1982) and Koptur (1992), respectively. To assess the importance of ant defense of plants in a tropical rain forest, Schupp and Feener (1991) surveyed 243 plant species on Barro Colorado Island, Panama, for evidence of extrafloral nectaries and food bodies. The sample included 109 canopy tree species (52% of the canopy tree flora). They concluded that at least 34% of these species had extrafloral nectaries and that 37% were ant-defended. Ants are also known to exploit floral nectar, but compared to extrafloral nectar it is of minor importance (Tobin, 1994).

Instead of exploiting nectar sources, certain ants obtain plant fluids by consuming foliage directly. Adult leafcutter ants derive a substantial portion of their energy from the sap of the leaves they cut and use as substrate to raise fungi (Quinlan and Cherrett, 1979). The main function of the fungi is to provide protein for the larvae. Ants also obtain nutrition from plants indirectly by tending arthropods such as homopterans (Buckley, 1987) or lepidopteran larvae (Pierce, 1987), with which they have evolved intricate and intimate symbioses. The tended insects tap into the vascular system of the plants, filtering out amino acids and other nutrients for their own use, and provide the ants with nutritional rewards in exchange for protection from predators and parasitoids. Aggregations of membracid homopterans on young shoots of tropical trees being tended by *Dolichoderus* ants are a common sight in Neotropical forest canopies.

Canopy ants find appropriate nesting sites in forest canopies in different ways. They build large carton nests, live in tree hollows or inside twigs, co-opt abandoned termite nests, dwell in leaf litter and humus accumulated on branches and in the inner crowns of trees, construct ant gardens, weave nests by joining leaves together with silk threads, and associate with domatia-bearing ant-epiphytes. The nesting strategies of ants associated with ant-plants have been summarized by Davidson and McKey (1993).

Nest-weaving has evolved at least four times in the subfamily Formicinae (in the genera *Dendromyrmex, Camponotus, Polyrhachis,* and *Oecophylla*) (Hölldobler and Wilson, 1990). *Oecophylla* has taken nest-weaving to its extreme. Ants in this genus are among the most abundant insects in tropical Africa, Asia, and Australia. Each nest consists of a chamber built by pulling together leaves from the trees they inhabit and stitching them together with silk threads dispensed by the larvae, which behave like living shuttles. Typical nests contain hundreds or thousands of workers and brood, and often include homopteran symbionts. A mature colony can consists of several hundred such nests, widely scattered throughout the extensive territories that such colonies defend (Hölldobler and Wilson, 1977; Hölldobler, 1983). The ability of weaver ants to build such nests with no other input than larval silk has freed them from the constraints on colony size imposed by limited nesting sites in the canopy (Hölldobler and Wilson, 1990).

Ant gardens are among the most abundant and distinctive ant–plant associations in Neotropical forest canopies (Kleinfeldt, 1978, 1986; Davidson, 1988). They were originally described by Ule (1902, cited in Hölldobler and Wilson, 1990). An ant garden is typically a small- to medium-sized carton nest associated with a particular set of specialized plants, often anchored at branching points in midcanopy trees. The plants, which belong to several families, appear to depend on the ants both for the nutrients with which ants enrich the substrate and for seed dispersal. They provide the ants with fruits and seed elaiosomes. A number of different myrmicine, dolichoderine, formicine, and ponerine genera have been found in ant gardens (Kleinfeldt 1986), and the usual situation involves two species of ants coexisting in a parabiotic relationship (Hölldobler and Wilson, 1990). In Peruvian Amazonia, the most common ant-garden species are *Camponotus femoratus* and a member of the *Crematogaster parabiotica* group (Davidson, 1988). Although ant-garden ants are known to disperse the seeds of ant-garden plants, relatively little is known about the roles that ants play in seed dispersal and recruitment of other canopy plants.

A large number of plant species belonging to several dozen families have domatia or other specialized structures that commonly house ant colonies. The taxonomic distribution of ant–plant associations has been summarized by Hölldobler and Wilson (1990) and Davidson and McKey (1993). Ant–plant associations include some of the most thoroughly studied animal–plant mutualisms. Terrestrial examples include *Pseudomyrmex–Acacia* (Janzen, 1966), *Pseudomyrmex–Triplaris* (Oliveira *et al.,* 1987), and *Azteca–Cecropia* (Longino, 1991) in the New World, and *Crematogaster–Macaranga* in Southeast Asia (Fiala *et al.,* 1991). Many epiphytic ant-plants have also been studied (Huxley, 1980). Although the degree of mutual dependence varies from one association to another, in extreme examples, the ants derive all their nutrition from their host plants in the form of nectar and protein-rich food bodies, in addition to obtaining nesting sites. The ants, in return, defend the plants against herbivores, provide nutrients in the form of refuse and excrement, and disperse seeds. In some cases, ants attack and destroy vegetation that may encroach and compete with the host plant (e.g., Janzen, 1966; Davidson *et al.,* 1988). In a few cases, ants serve plants as pollinators (Peakall *et al.,* 1991).

Studies on the distribution of arboreal ants in African agroecosystems have shown that dominant ant colonies partition the canopy into exclusive three-dimensional territories in the process of competing with one another. This pattern of aggressive species holding mutually exclusive foraging ranges is known as an *ant mosaic* (Room, 1971; Leston, 1973; Majer, 1976a, b, c; reviewed in Jackson, 1984). Colony members defend their territories against other conspecific colonies and against a series of codominant species. A distinctive fauna of subdominant ants positively or negatively associates with each of these codominant species. Room (1971) attributed these associations to the degree of specialization of the dominant species. Whether the observed ant mosaics were an artifact of the artificial simplicity of agroecosystems was not clear until Leston (1978) reported observations on ant mosaics for a secondary rain forest canopy near Bahia, Brazil. This pattern of competing colonies conducting ongoing, low-level harassment and aggression campaigns at their territorial borders against colonies of their own and of other species has been documented for natural forest canopies (Hölldobler, 1979, 1983; Majer and Camer-Pesci, 1991). Although several studies have documented multiple codominant species in dynamic equilibrium, at some Old World sites a single species, often an *Oecophylla* sp., is the sole dominant ant (Hölldobler, 1979, 1983).

Several ant species have important effects on the biology of forest canopies and are not true canopy species, in that they nest in the ground. Although most Neotropical canopy ants are arboreal nesters, a few species come into the canopy only to forage. The opposite pattern, however, is rarely observed in the New World: tree-nesting ants do not commonly forage on the ground. The flow of energy and materials (at least to the extent that ants are involved) is from the canopy to the ground. The most obvious ground-nesting, canopy-foraging ants in the neotropics are the leafcutter ants of the genera *Atta* and *Acromyrmex,* which build large subterranean nests and send out foraging columns into the trees. However, at least one leafcutter species, *Acromyrmex volcanus,* is also known to nest in trees (Wetterer, 1993). *Paraponera clavata* is another conspicuous ground-nesting ant that forages extensively in the canopy, although Breed and Harrison (1989) have shown that a few *P. clavata* nests are arboreal. In contrast to the neotropics, Wilson (1959) has documented extensive foraging movement from the canopy to the ground in New Guinea. Extensive ant migration between the ground and the trees in the seasonally flooded forests of central Amazonia has been documented by Adis and Schubart (1984). This movement appears to be closely tied to the inundation cycle and ostensibly serves to escape the flooded forest floor, though the phenomenon has not yet been thoroughly investigated.

III. Abundance

Although the dominance of ants in tropical forest canopies has long been clear, quantitative demonstrations of their abundance have had to await the development of appropriate sampling techniques. The relative and absolute abundance of ants in ground arthropod communities can be established with some certainty with litter sifting, sweeping, and other techniques (e.g., Pisarski, 1978). The inaccessibility of the canopy prevented such work until the development of insecticidal collecting methods, referred to variously as canopy fogging or insecticidal knockdown (Kikuzawa and Shidei, 1966; Martin, 1966; Roberts, 1973; Erwin and Scott, 1980; reviewed in Erwin, 1989). Some differences in the methodology and the compounds used by different researchers exist, but they all involve treating limited patches or columns of forest canopy with a short-lived insecticide, such as pyrethrum, and allowing the arthropods thus killed to fall into collecting sheets or funnels placed in the understory.

As with all methods, there are problems that confound the analysis of fogging data. The samples are not unbiased, as various taxa may be under- or overrepresented. Canopy fogging, however, is the single most effective technique currently available for obtaining samples of the arthropod communities of selected patches of forest canopy. Potential biases inherent in the method include:

1. Insecticide elicits a fleeing response in at least some arthropods, the effect of which is to flush them out of their nests, cavities, and webs; however, slow-moving arthropods such as wood-boring beetles and termites may die before they are flushed out (Adis *et al.,* 1984).

2. Insects that live inside or attached to vegetation, such as leaf miners and female scale insects, may fail to fall to the ground even after they are killed.
3. Vigorous fliers such as macrolepidoterans, odonates, and cicadas may flee before they are affected by the insecticide; these groups appear to be underrepresented in canopy fogging samples (T. L. Erwin, personal communication).
4. Other insect groups may be affected and eventually killed by the insecticide, but not before they have had a chance to flutter outside of the collection area, thus falling outside the range of the collectors.
5. Even at relatively high concentrations, insecticide may fail to penetrate deep inside the very large nests of certain species of ants and termites, potentially missing a significant segment of the population.
6. Some arthropods may get intercepted on the surface of leaves and branches during the process of falling to the ground; however, as most arthropods are still alive and moving when they fall, their movements should tend to dislodge them.
7. Minute arthropods, particularly immature forms, may be lost in the process of clearing the collecting trays and while sorting samples.
8. Some insects that were not originally in the fogged area may fly into the insecticide fog during the short active period and could end up in the collections; this bias would tend to overestimate the abundance of flying insects in the area.

Despite these problems, canopy fogging has allowed us to obtain virtual snapshots of canopy arthropod communities for the first time. We can now answer questions about the diversity and abundance of various taxonomic groups, positive and negative associations between taxa, and plant–host specificity, as this method allows us to assess the complete complement of arthropods in a patch of canopy at a particular time. One striking conclusion that has emerged from this work is the overwhelming abundance of canopy ants in tropical forests.

The first quantitative demonstration of the abundance of tropical canopy ants came from work near Manaus, Brazil, by Erwin (1983) and Adis *et al.* (1984). Erwin (1983) found ants to be the dominant arthropod group in terms of numbers of individuals in four different forest types. He sorted some 24,000 individual arthropods to major groups; of these, 43% were ants. Following in abundance were the orders Coleoptera (20% of the individuals), Homoptera (8%), and Diptera (7%).

Adis *et al.* (1984) fogged three forest types near Manaus to determine the density of all major canopy arthropod groups and the diversity of arboreal ants. They sampled several trees in várzea (a seasonally flooded white-water forest), igapó (a seasonally flooded black-water forest), and a terra firme forest. Their sample consisted of 9639 arthropods, of which 51% were ants. Ants also made up 29% of the total biomass of the sample. In the várzea forest, ants comprised 47% of the biomass, followed in importance by beetles and spiders. In the igapó forest, they were also the dominant group (36% of the biomass), followed in importance by beetles and termites. In the terra firme forest, however, saltatorian orthopterans and ants were equally abundant, each comprising about 25% of the arthropod biomass. The Hymenoptera (mostly ants), Orthoptera, Lepidoptera, and Coleoptera comprised about 80% of the total biomass in the sample. Dolichoderine ants of the genera *Azteca* and *Hypoclinea* were particularly abundant. Other important genera included *Crematogaster, Pseudomyrmex,* and *Camponotus.* Noticeably rare in the samples of Adis *et al.* (1984) were the arboreal ponerines, which were conspicuous in fogging samples from Peruvian Amazonia (Wilson, 1987; Tobin, 1991).

Other studies have confirmed the remarkable abundance of the ants elsewhere in the tropics (Table I). In Southeast Asia, Stork (1987, 1988, 1991) and Stork and Brendell (1990) found that ants vary in abundance between 20 and 40% at lowland sites, but are considerably less abundant in higher elevation forests. In dry tropical forests in Thailand, ants also comprised approximately 20 to 40% of the wet weight (Watanabe and Ruaysoongnern, 1989). In seasonally dry forests in Panama, J. E. Tobin (unpublished) estimated that ants comprise about 32% of the canopy arthropod biomass, which is lower at the beginning of the wet season due to a sudden increase in the abundance of other groups (J. E. Tobin, unpublished data; cf. Smythe, 1982).

In Africa, little canopy fogging has been done in natural forests, but data for tree plantations (Majer, 1990) and for other forests (Basset *et al.,* 1992) are consistent with patterns of ant abundance in other tropical areas. Although a few data are available for tropical Australian forests (Kitching *et al.,* 1993; cf. Majer, 1990), it appears that ants do not make as large a contribution to arthropod biomass as in other tropical forests. More studies are needed in that region. One area where ants appear to make an especially large contribution to arthropod abundance is in the forests of western Amazonia, near the foothills of the Peruvian Andes, where ants have been found to make up an absolute majority of the arthropod biomass and numbers (Erwin, 1989, personal communication; Tobin, 1991).

Just as arthropod diversity decreases from the tropics toward the temperate zones, so the absolute and relative abundances of ants decrease from the lowland tropics toward cooler areas (cf. Moran and Southwood, 1982). That ant abundance in seasonally dry forests in Central America (J. E. Tobin, unpublished data) and in dry forests in Southeast Asia (Watanabe and Ruaysoongnern, 1989) is comparable to their abundance in most lowland rain forests suggests that precipitation is not the determining factor. Also, the abundance of ants in some subtropical forests is comparable to that in

Table I   Relative Abundance of Ants in Canopy Arthropod Communities[a]

Source	Research site	Forest type	Percentage of total biomass	Percentage of individuals
Moran and Southwood (1982)	South Africa	St	0–45	0–58
Moran and Southwood (1982)	Britain	Te	0–<1	0–<1
Erwin (1983)	Manaus, Brazil	Tr	—	43
Hijii (1983, 1989)	Nagoya, Japan	Te[b]	0	0
Adis *et al.* (1984)	Manaus, Brazil	Tr	26–47	43–53
Stork (1987, 1991)	Borneo, Brunei	Tr	—	7–32
Majer and Recher (1988)	Western Australia	St[c]	—	26–48
Stork (1988)	Seram, Indonesia	Tr	—	42
Watanabe and Ruay-soongnern (1989)	Northeastern Thailand	Dt	17–46[d]	—
Majer (1990)	Western Australia	St[e]	6–12	—
Majer (1990)	Western Australia	St[f]	86	—
Majer (1990)	Ghana	Tr[g]	5–72	—
Stork and Brendell (1990)	Sulawesi, Indonesia	Tr[h]	—	1–32
Tobin (1991)	Manu, Peru	Tr	—	70
Kitching *et al.* (1993)	Queensland, Australia	Tr	—	Up to 26[i]
Harada and Adis (1994)	Central Amazonia	Tr	—	Up to 46[i]
J. E. Tobin (unpublished data)	Barro Colorado, Panama	Tr	16–57	—
J. E. Tobin *et al.* (unpublished data)	Tambopata, Peru	Tr	—	42–94

[a] Summary table of studies of relative abundance of ants in canopy arthropod samples collected using insecticidal knockdown techniques. Numbers are the percentage contribution of ants to the total arthropod biomass and to the total number of individual arthropods in the samples. Biomass is given as dry weight, except as noted. Most studies are of lowland tropical canopies, but several subtropical and temperate forest studies are included for comparison. The various types of lowland rain forest are not distinguished. Tr = lowland tropical rain forest; Dt = dry tropical forest; St = subtropical forest; Te = temperate forest.

[b] These two studies were conducted in plantations of Japanese cypress and Japanese cedar, respectively; although the associated canopy arthropod communities were large and diverse, ants were absent.

[c] Eucalypt forests at Dryandra State Forest, Western Australia.

[d] Wet weight.

[e] Eucalypt forests at Dryandra State Forest and at Karragullen, Western Australia.

[f] Mango plantation.

[g] Cocoa plantation; figures are for the dominant ant species only.

[h] Includes data from a variety of agricultural, swamp forest, and lowland and midelevation forest sites; in general, ants were less dominant at these sites than at other Southeast Asian and Neotropical sites; Diptera was the most abundant group.

[i] Authors do not give minimum figures.

tropical forests (Moran and Southwood, 1982; Majer and Recher, 1988; Majer, 1990; for estimates of arboreal ant abundance in subtropical forests using methods other than canopy fogging, see Majer and Recher, 1988; Yen, 1989). Even in tropical areas, canopy ants disappear quickly as elevation increases (Stork and Brendell, 1990; Kitching *et al.*, 1993). It appears that as long as the minimal requirements of insolation, heat, and moisture for the successful growth of colonies of dominant ant species are met, canopy ants make up approximately 20 to 40% of the arthropod biomass in forest canopies regardless of other biotic factors.

IV. Diversity

The recent development of canopy fogging and other methods have made it possible to sample canopy arthropods more extensively and more efficiently than ever before. Most ecological and taxonomic studies of canopy ants have involved collecting samples from trunks and low branches, fallen epiphytes, and felled trees. One of the first attempts to characterize the entire arboreal ant fauna in a rain forest was by Wilson (1959), who collected ants on vegetation near the ground and in the canopies of freshly felled trees in a logging area in New Guinea. He estimated colony sizes and described nesting conditions, observed foraging patterns, and produced a faunal list. Studies such as these, however, are limited by the enormous amount of effort required to sample extensively by hand and by having to work in logging areas. In addition, little can be said about the relative and absolute abundances of the component species from areas sampled by hand.

The development of insecticidal fogging has allowed biologists to carry out the large-scale, systematic sampling protocols that are required for meaningful ecological comparisons. One thing we have learned from faunistic studies using canopy fogging is that, although the local diversity of arboreal ants is by no means negligible, they are not nearly as diverse as they are abundant. Although ants comprise 20–40% or more of the individual arthropods and the arthropod biomass in many tropical canopies, nowhere does their contribution to arthropod species diversity approach this amount. This is illustrated by results of Stork (1991), who fogged 10 canopy trees at Ladan Hills Forest Reserve in Brunei, on the island of Borneo, and sorted nearly 24,000 individual specimens representing over 3000 arthropod species. The total arthropod richness of individual trees ranged between 288 and 1007 species, and only between 10 and 32 species in any tree were ants. The relative diversity of most major insect groups in Stork's study varied widely from one tree to another, but the ants comprised a relatively constant percentage of the species, ranging from 2.2 to 4.4% of the species in a tree, with an average of 3.1%.

The relative predictability of canopy ant species richness raises questions about the role that these ants play as energy sinks, as front-line antiherbivore defenses, and in structuring canopy arthropod communities. It also raises the interesting possibility that ants could serve as indicators of arthropod diversity. In estimating the species richness of a particular area, either for purposes of ecological work or for establishing conservation management areas, it would be useful to have one or a few taxa that serve as a proxy for ecosystem diversity. Ants, which are comparatively well-known taxonomically, easily collected, and comprise a manageable percentage of the species, are ideally suited if it can be established that their diversity is closely correlated to total diversity in particular areas. Whether the pattern documented by Stork (1991) is representative of other tropical forests requires further investigation.

Specificity of ants to habitats and substrate has also been investigated. In a lowland forest of Brazil, Adis *et al.* (1984) found that a high percentage (78%) of the 69 ant species was restricted to only one of the three types of forest studied, while only 4% of the ants were found in all forest types. There are at least two explanations for this pattern: either their sampling regime missed many canopy ants species in each forest type (and thus any species was likely to be collected only once), or that a large majority of ant species were restricted to certain forest types by specialized associations with other organisms. In the most comprehensive study published on the faunistics of a tropical canopy ant assemblage, Wilson (1987) suggested that the first explanation is likely to account for much of the habitat variation documented by Adis and his collaborators. Wilson classified over 100,000 ants collected by Erwin in four different forest types in Tambopata, Madre de Dios, Peru. He found 135 species belonging to 40 genera of ants and determined that 54% of the species were found in one forest type only, while 13% were found in all four forest types. Although Wilson found many fewer species restricted to one habitat than did Adis *et al.* (1984), his results suggest a fairly high degree of habitat specialization.

Considering how territorial many ant species are, it is surprising to find the numbers of species that we do coexisting in single trees. In a study of the ant assemblages associated with *Luehea seemannii* trees near Barro Colorado Island, Panama, Montgomery (1985) found an average of between 22 and 35 species of ants per tree, depending on the season. Wilson (1987) found 43 ant species belonging to 26 genera in a particularly diverse tree. J. E. Tobin (unpublished data; preliminary data on diversity and abundance summarized by Erwin, 1989; Tobin, 1991) found 79 species of ants in a small patch of canopy consisting of two adjacent trees and associated vegetation in Manu, Madre de Dios, Peru. Harada and Adis (1994) found 77 species in 20 genera in a single tree in the central Amazon region of Brazil. A closer examination of these samples reveals the overwhelming dominance of one or a few ant species and the relative scarcity of the majority of species. In Neotropical lowland forest canopies, the bulk of the energy flowing through ant assemblages is co-opted by some of the larger and more territorial species of *Azteca, Dolichoderus,* and *Camponotus.* Although these species patrol much of the canopy and exclude other potential competitors, they fail to exclude the small, predatory species with limited foraging ranges that constitute much of the diversity.

Although the application of canopy fogging has greatly expanded our ability to sample and study arboreal arthropod communities, it has also served to highlight how primitive our understanding is of the diversity of life. A canopy fogging sample, preserved in alcohol and placed under a dissecting microscope, can be a humbling experience: hundreds of unidentified and unidentifiable arthropods fill the field of view. Even some of the most abundant organisms in the samples are beyond our ability to reliably identify. The Neotropical ant genus *Azteca* is a good example. *Azteca* ants are extremely abundant and diverse in New World canopies. They have evolved prominent and (ecologically) well-studied symbioses with a variety of other species, and they are ubiquitous in canopy fogging samples. However, there are only a few overworked taxonomists in the world who are able to recognize major species groups within the genus and can identify some of the more distinctive species. Taxonomists that study the diversity of *Azteca* (e.g., Longino, 1989) have little option but to assign personal codes to the species with which they work. Ant workers in some species are indistinguishable on the basis of external morphology, and reproductives are needed to provide reliable identifications. An ecologist analyzing canopy samples may not know if a specimen belongs to a known species or is undescribed, and whether the ants in the sample belong to one or to several similar *Azteca* species. Unfortunately, *Azteca* is not atypical, as most other important canopy genera (e.g., *Camponotus, Crematogaster,* and *Solenopsis*) also defy our efforts at identification and classification. To make things worse, matching minor workers, major workers, and reproductives of the same species collected by canopy fogging can be difficult or impossible. Progress in ecology will be hindered as long as taxonomic information is lacking.

V. Canopy Arthropod Diversity and the Limitations of Taxonomy

Taxonomists have worked since the time of Linnaeus and have described between 1.4 and 1.8 million species of organisms (Stork, 1988; Wilson, 1988; Barnes, 1989). This represents an average of approximately 6000 to 7000 species per year since the modern system of classification was established. About half of these have been insects. Lane (1992) estimated that insect species are being described currently at a rate of 7000 per year. Plausible estimates of global species richness currently range between 3 million and 80 million species (Erwin, 1982, 1991; Stork, 1988; Stork and Gaston, 1990; Gaston, 1991; Hodkinson and Casson, 1991; estimates of global species richness summarized in Stork, 1993). If we estimate conservatively and assume that we have described about 50% of all species, it will take another century to finish the task of cataloguing the diversity of life on earth. Based on the higher estimates of species diversity, at the current rate, we could be describing new species well into the fourth millennium.

At present, there is no central repository or clearinghouse for taxonomic information. Until there is, we will not even be able to know how many species we have described, let alone how many remain undescribed. Taxonomic skills to identify and describe species are group-specific and vary enormously from taxon to taxon; they are difficult to teach or share except through tutorial instruction, and as they are stored in the minds of the practitioners, rather than in some more permanent and easily transferable medium, the knowledge is generally lost when taxonomists retire or die. Even some of the best-studied arthropod groups present intractable problems to taxonomists trying to make sense of their diversity. Canopy ants are a good example of these problems. Working taxonomists are able to quickly and reliably identify only a fraction of all known species. Most are too busy working on descriptions of new species and on general curatorial duties to dedicate much time to sharing their expertise with other scientists. Wilson (1985) estimated that it would take 25,000 scientists, each working for 40 years, to describe 10 million new species using traditional taxonomic methods. Given that the entire population of invertebrate taxonomists working today is a fraction of this number, the short-term prospects of making significant progress in the task of cataloguing life on earth are not good. This would involve a large increase in the amount spent on salaries and expenses for taxonomists, and a massive investment in laboratory space and collections.

The inadequacy of the present state of taxonomy and the need for a greatly increased effort directed at describing and cataloguing the richness of life are abundantly clear, especially in light of the huge numbers of new species found in tropical forest canopies and other recently explored habitats (Wilson, 1985; Wheeler, 1990; Lane, 1992; Raven and Wilson, 1992; Kim, 1993; Marshall, 1993). Compiling a comprehensive catalog of life on earth will involve more than simply increasing the number of taxonomists describing species. Even if our most optimistic hopes for increased funding became reality, we would still be hampered by the difficulties that large numbers of species and geographic variation present for taxonomy.

There may never be a large enough standing army of taxonomists to provide fast, reliable identifications of specimens or information on the general biology of any but a few conspicuous groups. The science of taxonomy will have to become more modern, more efficient, and more useful to non-taxonomists if it is to justify the increased funding it will need. The accumulation of taxonomic knowledge in computerized data bases to facilitate both its transfer and its analysis will be essential. Managing public taxonomic data bases and expert identification systems will have to become a curatorial duty. Scientists who need information on the identification, phylogeny, and general biology of particular taxa may one day be able to gain access to the appropriate data bases from their home institutions and perhaps even identify their own specimens.

The task of identifying, cataloguing, and studying the diversity of life, and putting it in useful forms, will be immense. Mustering the attention and the funds that will be needed to bring taxonomy up to the state of sophistication of other branches of science will not happen any time in the immediate future, and doing it well will be expensive (Lane, 1992) and will involve international coordination. A sizable portion of the investment will have to finance advanced technological tools with automated methods of species identification.

In the meantime, valuable habitats will be transformed throughout the world and many of their biological communities will be lost forever. What are we to do in the interim? One of our first priorities must be to support a greatly expanded effort at collecting and sampling in endangered ecosystems. There will always be time to describe and catalogue the diversity of life in the leisurely atmosphere of our laboratories if we have the specimens, but very little time remains before the last bits of Madagascan or western Ecuadorian forest are cleared and irreversibly transformed. Despite our best efforts, we are going to be unable to preserve much of the planet's diversity as populations expand into wilderness. Taxonomic field work in the disappearing forests, especially in the tropics, must receive immediate attention. We need to organize and support teams of collectors to move into the forest ahead of highway contractors and timber concessionaires with malaise traps, plant presses, Sherman traps, and canopy foggers in order to gather as complete a picture as possible of entire communities for later analysis.

Until now, museums have to a large extent been repositories of what is studied and what is known. Given the rate at which the biological wealth of the planet is being lost and the present inability of taxonomy to keep up with material that we must collect, museums must become temporary repositories of what is not known and cannot yet be studied. Later, we will be able to reconstruct much of the biology of these by then extinct communities from our collections. We need to partially reorder our priorities in taxonomic and ecological fieldwork and put much greater effort into obtaining samples of the life that now inhabits these areas for future study. Even if we are unable to preserve the biological diversity now existing in its natural state, it will be some consolation if we at least know what it is that we have lost.

VI. Summary

Ants comprise the single most abundant family of arthropods in lowland tropical forest canopies and are arguably the most pervasive in their effects on other canopy organisms. Because of the difficulty of access and systematic sampling in forest canopies, little information has been available until recently on the diversity and abundance of arboreal ants and other arthropods. With the development of insecticidal collection methods, a clearer understanding of the composition and assembly patterns of arboreal ant assemblages is unfolding. These methods have also highlighted the inadequacy of our understanding of the diversity of canopy arthropod diversity. This knowledge suggests the need for a dramatically increased effort at salvage collection work in threatened areas.

Acknowledgments

I am grateful to the following people for helpful discussions and/or for commenting on earlier versions of this manuscript: J. M. Carpenter, E. S. Chupaca, S. P. Cover, D. W. Davidson, T. L. Erwin, G. S. Gilbert, B. Hölldobler, M. Kaspari, J. T. Longino, M. D. Lowman, J. D. Majer, N. M. Nadkarni, N. E. Stork, and E. O. Wilson.

References

Adis, J., and Schubart, H. O. R. (1984). Ecological research on arthropods in central Amazonian forest ecosystems with recommendations for study procedures. *In* "Trends in Ecological Research for the 1980s" (J. H. Cooley and F. B. Golley, eds.), pp. 111–144. Plenum, New York.

Adis, J., Lubin, Y. D., and Montgomery, G. G. (1984). Arthropods from the canopy of inundated and terra firme forests near Manaus, Brazil, with critical considerations on the pyrethrum-fogging technique. *Stud. Neotrop. Fauna Environ.* 19, 223–236.

Barnes, R. D. (1989). Diversity of organisms: How much do we know? *Am. Zool.* 29, 1075–1084.

Baroni Urbani, C., Bolton, B., and Ward, P. S. (1992). The internal phylogeny of ants (Hymenoptera: Formicidae). *Syst. Entomol.* 17, 301–329.

Basset, Y., Aberlenc, H.-P., and Delvare, G. (1992). Abundance and stratification of foliage arthropods in a lowland rain forest of Cameroon. *Ecol. Entomol.* 17, 310–318.

Beattie, A. J. (1985). "The Evolutionary Ecology of Ant–Plant Mutualisms." Cambridge University Press, Cambridge, UK.

Bentley, B. L. (1977). Extrafloral nectaries and protection by pugnacious bodyguards. *Annu. Rev. Ecol. Syst.* 8, 407–427.

Breed, M. D., and Harrison, J. (1989). Arboreal nesting in the giant tropical ant, *Paraponera clavata* (Hymenoptera: Formicidae). *J. Kans. Entomol.* 62, 133–135.

Buckley, R. C. (1982). Ant–plant interactions: A world review. *In* "Ant–Plant Interactions in Australia" (R. C. Buckley, ed.), pp. 111–141. Dr. W. Junk Publ., The Hague, The Netherlands.

Buckley, R. C. (1987). Interactions involving plants, Homoptera, and ants. *Annu. Rev. Ecol. Syst.* 18, 111–135.

Davidson, D. W. (1988). Ecological studies of Neotropical ant gardens. *Ecology* 69, 1138–1152.

Davidson, D. W., and McKey, D. (1993). The evolutionary ecology of symbiotic ant–plant relationships. *J. Hymenop. Res.* 23, 13–83.

Davidson, D. W., Longino, J. T., and Snelling, R. R. (1988). Pruning of host plant neighbors by ants: An experimental approach. *Ecology* 69, 801–808.

Erwin, T. L. (1982). Tropical forests: Their richness in Coleoptera and other arthropod species. *Coleopterists Bull.* 36, 74–75.

Erwin, T. L. (1983). Beetles and other insects of tropical forest canopies at Manaus, Brazil, sampled by insecticidal fogging. *In* "Tropical Rain Forest: Ecology and Management" (S. L. Sutton, T. C. Whitmore, and A. C. Chadwick, eds.), pp. 59–75. Blackwell, Oxford.

Erwin, T. L. (1989). Canopy arthropod biodiversity: A chronology of sampling techniques and results. *Rev. Peru. Entomol.* 32, 71–77.

Erwin, T. L. (1991). How many species are there?: Revisited. *Conserv. Biol.* 5, 330–333.

Erwin, T. L., and Scott, J. C. (1980). Seasonal and size patterns, trophic structure, and richness of Coleoptera in the tropical arboreal ecosystem: The fauna of the tree *Leuhea seemannii* Triana and Planch in the Canal Zone of Panama. *Coleopterists Bull.* 34, 305–322.

Fiala, B., Maschwitz, U., and Pong, T. Y. (1991). The association between *Macaranga* trees and ants in South-east Asia. *In* "Ant–Plant Interactions" (C. R. Huxley and D. F. Cutler, eds.), pp. 263–270. Oxford Univ. Press, Oxford.

Gaston, K. J. (1991). The magnitude of global insect species richness. *Conserv. Biol.* 5, 283–296.

Harada, A. Y., and Adis, J. (1994). The ant fauna of tree canopies in central Amazonia: A first assessment. *Abst. Int. Cong. Ecol. (INTECOL), 6th*, Manchester, England.

Hijii, N. (1983). Arboreal arthropod fauna in a forest. I. Preliminary observation on seasonal fluctuations in density, biomass, and faunal composition in a *Chamaecyparis obtusa* plantation. *Jpn. J. Ecol.* 33, 435–444.

Hijii, N. (1989). Arthropod communities in a Japanese cedar (*Cryptomeria japonica* D. Don) plantation: Abundance, biomass and some properties. *Ecol. Res.* 4, 243–260.

Hodkinson, I. D., and Casson, D. (1991). A lesser predilection for bugs: Hemiptera (Insecta) diversity in tropical rain forests. *Biol. J. Linn. Soc.* 43, 101–109.

Hölldobler, B. (1979). Territories of the African weaver ant (*Oecophylla longinoda* [Latreille]): A field study. *Z. Tierpsychol.* 51, 201–213.

Hölldobler, B. (1980). Canopy orientation: A new kind of orientation in ants. *Science* 210, 86–88.

Hölldobler, B. (1983). Territorial behavior in the green tree ant (*Oecophylla smaragdina*). *Biotropica* 15, 241–250.

Hölldobler, B., and Wilson, E. O. (1977). Weaver ants: Social establishment and maintenance of territory. *Science* 195, 900–902.

Hölldobler, B., and Wilson, E. O. (1990). "The Ants." Harvard Univ. Press, Cambridge, MA.

Huxley, C. R. (1980). Symbiosis between ants and epiphytes. *Biol. Rev. Cambridge Philos. Soc.* 55, 321–340.

Jackson, D. A. (1984). Ant distribution patterns in a Cameroonian cocoa plantation: Investigation of the ant mosaic hypothesis. *Oecologia* 62, 318–324.

Janzen, D. H. (1966). Coevolution of mutualism between ants and acacias in Central America. *Evolution (Lawrence, Kans.)* 20, 249–275.

Kikuzawa, K., and Shidei, T. (1966). On the sampling technique to estimate the density and the biomass of the forest arthropods. *Jpn. J. Ecol.* 16, 24–28.

Kim, K. C. (1993). Biodiversity, conservation and inventory: Why insects matter. *Biodiversity and Conservation* 2, 191–214.

Kitching, R. L., Bergelson, J. M., Lowman, M. D., McIntyre, S., and Carruthers, G. (1993). The biodiversity of arthropods from Australian rainforest canopies: General introduction, methods, sites and ordinal results. *Aust. J. Ecol.* 18, 181–191.

Kleinfeldt, S. E. (1978). Ant-gardens: The interaction of *Codonanthe crassifolia* (Gesneriaceae) and *Crematogaster longispina* (Formicidae). *Ecology* 59, 449–456.

Kleinfeldt, S. E. (1986). Ant-gardens: Mutual exploitation. *In* "Insects and the Plant Surface" (B. E. Juniper and T. R. E. Southwood, eds.), pp. 283–294. Edward Arnold, London.

Koptur, S. (1992). Extrafloral nectary-mediated interactions between insects and plants. *In* "Insect–Plant Interactions" (E. Bernays, ed.), Vol. IV, pp. 81–129. CRC Press, Boca Raton, FL.

Lane, R. P. (1992). The 'new taxonomy'—Does it require new taxonomists or a new understanding? *Bull. Entomol. Res.* 82, 437–440.

Leston, D. (1973). Ecological consequences of the tropical ant mosaic. *Proc. Int. Cong. Int. Union Study Soc. Insects, 7th, 1973,* pp. 235–242.

Leston, D. (1978). A Neotropical ant mosaic. *Ann. Entomol. Soc. Am.* 72, 649–653.

Longino, J. T. (1989). Geographic variation and community structure in an ant–plant mutualism: *Azteca* and *Cecropia* in Costa Rica. *Biotropica* 21, 126–132.

Longino, J. T. (1991). *Azteca* ants in *Cecropia* trees: Taxonomy, colony structure, and behaviour. *In* "Ant–Plant Interactions" (C. R. Huxley and D. F. Cutler, eds.), pp. 271–288. Oxford Univ. Press, Oxford.

Longino, J. T., and Nadkarni, N. M. (1990). A comparison of ground and canopy leaf litter ants (Hymenoptera: Formicidae) in a Neotropical montane forest. *Psyche* 97, 81–93.

Majer, J. D. (1976a). The maintenance of the ant mosaic in Ghana cocoa farms. *J. Appl. Ecol.* 13, 123–144.

Majer, J. D. (1976b). The ant mosaic in Ghana cocoa farms: Further structural considerations. *J. Appl. Ecol.* 13, 145–155.

Majer, J. D. (1976c). The influence of ants and ant manipulation on the cocoa farm fauna. *J. Appl. Ecol.* 13, 157–175.

Majer, J. D. (1990). The abundance and diversity of arboreal ants in northern Australia. *Biotropica* 22, 191–199.

Majer, J. D., and Camer-Pesci, P. (1991). Ant species in tropical Australian tree crops and native ecosystems—Is there a mosaic? *Biotropica* 23, 173–181.

Majer, J. D., and Recher, H. F. (1988). Invertebrate communities on Western Australian eucalypts: A comparison of branch clipping and chemical knockdown procedures. *Aust. J. Ecol.* 13, 269–278.

Marshall, S. A. (1993). Biodiversity and insect collections. *Can. Biodiversity* 2, 16–22.

Martin, J. L. (1966). The insect ecology of red pine plantations in central Ontario. IV. The crown fauna. *Can. Entomol.* 98, 10–27.

Montgomery, G. G. (1985). Impact of vermilinguas (*Cyclopes, Tamandua:* Xenarthra = Edentata) on arboreal ant populations. *In* "The Evolution and Ecology of Armadillos, Sloths, and Vermilinguas" (G. G. Montgomery, ed.), pp. 351–363. Smithsonian Institution, Washington, D.C.

Moran, V. C., and Southwood, T. R. E. (1982). The guild composition of arthropod communities in trees. *J. Anim. Ecol.* 51 289–306.

O'Dowd, D. J. (1982). Pearl bodies as ant food: An ecological role for some leaf emergences of tropical plants. *Biotropica* 14, 40–49.

Oliveira, P. S., Oliveira-Filho, A. T., and Cintra, R. (1987). Ant foraging on ant-inhabited *Tri-*

plaris (Polygonaceae) in western Brazil: A field experiment using live termite-baits. *J. Trop. Ecol.* 3, 193–200.

Peakall, R., Handel, S. N., and Beattie, A. J. (1991). The evidence for, and importance of, ant pollination. *In* "Ant–Plant Interactions" (C. R. Huxley and D. F. Cutler, eds.), pp. 421–433. Oxford Univ. Press, Oxford.

Pierce, N. E. (1987). The evolution and biogeography of associations between lycaenid butterflies and ants. *In* "Oxford Surveys in Evolutionary Biology" (P. H. Harvey and L. Partridge, eds.), pp. 89–116. Oxford Univ. Press, Oxford.

Pisarski, B. (1978). Comparison of various biomes. *In* "Production Ecology of Ants and Termites" (M. V. Brian, ed.), pp. 326–331. Cambridge Univ. Press, Cambridge, UK.

Quinlan, R. J., and Cherrett, J. M. (1979). The role of fungus in the diet of the leaf-cutting ant *Atta cephalotes* (L.). *Ecol. Entomol.* 4, 151–160.

Raven, P. H., and Wilson, E. O. (1992). A fifty-year plan for biodiversity surveys. *Science* 258, 1099–1100.

Roberts, H. R. (1973). Arboreal Orthoptera in the rain forests of Costa Rica collected with insecticide: A report on the grasshoppers (Acrididae), including new species. *Proc. Acad. Nat. Sci. Philadelphia* 125, 49–66.

Room, P. M. (1971). The relative distributions of ant species in Ghana's cocoa farms. *J. Anim. Ecol.* 40, 735–751.

Schupp, E. W., and Feener, D. H. (1991). Phylogeny, lifeform, and habitat dependence of ant-defended plants in a Panamanian forest. *In* "Ant–Plant Interactions" (C. R. Huxley and D. F. Cutler, eds.), pp. 175–197. Oxford Univ. Press, Oxford.

Smythe, N. (1982). The seasonal abundance of night-flying insects in a Neotropical forest. *In* "The Ecology of a Tropical Forest: Seasonal Rhythms and Long-Term Changes" (E. G. Leigh, Jr., A. S. Rand, and D. M. Windsor, eds.), pp. 309–318. Smithsonian Institution, Washington, D.C.

Stork, N. E. (1987). Guild structure of arthropods from Bornean rain forest trees. *Ecol. Entomol.* 12, 69–80.

Stork, N. E. (1988). Insect diversity: Facts, fiction and speculation. *Biol. J. Linn. Soc.* 35, 321–337.

Stork, N. E. (1991). The composition of the arthropod fauna of Bornean lowland rain forest trees. *J. Trop. Ecol.* 7, 161–180.

Stork, N. E. (1993). How many species are there? *Biodiversity and Conservation* 2, 215–232.

Stork, N. E., and Brendell, M. J. D. (1990). Variation in the insect fauna of Sulawesi trees with season, altitude and forest type. *In* "Insects and the Rain Forests of South East Asia (Wallacea)" (W. J. Knight and J. D. Holloway, eds.), pp. 173–190. Royal Entomological Society of London, London.

Stork, N. E., and Gaston, K. J. (1990). Counting species one by one. *New Sci.* 1729, 43–47.

Tobin, J. E. (1991). A Neotropical rainforest canopy ant community: Some ecological considerations. *In* "Ant–Plant Interactions" (C. R. Huxley and D. F. Cutler, eds.), pp. 536–538. Oxford Univ. Press, Oxford.

Tobin, J. E. (1994). Ants as primary consumers: Diet and abundance in the Formicidae. *In* "Nourishment and Evolution in Insect Societies" (J. H. Hunt and C. A. Nalepa, eds.). Westview Press, Boulder, CO.

Watanabe, H., and Ruaysoongnern, S. (1989). Estimation of arboreal arthropod density in a dry evergreen forest in northeastern Thailand. *J. Trop. Ecol.* 5, 151–158.

Wehner, R. (1987). Spatial organization of foraging behavior in individually searching desert ants, *Cataglyphis* (Sahara Desert) and *Ocymyrmex* (Namib Desert). *Experientia Suppl.* 54, 15–42.

Wetterer, J. K. (1993). Foraging and nesting ecology of a Costa Rican leaf-cutting ant, *Acromyrmex volcanus. Psyche* 100, 65–76.

Wheeler, Q. D. (1990). Insect diversity and cladistic constraints. *Ann. Entomol. Soc. Am.* 83, 1031–1047.

Wilson, E. O. (1959). Some ecological characteristics of ants in New Guinea rain forests. *Ecology* 40, 437–447.

Wilson, E. O. (1985). The biological diversity crisis. *BioScience* 35, 700–706.

Wilson, E. O. (1987). The arboreal ant fauna of a Peruvian Amazon forest: A first assessment. *Biotropica* 19, 245–251.

Wilson, E. O. (1988). The current state of biological diversity. *In* "Biodiversity" (E. O. Wilson and F. M. Peters, eds.), pp. 3–18. National Academy Press, Washington, DC.

Yen, A. L. (1989). Overstory invertebrates in the Big Desert, Victoria. *In* "Mediterranean Landscapes in Australia: Mallee Ecosystems and Their Management" (J. C. Noble and R. A. Bradstock, eds.), pp. 285–299. CSIRO, Australia.

7

Lizard Ecology in the Canopy of an Island Rain Forest

Douglas P. Reagan

Life in both the forest and the sea is distributed in horizontal layers. Most life is near the top, because that is where the sunlight strikes and everything below depends on this surface.
—*Marston Bates, "The Forest and the Sea" (1960)*

I. Introduction

Information on animals in rain forest canopies consists almost exclusively of data on mammals and birds (Odum and Pigeon, 1970; Whitmore, 1975; Janzen, 1983; Leigh *et al.*, 1985). Although amphibians such as Wallace's flying frog (*Rhacophorus nigropalmatus*) and reptiles such as the paradise tree snake (*Chrysopelea pelias*) and flying dragon (*Draco volans*) are known to inhabit the upper parts of tropical rain forest, data on their populations in the canopy are sparse. The primary reasons for this are the difficulties in observing amphibians and reptiles (which are generally small or relatively immobile) more than a few meters above ground level, and the challenges of conducting studies within the forest canopy.

The anoline lizards (*Anolis* spp.) are the most abundant and conspicuous vertebrates that inhabit terrestrial ecosystems on islands in the Caribbean (Moermond, 1979; Williams, 1969, 1983). On large islands, several species occur syntopically in forest habitats and partition the habitat vertically (Williams, 1972). The habitat and vertical distribution of anoles in Puerto Rico have been investigated by ground-based observers (Rand, 1964; Schoener

Copyright © 1995 by Academic Press, Inc.
All rights of reproduction in any form reserved.

and Schoener, 1971; Moll, 1978; Lister, 1981), who documented the presence of anoles in the upper reaches of the forest, but provided few insights on their abundance or possible roles in the animal community of the forest canopy. Recent studies in Puerto Rico have produced quantitative information on the abundance and foraging of anoline lizards that shows that, at least in some rain forests, lizards play a significant role in the animal community of the forest canopy.

My work on canopy lizards began as the result of serendipitous observations made shortly after moving to Puerto Rico in 1979. I had accepted a position at the Center for Energy and Environment Research of the University of Puerto Rico and had made walking surveys through the forest near the El Verde Field Station. On one such visit, I climbed to the top of a tower that had been erected during the 1960s for irradiation studies. Looking out through the canopy, I noticed several *Anolis stratulus* (Fig. 1) on the upper-

Figure 1   Male *Anolis stratulus* with dewlap extended, defending its territory in the forest canopy. (Photograph by D. Reagan.)

most branches of nearby trees. Because I could see several individuals from a single point, I reasoned that they must be abundant and, therefore, it might be possible to study them from the tower. This chance observation of anoles in the canopy led me to investigate the distribution and abundance of *A. stratulus* and other anole species in the three-dimensional habitat of the forest canopy. I began my canopy research with vertical transects to document the vertical distribution and general abundance of anole species in the forest. The results of this study raised additional questions on population density, habitat use, and the role of *A. stratulus* in the forest food web, and started me on a course of research in the forest canopy that I have continued to pursue.

My objective in this chapter is to summarize what is known about anoline lizards that inhabit the canopy of rain forests in Puerto Rico and suggest directions for future research on canopy reptiles. Some of the insights have extended our knowledge of ecosystem structure and function in unexpected directions. The role of anoles in forest canopies elsewhere in the Caribbean and of lizards in other rain forest canopies is not yet known; however, the importance of these lizards in Puerto Rican rain forest canopies is now established.

II. The Forest

Most of the studies described in this chapter were conducted in tropical wet (tabonuco) forest between 250 and 450 m on the slopes of the Luquillo Mountains in eastern Puerto Rico (18°9′N, 65°45′W). The forest is exposed to the moisture-laden trade winds. Mean monthly temperatures are 21–24°C, annual rainfall is approximately 3.5 m, and there is a mild dry season from January to mid-April. Small streams dissect the forest along rocky channels, and the general topography varies from steep to gently sloping (Brown *et al.*, 1983).

The forest is diverse compared to temperate forests but depauperate compared to some mainland tropical forests (Richards, 1952; Gentry, 1990). More than 200 tree species occur in the Luquillo Mountains, most of which occur in the tropical wet forest near El Verde (Brown *et al.*, 1983). Tabonuco (*Dacryodes excelsa*), sierra palm (*Prestoea montana*), American muskwood (*Guerea trichilioides*), motillo (*Sloanea berteriana*), and granadillo (*Buchenavia capitata*) are common throughout the forest. The forest is evergreen and the canopy remains closed throughout the year. Canopy height is 20–25 m with few emergent trees and, unlike many other tropical rain forests, there is no well-defined subcanopy.

The animal community of the forest is characterized by a low species richness compared to similar mainland forests. The only native mammals are eight species of bats. Feral cats and a few Indian mongoose (*Herpestes auro-*

punctatus), both introduced species, also inhabit the forest. Introduction of the Indian mongoose contributed to the extinction of a few native ground-dwelling mammals on Puerto Rico and other Caribbean islands (Seaman, 1952). Only 49 bird species occur in the forest near El Verde (R. Waide, unpublished data). Eight anuran and nine lizard species inhabit the forest. Although some species attain high population densities (L. Woolbright, M. Stewart, and D. Reagan, unpublished data), species richness is low compared to similar mainland forests (Duellman, 1983). The Puerto Rican boa (*Epicrates inornatus*), red-tailed hawk (*Buteo jamaicensis*), and Puerto Rican screech owl (*Otus nudipes*) have the largest body sizes of the forest predators.

III. Canopy Methods

A. Canopy Access

1. Towers   Most of my observations of canopy lizards were conducted from scaffold towers and a canopy walkway. At El Verde, a 22-m tower (the "Odum tower") was erected during the 1960s for forest irradiation studies (Odum and Pigeon, 1970). Two other towers, 30 m apart and connected by a canopy walkway, were erected in 1984 (Fig. 2). A fourth tower (the "Bisley tower") was erected in similar forest type on the eastern side of the Luquillo Mountains. These towers provide safe and easy access to the canopy during all but the most severe weather conditions. Because towers are expensive and logistically difficult to construct, their use is recommended only where densities of canopy animals (e.g., anoline lizards) are sufficient to permit adequate numbers of observations from a few canopy locations.

2. Single-Rope Techniques   While investigating the food web relationships of canopy anoles, Dr. Roman Dial used a system of ropes and nylon harnesses to reach and move in the forest canopy. I made trips to the canopy with him and gained a firsthand appreciation of this access method. Initially, I felt insecure working 20 m above ground level, dangling from (or perched on) a limb that swayed over a distance of 2–3 m in the 10–20 km/hr trade winds that swept across the canopy. I recall wishing for an extra pair of limbs and a prehensile tail with which to anchor myself more securely to the precarious perch. This technique allows an observer the mobility to gain access to the forest up to the midcanopy of most large trees, but requires training and experience to reach and move about safely in the canopy. The system is described in Dial (1992).

3. Swiss Tree Grippers   Swiss tree grippers ("tree-bicycles") were used during initial canopy studies to determine if *A. stratulus* was distributed throughout the forest canopy. This device consists of a belt harness and pedals at-

Figure 2   Walkway through the forest canopy, 20 m above ground level and adjacent to tower in the forest near El Verde, Puerto Rico. (Photograph by D. Reagan.)

tached to metal hoops by metal brackets of unequal length so that the loops can be placed one above the other on the tree trunk. By alternately lifting one foot and raising it, and then stepping down to force a rubber brake against the trunk, one can walk up a tree trunk without causing visible damage. However, the method had three shortcomings. First, it worked only on trees of relatively narrow diameter and was difficult to install on trees with buttresses (of which there were many). Second, because the metal loops could not be easily adjusted, once the device was in use, ascent was limited to branchless boles. Although this allows some visual access to the upper canopy, observers still remain several meters below the top of the canopy. Third, because most climbable trees in the forest have smooth bark, when it rains (and sporadic, intense showers are typical), descent can become a rapid and uncontrolled slide down a slippery trunk. Dr. Alberto Sabat (University of Puerto Rico) has adapted this device for climbing palms by replacing the rubber brakes with scored metal plates to prevent sliding.

4. Other Methods   Other canopy methods have been considered but not implemented. Among these are balloons (too much wind and steep topographic relief), cranes (expensive to operate, difficult to install), canopy cable network (used in the 1960s, installation generally limited to lower canopy), cable or fixed ladders, and cable walkways (difficult to install).

B. Anole Studies

Because most of my canopy observations were made from towers, they are essentially point samples in areal terms. Frye's variable-width transect method (Overton, 1971) was adapted for this purpose. During initial vertical surveys, regular patterns of perch height and size emerged, which required the development of procedures to measure these habitat parameters. Detailed methods have been used to document the vertical distribution of lizards, population densities, habitat parameters, seasonal movements, and food habits (Reagan 1986, 1992; Reagan *et al.,* 1982; R. Garrison and D. Reagan, unpublished data). Additional methods for measuring anole densities and investigating canopy food webs are described in Reagan (1991) and Dial (1992).

The population density of *A. stratulus* was estimated using multiple mark and resight (recapture) methods of hot-branded individuals captured from towers. Ground surveys were conducted along 120- to 150-m transects to estimate the minimum density and relative abundance of each anole species using the Frye strip census method. Vertical transects were conducted by walking up towers and recording the species, distance from the central axis of the tower, and height above ground level for each lizard observed. Data on feeding habits were obtained from gut content analysis of all species except *A. cuvieri.* Because this latter species is uncommon in the forest, the stomachs were pumped and the individuals released at their place of capture.

IV. Vertical Stratification

In 1964, Stanley Rand published a landmark study describing the vertical stratification of anoline lizards in Puerto Rico. He related the differences in distribution to variations in structural habitat (perch height and perch diameter) and climatic habitat (shade conditions). Schoener and Schoener (1971) provided additional information on the structural habitat of Puerto Rican anoles. Building on this foundation, I surveyed anoles, with the difference that my surveys at towers were vertical and were limited to single locations (points) within the forest. Vertical surveys began in 1980 and consisted of slowly walking up the Odum tower and recording the species, perch diameter, and height above ground level for each lizard observed. These surveys, which were repeated at different times of day and during different seasons, undoubtedly involved repeated observations of some individuals. Following the construction of three additional towers, these surveys were repeated at other locations. Surveys at these four points

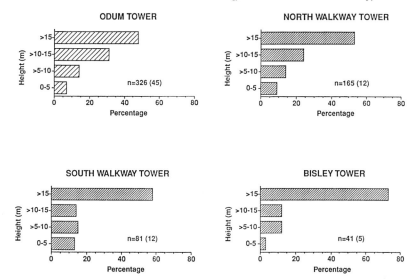

Figure 3 Occurrence (percentage) of *Anolis stratulus* in 5-m intervals at four tower locations within tabonuco forest. *n*, Total number of lizards observed. Number in parentheses indicates the number of transects conducted at each tower. (From Reagan, 1992.)

conducted between 1980 and 1989 (Fig. 3) show that *Anolis stratulus* is primarily a canopy species throughout the forest (Reagan, 1992). Using single-rope techniques, Dial (1992) obtained similar results in posthurricane forest.

The giant anole (*Anolis cuvieri*) also inhabits the forest canopy. With a mean adult body length of 125 mm and mean adult weight of about 50 g, it is a giant compared to *A. stratulus* (mean adult weight = 1.8 g). Little is known about the structural habitat requirements of giant anoles. Sightings in the canopy are rare and are usually of individuals sitting motionless on large branches. In the hundreds of hours I have spent in the forest canopy, I have seen only two giant anoles from towers; one on the upper portion of a tree trunk and the other on a large branch. Williams (1972, 1983) placed this species in the (tree) crown "ecomorph" category. Losos *et al.* (1990) monitored the movements of an adult male near El Verde and found it most often in the canopy during midday, 15–20 m above ground level. Juveniles are slow-moving and cryptically colored, and apparently spend much of the time near ground level (Rand and Andrews, 1975; Gorman, 1977; D. Reagan, personal observation).

Four other anole species inhabit tabonuco forest. *Anolis gundlachi* is a medium-sized trunk–ground species, generally limited to the lower 3 m of the forest (Rand, 1964), but occasionally ascending above this level to locate a resting location for the night (Reagan, 1992). The second species,

A. evermanni, is a true generalist. It is the only species that is regularly distributed from the canopy to ground level and also along rocky stream channels in the forest. Although it inhabits the canopy, it is most abundant in the lower canopy and near ground level (Reagan, 1992). Two other species, *A. krugi* and *A. occultus,* occur in forest openings or near streams.

V. The Canopy Habitat

The forest canopy habitat differs from that found near ground level in the forest interior. Whereas the forest interior is characteristically dark, humid, cool, and relatively open, the canopy is exposed to full sun, greater variability in humidity, higher midday temperatures, and a three-dimensional matrix of leaves and small branches. Canopy conditions resemble those found in shrubby habitats in drier areas at lower elevations on Puerto Rico and in the adjacent Virgin Islands, habitats also inhabited by *A. stratulus* (Rivero, 1978; D. Reagan, personal observation).

Perch diameter was identified as an important habitat variable for Puerto Rican anoles by Rand (1964). From the towers, I was able to record the vertical distribution of perches in different diameter classes within a 3-m radius of each tower throughout the vertical extent of the forest. Large stems predominate near ground level, stems of intermediate size are most frequent at intermediate height, and small stems are found primarily in the canopy (Fig. 4).

The strong association between the vertical distribution of *A. stratulus* (Fig. 3) and small stems (Fig. 4) suggests that lizards select for perch size rather than for perch height. To test this, I established ground transects in mature tabonuco forest and in a stand of *Eugenia jambos,* a small evergreen tree 5 to 10 m in height with a growth form of small stems branching from a single base. These trees resemble the structure of a tree crown minus the bole. Relative abundance estimates of anoles indicated that *A. stratulus* at

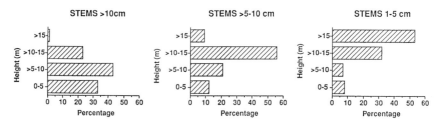

Figure 4 Vertical distribution of small, medium, and large perches (stems) within a 3-m radius of the Odum tower in tabonuco forest near El Verde, Puerto Rico. (From Reagan, 1992.)

ground level in mature tabonuco forest was only 9%, but in the *Eugenia* stand it was 53%, supporting the perch size selection hypothesis (Reagan, 1992).

VI. Population Density

Prehurricane population densities of *A. stratulus* along vertical transects at tower locations were estimated using multiple mark and resight techniques. Because transects were vertical, calculations were based on a point sample with the area based on the maximum reliable observation radius (2 m). Mean population estimates were 25,870 (SD = 7,005) and 21,333 (SD = 6,638) individuals/ha during the wet and dry seasons, respectively, or more than 2 individuals/m^2 (Reagan, 1992).

The highest population densities previously reported for any lizard species are for Caribbean anoles. Schoener and Schoener (1980) reported a maximum density of 0.97 individuals/m^2 for *A. sagrei* in the Bahamas, and Gorman and Harwood (1977) estimated up to 20,000 individuals/ha for *A. pulchellus* in the grassy lowlands of eastern Puerto Rico. The estimates for *A. stratulus* in the forest canopy are the highest reported for any lizard species.

VII. Home Range of *Anolis stratulus*

Home ranges were determined for 26 individually marked *A. stratulus* (Reagan, 1992). Data on height above ground level, distance from the centerline of the vertical transect, and compass direction were recorded in order to locate individual sightings in three dimensions. In both wet and dry seasons, individuals occupied home ranges 6.2 ± 1.2 m in diameter (all axes). All but one of the 12 home ranges had vertical axes substantially less than 10- to 14-m-thick canopy, and only one individual occupied a home range below the canopy.

Examination of individual home ranges, supplemented with observations during transect surveys, revealed that males defend three-dimensional territories. Females and subadults occupy home ranges that broadly overlap male territories. The resulting configuration is an assemblage of ellipsoids approximately 6 to 7 m in mean diameter, layered within the canopy. Individual home range shape is probably quite variable, depending on the distribution of available perches, territorial interactions, and other factors.

Three-dimensional home ranges have been reported for other animals. Meserve (1977) found that some congeneric mouse species (*Peromyscus* spp.) exhibited substantial arboreal behavior and speculated that use of three-dimensional home ranges provided a means of reducing interspecific competition. However, the layering of home ranges within a habitat volume as found in *A. stratulus* has not been reported for any other terrestrial vertebrate. The combination of small home range size, dispersed substrate (small branches), and thick canopy layer (10–14 m) account for this phenomenon and make it possible for this small territorial vertebrate to attain high population densities.

VIII. Canopy *Anolis* Foraging Behavior

The distribution of *A. stratulus* in the forest canopy indicates that they obtain most of their food there. Two other anoles, *A. evermanni* and *A. cuvieri*, also inhabit the canopy, but less is known about their foraging behavior. Information on the trophic relationships of these species is summarized in the following.

A. Foraging Rates

Foraging rates of *A. stratulus* were obtained from field observations of foraging behavior and extrapolations from gut content analyses. Reagan (1986), using direct observation methods, calculated foraging rates of 20.5 prey items/day during the wet season and 18.0 items/day during the dry season. These estimates were similar to those of Lister (1981), who found 25.6 and 15.4 prey items/lizard during wet and dry seasons, respectively, using stomach analysis methods.

An independent estimate of foraging rates was obtained from gut content analyses conducted on *A. stratulus* from El Verde (Reagan *et al.*, 1982). Stomachs of lizards contained 11.5 and 8.7 items/individual during wet and dry seasons, respectively. All lizards were collected during midday, approximately halfway through the available foraging time. Assuming that prey found in the stomach represented prey taken that day (Lister, 1981), doubling these numbers produced estimates of 23.0 and 17.4 items/individual for wet and dry seasons, respectively. Prey biomass was also lower during the dry season. A mean consumption rate of 9.5 items/juvenile was calculated from stomach content data (Reagan *et al.*, 1982). Combining mean daily consumption rate with population structure, adult and juvenile consumption rates, and mean population density produced an estimate of 3.4×10^5 prey items/ha/day taken by *A. stratulus*.

Reagan (1986) found that *A. stratulus* foraged chiefly in the canopy, but during the dry season extended its foraging range to lower levels of the forest. Reagan (1984) also noted competitive interactions between *A. stra-*

tulus and *A. evermanni* over food during the dry season. The expanded foraging range, greater foraging activity, competitive interactions, and reduced prey consumption during the dry season (prehurricane) suggest that this species may be food limited at times. Dial (1992) determined that increased population densities of both *A. stratulus* and *A. evermanni* in the canopy of isolated tabonuco trees under posthurricane conditions correlated with increased numbers of flying insects caught in canopy sticky traps. Assuming that the posthurricane anoles consume mostly flying insects, this provides additional evidence of food limitation under posthurricane conditions.

Data on the foraging behavior of *A. cuvieri* are limited because of their relatively low population density. Stomach content analyses (discussed in the next section) suggest that they eat sedentary or slow-moving prey that occur at all levels in the forest.

B. Prey

Anolis stratulus is an insectivore. Studies of gut contents prior to Hurricane Hugo showed less than 10% of noninsect materials in stomachs, in terms of number of items and volume (R. Garrison and D. Reagan, unpublished data). Numerically, ants were the most common prey, comprising nearly 60% of all items consumed. Homoptera (chiefly planthoppers) comprised 50% of food by volume, as compared to 10% for ants. No dietary differences were detected between male and female anoles. Other arthropod taxa included Diptera, Lepidoptera, Coleoptera, Orthoptera, Acarina, and Araneida. Small amounts of plant material were also found, but may have been ingested incidentally with arthropod prey. Anole skin was also found in the stomachs of some individuals, presumably ingested by individuals in the course of normal shedding. From a nutritional standpoint, prey volume is important, but prey number may be more relevant in terms of the population effects on insect prey.

Anolis stratulus search for prey along branch and leaf surfaces (Reagan, 1986), consuming prey of a mean length of 2–3 mm (R. Garrison and D. Reagan, unpublished data). Using sticky traps in the canopies of tabonuco trees isolated by Hurricane Hugo, Dial (1992) compared insect abundance in the canopies of trees with and without anoles. He showed that *A. stratulus* suppressed prey abundance in size categories greater than 2 mm and concluded that the species is an optimal forager.

The giant anole, *A. cuvieri,* is a slow-moving forager in the forest canopy in wet and dry forests of Puerto Rico. Rivero (1978) reported that the species feeds on large insects, snails, and smaller lizards. Examination of the regurgitated stomach contents of 11 individuals captured in tabonuco forest near El Verde indicated a similar diet: walkingsticks, snails, and lizards

were the largest portions of the diet by volume. A tail portion from a male
A. gundlachi was found in the stomach of one individual, suggesting that at
least some prey are consumed near ground level.

IX. Hurricane Effects

On September 18, 1989, the western part of Hurricane Hugo passed over
the Luquillo Mountains of eastern Puerto Rico. Maximum sustained winds
of 166 km/hr, with gusts to 194 km/hr, were recorded (U.S. National
Weather Service, 1989). Hurricanes of this intensity are estimated to pass
over the Luquillo Mountains on the average of once every 50–60 years (Scatena and Larsen, 1991). Because of their force and periodic occurrence,
they are a major influence on forest structure and dynamics on Puerto Rico
and throughout the Caribbean. Areas of forest directly exposed to Hugo's
winds were stripped of their branches; many trees remained upright, but
with broken trunks or missing most large branches. The immediate result for canopy anoles was a shift in available structural habitat from live
branches and leaves in the canopy to dead branches and other organic debris on or near the forest floor. This shifted the vertical habitat structure
from 10 to 25 m to < 5 m above the ground.

One month following the hurricane, *Anolis stratulus* was the most abundant anole at ground level. Estimates of relative abundance were similar
between the Bisley watershed site, an area severely affected by the hurricane
(48%), and El Verde, an area protected from the full force of the hurricane
and with much of the larger canopy branch structure intact (51%). Prehurricane surveys at both sites reported that only 2% of the anoles at ground
level were *A. stratulus*. The posthurricane increase in *A. stratulus* abundance
near ground level corresponded to a 300% increase in small perches in the
lower 5 m of the forest (Reagan, 1991).

A few *A. cuvieri* were seen near ground level at scattered locations in the
forest immediately after the hurricane (J. Wunderle, personal communication). In the three years following Hurricane Hugo, individuals have
been seen only rarely in highly disturbed areas of the forest. Dial (1992)
observed two individuals following Hurricane Hugo. Each was seen four
times during nearly 2000 hours of fieldwork (R. Dial, personal communication). The species may have suffered from lack of prey or, because of
their increased visibility and slow movements, they may have been exposed
to increased predation by birds.

From 1990 to 1992, Roman Dial (then a graduate student at Stanford
University) conducted intensive investigations of *A. stratulus* and *A. evermanni* in the canopy of the tabonuco forest that had suffered extensive
damage from Hurricane Hugo. Through a series of experiments and obser-

vations, he investigated aspects of the trophic relationships of these species in isolated and regenerating tabonuco tree canopies. In addition to the general results on food web relationships, he found that the two species that shared structural, climatic, and food niches (*A. stratulus* and *A. evermanni*) exhibited competitive exclusion at one spatial scale (isolated tree crowns) but not at larger spatial scales (Dial, 1992).

In October 1991, 25 months after the hurricane, differential forest regeneration between severely disturbed forest (Bisley) and moderately disturbed forest (El Verde) still showed differences in canopy habitat and use by *A. stratulus*. At El Verde, 71% of individuals were found in the canopy (above 10 m) compared to 46% at Bisley. Although rapid regrowth had restored small branches and leaves to most of the trees at Bisley, the upper canopy was still not closed three years after the hurricane.

X. Canopy Lizards in the Forest Community

Anoline lizards play a significant role in many Caribbean ecosystems because of their high abundance and their role as intermediate predators in food webs. The abundance and distribution of anoles throughout the vertical extent of the forest at El Verde indicate such a role for Puerto Rican anoles (Reagan, 1992). Dial (1992) showed that canopy anoles suppressed the abundance of large arthropods and spiders, demonstrating that anoles play a role in shaping food web structure in the canopy of tropical wet forests in Puerto Rico.

The rain forest canopy provides favorable conditions for numerous animal species. Primary productivity is high, and primary consumption by chewing and sucking insects occurs throughout the dense foliage. A three-dimensional matrix of leaves and small branches provides suitable substrate for perching, foraging, and other activities. Trade winds carry flying insects and ballooning spiders into the canopy that provide additional food for small consumers (Dial, 1992), including anoline lizards and birds.

Studies in the forest at El Verde have documented a number of anole predators. The colubrid snake *Alsophis portoricensis* preys on anoles, including *A. stratulus,* in the forest canopy (J. Gillingham, personal communication; D. Reagan, personal observation). The Puerto Rican lizard cuckoo (*Saurothera vieilloti*), pearly-eyed thrasher (*Margarops fuscatus*), and red-tailed hawk (*Buteo jamaicensis*) are major avian predators (R. Waide, unpublished data). The tailless whip scorpion (*Phrynus longipes*), tarantula (*Avicularia laeta*), and centipede (*Scolopendra alternans*) all prey on anoles (D. Reagan, unpublished data). Intrageneric predation has also been observed; *A. evermanni* has been observed consuming juvenile *A. stratulus,* and a piece of tail from a male *A. gundlachi* was found in the stomach of an adult

A. cuvieri (R. Garrison and D. Reagan, unpublished data). Introduced mammals, feral cats, and the Indian mongoose also feed on anoles, but are not arboreal and probably consume few, if any, *A. stratulus.* The abundance and species richness of birds is low in the forest canopy at El Verde, as in forests on other Caribbean islands. Competition with lizards and predation have been suggested to account for this (Wright, 1981; Waide and Reagan, 1983).

XI. Summary and Recommendations for Future Research

The abundance and observability of anoles have made them ideal subjects for investigating ecological questions. The importance of anoles in the canopy of Puerto Rican rain forests has been established, but similar studies are lacking elsewhere in the Caribbean. Even on Puerto Rico, studies to date have been limited to only one forest type (tabonuco forest). Although studies in Puerto Rico have produced a wealth of information, almost nothing is known about canopy lizards elsewhere in the tropics.

Basic population information is still lacking for some common species of Puerto Rican forest anoles. Recent studies focused on *A. stratulus. Anolis evermanni* also inhabits the canopy of tabonuco forest, but occurs throughout the vertical extent of the forest. Dial (1992) estimated the populations of *A. evermanni* at two sites that differed in the degree of forest damage from Hurricane Hugo and found a high degree of variation in relative abundance among individual trees sampled and between sites. It is the only anole species that regularly forages on rocks adjacent to stream channels in the forest (Reagan *et al.,* 1982), and remains the least known and perhaps the most interesting of the common anoles in Puerto Rican forests.

Long-term studies of anole populations at forest sites in Puerto Rico indicate that the species composition, relative abundance, and vertical distribution of anoles may still be responding to Hurricane Hugo, four years after it struck Puerto Rico (D. Reagan, unpublished data). Long-term investigations of anole responses to changes in structural habitat and food sources as a result of Hurricane Hugo continue, but these studies must be supplemented by experiments. Field studies conducted by Dial (1992) illustrate that much can be learned in a short period of time using properly designed manipulations.

The small size of *A. stratulus* and abundance of small insects allow this anole to occupy small home range volumes layered within the canopy and to attain the highest population densities documented for any lizard species. Although habitat stratification is known for other small vertebrates, *A. stratulus* remains the only terrestrial vertebrate for which this home range layering has been demonstrated. This phenomenon should be investigated in similar structural habitats and animal communities elsewhere in the Caribbean. Observers in mainland tropical forests have not noted similarly high abundances of anoles in the canopy, possibly because the presence of competitors (insectivorous birds) and numerous lizard predators (particularly snakes and birds) make it unlikely that lizards in mainland forests attain densities similar to that of *A. stratulus* in Puerto Rico.

Interactions among canopy anoles also require additional investigation. Competitive behavior has been noted between *A. stratulus* and *A. evermanni* (Reagan, 1984), and Dial (1992) demonstrated competitive exclusion between these species. In spite of the extensive literature on Caribbean anoles, these and many other ecological principles should be investigated using these abundant organisms.

References

Bates, M. (1960). "The Forest and the Sea." Vintage Books, New York.

Brown, S., Lugo, A., Silander, S., and Liegel, L. (1983). Research history and opportunities in the Luquillo Experimental Forest. *U.S., For. Serv. South. For. Exp. St., Gen. Tech. Rep.* SO-44.

Dial, R. (1992). A food web for a tropical rain forest: The canopy view from *Anolis*. Ph.D. Dissertation, Stanford University, Stanford, CA.

Duellman, W. E. (1990). Herpetofaunas in Neotropical rainforests: Comparative composition, history, and resource use. *In* "Four Neotropical Rainforests" (A. H. Gentry, ed.), pp. 455–487. Yale Univ. Press, New Haven, CT.

Gentry, A. H., ed. (1990). "Four Neotropical Rainforests." Yale Univ. Press, New Haven, CT.

Gorman, G. C. (1977). Comments on the ontogenetic color change in *Anolis cuvieri* (Reptilia, Lacertilia, Iguanidae). *J. Herpetol.* 11, 221.

Gorman, G. C., and Harwood, R. (1977). Notes on the population density, vagility, and activity patterns of the Puerto Rican grass lizard, *Anolis pulchellus* (Reptilia, Lacertilia, Iguanidae). *J. Herpetol.* 11, 363–368.

Janzen, D. H. (1983). "Costa Rican Natural History." Univ. of Chicago Press, Chicago.

Leigh, E. G., Jr., Rand, A. S., and Windsor, D. M. (1985). "The Ecology of a Tropical Forest: Seasonal Rhythms and Long-Term Changes." Smithsonian Institution Press, Washington, D.C.

Lister, B. C. (1981). Seasonal niche relationships of rain forest anoles. *Ecology* 62, 1548–1560.

Losos, J. B., Gannon, M. R., Pfeiffer, W. J., and Waide, R. B. (1990). Notes on the ecology and behavior of *Anolis cuvieri* (Lacertilia: Iguanidae) in Puerto Rico. *Caribb. J. Sci.* 26, 65–66.

Meserve, P. L. (1977). Three-dimensional home ranges of cricetid rodents. *J. Mammal.* 58, 549–558.

Moermond, T. C. (1979). Habitat constraints on the behavior, morphology, and community structure of *Anolis* lizards. *Ecology* 60(1), 152–164.

Moll, A. G. (1978). Abundance studies on the *Anolis* lizards and insect populations in altitudinally different tropical forest habitats. *Occas. Pap., Cent. Energy Environ. Res.* CEER-11.

Odum, H. T., and Pigeon, R. F., eds. (1970). "A Tropical Rain Forest." U.S. Atomic Energy Commission, Oak Ridge, TN.

Overton, W. S. (1971). Estimating the numbers of animals in wildlife populations. *In* "Wildlife Management Techniques" (R. H. Giles, ed.), pp. 403–456. The Wildlife Society, Washington, D.C.

Rand, A. S. (1964). Ecological distribution in anoline lizards of Puerto Rico. *Ecology* 45, 745–752.

Rand, A. S., and Andrews, R. (1975). Adult color dimorphism and juvenile pattern in *Anolis cuvieri. J. Herpetol.* 9, 257–260.

Reagan, D. P. (1984). Competitive interactions between rain forest lizards: Field observations and experimental evidence. *Bull. Ecol. Soc. Am.* 65, 233.

Reagan, D. P. (1986). Foraging behavior of *Anolis stratulus* in a Puerto Rican rain forest. *Biotropica* 18, 157–160.

Reagan, D. P. (1991). The response of *Anolis* lizards to hurricane-induced habitat changes in a Puerto Rican rain forest. *Biotropica* 23, 468–474.

Reagan, D. P. (1992). Congeneric species distribution and abundance in a three-dimensional habitat: The rain forest anoles of Puerto Rico. *Copeia,* pp. 392–403.

Reagan, D. P., Garrison, R. W., Martinez, J. E., Waide, R. B., and Zucca, C. P. (1982). "Tropical Rain Forest Cycling and Transport Program," Phase I Report, CEER-T-137. Terrestrial Ecology Division, San Juan, PR.

Richards, P. W. (1952). "The Tropical Rain Forest." Cambridge Univ. Press, Cambridge, UK.

Rivero, J. A. (1978). "Los Anfibios y Reptiles de Puerto Rico." Universidad de Puerto Rico, Rio Piedras.

Scatena, F. N., and Larsen, M. C. (1991). Physical aspects of Hurricane Hugo in Puerto Rico. *Biotropica* 23, 317–323.

Schoener, T. W., and Schoener, A. (1971). Structural habitats of West Indian *Anolis* lizards. II. Puerto Rico uplands. *Breviora* No. 375.

Schoener, T. W., and Schoener, A. (1980). Densities, sex ratios, and population structure in four species of Bahamian *Anolis* lizards. *J. Anim. Ecol.* 49, 19–53.

Seaman, G. A. (1952). The mongoose and Caribbean wildlife. *Trans. North Am. Wildl. Conf.* 17, 188–197.

U.S. National Weather Service (1989). Forest Office in San Juan, 1989. October 24, 1989, Statement on Hurricane Hugo. U.S. Natl. Weather Serv., San Juan, PR.

Waide, R. B., and Reagan, D. P. (1983). Competition between West Indian anoles and birds. *Am. Nat.* 121, 133–138.

Whitmore, T. C. (1975). "Tropical Rain Forests of the Far East." Clarendon Press, Oxford.

Williams, E. E. (1972). The origin of faunas: Evolution of lizard congeners in a complex island fauna—A trial analysis. *Evol. Biol.* 6, 47–89.

Williams. E. E. (1983). Ecomorphs, faunas, island size, and diverse endpoints in island radiations of *Anolis. In* "Lizard Ecology: Studies of a Model Organism" (R. B. Huey, E. R. Pianka, and T. W. Schoener, eds.), pp. 326–370. Harvard Univ. Press, Cambridge, MA.

Wright, S. J. (1981). Extinction-mediated competition: The *Anolis* lizards and insectivorous birds of the West Indies. *Am. Nat.* 117, 181–192.

8

Canopy Access Techniques and Their Importance for the Study of Tropical Forest Canopy Birds

Charles A. Munn and Bette A. Loiselle

I. Introduction

Most studies of birds in the canopy of tropical forests using specialized techniques have included laborious mist-netting and banding of canopy birds (Lovejoy, 1974; Munn, 1985, 1991), observation of these marked birds from the ground (Munn, 1985; Robinson *et al.,* 1990; Terborgh *et al.,* 1990) and canopy tower or tree platform observation of unmarked birds (McClure, 1963; Greenberg, 1981; Loiselle, 1987, 1988; Nadkarni and Matelson, 1989). Working near Belém in the eastern Amazon, Lovejoy (1974) netted from the ground level to a height of 23 m (Humphrey *et al.,* 1968), banded the birds, and used net capture data to calculate species diversity indices and other features of the forest bird community. He did not observe the marked birds after initial banding. Munn (1985, 1991) canopy-netted from the ground to a height of 50 m, banded, and observed from the ground permanent mixed-species canopy flocks in the Western Amazon in southeastern Peru. Robinson *et al.* (1990) and Terborgh *et al.* (1990) included analyses of the structure and organization of the entire forest bird community at Cocha Cashu in southeastern Peru. Greenberg's (1981) canopy tower censuses on Barro Colorado Island in Panama showed that most of the common canopy birds were seasonal in abundance and that two notable sources of this seasonal fluctuation were omnivorous tanagers and temperate-zone migrants. Loiselle (1987, 1988) recorded bird abundance and seasonality from observation points in the tops of two tall canopy emer-

Copyright © 1995 by Academic Press, Inc.
All rights of reproduction in any form reserved.

gents at La Selva Biological Station in Costa Rica. As with the Panama site, Loiselle (1988) reported significant seasonal variation in small frugivores and insectivores, as well as parrots, but seasonal patterns occasionally varied between census sites. Nadkarni and Matelson (1989) recorded a surprisingly large amount of use of treetop epiphytes by a wide variety of small foraging canopy birds in Costa Rica. These results emphasize the importance of considering the spatial scale at which canopy studies are done.

The tall stature and vertical complexity of tropical humid forests have greatly limited our abilities to conduct in-depth studies of bird biology and ecology. Despite the accomplishments of the studies cited here, our knowledge of the biology of canopy birds is still rudimentary. From a conservation perspective, this is clearly unwelcome news as canopy birds, especially large canopy frugivores and raptors, are declining or are recognized as being particularly vulnerable in tropical forests (Terborgh, 1974; Willis, 1979; Hilty, 1985; Levey and Stiles, 1994). Given the current rates of tropical deforestation (World Resources Institute, 1993) and the likelihood that large tracts of mature forests will remain in only a few regions of the world, conservation of many of these vulnerable canopy bird species will depend on an adequate understanding of the species' habitat requirements, breeding and nesting cycles, and other life-history attributes (Soulé, 1987). In many cases, collection of such crucial data may be possible only by using specialized canopy access techniques.

In this chapter, we review the advantages and disadvantages of canopy access methods and describe two promising new methods for studying canopy birds—remote-sensing and radio-tracking from ultralight aircraft. Such methods are reviewed in light of important directions for future research in canopy bird studies.

II. Methods

A. Canopy Netting

Past methods of canopy netting have varied from destructive and expensive to tedious and laborious. Humphrey *et al.* (1968) netted canopy birds in the Eastern Amazon by paying skilled local assistants to climb trees and clear all vegetation from entire 12-m-long corridors from the ground all the way up to the canopy. These assistants then hung pulley and rope systems in tree branches to permit a series of nets to be raised in that corridor. As the nets were raised, they resembled a diaphanous theater curtain sliding up into the canopy. This method caught birds effectively, but at considerable cost in terms of labor, danger to assistants (in tree climbing), and destruction to vegetation. Also, it is likely that some birds learned to avoid the open net-corridors that resulted in the nets being well-lit and conspicuous.

Munn (1985; 1991) improved the canopy-netting method developed by Greenlaw and Swinebroad (1967) and found it to be less expensive, less damaging to vegetation, less conspicuous, more mobile, and more adaptable than of Humphrey *et al.* (1968). Munn's method involved removing the nets' horizontal shelf strings and restringing them vertically, that is, perpendicular to their original orientation. A pole was then tied to each end of the net, and the net was hung from one of the poles like a tall, vertical sail. The rope used to raise and lower the net passed through one fixed pulley, which itself was hung from a separate support rope placed over a high branch. Starting with shooting a thin monofilament line over a branch, each support rope was placed in position without anyone leaving the ground. The monofilament line was used to pull over progressively heavier lines. This netting technique permitted tall, saillike nets to be hung both vertically and, in some cases when vegetation configuration dictated, even diagonally in natural openings in the forest canopy and midstory. Little or no vegetation had to be cleared to use this method, which probably made it harder for birds to detect and avoid the nets. With practice, one person working alone from the ground could erect and operate these nets.

B. Observation of Canopy Birds

In addition to high-quality binoculars, spotting scopes, and small mirror telescopes used to observe canopy birds, other equipment to gain access to tall forest canopies and enhance canopy bird observations are rope and tower systems (e.g., McClure, 1963; Lovejoy, 1974; Greenberg, 1981; Loiselle, 1988; Nadkarni and Matelson, 1989; Munn, 1991; Chapter 1). Many ingenious techniques have been developed by ornithologist and artist Paul Donahue for climbing tropical trees with webbing loops rather than with spikes and for rappelling down ropes that are wrapped rather than tied around trunks. From 1989 through 1991, he installed canopy platforms near Manu Lodge in southeastern Peru, and in 1991–1992 he designed and constructed a 400-m-long canopy walkway at the Amazon Biosphere Reserve in northeastern Peru in conjunction with the Amazon Center for Environmental Education and Research Foundation (ACEER). Canopy researchers will benefit from published descriptions and safety evaluations of Donahue's new techniques of moving through the canopy, which are elegantly minimalist, versatile, and inexpensive.

III. Directions and Methods for Future Research

Past research on canopy birds has been restricted mostly to basic censuses, principally of small passerines. Experienced observers can conduct visual and auditory censuses of canopy birds from ground level at much lower cost and effort and at many more locations (e.g., Hilty, 1985; Parker *et al.*, 1985; Robbins *et al.*, 1985; Remsen, 1985; Stiles, 1985a; Wiedenfeld *et al.*, 1985; Terborgh *et al.*, 1990; Parker and Bailey, 1991). Detailed observations of canopy birds [e.g., foraging behavior and specialization, nesting biology (cf. Rettig, 1978), ranging patterns of individual birds], however, can be extremely difficult and even biased from ground locations, especially in forests with heavy epiphyte loads. Few ground-based studies have provided detailed observations on canopy birds in such forests (but see Stiles, 1985a; Fitzpatrick, 1985; Munn, 1985). From well-situated positions within or above the forest canopy (see the following), the possibilities for canopy research on birds is greatly enhanced. Moreover, population studies require individual marking of canopy birds and further observations of such individuals over time. In tall forests, capturing and marking (with radio-collars or color bands) will necessitate specialized techniques (Munn, 1985, 1991).

Perhaps the only example of a detailed study of ecological interactions between canopy birds and canopy vegetation is Nadkarni and Matelson's (1989) investigation of small passerines and hummingbirds foraging in canopy epiphytes in Costa Rica's Monteverde cloud forest. They documented eight species of birds foraging in canopy epiphytes for more than 40% of foraging visits; 33 of the 56 recorded bird species foraged in epiphytes to some extent. They conclude by correctly pointing out that epiphyte use by canopy birds is largely unstudied and overlooked. In particular, they argue forcefully that epiphyte resources provide birds with a host of unusual foraging opportunities and niches, which may in turn help explain the high species diversity of tropical birds.

Two fundamental questions in ecology are what determines species numbers in ecological communities and how such diversity is maintained (Ricklefs and Schluter, 1993). In tropical bird communities, high species diversity of birds has been attributed to historical and biogeographical factors, and to availability of "new" resources and increased species packing (Karr, 1971; Haffer, 1974; Terborgh, 1980; Parker, 1982; Remsen and Parker, 1984). Many of these "new" resources (e.g., epiphytes, fruits, flowers) are concentrated in forest canopies. As Remsen (1985) noted, our ability to recognize foraging specialization by canopy birds, and thus address some of the hypotheses to explain increased species diversity in the tropics, is hampered by our inability to conduct detailed and replicated observations on canopy birds. These are essential lines of future research, as birds play critical ecological roles as flower pollinators and seed dispersers in tropical forests (Stiles, 1985b). Additionally, raptors are important predators in these forests and although they frequently occur at low population densities (Thiollay, 1989), their impact on populations of small mammals may be large (Karr, 1982; Loiselle and Hoppes, 1983). Because local extinction resulting from habitat alteration is not a random process (i.e., certain groups are more prone to extinction than others), understanding the ecological role of canopy bird species is essential to predict and mitigate impacts to forest structure and function that result from the loss of these canopy birds.

Another fundamental question in the study of ecological communities is the stability of populations and communities over time. Much of this discussion has been hampered by problems concerning issues of spatial and temporal scale (e.g., Wiens, 1990); what may be reported as a variable population at the scale of a study plot (or canopy platform) may not be so at the scale of a forest reserve. Tropical forest canopies undergo daily, seasonal, and annual variation in rainfall, wind, temperature, humidity, and insolation. These daily and seasonal patterns in the canopy environment likely influence canopy birds directly and indirectly through effects on canopy leaf flush, arthropod availability and abundance, and fruit and nectar supply. It is likely that the illumination and accessibility of arthropod, lizard, and frog prey items on the tops and bottoms of leaves in the canopy vary sufficiently throughout the day and seasonally to affect arthropod abundance, and hence the behavior of canopy birds. In the understory, the vast majority of palatable, visible arthropods hide during the day on the undersides of leaves (Munn, 1984), which are more poorly illuminated and harder to search than the leaftops. Conversely, the uppermost and outermost leaves on canopy trees may be well-lit, but birds may have difficulty examining them from comfortable perched positions.

Understanding how birds respond to such environmental variation is an important step in understanding the species' biology, as well as predicting a species' vulnerability to extinction following habitat alteration. For example, selective logging of forests, or habitat alteration that leads to forest fragmentation and increased edge, is expected to accentuate the environmental variation in the remaining forest (e.g., Lovejoy *et al.,* 1986).

General questions of how canopy birds use and benefit from the vegetation and animal resources of the tropical canopy are a virtually unexplored area of tropical research. When one considers that birds are the best known of tropical organisms, this ignorance of the details of the biology and adaptations of canopy birds is sobering, and suggests how profoundly ignorant we are of most canopy biology for all other groups of more inconspicuous and speciose organisms such as bats, ants, and beetles. How do the daily activity patterns of canopy birds compare with those of understory birds? How do canopy regimes of sunlight, wind, temperature, and rain affect bird behavior and ecology? What adaptations do canopy birds demonstrate during nesting season? Is the nesting season of canopy birds more or less concentrated in one part of the year than that of understory species? How do the insect and fruit resources of the canopy differ from those of the understory, both in terms of physical and nutritional characteristics and in terms of daily, seasonal, and annual availability? How does the presentation of fruit, nectar, and pollen resources differ between the canopy and the understory, and what adaptations do canopy birds exhibit to take advantage of these plant resources? Do canopy arthropods, frogs, and lizards demonstrate special adaptations to canopy living? What techniques do canopy birds use to capture these prey? The list of unexplored research topics is endless, and we still lack even basic counts and censuses of canopy birds in most tropical regions.

For canopy research of birds to advance and address these fundamental ecological questions and conservation concerns, it is essential that more intensive netting and marking of canopy birds of all sizes be undertaken. Expanding on Munn's (1991) netting methods, it should be possible for one or two persons working from the ground to erect and to simultaneously operate 20 or more canopy nets in natural openings in the tropical forest. A systematic marking effort combined with good optical equipment and judicious radio-tagging would permit researchers to understand the ranging patterns, ecology, and perhaps even the population dynamics of most canopy species. Large and small frugivores could be netted at fruit trees by hanging restrung nets vertically from pulleys that are hung in midair in the gaps between adjacent canopy trees (Munn, 1991).

Birds that fly above the forest could be captured in nets suspended horizontally in flyways between canopy emergents or in gaps in hills and mountains. It may also be possible to net birds above the canopy by erecting net poles that jut 6–15 m or more above the forest canopy. The simplest way to mount such poles might be to guy or fix them in position with their bases in crotches or on small platforms near the tops of canopy trees. Extracting birds from such nets would require special ropework or other access to the nets, perhaps including features that allow one to reel the nets in horizontally, like clothes on a tenement building clothesline, until the bird is within reach from a stable spot just above the canopy of one of the support trees. However, beware of placing extreme stresses on horizontally deployed climbing or support ropes (Whitacre, 1981; MacInnes, 1984).

Radio-tagging methods are now available to allow distant monitoring of location, activity, vocalizations, body temperature, and heart rate of birds (Bloom *et al.,* 1993; Bull and Holthausen, 1993; Clark *et al.,* 1993; Conway *et al.,* 1993; Fenton *et al.,* 1993; Schneil and Wood, 1993; Wuethrich, 1993; Guetterman, 1991; Kenward, 1987). Radios can be fixed to birds in many ways, including tying to tail feathers, gluing to back feathers, attaching to neck collars or to leg bands, or surgically implanting them. For tropical canopy birds, particularly the large ones that fly great distances daily or seasonally, it might be possible to catch and radio-tag them, but until now it has been difficult and prohibitively expensive to follow them with small planes. Smaller birds with small home ranges have proven feasible to radio-tag and follow on foot or by car or boat, and we shall not discuss these methods here.

Studies of far-ranging, typically larger species of canopy birds, however, present different challenges and difficulties that require more complex and more aggressive techniques of canopy observation and access. New techniques for gaining access to the canopy are not all promising for the study of canopy birds. Although huge canopy cranes (G. G. Parker *et al.,* 1992) could provide "sky-hooks" from which to hang nets to catch birds flying above the canopy, in general these cranes do not offer much improvement in canopy observation over tall emergent trees. Likewise, the expensive French canopy raft made famous in botanical and pollination studies in the forest canopy in French Guyana (Hallé, 1990) offers little or no improvement in methods for study or netting canopy birds.

Among all the new methods of canopy study, however, two are especially promising for research on large and far-ranging canopy birds: satellite radiotracking (remote-sensing) and searching, following, and radio-tracking from ultralight aircraft (ultralights). These two methods are vast improvements over expensive and dangerous small planes and fixed treetop or ridgetop observation points.

Satellites and ultralights are excellent new methods for different reasons. Satellites provide long-term coverage of the few far-ranging canopy birds large enough to carry a satellite transmitter. In 1993, the smallest satellite transmitters with batteries weighed 48 g (W. Berger of Telonics, Inc., personal communication). Using the general rule that no flying animal should carry radio-tags that weigh more than 3% of its body weight, a 48-g tag cannot be worn by a bird weighing less than 1584 g.

The only canopy or above-canopy birds from the tropics that weigh more than this limit are eagles, hawk-eagles, vultures, some guans, some hornbills, and some huge pigeons from the New Guinea region. Tropical forest ground birds that are large enough to carry such transmitters include currasows, trumpeters, certain pheasants and jungle fowl, and cassowaries, but such birds might move slowly enough and in small enough areas to permit cost-effective following on foot, a much more economical alternative to satellite telemetry. With the possible exception of the nearly extinct Kakapo of New Zealand, all other parrot species of the world, including the large macaws, cockatoos, and Amazon parrots, are too light to carry satellite transmitters. Disadvantages of satellite-tracking are the high cost of transmitters and of data collection (approximately $3500 per animal per year), the risk of loss of contact with such a costly bird, and lack of detailed, hour-by-hour or even day-by-day ground truth information about habitat and local weather conditions.

To track far-ranging tropical birds that are too small to carry satellite transmitters, the best option may be to mark birds with conventional small transmitters and follow them with ultralights. Ultralights are much safer and more economical to operate than small planes. With the recent development of more reliable engines, ballistic parachute safety systems (Gorman, 1993), better-engineered wings and tails, and inexpensive, hand-held Geo-Positioning Systems (GPSs), ultralights offer excellent opportunities for canopy biologists working in relatively wind-free tropical forests. Safe, very-slow-flying ultralights with aerodynamic stall speeds close to or just above 32 km/hr (20 mph) are now available for less than $6000. A two-seat, float-mounted ultralight for landing on water costs approximately $8000 (1993 estimate). Ballistic parachute systems for safely bringing down the entire ultralight with the pilot or pilots are available for $1300–2000. Walkman-sized GPSs cost approximately $800. A good source of information about ultralight models, specifications, capabilities, and prices is the U.S. Ultralight Association.[1]

The New York Zoological Society (NYZS)/Wildlife Conservation Society (WCS) and World Wildlife Fund have used ultralights successfully in three studies in tropical ecosystems: radio-tracking jaguars in the Pantanal savannah–forest mosaic of west-central Brazil (Quigley and Crawshaw, 1989), radio-tracking white-lipped peccaries in the Eastern Amazon forest of Brazil (J. Fragoso, personal communication), and visually following large macaws in the Amazon forest of southeastern Peru (C. Munn, personal observation). The latter project used a float-mounted model that landed on rivers and lakes, and permitted slow, nonintrusive following of large macaws flying across the 5-km-wide floodplain of the Manu River. When tracking macaws with an ultralight, I (C.M.) could see and follow the birds easily with the naked eye while flying several hundred meters above them. The birds showed no fear of the plane and did not modify either their flight directions or flight speeds as I approached and followed from a higher altitude. Macaws fly slightly more slowly than the ultralight, so we had to make slight "S-turns" to avoid getting ahead of them.

Ultralights have proven to be adaptable tools for many novel applications. In the western United States, ultralights have been used to locate raptor nests (Looman *et al.*, 1985) and to herd cattle and check fencelines (V. Vitollo, personal communication). In New Jersey's Pinelands, ultralights have been used to relocate radio-tagged pine snakes (V. Vitollo, personal communication). In Bimini, they been used to observe lemon sharks in clear water in a lagoon and in deeper, off-shore water (Gruber, 1988). In Namibia, an ultralight has been used to locate elephants in semiarid and

---

[1]P.O. Box 667, Frederick, MD 21705, telephone: (301) 695-9100. For special features and custom designs of inexpensive, reliable, float-mounted, hands-free models, we recommend V. Vitollo of the Ultralight Flight Center, 545 Whitesville Rd., Jackson, NJ 08527, telephone: (908) 363-9888.

arid landscapes and sand dunes. In Kenya, an ultralight has been used intensively for low-level nature photography (Rosane, 1993). Two recent documentary films for television (Discovery Channel, 1991, 1993) used float-mounted ultralights to film the rainforest of Peru's Manu National Park and whales off the coast of Patagonia.

We believe that biologists and wildland managers fortunate enough to work in large expanses of tropical forest will find ultralights to be the tool of choice for much of their research and monitoring. In 1983, WCS's operating costs for the ultralight jaguar research were only $17/hr, much less than the minimal charter costs of $130–200/hr for conventional small planes. In the case of remote sites, small planes are rarely available, which makes it expensive to bring a plane to the study site for radio-tracking. Ultralights based at the study site solve this problem.

Finally, ultralights fly so slowly and weigh so little that they can land easily and safely in the canopy of a large or medium-sized tropical tree. In New Jersey, V. Vitollo intentionally landed a float-mounted ultralight successfully in the canopy of a medium-sized tree. He selected a tree near a road, from which a crane was able to rescue the ultralight, which suffered no significant damage.

For ten years, ultralights with in-flight emergencies have routinely floated safely to earth thanks to small, rocket-deployed safety parachutes (Gorman, 1993; Kenner-Smith, 1984). The manufacturer, who has sold 10,000 parachute systems, says they have saved 73 lives (Gorman, 1993). When emergencies force conventional small planes to land in the forest, they weigh so much more and land so much faster than ultralights that the results are often fatal. Only recently have larger, more expensive parachutes been developed to allow small planes and passengers to float down to the ground safely (Gorman, 1993).

After flying the float-mounted WCS ultralight over the Peruvian rain forest in 1993, V. Vitollo stated, "I believe that it would be nearly impossible to injure yourself seriously if you were forced to land this ultralight without a safety parachute in the top of a relatively large tree." The 70-year-old Vitollo has been a professional flying instructor for 52 years, longer than any other currently licensed instructor in the United States. He also was the first person to successfully use ultralights on floats and has piloted small planes for over 30,000 hours and ultralights for over 10,000 hours. Because of his tremendous amount and variety of flying experience, Vitollo's opinion about the safety of ultralights carries considerable authority.

In the event of a treetop landing, one must be prepared to use thin safety ropes with seat harnesses to descend to the ground. Also, prior to piloting float-mounted ultralights over forest beyond the safe glide range back to rivers or lakes, one should plan and test reliable survival and search-and-rescue protocols. Otherwise, a safe emergency landing in distant treetops could be followed by panic, disorientation, confusion, and even death.

The ecologically most significant species of canopy birds are the large frugivores and carnivores, for example, eagles and other large raptors, macaws, Amazon parrots, cockatoos, large cotingas, guans, hornbills, and toucans. Logistical obstacles have prevented detailed studies of these birds. Ultralights would be effective in surveying, following, or radio-tracking these canopy birds. So far, macaws have proven more difficult to radio-tag than elephants, whales, large cats, wolves, manatees, and many other large vertebrates, because the birds use their powerful beaks to chew and destroy their mates' radio collars (N. Snyder, personal communication).

IV. Conclusions

Observation from the ground is not enough, as Nadkarni and Matelson (1989) showed in their study of birds foraging in treetop epiphytes. The possibilities for canopy research are endless and the encouraging message for aspiring tropical ornithologists is that with safe, simple ropework, one can gain access to observation points in medium-sized, large, and huge trees and thereby investigate canopy birds from an appropriate perspective (Whitacre, 1981; Nadkarni, 1988).

We particularly encourage the initiation of canopy research on large canopy birds, especially those identified as ecologically important to the maintenance of biodiversity and ecosystem function in tropical forests. These birds are consistently among the most predictable, attractive, photogenic, and manageable of tropical forest animals. Therefore, understanding and conserving these large canopy birds will be critical to the success of ecotourism projects that are attempting to save the world's finest tropical wilderness sites (Munn, 1992, 1994; Munn *et al.*, 1991; Groom *et al.*, 1991).

Acknowledgments

C.M. thanks the Peruvian government's Ministry of Agriculture for permission to research macaws in Manu National Park. The macaw research in Peru was funded entirely by NYZS/WCS. V. Vitollo selflessly donated his time and expertise to assembling the WCS ultralight in Manu and to teaching C. Munn and his co-investigator and wife Mariana Valqui Munn how to fly it. Superflow Corporation of Colorado Springs, Colorado, and the Discovery Channel of Bethesda, Maryland, generously permitted WCS to buy their customized two-seater ultralight, which had originally been brought to Peru for film-making in Manu, at a deep discount. Finally, we are grateful to N. Snyder of Jersey Wildlife Preservation Trust International, G. Powell of RARE Center for Tropical Conservation, A. Taber of WCS, and W. Berger of Telonics Corporation (Mesa, Arizona), who offered information about new options in conventional and satellite radiotelemetry.

References

Bloom, P. H., McCrary, M. D., and Gibbons, M. J. (1993). Red-shouldered Hawk home-range and habitat use in southern California (*Buteo lineatus elegans*). *J. Wildl. Manage.* 57, 258–265.

Bull, E. L., and Holthausen, R. S. (1993). Habitat use and management of pileated woodpeckers in northeastern Oregon (*Dryocopus pileatus*). *J. Wildl. Manage.* 57, 335–345.

Clark, B. S., Leslie, D. M., Jr., and Carter, T. S. (1993). Foraging activity of adult female Ozark big-eared bats (*Plecotus townsendii ingens*) in summer. *J. Mammal.* 74, 422–427.

Conway, C. J., Eddleman, W. R., and Anderson, S. H. (1993). Seasonal changes in Yuma Clapper Rail vocalization rate and habitat use (*Rallus longirostris yumanensis*). *J. Wildl. Manage.* 57, 282–290.

Discovery Channel (1991). "In the Company of Whales." Discovery Communications, Bethesda, MD.

Discovery Channel (1993). "Spirits of the Rainforest." Discovery Communications, Bethesda, MD.

Fenton, M. B., Audet, D., and Dunning, D. C. (1993). Activity patterns and roost selection by *Noctilio albiventris* (Chiroptera: Noctilionidae) in Costa Rica. *J. Mammal.* 74, 607–613.

Fitzpatrick, J. W. (1985). Form, foraging behavior, and adaptive radiation in the Tyrannidae. *Ornithol. Monogr.* 36, 447–470.

Gorman, C. (1993). A parachute—But no jump. *Time* 142, 76.

Greenberg, R. (1981). The abundance and seasonality of forest canopy birds on Barro Colorado Island, Panama. *Biotropica* 13, 241–251.

Greenlaw, J. S., and Swinebroad, J. (1967). A method for constructing and erecting aerial-nets in a forest. *Bird-Banding* 38, 114–119.

Groom, M. J., Podolsky, R. D., and Munn, C. A. (1991). Tourism as a sustained use of wildlife: A case study of Madre de Díos, southeastern Peru. *In,* "Neotropical Wildlife Use and Conservation" (J. G. Robinson and K. H. Redford, eds.), pp. 393–412. Univ. of Chicago Press, Chicago.

Gruber, S. H. (1988). Sharks of the shallows. *Nat. Hist.* 97, 50–59.

Guetterman, J. H. (1991). "Radio Telemetry Methods for Studying Spotted Owls in the Pacific Northwest." U.S. Department of Agriculture, For. Serv. Pac. Northwest Res. St., Portland, OR.

Haffer, J. (1974). Avian speciation in tropical South America. *Publ. Nuttall Ornithol. Club* No. 14.

Hallé, F. (1990). A raft atop the rainforest. *Natl. Geogr.* 178, 128–138.

Hilty, S. L. (1985). Distributional changes in the Colombian avifauna: A preliminary blue list. *Ornithol. Monogr.* 36, 1000–1012.

Humphrey, P., Bridge, D., and Lovejoy, T. E. (1968). A technique for mist-netting in the forest canopy. *Bird-Banding* 39, 43–50.

Karr, J. R. (1971). The structure of avian communities in selected Panama and Illinois habitats. *Ecol. Monogr.* 41, 207–233.

Karr, J. R. (1982). Avian extinctions on Barro Colorado Island, Panama: A reassessment. *Am. Nat.* 119, 220–239.

Kenner-Smith, S. (1984). Safety parachutes for disabled ultralights. *Pop. Sci.* 224, 75.

Kenward, R. (1987). "Wildlife Radio Tagging: Equipment, Field Techniques, and Data Analysis." Academic Press, London.

Levey, D. J., and Stiles, F. G. (1994). Birds: Ecology, Behavior, and Taxonomic Affinities. *In* "La Selva: Ecology and Natural History of a Lowland Forest" (L. McDade, K. Bawa, G. S. Hartshorn, and H. Hespenheide, eds.), pp. 217–228. Univ. of Chicago Press, Chicago.

Loiselle, B. A. (1987). Migrant abundance in a Costa Rican lowland forest canopy. *J. Trop. Ecol.* 3, 163–168.

Loiselle, B. A. (1988). Bird abundance and seasonality in a Costa Rican lowland forest canopy. *Condor* 90, 761–772.

Loiselle, B. A., and Hoppes, W. G. (1983). Nest predation in insular and mainland lowland rainforest in Panama. *Condor* 85, 93–95.

Looman, S. J., Boyce, D. A., Jr., White, C. M., Shirley, D. L., and Mader, W. J. (1985). Use of ultralight aircraft for raptor nest surveys. *Wildl. Soc. Bull.* 13, 539–543.

Lovejoy, T. E. (1974). Bird diversity and abundance in Amazon forest communities. *Living Bird* 13, 127–191.

Lovejoy, T. E., Bierregaard, R. O., Jr., Rylands, A. B., Malcolm, J. R., Quintela, C. E., Harper, L. H., Brown, K. S., Jr., Powell, A. H., Powell, G. V. N., Schubart, H. O. R., and Hays, M. B. (1986). Edge and other effects of isolation on Amazon forest fragments. *In,* "Conservation Biology: The Science of Scarcity and Diversity" (M. E. Soulé, ed.), pp. 257–285. Sinauer Assoc., Sunderland, MA.

MacInnes, H. (1984). "International Mountain Rescue Handbook." Constable, London.

McClure, H. E. (1963). Flowering, fruiting and animals in the canopy of a tropical rain forest. *Malays. For.* 29, 182–203.

Munn, C. A. (1984). The behavioral ecology of mixed-species bird flocks in Amazonian Peru. Doctoral Dissertation, Princeton University, Princeton, N.J.

Munn, C. A. (1985). Permanent canopy and understory flocks in Amazonia: Species composition and population density. *Ornithol. Monogr.* 36, 683–712.

Munn, C. A. (1991). Tropical canopy netting and shooting lines over tall trees. *J. Field Ornithol.* 62, 454–463.

Munn, C. A. (1992). Macaw biology and ecotourism, or "When a bird in the bush is worth two in the hand." *In,* "New World Parrots in Crisis: Solutions from Conservation Biology" (S. R. Beissinger and N. F. R. Synder, eds.), pp. 47–72. Smithsonian Institution Press, Washington, DC.

Munn, C. A. (1994). Macaws: Winged rainbows. *Natl. Geogr.* 184.

Munn, C. A., Blanco Z., D. H., Nycander v.M., E., and Ricalde R., D. G. (1991). Prospects for sustainable use of large macaws in southeastern Peru. *In* "Proceedings of the First Mesoamerican Workshop on the Conservation and Management of Macaws" (J. Clinton-Eitniear, ed.), pp. 42–47. Misc. Publ. No. 1. Center for the Study of Tropical Birds. San Antonio, TX.

Nadkarni, N. M. (1988). The use of a portable platform to observe bird behavior in tropical tree crowns. *Biotropica* 20, 350–351.

Nadkarni, N. M., and Matelson, T. J. (1989). Bird use of epiphyte resources in neotropical trees. *Condor* 91, 891–907.

Parker, G. G., Smith, A. P., and Hogan, K. P. (1992). Access to the upper forest canopy with a large tower crane: Sampling the treetops in three dimensions. *BioScience* 42, 664–670.

Parker, T. A., III (1982). Observations of some unusual rainforest and marsh birds in southeastern Peru. *Wilson Bull.* 94, 477–493.

Parker, T. A., III, and Bailey, B., eds. (1991). "A Biological Assessment of the Alto Madidi Region," Rapid Assessment Program Working Papers No. 1. Conservation International, Washington, DC.

Parker, T. A., III, Schulenberg, T. S., Graves, G. R., and Braun, M. J. (1985). The avifauna of the Huancabamba region, northern Peru. *Ornithol. Monogr.* 36, 169–197.

Quigley, H. B., and Crawshaw, P. G., Jr. (1989). Use of ultralight aircraft in wildlife radio telemetry. *Wildl. Soc. Bull.* 17, 330–334.

Remsen, J. V., Jr. (1985). Community organization and ecology of birds of high elevation humid forest of the Bolivian Andes. *Ornithol. Monogr.* 36, 733–756.

Remsen, J. V., Jr., and Parker, T. A., III (1984). Arboreal dead-leaf searching birds of the neotropics. *Condor* 86, 36–41.

Rettig, N. L. (1978). Breeding behavior of the Harpy Eagle (*Harpia harpyja*). *Auk* 95, 629–643.

Ricklefs, R. E., and Schluter, D., eds. (1993). "Species Diversity in Ecological Communities." Univ. Chicago Press, Chicago.

Robbins, M. B., Parker, T. A., III, and Allen, S. E. (1985). The avifauna of Cerro Pirre, Darién, eastern Panama. *Ornithol. Monogr.* 36, 198–232.

Robinson, S. K., Terborgh, J., and Munn, C. A. (1990). Lowland tropical bird communities of a site in Western Amazonia. *In* "Biogeography and Ecology of Forest Bird Communities" (A. Keast, ed.), pp. 229–258. S.P.B. Academic Publishing, The Hague, The Netherlands.

Rosane, D. (1993). Kenya from the clouds. *Int. Wildl.* 23, 18–23.

Schneil, C. R., and Wood, J. M. (1993). Measurement of blood pressure and heart rate by telemetry in conscious, unrestrained marmosets. *Am. J. Physiol.* 264, H1509–H1518.

Soulé, M. E., ed. (1987). "Viable Populations for Conservation." Cambridge Univ. Press, Cambridge, UK.

Stiles, F. G. (1985a). Seasonal patterns and coevolution in the hummingbird–flower community of a Costa Rican subtropical forest. *Ornithol. Monogr.* 36, 757–787.

Stiles, F. G. (1985b). On the role of birds in the dynamics of neotropical forests. *In* "Conservation of Tropical Forest Birds" (A. W. Diamond and T. E. Lovejoy, eds.), pp. 49–59. International Council of Bird Preservation, Cambridge, UK.

Terborgh, J. (1974). Preservation of natural diversity: The problem of extinction of species. *BioScience* 24, 715–722.

Terborgh, J. (1980). Causes of tropical species diversity. *Proc. Int. Ornithol. Congr. 17*, 955–961.

Terborgh, J., Robinson, S. K., Parker, T. A., III, Munn, C. A., and Pierpont, N. (1990). Structure and organization of an Amazonian bird community. *Ecol. Monogr.* 60, 213–238.

Thiollay, J. M. (1989). Area requirements for the conservation of rain forest raptors and game birds in French Guiana. *Conserv. Biol.* 3, 128–137.

Whitacre, D. F. (1981). Additional techniques and safety hints for climbing tall trees, and some equipment and information sources. *Biotropica* 13, 286–291.

Wiedenfeld, D. A., Schulenberg, T. S., and Robbins, M. B. (1985). Birds of a tropical deciduous forest in extreme northwestern Peru. *Ornithol. Monogr.* 36, 305–315.

Wiens, J. A. (1990). "The Ecology of Bird Communities," Vols. I and II. Cambridge Univ. Press, Cambridge, UK.

Willis, E. O. (1979). The composition of avian communities in remanescent woodlots in southern Brazil. *Pap. Avulsos Dep. Zool., Secr. Agric., Ind. Comer. (São Paulo)* 33, 1–25.

World Resources Institute (1993). "World Resources 1992–1993." World Resources Institute, Oxford Univ. Press, New York.

Wuethrich, B. (1993). Tracking turtles' ocean highways. *New Sci.* 138, 8.

9

Forest Structure and the Abundance and Diversity of Neotropical Small Mammals

Jay R. Malcolm

I. Introduction

Of the diverse array of nonflying mammalian species known from the tropics, approximately one-half can be classified as arboreal or scansorial, a proportion that appears to be substantially higher than that in temperate forests. Estimates of the percentage of arboreal plus scansorial species from rain forest sites in Malaysia, Australia, Borneo, Panama, and Gabon range between 45 and 61% (Harrison, 1962; Davis, 1962, cited in Bourlière, 1973; Fleming, 1973; Emmons *et al.,* 1983), whereas the corresponding figure was only 15% in temperate woodlands in Virginia (Davis, 1962, cited in Bourlière, 1973). In a comparison of the mammal fauna across sites located between Alaska and Panama, Fleming (1973) found that as latitude decreased, the proportion of terrestrial plus scansorial species decreased, and the proportion of arboreal plus aerial species increased. Biomass calculations also highlight the importance of the forest canopy to tropical mammals: estimates of the mammalian biomass in the canopy were 60–70% of the total mammalian biomass (Eisenberg and Thorington, 1973; Terborgh, 1986). The richer arboreal fauna in tropical forests has been attributed to the greater opportunities for subdivision of a structurally complex habitat and the resources it provides, particularly the abundant fruit crop that is available to varying degrees throughout the year (Lein, 1972; Fleming, 1979; Emmons, 1980; August, 1983).

Ecological information on canopy-dwelling mammals is restricted pri-

Copyright © 1995 by Academic Press, Inc.
All rights of reproduction in any form reserved.

marily to species that are diurnal and relatively large-bodied. Relatively detailed ecological information is now available for almost all Neotropical primate species (C. Peres, personal communication), whereas similar information is available for only a few Neotropical arboreal rodents and marsupials, most of which comes from one research site (Charles-Dominique *et al.*, 1981; Atramentowicz, 1982). The paucity of information for small nocturnal species reflects, at least in part, methodological difficulties. Some arboreal species rarely frequent the ground (Malcolm, 1991a) and are rarely observed unless they are trapped in the canopy. Direct observation of more than a few individuals, as around a tower or platform, is virtually impossible unless the individuals have been radio-tagged. Indirect methods such as spool-and-line devices and fluorescent powder are unlikely to meet with great success (e.g., Miles *et al.*, 1981). Most studies used only terrestrial and understory trapping (e.g., Delany, 1971; Fleming, 1972; Rahm, 1972; Dosso, 1975; August, 1984; Fonseca, 1988; Stallings, 1988; O'Connell, 1989); only a few used traps at heights greater than 5 m (e.g., Davis, 1945; Adam, 1977; Charles-Dominique *et al.*, 1981; Stallings, 1988; Malcolm, 1991a).

In this chapter, I examine the problem of how arboreal and terrestrial small mammal abundance and diversity vary from location to location within tropical rain forests. The number of species in a particular area depends on the interaction between ecological, biogeographic, and evolutionary processes (Brown, 1973). I will emphasize the importance of ecological processes by examining a set of habitats potentially open to colonization by all members of a local species pool (Brown, 1973; August, 1983).

Two habitat characteristics of possible importance to small animals are the vertical and horizontal diversity of a habitat, variables that correlate with measures of abundance and diversity for a variety of taxa (August, 1983). August (1983) found that diversity and guild structure of the total nonvolant mammalian fauna at a site in the llanos of Venezuela varied predictably with the vertical complexity of a habitat (in agreement with Eisenberg, 1980), but was unable to correlate small mammal richness, diversity, or abundance with either the vertical complexity or horizontal heterogeneity of a habitat. I repeated the test at a site in the central Amazon with a larger sample size and included an intensive survey of the canopy-dwelling fauna. I also included another habitat variable: the grain (*sensu* Levins, 1968) or "patch size" of habitat heterogeneity. Correlations between habitat structure and characteristics of the small mammal community were examined for three forest types: (a) primary forests with varying proportions of edge-modified habitat, (b) undisturbed primary forest, and (c) secondary habitats including pasture and young secondary forest.

II. Materials and Methods

A. Study Site

Research was conducted at the Biological Dynamics of Forest Fragments Project study site, approximately 80 km north of Manaus, Brazil. Primary forest in the area is upland, or *terra firme,* and is far from large rivers and associated riverine habitats. Soils are nutrient-poor latosols, and annual rainfall averages about 2200 mm, with a dry season of <100 mm per month from July to September. Details on the study site are in Lovejoy and Bierregaard (1990).

In each of four areas (blocks), characteristics of the vegetation and of the small mammal community were measured in five major habitat types: (a) continuous forest (CF), (b) edge of continuous forest where it abutted pasture or young secondary forest (CF edge), (c) a 10-ha forest fragment, (d) a 1-ha forest fragment, and (e) the matrix of pasture and/or second growth surrounding the forest fragments. Habitat-specific nearest-neighbor distances between blocks ranged from 2 to 20 km. In each block, I sampled four 1-ha subsampling units in CF (each unit was at least 400 m from the nearest forest edge), two or three units abutting the edge of continuous forest, four units in the 10-ha fragment (these units included fragment corners, edge, and, in a few cases, fragment interior), one unit in the 1-ha fragment, and two or three units in the matrix (units were at least 150 m from primary forest). Matrix habitat included pastures (two blocks) established in 1980 or 1984 and secondary forests (two blocks) in large areas where primary forest was clear-cut in 1983, but was never burned or recut.

B. Small Mammal Trapping

Each block was sampled once during each of three censuses: (a) September 1987–February 1988, (b) March 1988–September 1988, and (c) October 1988–March 1989. To maximize coverage of each 1-ha unit with as few traps as possible, I divided each unit into two 50 × 100-m halves and established traps along a 100-m-long transect in the center of each half. Terrestrial and arboreal trap-stations (one Sherman and one Tomahawk trap per station) were placed at 20-m intervals along the transects to provide 12 terrestrial and 12 arboreal trap-stations per unit, which were run for eight consecutive nights per census. Terrestrial and arboreal trap-stations were set following Malcolm (1991a). Arboreal traps in primary forest were set at average heights of 12–14 m using the "pulley" method; in secondary forest, they were set approximately 2 m high using the "V" method. Arboreal traps were not set in pastures.

Three variables were used to characterize the small mammal community

Table I Abundance, Richness, and Diversity of Small Mammals in Six Major Habitats in the Central Amazon[a]

	CF	CF edge	10-ha fragment	1-ha fragment	Secondary forest	Pasture
No. of sites	4	4	4	4	2	2
No. of 1-ha units per site	4	2 or 3	4	1	2 or 3	3
Terrestrial abundance	3.56 ± 1.74	6.95 ± 3.37	7.18 ± 4.62	12.00 ± 8.04	16.16 ± 0.23	11.66 ± 8.48
richness	2.12 ± 0.66	4.16 ± 1.37	4.12 ± 1.65	5.25 ± 2.36	4.16 ± 0.23	4.00 ± 1.41
diversity	0.61 ± 0.26	1.24 ± 0.32	1.20 ± 0.39	1.41 ± 0.48	1.16 ± 0.07	1.14 ± 0.26
Arboreal[b] abundance	5.75 ± 1.24	4.20 ± 2.78	9.62 ± 6.33	8.25 ± 5.31	18.08 ± 0.82	
richness	2.50 ± 0.54	2.25 ± 1.10	3.43 ± 0.89	2.25 ± 0.95	2.91 ± 0.58	
diversity	0.76 ± 0.26	0.64 ± 0.47	0.99 ± 0.18	0.53 ± 0.42	0.87 ± 0.15	
Total abundance	8.87 ± 2.50	10.41 ± 5.10	15.81 ± 9.52	19.75 ± 13.40	32.33 ± 0.47	11.66 ± 8.48
richness	4.18 ± 0.96	5.70 ± 1.72	6.62 ± 1.78	6.50 ± 3.00	5.08 ± 0.82	4.00 ± 1.41
diversity	1.20 ± 0.28	1.49 ± 0.45	1.61 ± 0.21	1.49 ± 0.60	1.38 ± 0.13	1.14 ± 0.26

[a] Subsampling units were 1-ha plots with 12 terrestrial and 12 arboreal trap-stations and were censused three times for eight consecutive nights each. "Abundance" is the total number of individuals captured during the three censuses in a plot, "richness" is the total number of species, and "diversity" is the Shannon–Weiner information index (H'). Each habitat was censused at different sites by use of one or more subsampling units per site, and average site-specific means (± SD) are listed below.

[b] Average trap height in primary forest habitats was 12–14 m and in secondary forest was 2 m. Arboreal traps were not set in pasture.

in each unit: overall species richness (the total number of species encountered during the three censuses), overall species diversity (the Shannon–Wiener information index, H′), and total abundance (the total number of individuals captured during the three censuses). The three measures were calculated for terrestrial captures, arboreal captures, and combined terrestrial and arboreal captures (Table I). The list of small mammal species in the area is in Malcolm (1990).

C. Vegetation Measurements

I modified Hubbell and Foster's (1986) method to measure vertical stratification of foliage. In 1-ha units in primary forest, a 2.5-m pole was used to make a vertical sighting at each point on a 10 × 10-m grid extending 10 m outside the unit (169 points). Along the sighting, foliage density was scored in six height intervals: 0–2, 2–5, 5–10, 10–20, 20–30, and 30–40 m. In matrix units, I scored foliage density at 36 points: points on the fourth and tenth rows of the 13 × 13 grid, and points between these rows on the second and twelfth columns. Height estimates were periodically checked with a range finder. The first three height intervals were scored as 0 (<25% coverage), 1 (25–<50% coverage), 2 (50–<75% coverage), or 3 (>75% coverage). The last three were scored as 0 (<10% coverage), 1 (10–<50% coverage), 2 (50–<75% coverage), or 3 (>75% coverage). To provide a measure of foliage "thickness" for each interval, I recoded nonzero scores by multiplying the number of meters in the interval by the mean percentage corresponding to the score, for example, a score of 1 in the first height interval (0–2 m) was recoded as $2 \times 0.375 = 0.75$. Because vegetation samples differed between primary forest (CF, CF edge, 10-ha fragments, and 1-ha fragments) and secondary forest (matrix), I analyzed the two habitats separately.

D. Vegetation Complexity and Heterogeneity

These six vegetation measurements represented a potentially voluminous habitat description (if one calculates the mean, variance, and grain for each stratum) and very likely a redundant one, as only a few major patterns of variation were likely to exist. A comparison of the means of strata among the four primary forest habitats revealed two major patterns of variation (Malcolm, 1991b). As the proportion of edge-modified forest in a habitat increased (i.e., in the sequence CF, CF edge, 10-ha forest fragment, 1-ha forest fragment), understory (0–2 and 2–5 m) density increased and overstory (10–20 and 20–30 m) density decreased. Differences among habitats were not as pronounced for the other two strata. In contrast to the understory and overstory strata, vegetation thickness in the strata of 5–10 and 30–40 m showed little relationship with distance from the edge (Malcolm, 1991b). To reduce the number of vegetation variables, I defined just two strata in primary forest: understory [the sum of the thickness scores in strata 1 and 2 (0–5 m)] and overstory [the sum of the thickness scores in strata 4 and 5 (10–30 m)].

As a measure of vertical complexity in each of these derived strata, I used the hectare-specific mean. As a measure of heterogeneity, I used the hectare-specific variance. The relationship between the mean and variance was strong and linear. To derive a measure of variability that was "independent" of the mean, I regressed the variance against the mean and used the residuals ($n = 47$ 1-ha units). I also calculated a simple measure of understory and overstory "grain" with the relationship between surface area and volume. A small relative surface area indicates a coarser grain and vice versa. Each of the two derived strata was visualized as a square array of 169 blocks, where the height of each block was the vegetation thickness at the corresponding grid point. Blocks were 10×10 m, hence the volume of a block was 100 times the stratum thickness. Surface area included any surfaces of the blocks not in contact with other block surfaces within the stratum. The relationship between log-surface area and log-volume was linear. To render the surface area "independent" of the mean and variance, I used residuals from the multiple regression of surface area on log-mean and variance. In summary, six variables were calculated for each 1-ha unit: mean thickness, residual variance, and residual surface area for understory and overstory. Because vegetation structure in secondary habitats was measured at only a few points in each 1-ha unit, I calculated only vertical complexity, the sum of the vegetation thicknesses in the six height strata.

E. Data Analysis

To compare primary forest habitats (CF, CF edge, 10-ha fragment, and 1-ha fragment), I combined data from the 1-ha units in each habitat-by-block combination and calculated the mean. Sample size was 16 (four blocks for each of the four forest types). Secondary habitats and sites within continuous forest were compared with the data from each 1-ha unit, hence sample size was 11 for secondary habitats and 16 for continuous forest. Spearman's correlations were calculated between small mammal community measurements and habitat structure variables.

III. Results

A. Relationships in Edge-Modified Primary Forest

Two major patterns of correlation were evident between vegetation structure in primary forest habitats and characteristics of the small mammal community (Table II). First, as understory density increased and overstory density decreased, the abundance, richness, and diversity of the terrestrial

Table II   Correlations between Characteristics of Small Mammal Populations at 16 Primary Forest Sites and Measurements of Vegetation Structure[a]

| | | Vegetation thickness | | | | | |
| | | Mean | | Residual variance | | Residual surface area | |
Small mammal		Understory	Overstory	Understory	Overstory	Understory	Overstory
Terrestrial	abundance	0.721[b]	−0.657	−0.154	0.295	−0.101	−0.490
		<0.01[c]	<0.01	0.57	0.27	0.71	0.05
	richness	0.714	−0.754	−0.148	0.224	−0.128	−0.523
		<0.01	<0.01	0.58	0.40	0.64	0.04
	diversity	0.682	−0.782	−0.155	0.167	−0.097	−0.505
		<0.01	<0.01	0.56	0.53	0.72	<0.05
Arboreal	abundance	0.520	−0.594	−0.402	0.020	0.058	−0.423
		0.04	0.02	0.12	0.94	0.83	0.10
	richness	0.272	−0.319	−0.055	0.032	0.506	−0.740
		0.31	0.23	0.84	0.90	<0.05	<0.01
	diversity	0.017	−0.098	0.069	−0.100	0.547	−0.618
		0.95	0.72	0.80	0.71	0.03	0.01
Total	abundance	0.652	−0.632	−0.326	0.200	−0.017	−0.558
		<0.01	<0.01	0.22	0.46	0.95	0.02
	richness	0.631	−0.730	−0.274	0.118	−0.014	−0.585
		<0.01	<0.01	0.30	0.66	0.96	0.02
	diversity	0.500	−0.535	−0.105	0.102	0.044	−0.441
		<0.05	0.03	0.70	0.70	0.87	0.09

[a] Four major forest types were censused: continuous forest, the edge of continuous forest, 10-ha forest fragments, and 1-ha forest fragments. See text for definitions of the mammal and vegetation variables.
[b] First row is the Spearman correlation coefficient.
[c] Second row is the probability.

small mammal fauna increased (in all cases, $P < 0.01$; Fig. 1A). The three measures of the combined small mammal sample were similarly correlated with vegetation density (in all cases, $P < 0.05$). This was in large part due to the terrestrial sample, because neither the richness nor diversity of the arboreal fauna was correlated with vegetation density. Overstory residual surface area showed a similar pattern; correlations with the same set of small mammal variables were relatively high ($P \le 0.10$). I suspect that these correlations represented a single pattern of variation, because the correlation between understory and overstory density was itself relatively high ($r_s = -0.45$, $P = 0.08$). Indeed, when the various small mammal variables were regressed against understory density, and the resulting residuals were regressed against the other vegetation variables, these correlations all declined in magnitude. They were significant (and negative) only for overstory density versus terrestrial abundance, richness, and diversity ($0.01 < P < 0.05$).

Second, as relative grain size in the overstory increased (i.e., it became more continuous) and relative grain size in the understory decreased, the

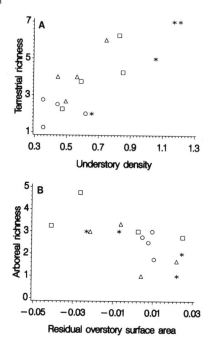

Figure 1   Mean species richness of small mammals versus characteristics of the vegetation at 16 primary forest sites. Four sites were sampled in each of four major forest types: continuous forest (circles), the edge of continuous forest (triangles), 10-ha forest fragments (squares), and 1-ha forest fragments (stars). Species richness is the total number of species encountered during three censuses in a 1-ha subsampling unit; points represent the mean richness across the 1-ha units at a site. In (A), $r_s = 0.71$, $P < 0.01$; in (B), $r_s = 0.74$, $P < 0.01$. See text for definitions of vegetation variables.

richness of the arboreal fauna increased (Fig. 1B). Both understory and overstory residual surface area were correlated (positively and negatively, respectively) with arboreal richness and diversity (Table II). When residuals from these regressions were used, the correlations remained approximately the same magnitude (in two cases $P < 0.05$, and in two cases $P < 0.01$).

B. Relationships in Undisturbed Primary Forest

Only three correlations between vegetation structure in continuous forest and characteristics of the small mammal community were significant; understory residual variance was positively correlated with terrestrial abundance, richness, and diversity (Table III). Understory residual variance was negatively correlated with residual understory surface area ($r_s = -0.58$, $P = 0.02$). Thus, neither of the two patterns of variation described in the previous section was found among the sampling units in continuous forest. This contrasted to the other primary forest habitats, where correlations within habitats were similar to those among habitats (Fig. 1).

Table III   Correlations between Characteristics of Small Mammal Populations and Measurements of Vegetation Structure in 16 1-ha Plots in Continuous Primary Forest

Small mammal		Vegetation thickness					
		Mean		Residual variance		Residual surface area	
		Understory	Overstory	Understory	Overstory	Understory	Overstory
Terrestrial	abundance	0.065[a]	0.242	0.552	0.419	−0.326	−0.107
		0.81[b]	0.36	0.03	0.11	0.22	0.69
	richness	−0.180	0.126	0.508	−0.016	−0.356	−0.229
		0.50	0.64	0.04	0.95	0.18	0.39
	diversity	−0.308	0.037	0.507	−0.104	−0.382	−0.235
		0.24	0.89	0.04	0.70	0.14	0.38
Arboreal	abundance	−0.343	0.142	−0.133	0.068	0.112	0.144
		0.19	0.60	0.62	0.80	0.68	0.59
	richness	−0.132	−0.296	0.100	−0.201	0.177	0.327
		0.62	0.27	0.71	0.45	0.51	0.22
	diversity	−0.104	−0.315	0.165	−0.141	0.088	0.331
		0.70	0.23	0.54	0.60	0.74	0.21
Total	abundance	−0.240	0.331	0.141	0.316	−0.078	0.034
		0.37	0.21	0.60	0.23	0.77	0.90
	richness	−0.318	−0.180	0.312	−0.358	−0.237	−0.055
		0.23	0.50	0.24	0.17	0.35	0.84
	diversity	−0.328	−0.237	0.289	−0.339	−0.348	−0.060
		0.21	0.38	0.28	0.20	0.19	0.82

[a] First row is the Spearman correlation coefficient.
[b] Second row is the probability.

C. Relationships in Secondary Habitats

The relationship between overall thickness of arborescent vegetation in secondary habitats and the abundance of the combined small mammal samples was highly significant ($P < 0.01$) and close to significant for only the terrestrial subset ($P = 0.07$, Table IV). Abundance increased with in-

Table IV   Correlations between Characteristics of Small Mammal Populations in Eleven 1-ha Units in Secondary Habitats and the Density of Arborescent Vegetation[a]

	Terrestrial			Combined terrestrial and arboreal		
	Abundance	Richness	Diversity	Abundance	Richness	Diversity
Density of arborescent vegetation	0.570[b]	0.345	0.291	0.952	0.478	0.418
	0.07[c]	0.30	0.39	<0.01	0.14	0.20

[a] Of the eleven units, six were in pasture (three at each of two sites) and five were in young secondary forest (three units at one site and two units at another). Units in pasture were censused using terrestrial traps only.
[b] First row is the Spearman correlation coefficient.
[c] Second row is the probability.

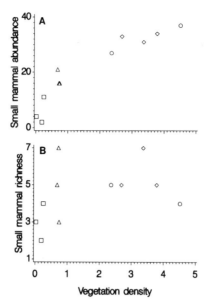

Figure 2   Small mammal abundance and species richness versus arborescent vegetation density in secondary habitats. Two or three 1-ha sampling units were censused at two sites in pasture (squares and triangles) and two in young secondary forest (circles and diamonds). Abundance is the total number of individuals captured during three censuses; species richness is the total number of species encountered during the three censuses. In (A), $r_s = 0.95$, $P < 0.01$; in (B), $r_s = 0.48$, $P = 0.14$. See text for definition of vegetation density.

creasing density of arborescent vegetation (Fig. 2A), whereas richness appeared to plateau at relatively low vegetation densities (Fig. 2B).

IV. Discussion

A. General Considerations

Brown (1973) argued that if a group of species has the opportunity to colonize a certain set of habitats, changes in community characteristics between the habitats are expected to reflect ecological characteristics that allow the species to exist and coexist. August (1983) suggested that biogeographic and historical biases were limited among a set of habitats in the Venezuelan llanos because the study sites were present in a small area and the habitats were broadly contiguous. In this study, I examined two habitats (continuous forest and its edge) that were contiguous with each other and two habitats (1- and 10-ha forest fragments) that were isolated from continuous forest by secondary habitats. Thus, part of the variation in small mammal community structure that I observed among habitats could be attributable to characteristics of the sites other than the nature of the forest, for example, the size of a forest fragment or its degree of isolation from continuous forest (as predicted by island biogeography theory).

Two lines of evidence, however, suggest that this was not the case (Malcolm, 1991b). First, vegetation structure and the abundance of two common small mammal species in forest fragments could be predicted by use of a model of edge effects and data from the edge of continuous forest. Second, relationships between resource variables and small mammal abundance among "nonisolated" sites (continuous forest and its edge) were the same as those among "isolated" sites (forest fragments). This is indicated by analysis of covariance and canonical correlation (e.g., there is little difference between isolated and nonisolated habitats in the relationship between vegetation structure and species richness in Fig. 1). Thus, with a given resource base, it is possible to predict characteristics of the small mammal community regardless of habitat isolation; insularization alone was unimportant. Therefore, it seems that the primary forest habitats examined here were potentially open to colonization by any species. In fact, most species in the area were recorded in all the habitats during nearly six years of study. Of the 16 species trapped in continuous forest, all were trapped in either 1- or 10-ha fragments. The four species recorded only from fragments and/or secondary habitats (*Caluromys lanatus, Oecomys regalis, Philander opossum,* and *Neacomys* sp.) were either represented by only a few individuals or had restricted distributions in the area.

August (1983) failed to find any significant correlations between habitat complexity or heterogeneity in Venezuela and the diversity and abundance of small mammals. He attributed this result to either an inadequate sample (only five 1-ha grids were examined in less than two years) or to a general lack of relevance of the measured habitat features to processes that structure the community. The results presented here suggest that the latter is unlikely. First, the correlations I observed between vertical complexity and abundance and diversity of small mammals were high in both primary and secondary habitats, and were evident even without the arboreal sample (*contra* Malcolm, 1990). Second, assembly rules probably do not differ substantially between the central Amazon and forests of the llanos as a majority of the small mammal genera in the llanos were also trapped in the central Amazon.

B. Vertical Complexity

As vertical complexity of a habitat increases, increased opportunities for niche subdivision are presented. Increased complexity of vegetation may also correlate with a more abundant and diverse resource base (Brown, 1973; August, 1983). Both Fleming (1973) and August (1983) argued that the increased resource base that accompanies increasing vertical complexity is the most relevant variable in tropical systems because as forests increase in vertical complexity, terrestrial richness increases more than arboreal richness.

In this study, correlations between understory density and terrestrial small mammal abundance in both primary and secondary habitats may have resulted in part from variation in the resource base, in particular from variation in the abundance of insect prey. Diets are not well known, but insects appear to be an important dietary element for most tropical small mammal species (e.g., Cole, 1975; Genest-Villard, 1980; Charles-Dominique *et al.*, 1981; Duplantier, 1982; Robinson and Redford, 1986). Creation of edges at this site resulted in pervasive environmental changes. Understory density increased, presumably due to increased light levels and decreased root competition, and canopy density decreased due to wind damage (and as a result, the quantity of fallen timber along edges increased; J. R. Malcolm, unpublished data). Presumably in response to the increased volume and productivity of the understory vegetation, and to an increased proportion of actively growing plant tissues (Janzen, 1973), an effect of edge creation was to increase insect biomass close to the ground, both absolutely and relative to insect biomass in the canopy (Malcolm, 1991b). An increase in overall insect abundance is also expected to result in the reduced temporal and spatial variation of insect prey (Winnett-Murray, 1986). Thus, as Brown (1973) suggested for desert rodents, a more diverse and abundant small mammal fauna in these edge-modified primary habitats may have been the result of a more abundant and predictable resource base.

The extent to which terrestrial small mammals at this site partition microhabitats is unknown. Dickman (1988) suggested that ecological separation of insectivores may be achieved by specialization for foraging in different microhabitats, and that variation in body size itself is irrelevant. Emmons (1980) noted a greater size range among arboreal squirrel species than among terrestrial species; in accordance with Dickman's (1988) hypothesis, she attributed the result to greater insectivory among terrestrial species. Preliminary information (J. R. Malcolm, unpublished data) from spool-and-line devices suggested substantial differences in the use of space in relation to fallen timber, a habitat characteristic that also varied systematically with edge effects. For example, short-tailed opossums (*Monodelphis brevicaudata*) frequently foraged under the leaf litter and under any fallen timber, whereas spiny rats (*Proechimys guyannensis*) usually traveled on top of fallen timber. Scansorial species (e.g., *Marmosa parvidens* and *M. murina*) apparently frequently forage in the understory, and possibly partition foraging sites by support size (Charles-Dominique *et al.*, 1981). Spatial complexity along edges may thus directly contribute to coexistence.

Increased fruit production in the understory was probably unimportant in these young (4–8 years of age) primary forest edges. The frugivorous *Caluromys philander* was significantly less abundant in edge-modified forest than in continuous forest (Malcolm, 1991b). As edges age and colonizing perennial species begin to fruit, however, fruit productivity in the understory may increase. A brief arboreal trapping effort (360 station-nights) along a 10-year-old edge during a time when *Bellucia* sp. was in fruit yielded approximately three times more individuals of *C. philander* than would be expected given the same effort in continuous forest. In contrast to observations of a less diverse terrestrial fauna with a less developed forest canopy (August, 1983), decreased development of the canopy correlated with an increase in terrestrial diversity in the present study. If fruit production per tree stayed the same or declined along edges, then the net result would be lower fruit fall in the understory along edges (because overstory density declined along edges). However, individual trees could increase their fruit production along edges.

In secondary habitats, fruit production (except for *Cecropia* spp.) appeared to be very low (J. R. Malcolm, unpublished data), and small mammals rarely use *Cecropia* fruits (Charles-Dominique, 1986). Thus, the high density and diversity of small mammals in young secondary forest may also have been due to an abundant insect fauna resulting from high vegetation productivity and to increased spatial complexity.

The observation of increased small mammal diversity and abundance as understory density increased raises the question of why arboreal diversity and abundance did not increase with increased foliage density in the overstory. Possible explanations include: (a) arboreal species tend to be more frugivorous than terrestrial species (as with other mammalian groups; Emmons, 1980; Gautier-Hion *et al.*, 1980) and fruit productivity does not correlate with vertical complexity; (b) a less diverse or less predictable array of foraging microhabitats is available in the canopy than on the ground; and (c) measurements of the canopy were too crude to reflect any relationship. Concerning the latter possibility, understory measurements were more detailed than overstory measurements (in the understory, two measurements represented a 5-m-thick stratum, whereas in the overstory, two measurements represented a 20-m-thick stratum) and, because understory measurements were closer to the observer, they were likely more accurate. Emmons suggests that vine tangles are important in determining the abundance of canopy small mammals (Chapter 10), a habitat feature that my method measured only indirectly.

C. Horizontal Heterogeneity

In patchy habitats, potential niches are distributed both horizontally and vertically (MacArthur *et al.*, 1962; Levins, 1968). In the present study, two correlations suggested a possible role of habitat dispersion in determining small mammal abundance and diversity. First, richness of the arboreal fauna increased with increased relative grain size in the overstory. This correlation may reflect the importance of connectivity in constraining the movements of arboreal mammals. Arboreal species would spend considerable energy and time traveling on the ground between forest patches (August, 1983) and perhaps incur increased risk of predation (some arboreal species are not adept at locomotion on the ground). As a result, individuals may avoid areas of low connectivity.

Second, terrestrial small mammal abundance and diversity in continuous forest increased with increasing variability and relative grain in the understory. The distribution of terrestrial small mammals may be correlated with the distribution of treefall gaps, which are the most obvious reason for variation in the density and grain size of understory vegetation in undisturbed forest. Correlations between small mammal abundance and the density of treefall gaps in continuous forest have been noted for spiny rats (*Proechimys* spp.) (Emmons, 1982; J. R. Malcolm, unpublished data) and a species of rice rat (*Oryzomys macconnelli*) (J. R. Malcolm, unpublished data). Some gap features vary with physical gap characteristics, including mircoclimate and characteristics of the community of colonizing plant species (Harshorn, 1978; Brokaw, 1985). Certain of these features may be relevant to small mammals, for example, fallen timber (Guillotin, 1982; J. R. Malcolm, unpublished data) and the protection afforded by the tangle of vegetation (Putz, 1984). A mosaic of different-sized gaps may provide a diverse array of food types at different times and thus contribute to the coexistence of competing species (Levin, 1974; Wiens, 1985). It seems unlikely that particular species of small mammals specialize on certain types of gaps, given the unpredictability of the occurrence of gaps in space and time and the limited mobility of small mammal species. However, the characteristic distribution of gaps in a forest may determine small mammal abundance and diversity indirectly by determining understory productivity and the availability of foraging microhabitats.

D. Future Research

A key goal of future research is to determine the extent to which tropical small mammal populations, particularly those in the forest canopy, are resource-limited. Resource limitation is a key assumption of equilibrium theory (Wiens, 1977; Grant, 1986). According to this theory: (a) populations and communities are at an equilibrium determined by resource limitation; (b) selection of resource-exploiting attributes is continuous and intense; and (c) competition is the major selective force acting upon resource utilization trends. Grant (1986) provided a discussion of the kinds of evidence that provide support for equilibrium theory. Experiments that provide direct evidence, such as manipulation of resource levels, have not yet been undertaken in tropical rain forests.

Another line of evidence includes correlations between population density and resource abundance. Variation in faunal diversity and abundance among Neotropical sites has been attributed to variation in resource productivity, which in turn may be related to soil richness (Emmons, 1984; Gentry and Emmons, 1987). Several studies have suggested that variation in the quantity or pattern of rainfall among years results in variation in patterns of resource availability (Karr, 1976; Foster, 1982; Leighton and Leighton, 1983), which in turn can affect mammal populations (Foster, 1982; Malcolm, 1991b).

A second line of evidence concerns population responses to resource variability within years. Resource production in a variety of Neotropical forests is seasonal, and apparently in response, reproduction of most small mammal species peaks in the period of maximum resource abundance. Terborgh (1986) suggested that a few keystone plant species may support a significant portion of the frugivore community during the period of resource scarcity. As a result, species-specific responses to the set of keystone species in an area may determine α diversity and variation between habitats in the distribution of keystone species could be a major determinant of β diversity (Terborgh, 1986).

The final tenet of equilibrium theory, the importance of competition, is perhaps the most difficult to test. Indirect evidence for tropical mammals includes reduced niche overlap during seasons of scarcity (Emmons, 1980; Terborgh, 1986) and overdispersion of body weights within guilds (Emmons, 1980; Emmons *et al.*, 1983; Emmons, 1984). Ecological separation by body size may hold true for some guilds and not others; it will be of particular interest to compare patterns of variation among terrestrial species with those among arboreal species.

Further advances in the understanding of canopy populations of tropical small mammals will come only with increased efforts to work in the forest canopy. Much more research using canopy-based sampling is required, and canopy trapping should become an integral part of inventories and ecological studies. I have focused here on the ecology of the small mammal fauna; the paucity of information is equally apparent concerning their biogeography and systematics. Much remains to be done, and given the current rates of tropical forest destruction, we may have little time to do it.

V. Summary

Correlations between measures of vegetation structure and characteristics of the small mammal fauna were calculated for three sets of habitats in the central Amazon: (a) primary forest habitats with varying proportions of edge-modified habitat, (b) undisturbed primary forest, and (c) secondary habitats (pasture and young secondary forest). Average density, variability, and patchiness of understory and overstory vegetation were qualified by measuring the vertical distribution of foliage. Estimates of small mammal abundance, richness, and diversity (H′) were obtained by live-trap censuses of the terrestrial and the canopy fauna. In primary forest with varying proportions of edge-modified habitat, two major patterns of correlation were observed. First, as understory density increased and overstory density decreased, the abundance, richness, and diversity of the terrestrial small mammal fauna increased. Second, as relative grain size in the overstory increased, richness of the arboreal fauna increased. I hypothesize that the former relationship was the result of a more abundant and predictable insect prey base along forest edges, whereas the latter relationship reflected locomotory constraints imposed by the degree of connectivity in the canopy. In undisturbed primary forest, abundance and diversity of the terrestrial small mammal fauna increased with understory variability and grain size, indicating the potential importance of treefall gaps in structuring small mammal communities. As the overall density of arborescent vegetation in secondary habitats increased, small mammal abundance increased, presumably in response to a more abundant arthropod fauna.

Acknowledgments

I extend special thanks to A. Cardoso, R. Cardoso, J. Santos, C. Martins, D. Oliveira, and J. Voltolini for assistance in the field, and to R. Bierregaard, J. Eisenberg, and T. Lovejoy, who provided me with the opportunity to conduct the research. L. Emmons and J. Ray provided helpful comments on the manuscript. Funding was from the World Wildlife Fund - US, the Instituto Nacional de Pesquisas da Amazônia, the National Geographic Society, the Tinker Foundation, and Sigma-Xi, a postgraduate scholarship from the Natural Sciences and Engineering Research Council of Canada, and graduate assistantships from the Department of Wildlife and Range Sciences and the Katharine Ordway Chair of Ecosystem Conservation of the University of Florida. This is publication Number 20 in the Biological Dynamics of Forest Fragments Project Technical Series.

References

Adam, F. (1977). Données préliminaires sur l'habitat et la stratification des ronguers en forêt de Basse Côte d'Ivoire. *Mammalia* 41, 283–290.

Atramentowicz, M. (1982). Influence du milieu sur l'activité locomotrice et la reproduction de *Caluromys philander* (L.). *Rev. Ecol. (Terre Vie)* 36, 376–395.

August, P. V. (1983). The role of habitat complexity and heterogeneity in structuring tropical mammal communities. *Ecology* 64, 1495–1513.

August, P. V. (1984). Population ecology of small mammals in the llanos of Venezuela. *Spec. Publ.—Mus. Tex. Tech Univ.* 22, 1–234.

Bourlière, F. (1973). The comparative ecology of rain forest mammals in Africa and tropical America: Some introductory remarks. *In* "Tropical Forest Ecosystems in Africa and South

America: A Comparative Review" (B. J. Meggers, E. S. Ayensu, and W. D. Duckworth, eds.), pp. 279–292. Smithsonian Institution Press, Washington, DC.

Brokaw, N. V. L. (1985). Treefalls, regrowth, and community structure in tropical forests. *In* "The Ecology of Natural Disturbance and Patch Dynamics" (S. T. A. Pickett and P. S. White, eds.), pp. 53–69. Academic Press, New York.

Brown, J. H. (1973). Species diversity of seed-eating desert rodents in sand dune habitats. *Ecology* 54, 775–787.

Charles-Dominique, P. M. (1986). Inter-relations between frugivorous vertebrates and pioneer plants: *Cecropia*, birds and bats in French Guyana. *In* "Frugivores and Seed Dispersal" (A. Estrada and T. H. Fleming, eds.), pp. 119–135, Dr. W. Junk Publ., Dordrecht, The Netherlands.

Charles-Dominique, P. M., Atramentowicz, M., Charles-Dominique, M., Gerard, H., Hladik, C. M., and Prévost, M. F. (1981). Les mammifères frugivores arboricoles nocturnes d'une forêt Guyanaise: inter-relations plantes-animaux. *Rev. Ecol. (Terre Vie)* 35, 341–435.

Cole, L. R. (1975). Foods and foraging places of rats (Rodentia, Muridae) in the lowland evergreen forest of Ghana. *J. Zool.* 175, 453–471.

Davis, D. E. (1945). The annual cycle of plants, mosquitos, birds and mammals in two Brazilian forests. *Ecol. Mongr.* 15, 244–295.

Delany, M. J. (1971). The biology of small rodents in Mayanja Forest, Uganda. *J. Zool.* 165, 85–129.

Dickman, C. R. (1988). Body size, prey size, and community structure in insectivorous mammals. *Ecology* 69, 569–580.

Dosso, H. (1975). Liste preliminaire des Rongeurs de la forêt de Tai (5°53'N, 7°25'W), Côte D'Ivoire. *Mammalia* 39, 515–517.

Duplantier, J. M. (1982). Les rongeurs myomorphes forestiers du nord-est du Gabon. Ph.D. Dissertation, Université des Sciences et Techniques du Languedoc, France.

Eisenberg, J. F. (1980). The density and biomass of tropical mammals. *In* "Conservation Biology" (M. E. Soulé and B. A. Wilcox, eds.), pp. 35–56. Sinauer Assoc., Sunderland, MA.

Eisenberg, J. F., and Thorington, R. W., Jr. (1973). A preliminary analysis of a Neotropical mammal fauna. *Biotropica* 5, 150–161.

Emmons, L. H. (1980). Ecology and resource partitioning among nine species of African rain forest squirrels. *Ecol. Monogr.* 50, 31–54.

Emmons, L. H. (1982). Ecology of *Proechimys* (Rodentia, Echimyidae) in south-eastern Peru. *Trop. Ecol.* 23, 280–290.

Emmons, L. H. (1984). Geographic variation in densities and diversities of non-flying mammals in Amazonia. *Biotropica* 16, 210–222.

Emmons, L. H., Gauteir-Hion, A., and Dubost, G. (1983). Community structure of the frugivorous-folivorous forest mammals of Gabon. *J. Zool.* 199, 209–222.

Fleming, T. H. (1972). Aspects of the population dynamics of three species of opossums in the Panama Canal zone. *J. Mammal.* 53, 619–623.

Fleming, T. H. (1973). Numbers of mammal species in North and Central American forest communities. *Ecology* 54, 555–563.

Fleming, T. H. (1979). Do tropical frugivores compete for food? *Am. Zool.* 19, 1157–1172.

Fonseca, G. A. B. (1988). Patterns of small mammal species diversity in the Brazilian Atlantic forest. Ph.D. Dissertation, University of Florida, Gainesville.

Foster, R. B. (1982). Famine on Barro Colorado Island. *In* "The Ecology of a Tropical Forest: Seasonal Rhythms and Long-Term Changes" (E. G. Leigh, Jr., A. S. Rand, and D. M. Windsor, eds.), pp. 201–212. Smithsonian Institution Press, Washington, DC.

Gautier-Hion, A., Emmons, L. H., and Dubost, G. (1980). A comparison of three major groups of primary consumers of Gabon (primates, squirrels and ruminants). *Oecologia* 45, 182–189.

Genest-Villard, H. (1980). Régime alimentaire des rongeurs myomorphes de forêt équatoriale (Région de M'Baiki, République Centrafricaine). *Mammalia* 44, 423–484.

Gentry, A. H., and Emmons, L. H. (1987). Geographic variation in fertility, phenology, and composition of Neotropical forests. *Biotropica* 19, 216–227.

Grant, P. R. (1986). Interspecific competition in fluctuating environments. *In* "Community Ecology" (J. Diamond and T. J. Case, eds.), pp. 173–191. Harper & Row, New York.

Guillotin, M. (1982). Rhythmes d'activité et regimes alimentaires de *Proechimys cuvieri* et d'*Oryzomys capito velutinus* (Rodentia) en forêt Guyanaise. *Rev. Ecol. (Terre Vie)* 36, 337–371.

Harrison, J. L. (1962). The distribution of feeding habits among animals in a tropical rain forest. *J. Anim. Ecol.* 31, 53–63.

Hartshorn, G. S. (1978). Treefalls and tropical forest dynamics. *In* "Tropical Trees as Living Systems" (P. B. Tomlinson and M. H. Zimmerman, eds.), pp. 617–638. Cambridge Univ. Press, Cambridge, UK.

Hubbell, S. P., and Foster, R. B. (1986). Canopy gaps and the dynamics of a Neotropical forest. *In* "Plant Ecology" (M. J. Crawley, ed.), pp. 77–95. Blackwell, Oxford.

Janzen, D. H. (1973). Sweep samples of tropical foliage insects: Effects of seasons, vegetation types, elevation, time of day, and insularity. *Ecology* 54, 687–708.

Karr, J. R. (1976). Seasonality, resource availability, and community diversity in tropical bird communities. *Am. Nat.* 110, 973–994.

Leighton, M., and Leighton, D. R. (1983). *In* "Tropical Rain Forest: Ecology and Management" (S. L. Sutton, T. C. Whitmore, and A. C. Chadwick, eds.), pp. 181–196. Blackwell, Oxford.

Lein, M. R. (1972). A trophic comparison of avifaunas. *Syst. Zool.* 21, 135–150.

Levin, S. A. (1974). Dispersion and population interactions. *Am. Nat.* 104, 413–423.

Levins, R. (1968). "Evolution in Changing Environments." Princeton Univ. Press, Princeton.

Lovejoy, T. E., and Bierregaard, R. O., Jr. (1990). Central Amazonian forests and the Minimum Critical Size of Ecosystems Project. *In* "Four Neotropical Rainforests" (A. H. Gentry, ed.), pp. 60–71. Yale Univ. Press, New Haven, CT.

MacArthur, R. H., MacArthur, J. W., and Preer, J. (1962). On bird species diversity. II. Prediction of bird census from habitat measurements. *Am. Nat.* 96, 167–174.

Malcolm, J. R. (1990). Mammalian densities in continuous forest north of Manaus, Brazil. *In* "Four Neotropical Rainforests" (A. Gentry, ed.), pp. 339–357. Yale Univ. Press, New Haven, CT.

Malcolm, J. R. (1991a). Comparative abundances of Neotropical small mammals by trap height. *J. Mammal.* 72, 188–192.

Malcolm, J. R. (1991b). The small mammals of Amazonian forest fragments: Pattern and process. Ph.D. Dissertation, University of Florida, Gainesville.

Miles, M. A., Souza, D. A. A., and Povoa, M. M. (1981). Mammal tracking and nest location in Brazilian forest with an improved spool-and-line device. *J. Zool.* 195, 331–347.

O'Connell, M. A. (1989). Population dynamics of Neotropical small mammals in seasonal habitats. *J. Mammal.* 70, 532–548.

Putz, F. E. (1984). The natural history of lianas on Barro Colorado Island, Panama. *Ecology* 65, 1713–1724.

Rahm, U. (1972). Zur oekologie der muriden in regenwaldgebiet des östlichen Kongo (Zaire). *Rev. Suisse Zool.* 79, 1121–1130.

Robinson, J. G., and Redford, K. H. (1986). Body size, diet, and population density of Neotropical forest mammals. *Am. Nat.* 128, 665–680.

Stallings, J. R. (1988). Small mammal communities in an eastern Brazilian park. Ph.D. Dissertation, University of Florida, Gainesville.

Terborgh, J. (1986). Community aspects of frugivory in tropical forests. *In* "Frugivores and Seed Dispersal" (A. Estrada and T. H. Fleming, eds.), pp. 371–384. Dr. W. Junk Publ., Dordrecht, The Netherlands.

Wiens, J. A. (1977). On competition and variable environments. *Am. Sci.* 65, 590–597.

Wiens, J. A. (1985). Vertebrate responses to environmental patchiness in arid and semi-arid ecosystems. *In* "The Ecology of Natural Disturbance and Patch Dynamics" (S. T. A. Pickett and P. S. White, eds.), pp. 169–193. Academic Press, New York.

Winnett-Murray, K. (1986). Variation in the behavior and food supply of four Neotropical wrens. Ph.D. Dissertation, University of Florida, Gainesville.

10

Mammals of Rain Forest Canopies

Louise H. Emmons

I. Introduction

The forest canopy is one of the most constraining habitats for mammals. Large canopy mammals (those above 1 kg mass that acquire most of their food in the trees) are quite narrowly restricted geographically, but small mammals (below 1 kg mass) have wider distributions. Although many mammals use the canopy, there are no large strictly canopy mammals in North America, in the Palearctic, or south of 30°S in Africa or South America. With the exception of Australia, almost all canopy mammals are found in tropical forests, and mainly in rain forests. The reason for this pattern is simple: to feed only in the canopy, an animal must find food in it year-round. North temperate and tropical deciduous forests lack year-round canopy resources for mammals. Temperate and dry forest arboreal species (e.g., mice, squirrels, porcupines, monkeys, and martens) can forage on the ground much of the year or hibernate (e.g., dormice, some small lemurs). In the vast evergreen boreal forests, only one genus (2 spp.) of tiny mice (*Arborimus*, 30 g) is able to thrive by feeding on the needles of conifers, and these mice are partly terrestrial (Burt and Grossenheider, 1976; Nowak, 1991).

Australia has a unique type of flora and associated arboreal mammal complement. Its temperate and "dry" forests are dominated by evergreen eucalypts, and several lineages of marsupials have evolved the astonishing ability to use eucalypt leaves as a primary diet. This gives them a year-round resource in a climatic regime that elsewhere does not support a significant

Copyright © 1995 by Academic Press, Inc.
All rights of reproduction in any form reserved.

canopy mammal fauna. Many smaller canopy mammals inhabit temperate and dry tropical Australian forests, but there are only four genera of large species (the koala, *Phascolarctos cinereus;* greater glider, *Petauroides volans;* brushtail possums, *Trichosurus* spp.; and scaly-tailed possum, *Wyulda squamicaudata*). Other large canopy mammals of the Australian region (e.g., tree kangaroos, gliders, and cuscuses) are found only in the tropical rain forests (Strahan, 1983). Thus, almost all large canopy mammals, and also the majority of small ones, are found in evergreen tropical forests.

This chapter focuses on the reciprocal issues of (1) the characters that mammals have for canopy life and (2) the characters the canopy has for mammal life. I first discuss the taxonomic distribution and physical characteristics of canopy mammals, and the canopy as a special environment. I then describe the feeding categories of canopy species and the correlation between diet and physical adaptations. Finally, I compare canopy mammal assemblages and canopy fruit resources of different continents and speculate on the association of intercontinental vegetation differences with intercontinental mammal community differences.

A. Taxonomic Distribution

Twelve of the 26 orders of mammals include species that use the forest canopy. Of the 79 families of nonflying (nf) mammals within these orders, 39 (49%) include canopy-feeding taxa. In these families, 195 of 252 genera (77%) include arboreal or scansorial species (classification of higher taxa follows Wilson and Reeder, 1993). The numbers are approximate, due to unsettled taxonomic questions and uncertain natural history, but their magnitude shows that arboreality has been one of the chief axes of mammalian diversification, comprising nearly half of higher taxa. Nevertheless, only four orders include the majority of nf arboreal mammal species. Rodents dominate the scene, with 85 specialized climbing genera, followed by primates (58 genera), marsupials (23), and carnivora (18). In the primates alone among large orders, most genera are arboreal (94%). Virtually the entire living primate radiation, and 40% of that of marsupials, is directed toward exploitation of arboreal resources. The second largest order of mammals, the Chiroptera (bats), with close to a thousand species, is the largest and probably the most ecologically important group of mammals that use the forest canopy. The exploitation of the canopy by bats remains a frontier for basic research. Because it would require a whole chapter to do them justice and because knowledge is so scarce, I reluctantly exclude them from this review, apart from general comments.

B. Morphologies of Canopy Nonflying Mammals

The infinitely diverse irregularities, discontinuities, and instabilities of these [forest canopy] support surfaces create barriers to direct, point-to-point movements for all non-

flying canopy dwellers. Moreover, food, in the form of leaves, fruit, and insects, is the most important motivation for movement and is both dispersed and available only seasonally within this complex space.
    —T. I. Grand (1984, p. 53)

The locomotion of arboreal nf mammals has been well studied (e.g., Rodman and Cant, 1984), so I will only summarize some general principles. To a mammal, canopy travel presents three separate challenges: (1) moving up and down large vertical trunks; (2) balancing on thin, unstable, horizontal and vertical branches; and (3) crossing open spaces.

The first challenge requires little specialization, because the only tools needed are sharp, hooked claws. These are remarkably efficient. Among the carnivores, many small species and heavy, largely terrestrial species, including bears, big cats, and raccoons, quite easily climb up large tree trunks. However, to climb down again as easily, a second level of adaptation is needed: ankle joints that swivel to allow head-first descent. Only truly arboreal species have these (squirrels, margays), and without them, other species have to back down awkwardly, craning their necks over their shoulders to see where they are going. Kittens of domestic cats get stuck in trees because backing down is a difficult maneuver that must be learned.

The second challenge, balancing on thin, flexible branches, requires an evolutionary commitment consisting of toes that can flex against some other part of the hand or foot to grip a branch tightly. Many independent solutions to this have evolved, such as the opposable hallux of opossums; the pincerlike grips of porcupines, tamanduas, and pangolins; and the long grasping fingers of Amazon bamboo rats, opossums, and primates. The heavier an animal is, the stronger or more specialized its grip must be. A lightweight squirrel or rat can climb a branch by gripping with sharp claws applied by strong toes. Arboreal rodents [e.g., South American tree rats (Echimyidae; L. H. Emmons, personal observation) and Asian and African tree mice (Muridae; Nowak, 1991)] and mouse opossums (L. H. Emmons, personal observation) grip slender twigs between the digits of hind- and/or forefeet. The digits used differ among taxa, and they are often modified by having a space between them.

The third and most difficult challenge for arboreal species is crossing gaps. The three basic solutions are: reaching across, jumping across, and gliding across. Many totally or highly arboreal species can cross only those gaps that they can reach across, so they must descend to the ground to travel between widely spaced trees. These species include all the "pincer-grip" species (e.g., sloths, porcupines, anteaters, and pangolins) and also many "finger-grip" species (e.g., lorises, cuscuses, common opossums, Amazon bamboo rats, orangutans, and binturongs (L. H. Emmons, personal observation; M. Roberts, personal communication).

Arboreal adaptations can be conceptualized as a series of levels (Table I).

Table I Classification of Canopy Locomotor Adaptations of Larger Mammals (Excluding Tiny Scampering Species) and Their Ecological Associations

Adaptation	Capability	Species examples	Ability in canopy	Canopy diet
TYPE I: Claws only	Up wide or narrow vertical trunks; along large main branches	Bears, jaguars, raccoons	Escape terrestrial predators, sleep, lookout, feed on main-branch resources	Minor part of diet, honey, some fruit, birds eggs, mammal nestlings
TYPE II: Tightly gripping hands/feet, does not jump	Up wide (if with claws) or narrow (without claws) vertical trunks; along thin, unstable branches	Porcupines, anteaters, lorises, sloths, cuscuses, Amazon bamboo rats, binturongs	Can feed on branch tips, locally confined foraging	Up to whole diet, often specialized, leaves, social insects, noxious insects, fruits, vertebrates
Type III: Leapers	Can move easily between unconnected trees	Squirrels, monkeys, lemurs, kinkajous, martens	Can travel quickly to dispersed resources	Whole diet, mixed fruit, random insects, nectar, leaves
TYPE IV: Gliders	Can move between widely isolated trees	Flying lemurs, flying squirrels, marsupial gliders, anomalurids	Can travel widely very quickly, with little energy, but slow or awkward running	Leaves, nectar/exudates, fruit/insects
TYPE V: Flyers	Access to entire canopy	Bats	Can travel over wide regions	Fruit, nectar, and/or insects or vertebrates

This arrangement reveals ecological correlates of broad classes of morphological types. Each type represents an increased degree of mobility within the canopy and, consequently, increased foraging opportunities. In addition to these correlates, the structure of the canopy affects animals of different types in different ways. For example, many Type II species that can easily climb slender branches, but cannot jump, favor dense, viny habitats where all the vegetation is connected.

There is a strong phylogenetic component to types of arboreal locomotion. Three of the large orders that include arboreal nf mammals (primates, carnivores, and rodents) include separate lineages of Type II and Type III species. One of them (rodents) also includes Type IV (Table I). Surprisingly, the marsupials, one of the largest arboreal radiations, includes no real leapers; all arboreal species are Types II and IV. Likewise, the small orders with highly specialized arboreal taxa have species only of Type II (anteaters, sloths, pangolins, hyraxes) and IV (flying lemurs). Arboreal leaping is often associated with speciose radiations (squirrels, primates, palm civets), as if this proficiency launched a range of new ecological opportunities. There are a few exceptions to these generalizations among lightweight species: a few marsupials can at least jump from branch to branch while foraging, including a rare Neotropical opossum (*Glirona venusta;* Emmons and Feer, 1990) and Australian pygmy-possums (*Cercartetus* spp.; Strahan, 1983). The squirrellike tree shrews (order Scandentia) are scansorial leapers but have failed to diversify, and only two or three species use the canopy (L. H. Emmons, personal observation). Because tiny mammals such as mice or mouse opossums can jump only short distances, they cannot cross significant gaps between trees.

C. The Canopy Environment

The rain forest canopy fauna must cope with special environmental conditions. The most important of these are dehydration, because the canopy in drier seasons or regions has few drinking water sources, and exposure to wind, rain, sun, and temperature extremes. It is hotter by day and colder by night within the canopy than under it (Kira and Yoda, 1989). Mammals have behavioral ways to cope with these extremes, (e.g., resting during midday heat, sheltering from the elements below dense foliage), but they also have structural and physiologial features that are associated with arboreal life. I have noted that many large arboreal mammals in rain forests (e.g., sloths, binturongs, and most large primates) have long, coarse hair. Such hair probably sheds rain, does not clump when wet, and dries easily. Similarly, arboreal mammals often have thick, tough skin, which I assume protects them from dehydration and the elements and shields them as they make crash-landings onto branches when they leap. Small, nocturnal arboreal mammals can almost always be distinguished from their terrestrial counterparts by their denser, woollier fur (e.g., in the canopy-living woolly mouse opossums, *Micoureus,* and the woolly opossums, *Caluromys*). Some arboreal mammals do not drink, such as koalas and greater gliders (Strahan, 1983). For obvious reasons (branches break), canopy mammals are limited in body weight to under about 50 kg, but most are below 15 kg.

D. Diet

In a review of insectivory among arboreal nf mammals (Emmons, 1995), I found that only five genera with 11 species feed almost entirely on arboreal insects. All are small, nocturnal forms (<400 g). There are no reports of mammalian carnivores that feed exclusively in the trees on vertebrate prey. The most carnivorous arboreal Carnivora, linsangs (*Prionodon* and *Poiana,* Viverridae), martens (*Martes* spp., Mustelidae), and margays (*Leopardus wiedii,* Felidae), do much of their foraging on the ground. Similarly, the most highly specialized large arboreal insectivores, anteaters and pangolins (*Tamandua* spp., Myrmecophagidae; *Manis* spp., Manidae), also forage terrestrially, although the small ones, such as the tiny anteater (*Cyclopes didactylus,* Myrmecophagidae), may not.

All of the truly arboreal Carnivora, such as olingos and kinkajous (*Bassaricyon* and *Potos,* Procyonidae), and palm civets and binturongs (*Nandinia* and *Arctictis,* Viverridae), are the most frugivorous members of their orders (Emmons *et al.,* 1983; Payne *et al.,* 1985). The least terrestrial climbing opossums (*Caluromys* and *Micoureus* spp., Didelphidae) are the most frugivorous (Charles-Dominique *et al.,* 1981), as is the case among rain forest squirrels of Gabon (Emmons, 1980) and Malaysia (Medway, 1978; Emmons, 1995). The pursuit of vertebrate and invertebrate prey in the latticework of the canopy does not appear productive enough to fully support mammalian quadrupeds. For nf mammals, the forest canopy is thus a realm for consumers of the primary production of the canopy itself: fruit, leaves, and nectar.

Entirely folivorous arboreal mammals that eat no fruit or insects (koalas and three-toed, but not two-toed, sloths) are as rare as completely arboreal insectivores. Moreover, only one species, the tiny (10–20 g) Australian honey-possum (*Tarsipes rostratus,* Tarsipedidae), probably feeds on arboreal nectar alone, but it must enter torpor to survive food shortages (Strahan, 1983). I find no reports of any nf mammal that feeds entirely on fruit. However, some megachiropteran fruit bats and some phyllostomid bats seem to do so. These must overcome the nutritional deficiencies of fruit by eating up to 250% in excess calories to obtain scarce nutrients (Thomas, 1984).

The living array of arboreal nf mammals overwhelmingly suggests that the forest canopy is a place where only species with mixed diets of fruit, leaves, invertebrates, and sometimes nectar, exudates, or pollen have achieved significant diversity. Nutritional balance is readily achieved in the physically restricted canopy environment only by combining plant and animal resources. This superficially similar cafeteria at which all arboreal nf mammals dine can nonetheless be exploited in a number of distinctive ways by members of sympatric assemblages (Emmons, 1995). Mammals specialize in the types and proportions of fruit and foliage that they eat and, much like birds, they differ in how they prey upon arthropods.

E. Correlates of Locomotion and Diet

Mammals that can climb but not jump (Type II in Table I) must travel arboreal paths with many detours to where trees interdigitate or are linked by lianas. Alternatively, the animals must descend to the ground to travel between trees. This takes more time and energy than direct routes available to species that can jump gaps. Travel on the ground also entails increased risk of predation by felids or canids. One might therefore expect that Type II locomotion would be associated with species that use concentrated resources that entail little arboreal travel. Conversely, species that leap may use more spatially scattered resources. To a large extent, this is the case. Type II species include most of the specialized folivores that eat mature leaves (koalas, sloths, porcupines, tree kangaroos), as well as mammals that eat sedentary and/or aggregated social or noxious insects (anteaters, pangolins, slow-lorises, pottos). These food items are often chemically or physically defended from general predators, and thus there are few competitors for them in any assemblage (competition depletes resources and thereby increases the need for consumer travel). Resources are also relatively more concentrated for species with small absolute needs either because their body sizes are absolutely small (rats, mice, mouse opossums) or because their metabolic rates are low (pottos, lorises, tree hyraxes, pangolins, porcupines, sloths, marsupials) (Hildewein, 1972; McNab, 1978).

Only animals that can launch themselves into space across a gap are able to travel long distances at great speed through the canopy. The increased range, and the presumed "cheapness" of travel by arboreal leaping between separated trees, gives mammals access to large areas to search for particular fruits or hunt for solitary invertebrates. Because of their high tree species diversity/low density of conspecific individuals, tropical rain forest fruit resources are almost always scattered. To avoid toxic secondary chemicals and/or maximize nutrients, many arboreal mammals feed on young leaves of only certain species (Hladik, 1978). Like fruit, these are spatially scattered. The primates, squirrels, marsupials, and a few Carnivora have diversified in exploiting these resources. The most active rain forest canopy monkeys worldwide have daily maximum pathlengths of about 1.5–2.5 km (Smuts *et al.*, 1987). The surprising uniformity, from marmosets (400 g) to gibbons (6000 g), of (leaping) primate daily travels through a wide range of seasons, habitats, and locomotory adaptations suggests that either tropical canopy resources have roughly equivalent dispersions worldwide or primates are energetically limited to areas where resources can be acquired within this range (terrestrial primates can travel much farther).

A detailed study addresses this question for a rain forest monkey at Makokou, Gabon (*Cercopithecus cephus*). Gautier-Hion *et al.* (1981) determined that the daily pathlength of 1300 m (on 11 ha) gave a 100% probability for encounter of the 3 most common fruit species, of 7 available in the dry season, and over 60% probability of finding 4. In the rainy season, the same pathlength/area allowed monkeys to find at least 8 of 14 available species. In both seasons, a huge further increase (almost double) in daily range would have been needed to encounter additional fruit species. These data vividly attest to the need for extensive travel by tropical rain forest canopy frugivores and may help explain that most species have mixed diets to alleviate dependence on rare food items. This study also addressed the question of why the particular distance of about 1.5 km of daily travel is commonly found among mammals that have little but frugivory in common.

A cost of travel by leaping is the risk of falling. Small species (e.g., squirrels) can fall 20 m without apparent harm, and lightweight taxa often have a predator defense of letting go and dropping from the canopy (L. H. Emmons, personal observation). However, falls seem to be a major cause of injury and death among large primates (Milton, 1982; Symington, 1988). I have watched a huge male orangutan climb to 40 m along an attached, vertical fig root, carefully testing each successive handhold with a strong tug before transferring body weight to that arm, exactly as a nervous human would have climbed it.

Gilders can cross gaps of scores of meters with little energy cost or physical risk. Gliding has evolved independently at least five times among the living mammals. In contrast to arboreal leaping and flying, it has spawned no great mammalian radiations, except the presumed gliders that led to the bats. The number of gliding species in all but one of these lineages is small: Petauridae (5); Burramyidae (1); Dermoptera (2); Anomaluridae (6); and Sciuridae (30). All gliders are nocturnal and limited in body weight to a maximum of about 2 kg. The largest, including six species of squirrels, one marsupial, and the two flying lemurs, weigh between 1 and 2 kg. All the larger species appear to eat mainly leaves and/or plant exudates or nectar, and a few insects (Eisenberg, 1978; Strahan, 1983). The smaller rodent-gliders feed on these plus fruits and seeds, and the marsupials on plant exudates (Strahan, 1983). Although gliding facilitates travel between trees, the membranes appear to hamper quadrupedal movements, and gliders are not as agile as nongliders when climbing within a tree (L. H. Emmons, personal observation of all families except marsupial gliders).

The diets of gliders seem more like those of Type II than of Type III species: the "giant" flying squirrels are folivores, but the "giant" nongliding squirrels feed on seeds and fruits (Emmons, 1980, and personal observation; Payne *et al.*, 1985). It is unclear why gliding is associated with folivory, but arboreal folivores tend to have low metabolic rates (McNab, 1978) because microbial fermentation of tree leaves in the gut can provide nutrients at only limited rates (Parra, 1978). Nectar or exudates may also be energy- and nutrient-poor resources for foot-travelers, with a low caloric return per flower or feeding site. Small fliers (insects, birds, and bats) with small total energy needs and quick progression from one flower to the next can exploit nectar efficiently, and gliding may be another way to achieve this. Gliding may thus be a means to reduce energy expenditure and enable use of otherwise marginal foods. The extra weight and volume of fermentation chambers for folivory may also limit jumping agility.

The ultimate means of canopy travel is by powered flight. Bats have the greatest access to the canopy and maximum range of daily movements. Flight has also imposed many limitations, such as on body size and litter size. The digestive tracts of bats are mere tubes of the simplest kind (Mitchell, 1916). Bats are limited to feeding only on other animals, or on the most readily assimilated plant products, juicy fruits and nectar. Perhaps digestive fermentation by microorganisms has never developed in Chiroptera because of the extra weight involved (although it has evolved in one small bird and many large ones). With flight, some bats are able to commute far enough to feed entirely on fruit (e.g., over 10 km each way; Bradbury, 1977), while primates in the same habitats switch to other resources during dry-season fruit shortages and supplement their diets with leaves and insects (Gautier-Hion, 1980).

II. Intercontinental Comparisons

A. Species Numbers

All tropical rain forests, except those on some far islands, have canopy mammal faunas. In this section, I discuss the faunas of three continental sites, and compare them to two geographically isolated island faunas (Table II). Two of these forests (Makokou, Gabon, and Cocha Cashu, Peru) have been subjects of decades of research by many investigators. Two other sites (Ranomafana, Madagascar, and Danum Valley, East Malaysia) are new research stations that are much less well known. West Sepik, Papua New Guinea, was the site of brief mammal survey work. As part of Sundaland, I consider Borneo (Danum Valley) to have a continental fauna, although it is an island smaller than New Guinea. New Guinea is on the Australian continental shelf, but its tropical rain forest habitat has small extent in Australia, so I consider it an island fauna. The mammal fauna of Makokou has been analyzed in detail elsewhere (Gautier-Hion *et al.*, 1980; Emmons *et al.*, 1983) and compared to another Neotropical fauna (Dubost, 1987; Skalli

Table II Mammal Faunas of Five Rain Forest Localities

| Locality[a] | Total nonflying species | Nonflying arboreal species | Total bats | "Fruit" bats[b] | Arboreal/scansorial | | | |
					Marsu-pials	Primates	Carni-vores	Rodents
Cocha Cashu	78	40	42	27	10	13	3	12
Makokou	89	38	33	7	—	14	2	17
Danum	75[c]	39	28	7	—	10	6	16
West Sepik	32	16	19	10	10	—	—	6
Ranomafana	33	16	?	?	—	12	1	3

[a] Estación Biológica de Cocha Cashu and Pakitza (350 m), Manu National Park, Peru; fauna not completely known, especially for bats (Pacheco *et al.*, 1993). Makokou, Gabon (500 m); fauna probably completely known (Emmons *et al.*, 1983; Skalli and Dubost, 1986). Danum Valley Conservation Area, Sabah, Malaysia (Borneo, 150 m); fauna incompletely known, bats and shrews not inventoried, rodents incompletely (Anonymous, 1993; L. H. Emmons, personal observation). West Sepik Province, Papua New Guinea, (<1000 m only); completeness of inventory not known (Flannery and Seri, 1990). Ranomafana National Park Biological Station, Madagascar (700–1200 m); bats not inventoried (K. Creighton, L. H. Emmons, J. Ryan, and P. Wright, unpublished data). Tallies do not include terrestrial species that occasionally climb, but do include scansorial understory species.

[b] Including nectar feeders. Note that only the list from Makokou is nearly complete.

[c] I exclude seven unreliable species records from Anonymous (1993).

and Dubost, 1986). In the accompanying figures and tables, the data are in a few cases approximations based on few individuals. Where mammal weights were available as ranges and not means, I have used the mean of the extremes; likewise for sexual dimorphisms.

A summary of the species composition of the five faunas (Table II) shows striking features. When compared at the crudest level (numbers of species in different categories, such as arboreal), there is such symmetric occupancy of the ecological sets that it is hard to escape the conclusion that these faunas are structured. If so, mammal species must interact with each other in the context of how they assemble in communities (Gautier-Hion *et al.*, 1980; Emmons *et al.*, 1983; Dubost, 1987). Considering that the faunas of Danum Valley and Cocha Cashu are still incompletely known, the three large continental nf mammal faunas are very close in total species numbers. Likewise, the species richness of the two large island sites is identical, at just under half the size of the former (but both may have had anthropogenic extinctions of large species). Even more dramatic are the virtually identical numbers of arboreal species in the three large faunas and two small faunas.

B. Taxonomic History

The arboreal species of the two Old World continental sites are derived largely from the same families (Cercopithecidae, Viverridae, Sciuridae, and Muridae) and are thus phylogenetically close, with only a quarter of the species from regionally endemic families. One might therefore expect a high level of similarity between their faunas, which is indeed the case. The most prominent unique feature of the Asian arboreal mammal fauna is the large radiation of frugivorous climbing civets: the African site has only one palm civet, whereas there are at least three, and probably four, at Danum Valley.

The New World arboreal fauna at Cocha Cashu, in contrast, shares only two major families with the Old World sites (Sciuridae and Muridae, with nine arboreal species). The fauna thus largely independently evolved into the rain forest habitat. The faunivorous-frugivorous platyrrhine monkeys, procyonids, marsupials, and echimyid rodents of the New World forest canopy to some extent fill roles equivalent to those of catarrhine monkeys, civets, tree shrews, prosimians, and flying squirrels in Asia, and to rodents, prosimians, and anomalurids in Africa. Although the numbers of species in the three localities are equivalent, the resemblances are inexact except in a few startling cases. The Neotropical mouse opossums, Madagascar mouse lemurs, Malaysian pentail tree shrews, and Australian pygmy-possums are perfectly convergent, even to color pattern. Neotropical kinkajous (*Potos flavus*) and olingos (*Bassaricyon* spp., Procyonidae) are almost exact counterparts of Paleotropical palm civets (Viverridae). The scansorial Neotropical tayra (*Eira barbara,* Mustelidae) is (apart from eating some fruit) the equivalent of the Asian yellow-throated marten (*Martes flavigula*, Mustelidae). As the latter are members of the same family, the case for physical convergence is weak, but that each habitat has one is evidence of community-level similarity.

The salient difference between New World tropical faunas and all others is the great radiation of echo-locating leaf-nosed bats (Phyllostomidae) into fruit- and nectar-feeding roles (Table II). The Old World megachiropteran fruit bats nowhere achieve the sympatric trophic or species richness of the Phyllostomidae, nor their biomass (L. Emmons, personal observation). In a well-inventoried Neotropical forest in French Guyana, bats comprise 46% of the total mammal species, but they are only 26% of the fauna at Makokou, Gabon (Skalli and Dubost, 1986; Dubost, 1987). The ecological prominence of leaf-nosed bats is thus unique to the New World.

The Ranomafana and West Sepik arboreal faunas, derived totally from regionally endemic families or subfamilies (except for murid rodents in New Guinea), display a curious phenomenon. The Ranomafana fauna, composed mostly of primates, has the same number of primates as in each of the three continental rain forest faunas, and the marsupial fauna of West Sepik has the same number of marsupials as a Neotropical forest (Table II). However, the 10 climbing marsupials of Cocha Cashu are a morphologically uniform set from a single family, whereas the 10 from West Sepik are an extraordinarily diverse array (kangaroos to gliders) from six families. Similarly, the 13 primates of Cocha Cashu belong to two or three families, but the 12 from Ranomafana belong to five. Thus, the island assemblages that have independently arisen from isolated orders show a greater morphological diversity than do the same orders in continental assemblages with greater overall ordinal richness, but only the same number of sympatric species within dominant orders.

The weight distribution of the subset of arboreal mammals that forage in the forest canopy (excluding scansorial understory species) is nonuniform (Fig. 1). Weight classes are in powers of two, the scale chosen ad hoc to emphasize evident groupings (see Dubost, 1987, for similar representations of entire faunas). Weight classes are not equally occupied, which gives rise to quite similar, trimodal distributions at different localities. In general, the peaks represent different modal weights for families or orders. The first peak (Fig. 1) includes small rodents. The depauperate rodent fauna of Ranomafana is evident, but in all other regions, canopy nf mammals tend to be large. Murid rodents, the most speciose family of mammals (1326 species), have few sympatric canopy species (usually only 3–4 in the richest habitats). The same is true of marsupials. Both orders have many tiny scansorial understory species, but species whose major activity and foraging center is the upper canopy layer tend to be few and larger.

Small primates (galagos, marmosets), marsupials, and squirrels make up

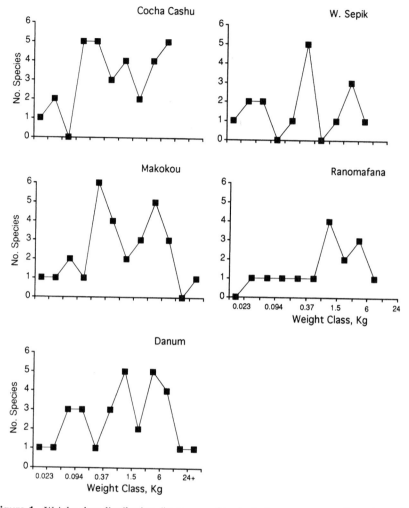

Figure 1   Weight class distribution (in powers of two) of all canopy mammal species at the five localities listed in Table II. This figure excludes arboreal species restricted to the understory and those that only occasionally climb.

most of the second peak (Fig. 1), and primates and carnivores form the third. Primate communities are bimodal, with peak numbers of small, highly insectivorous prosimians or tamarins and large frugivorous or folivorous monkeys (Fig. 2). The slight bimodality in Madagascar does not correspond to the dietary difference of the clusters of species of continental faunas and is probably not significant. The high body mass tails of the

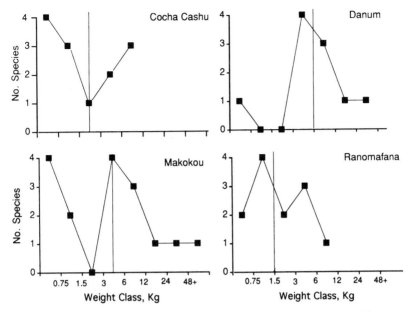

Figure 2 Weight distribution of all primates. Vertical line is median weight.

curves for Makokou and Danum (Fig. 2) include the great apes, which are absent from the other sites. The median (and mean) body mass of primates is higher in Old World than in New World continental sites (Fig. 2) (Terborgh, 1986). Rain forest canopy mammal faunas worldwide therefore show strong convergences in species numbers, weights, and taxonomic balance. This suggests similarity in basic resources.

C. Fruit Size

Fruit (including seeds) is the chief resource used by canopy mammals. In each of the continental sites, fruit was collected at ground level to evaluate local production on study areas. At Makokou, fallen fruit was collected biweekly on a 6-km transect of trails (Gautier-Hion *et al.*, 1985). At Danum Valley, fallen fruit was collected monthly for 15 months on 4.5 km of trails (L. H. Emmons and E. Gasis, unpublished data). Our fruit collection at Danum overlapped both a masting and a nonmasting year. At Cocha Cashu, fruiting trees were monitored biweekly with fruit-traps and a general collection of the flora (Janson, 1983; Foster and Janson, 1985; Janson and Emmons, 1990). The data base collected at Cocha Cashu is the most comprehensive and includes information on 256 fruit species dispersed by vertebrates (Foster and Janson, 1985). Separating this sample into canopy

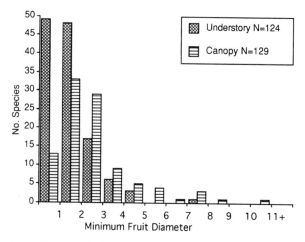

Figure 3   Minimum diameter of whole fruits from canopy (≥10 m) and understory (≤10 m) at Cocha Cashu, Peru.

versus understory species (Fig. 3) shows that understory fruits are generally much smaller. To make the data more comparable, I use only the canopy portion of the Cocha Cashu sample, because the Old World data sets include few understory species, and the collection method may be biased against tiny fruits. Each fruit list is only a partial sample of the species in the habitat, but the most important vertebrate food species in the study areas are all likely to be included. I include only fruit potentially eaten by vertebrates and exclude wind-dispersed and exploding capsule seeds and inedible legumes. Both absolute and relative fruit species numbers in a size class are shown (Fig. 4), but from an animal's viewpoint, the former are more relevant.

The canopy fruit size spectrum differs in each of the three study sites (Fig. 4). Three differences are evident: (1) Cocha Cashu has a large number of fruit species in the 2- to 3-cm size class (38 species), and Danum Valley few (11 species), with Makokou intermediate (18 species); (2) Makokou has more very large fruits (15 species ≥6 cm versus 7 species in each other site); and (3) Danum, with no very large fruits over 8 cm, and very few 2- to 3-cm fruits, is dominated by a single size class, 1–2 cm. A comparison of the mass of individual fruits (Fig. 5) also shows the shift toward large fruits in the African forest. Nevertheless, on all continents, the modal fruit size class is the same (1–2 cm). I believe that the differences in fruit size spectra at the three study sites are linked to faunal differences.

On the basis of known bird and mammal consumption of the fruit in our sample at Makokou (Gautier-Hion *et al.,* 1985), we concluded that in this forest, monkeys dispersed the most species (59), followed by birds (32) and

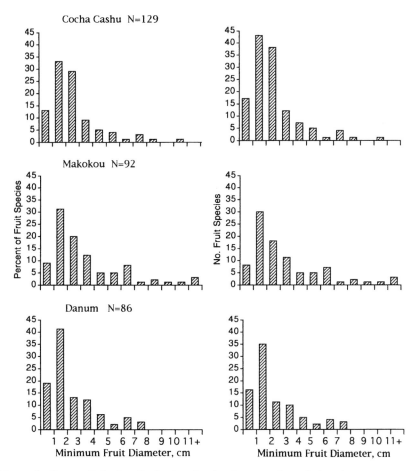

Figure 4 Canopy fruit size distribution for the three continental sites, in percentage of sample (left) and total species (right).

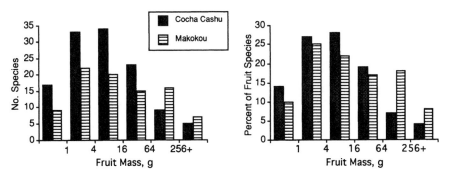

Figure 5 Canopy fruit species mass distributions for Cocha Cashu and Makokou study areas, for total species (left) and percentage of sample (right).

elephants (30), while other taxa (rodents, artiodactyla) tended to be predators rather than dispersers. There was extensive overlap in fruit species eaten by consumer taxa at Makokou, especially by birds and monkeys (42% of the total species eaten by both taxa were shared). Virtually all the large-sized fruit species at Makokou are dispersed by elephants. I attribute the paucity of fruits of this size at the other sites to lack of elephant, or equivalent, dispersers. Asian elephants occur at Danum Valley, where there is a small, regionally restricted population (perhaps feral domestics), but on the rare occasions when they entered my study plot, their droppings contained no seeds, suggesting that this species eats far fewer fruit than African elephants.

The New World tropics has a frugivorous bat fauna unparalleled in the Old World (Section II, B). If we isolate the sample of 20 bat-dispersed fruits at Cocha Cashu (from Foster *et al.*, 1986) that occur in the same fruit data base used in Fig. 4, they show a different distribution from the overall pattern, with 50% of the species in the 2- to 3-cm class and 25% in the 3- to 4-cm size class (Fig. 6). Although I cannot identify all bat-dispersed fruits in the whole fruit data base, it is likely that the large numbers of 2- to 3-cm fruits at Cocha Cashu, compared to Old World sites, represents increased numbers of bat-dispersed species. Unfortunately, we know little about which fruit species are bat-dispersed, or the characters of fruits eaten by Megachiroptera, in the two Old World sites.

Finally, my own observations in Sabah during 18 months of fieldwork suggested that large birds (e.g., barbets, pigeons, broadbills, and hornbills) fed together with mammals on the majority of the vertebrate-dispersed species. Hornbills ate fruits up to 8.5 cm in diameter (*Diospyros macrophyllum*, pecked open while attached). Two species generally eaten only by mammals were small, 1.5 and 2.0 cm in diameter (*Diallium indum* and *Alangium eben-*

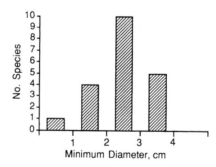

Figure 6   Size distribution of 20 species of bat-dispersed fruits at the Cocha Cashu site (data from Foster *et al.*, 1986) from the same sample as Fig. 3.

aceum). My impression was that in comparison to Africa and the neotropics, bird dispersal has greater importance on Borneo. Mammals there are not less frugivorous, but large frugivorous canopy birds use most available species. For example, in Neotropical forests, there are many species of large, bat-dispersed figs. These are green when ripe and are also eaten by a few monkeys. In Sabah, canopy figs are all brightly colored, small, and eaten by both birds and all larger mammals. I hypothesize that this accounts for the very high percentage of 1- to 2-cm fruits at Danum, which are eaten by both birds and mammals. The civets eat the same fruits as birds and monkeys and are excellent dispersers because they gulp down masses of large seeds and then range widely, defecating them about.

Thus, although rain forest fruits are not tightly coevolved with disperser species (Gautier-Hion *et al.*, 1985), except perhaps African elephants, intercontinental comparisons of canopy fruit arrays show differences that seem linked to major taxonomic differences in the faunas.

D. Canopy Structure

The most apparent difference between the dipterocarp forests of Sabah and the forests of either Peru or Gabon is in physical structure. Each geographic region has a wide range of forest structures, depending on age, soil, climate, drainage, disturbance, and species pools. Nevertheless, mature, terra firme, evergreen equatorial forest on well-drained, medium-fertility soil has a fairly typical regional physiognomy. Both African and Amazonian forests have closed canopy structure formed by a dense tree population of quite uniform height (Figs. 7A and 7B). In these forests, leafless liana stems climb to the canopy and, once there, branch into foliage and clamber above the host tree foliage and from tree to tree, which further closes the canopy. This allows canopy mammals to travel horizontally with relatively few detours. Connecting lianas make canopy travel easier for species that cannot jump (Chapter 14).

In the magnificent dipterocarp forests at Danum, large numbers of tall (50–60 m) emergents are widely separated from other tree crowns (Fig. 7C), so that there is no continuous upper-level canopy. Lianas densely wrap some emergent trunks, but rarely occupy distal canopy branches or connect canopy crowns. Instead, the true light-screening canopy is often far below, at 10–15 m, where lianas can connect a closed subcanopy of smaller trees. Only gliding or volant mammals can cross from emergent to emergent, and they regularly do so; other species generally travel in the subcanopy layer. The macaque monkeys and palm civets often travel on the ground. I have seen langurs doing this, but langurs and orangutans usually travel in the low subcanopy, and from there climb up into food trees. Only gibbons often move at a higher level, by flinging themselves with breath-

Figure 7   Aerial photographs of forest canopy. (A) Africa, Zaire, similar to Gabon (photo by Frans Lanting); (B) South America, Pando, Bolivia (Amazonia); (C) Danum Valley study area, Sabah (dipterocarp forest). The paler crowns are emergents in brighter light, the darker crowns are the subcanopy many meters below. Note the uniform, closed canopies in A and B, and the tall, exposed emergents, large gaps, and a few vine-wrapped trunks in C.

Figure 7—*Continued*

taking skill across large gaps. All of the larger monkeys, when alarmed by terrestrial threats, drop downward into the denser subcanopy to disappear, a behavior I have not seen in Africa nor in the neotropics.

III. Canopy Structure and Animal Locomotion

In an earlier publication, the prevalence of gliding mammals in Asia was attributed to a lack of connecting lianas (Emmons and Gentry, 1983). I modify this conclusion to hypothesize that lack of a high, closed-canopy connection between tree crowns is a factor promoting evolution of gliding taxa, but that it is not lack of lianas (which are often abundant at low levels) but rather the height distribution of connections of both tree and liana canopies that is critical.

The other morphological feature that shows regional concentration in canopy animals is the prehensile tail, which is predominant in Neotropical species (Emmons and Gentry, 1983). It is still not clear which, if any, specific continental aspects of the vegetation structure may have influenced evolution of prehensile tails in South America but not in Africa. A characteristic of prehensile tails is that their musculature makes them heavy (almost 6% of the body mass in howler monkey, or more than a hindleg; Grand, 1978). This extra mass should hinder arboreal leaping compared to the same body without it. This may be why gibbons have no tails. The only short-tailed and non-prehensile-tailed larger monkeys in the neotropics are the pithecines (sakis and uakaris), which are also the most accomplished nonbrachiating leapers. My impression is that the tailless gibbons leap (from brachiation) much farther than do the equivalent, but prehensile-tailed, American spider monkeys. Prehensile tails would be most useful in canopies connected enough so that the tail can be used for foraging and to bridge gaps, but where travel usually does not necessitate frequent, ballistic leaping of wide spaces. Most prehensile-tailed mammals (other than primates and kinkajous) are not leapers (tamanduas, pangolins, binturongs, prehensile-tailed porcupines, marsupials). Prehensile-tailed mammals therefore may use resources that either do not require much travel, and/or they inhabit well-connected canopy environments, or they travel on the ground.

On a microhabitat scale, resources and connectedness may often be related: where vegetation is dense and viny, invertebrates, leaves, and some fruits can be concentrated, and such vegetation offers shelter from the elements and refuge from birds of prey. Lianas also form large festoons that increase both the volume of the canopy space that is occupied by plants and the surface area of substrate for insect-foraging. Worldwide, almost all smaller arboreal mammals, including prosimians, tamarins, squirrels, tree shrews, opossums, fruit bats, and rodents, are aggregated in vinier sections of habitat (Charles-Dominique *et al.,* 1981; Emmons and Feer, 1990; L. H. Emmons, personal observation; Chapter 9). A number of larger mammals such as titi monkeys and prehensile-tailed porcupines also favor vine tangles (Emmons and Feer, 1990). Canopy vines and lianas therefore increase faunal abundance and richness.

IV. Conclusions

The forest canopy is home for a major fraction of all mammals. In most temperate and subtropical forests, the canopy is a temporary resource base for many species that during nonproductive seasons can sleep, hoard, forage terrestrially (sometimes on fallen canopy products such as mast), or migrate to forests in other production regimes. In the tropics, however, the tremendous species richness and year-round production of edible materials in the rain forest canopy allow assemblages of about 40 arboreal nf mammal species to coexist.

The rain forest canopy is the primary production layer of the habitat (Chapter 3), and nf mammals for the most part have radiated into the canopy as users of its primary products—fruit and leaves. Most also capture some invertebrates, but only flying vertebrates (bats, birds) seem highly successful at this in terms of fraction of diet and species numbers. Four-footed pursuit of animal prey, much of which can drop, fly, or sting to avoid consumption, may be more difficult within the thin network of aerial routes than on the ground, where most faunivorous mammals live (Dickman, 1995). The most insectivorous canopy nf mammals are small species that concentrate their activity in the densest arboreal substrate, vine tangles (e.g., small galagos, tamarins, tree shrews, mice, mouse opossums, pygmy anteaters; L. H. Emmons, personal observation). Assemblages of rain forest canopy nf mammals are in general poor in small species (Fig. 1, Dubost, 1987; Emmons *et al.*, 1983). I believe this reflects the greater success of frugivory/folivory than of insectivory as a canopy life-style.

When whole faunas are compared (Dubost, 1987), there is strong resemblance or convergence between canopy mammal faunas on different continents, in numbers, sizes, and diets of species. This implies that interactions between species regulate some community properties. The convergences seen in mammal assemblages are in strong contrast to assemblages of snakes described by Cadle and Green (1993), where communities within similar ecosystems in different regions lack similarity and appear to be the result of chance historical grouping of lineages with stable ecological characters (e.g., size). I conjecture that competitive interactions may be stronger among nf mammals than among snakes and many other animal taxa because: (1) their endothermy and large body size require large amounts of food and (2) the broad ("generalized") feeding habits of many mammals causes large overlap in resource use, inducing direct competition. For example, most of the arboreal nf mammals feed on fruits, seeds, leaves, and invertebrates. A phenomenon relevant to this idea is that in rich assemblages worldwide, the maximum number of sympatric species in a taxon is 10–15, usually 12 or 13 (e.g., Table II; Gautier-Hion *et al.*, 1980). This may reflect the maximum size spread for a given morphological type, coupled with the breadth of resource use of a typical mammal (maximum species packing). When there are more than a dozen sympatric species (such as the 17 primates at Makokou), they often include a separate ecological set (in that case, nocturnal prosimians) or habitat. Obviously, most localities have fewer than the maximum species number. For most bat assemblages, the ecological data are still too poor to understand their community characteristics (year-round resources, exact position, and manner of food item acquisition).

Because most of the fruit biomass of forests is produced in the canopy, arboreal animals have first choice of that production. If animal dispersal has been a force in the evolution of fruit characters, then canopy animals are the primary receptors for which fruit qualities are directed, especially those that both swallow seeds and travel broadly: birds, bats, and monkeys. Nonetheless, a large set of African fruits seems to have characters directed toward elephant dispersal. The wide overlap in fruit species choice between distant taxa (Gautier-Hion *et al.*, 1985) makes it difficult to assess the influence of individual taxa on the evolution of fruit characters from work at one location. The intercontinental comparisons given here, however, show differences in the fruit-size spectra that I conjecture are due to the faunal differences at the same sites. This hypothesis needs testing by comparing complete data sets of the fruits and which animals disperse each fruit. Unfortunately, complete mammal faunas have been documented for few tropical localities, and the feeding ecology of most species in an assemblage is known even more rarely. Just as uncommon are data on the properties of a whole array of local fruits, other than just a subset eaten by primates. There seem to be virtually no data on feeding habits of different species within large communities of megachiropteran fruit bats in equatorial forests of Africa, Southeast Asia, or New Guinea. New methods for observing the high canopy (Chapter 1), where almost all the megachiropterans are active (C. Francis, personal communication), should allow detailed studies of the foraging ecology of these and all other canopy bats. I believe that bats will be found to be far more important actors in the ecology of the canopy than is currently recognized, perhaps nearly equivalent to birds.

V. Summary

Most obligate forest canopy mammals are restricted to evergreen tropical forests. The chief exceptions are in Australia, where subtropical and temperate forests of evergreen eucalypts support specialized folivores and exudate-feeders. Globally, nearly 200 genera of nonflying mammals feed arboreally. Few species have narrow diets of only animals, fruits, or leaves; almost all canopy mammals combine all of these in their diets. Several classes of locomotory adaptation confer different levels of mobility within the canopy and, consequently, access to resources. The rain forest canopy mammal faunas of three continental sites and two large island sites that I compared show both notable convergences in numbers of species in different roles and some differences. These differences are largely due to phylogenetic differences between continental faunas, such as the dominance of echo-locating frugivorous bats in the neotropics or of folivorous marsupials in the Australian region. Some features of the size spectrum of canopy fruit at the same sites appear related to differences in the taxonomic composition of the frugivore faunas.

Acknowledgments

The fruit data from Makokou were collected as a joint effort by all the authors of Gautier-Hion *et al.* (1985), and I am indebted to Annie Gautier-Hion for sending me the original data for use here. I am likewise indebted to Elaine Gasis, who was responsible for collecting the fruit data at Danum Valley, and to Charles Janson for the use of his fruit data base for Cocha Cashu. Robin Foster and Ted Grand made helpful comments on the work in progress. To Frans Lanting, I owe thanks for his superb photo. Many persons and organizations supported portions of the research described here, but I particularly wish to thank The Smithsonian Institution, The National Geographic Society, The Douroucouli Foundation, and Yayasan Sabah.

References

Anonymous (1993). "Danum Valley Conservation Area: A Checklist of Vertebrates." Innoprise Corp, Kota Kinabalu, Malaysia.

Bradbury, J. W. (1977). Lek mating behavior in the hammer-headed bat. *Z. Tierpsychol.* 45, 225–255.

Burt, W. H., and Grossenheider, R. P. (1976). "A Field Guide to the Mammals." Houghton Mifflin, Boston.

Cadle, J. E., and Greene, H. W. (1993). Phylogenetic patterns, biogeography, and the ecological structure of Neotropical snake assemblages. *In* "Species Diversity in Ecological Communities: Historical and Geographical Perspectives" (R. E. Rickleffs and D. Schluter, eds.), pp. 281–293. Univ. of Chicago Press, Chicago.

Charles-Dominique, P., Atramentowicz, M., Charles-Dominique, M., Gérard, H., Hladik, A., Hladik, C. M., and Prévost, M. F. (1981). Les mamifères frugivores arboricoles nocturnes d'une forêt guyanaise: inter-relations plantes–animaux. *Rev. Ecol.* 35, 341–435.

Dickman, C. R., ed. (1995). "Insect-Eating Mammals: Evolution and Adaptations for Insectivory." Cambridge Univ. Press, Cambridge, UK. (In press.)

Dubost, G. (1987). Une analyse écologique de deux faunes de mamifères forestiers tropicaux. *Mammalia* 51, 415–436.

Eisenberg, J. F. (1978). The evolution of arboreal herbivores in the class Mammalia. *In* "The Ecology of Arboreal Folivores" (G. G. Montgomery, ed.), pp. 135–152. Smithsonian Institution Press, Washington, DC.

Emmons, L. H. (1980). Ecology and resource partitioning among nine species of African rain forest squirrels. *Ecol. Monogr.* 50, 31–54.

Emmons, L. H. (1995). Mammalian arboreal insectivores. *In* "Insect-Eating Mammals: Evolution and Adaptations for Insectivory" (C. R. Dickman, ed.). Cambridge Univ. Press, Cambridge, UK. (In press.)

Emmons, L. H., and Feer, F. (1990). "Neotropical Rainforest Mammals: A Field Guide." Univ. of Chicago Press, Chicago.

Emmons, L. H., and Gentry, A. H. (1983). Tropical forest structure and the distribution of gliding and prehensile-tailed vertebrates. *Am. Nat.* 121, 513–524.

Emmons, L. H., Gautier-Hion, A., and Dubost, G. (1983). Community structure of the frugivorous-folivorous forest mammals of Gabon. *J. Zool.* 199, 209–222.

Flannery, T. F., and Seri, L. (1990). The mammals of southern West Sepik Province, Papua New Guinea: Their distribution, abundance, human use and zoogeography. *Rec. Aust. Nat. Mus.* 42, 173–208.

Foster, R. B., Arce, B. J., and Wachter, T. S. (1986). Dispersal and the sequential plant communities in Amazonian Peru floodplain. *In* "Frugivores and Seed Dispersal" (A. Estrada and T. H. Fleming, eds.), pp. 357–370. Dr. W. Junk Publ., Dordrecht, The Netherlands.

Foster, S. A., and Janson, C. H. (1985). The relationship between seed size and establishment conditions in tropical woody plants. *Ecology* 66, 773–780.

Gautier-Hion, A. (1980). Seasonal variations of diet related to species and sex in a community of *Cercopithecus* monkeys. *J. Anim. Ecol.* 49, 237–269.

Gautier-Hion, A., Emmons, L. H., and Dubost, G. (1980). A comparison of the diets of three major groups of primary consumers of Gabon (primates, squirrels and ruminants). *Oecologia* 45, 182–189.

Gautier-Hion, A., Gautier, J. P., and Quris, R. (1981). Forest structure and fruit availability as complementary factors influencing habitat use by a troop of monkeys (*Cercopithecus cephus*). *Rev. Ecol.* 35, 511–536.

Gautier-Hion, A., Duplantier, J.-M., Quris, R., Feer, F., Sourd, C., Decoux, J.-P., Dubost, G., Emmons, L., Erard, C., Hecketsweiler, P., Moungazi, A., Roussilhon, C., and Thiollay, J.-M. (1985). Fruit characters as a basis of fruit choice and seed dispersal in a tropical forest vertebrate community. *Oecologia* 65, 324–337.

Grand, T. I. (1978). Adaptations of tissue and limb segments to facilitate moving and feeding in arboreal folivores. *In* "The Ecology of Arboreal Folivores" (G. G. Montgomery, ed.), pp. 231–241. Smithsonian Institution Press, Washington, DC.

Grand, T. I. (1984). Motion economy within the canopy: Four strategies for mobility. *In* "Adaptations for Foraging in Nonhuman Primates" (P. S. Rodman and J. G. H. Cant, eds.), pp. 54–72. Columbia Univ. Press, New York.

Hildewein, G. (1972). Métabolisme énergétique de quelques mamifères et oiseaux de la forêt équatoriale. II. Résultats expérimentaux et discussion. *Arch. Sci. Physiol.* 26, 387–340.

Hladik, A. (1978). Phenology of leaf production in rain forest of Gabon: Distribution and composition of food for folivores. *In* "The Ecology of Arboreal Folivores" (G. G. Montgomery, ed.), pp. 51–71. Smithsonian Institution Press, Washington, DC.

Janson, C. H. (1983). Adaptation of fruit morphology to dispersal agents in a Neotropical forest. *Science* 219, 187–189.

Janson, C. H., and Emmons, L. H. (1990). Ecological structure of the non-flying mammal community at Cocha Cashu Biological Station, Manu National Park, Peru. *In* "Four Neotropical Rainforests" (A. H. Gentry, ed.), pp. 314–338. Yale Univ. Press, New Haven, CT.

Kira, T., and Yoda, K. (1989). Vertical stratification in microclimate. *In* "Tropical Rain Forest Ecosystems" (H. Lieth and M. J. A. Werger, eds.), pp. 55–71. Elsevier, Amsterdam.

McNab, B. K. (1978). Energetics of arboreal folivores: Physiological problems and ecological consequences of feeding on an ubiquitous food supply. *In* "The Ecology of Arboreal Folivores" (G. G. Montgomery, ed.), pp. 153–162. Smithsonian Institution Press, Washington, DC.

Medway, Lord (1978). "The Wild Mammals of Malaya (Peninsular Malaysia) and Singapore." Oxford Univ. Press, Kuala Lumpur.

Milton, K. (1982). Dietary quality and demographic regulation in a howler monkey population. *In* "The Ecology of a Tropical Forest: Seasonal Rhythms and Long-Term Changes" (E. G. Leigh, Jr., A. S. Rand, and D. M. Windsor, eds.), pp. 273–289. Univ. of Chicago Press, Chicago.

Mitchell, P. C. (1916). Further observations on the intestinal tract of mammals. *Proc. Zool. Soc. London*, pp. 183–251.

Nowak, R. M. (1991). "Walker's Mammals of the World," 5th ed. Johns Hopkins Univ. Press, Baltimore, MD.

Pacheco, V., Patterson, B. D., Patton, J. L., Emmons, L. H., Solari, S., and Ascorra, C. F. (1993). List of mammal species known to occur in Manu Biosphere Reserve, Peru. *Publ. Mus. Hist. Nat. Lima, Ser. A: Zool.* 44, 1–12.

Parra, R. (1978). Comparison of foregut and hindgut fermentation in herbivores. *In* "The Ecology of Arboreal Folivores" (G. G. Montgomery, ed.), pp. 205–229. Smithsonian Institution Press, Washington, DC.

Payne, J., Francis, C. M., and Phillipps, K. (1985). "A Field Guide to the Mammals of Borneo." The Sabah Society, Kota Kinabalu, Malaysia.

Rodman, P. S., and Cant, J. G. H., eds. (1984). "Adaptations for Foraging in Nonhuman Primates." Columbia Univ. Press, New York.

Skalli, A., and Dubost, G. (1986). Comparison de deux faunes de mamifères tropicaux par l'analyse des charactères de leur mode de vie. *Cah. Anal. Données* 11, 403–440.

Smuts, B. B., Cheney, D. L., Seyfarth, R. M., Wrangham, R. W., and Struhsaker, T. T., eds. (1987). "Primate Societies." Univ. of Chicago Press, Chicago.

Strahan, R. (1983). "The Complete Book of Australian Mammals." Angus & Robertson, London.

Symington, M. (1988). Demography, ranging patterns, and activity budgets of black spider monkeys (*Ateles paniscus chamek*) in the Manu National Park, Peru. *Am. J. Primatol.* 15, 45–67.

Terborgh, J. (1986). Community aspects of frugivory in tropical forests. *In* "Frugivores and Seed Dispersal" (A. Estrada and T. H. Fleming, eds.), pp. 371–384. Dr. W. Junk Publ., Dordrecht, The Netherlands.

Thomas, D. W. (1984). Fruit intake and energy budgets of frugivorous bats. *Physiol. Zool.* 57, 457–467.

Wilson, D. E., and Reeder, D. M. (1993). "Mammal Species of the World," 2nd ed. Smithsonian Institution Press, Washington, DC.

11

Vascular Epiphytes

David H. Benzing

I. Introduction

Forest canopies support extensive flora that include well over 20,000 species or about 10% of all the tracheophytes (Kress, 1986, but see Atwood, 1986, about Orchidaceae). Vascular epiphytes differ greatly by structure, function, and fidelity to canopy versus terrestrial substrates. At one extreme are the accidental epiphytes, plants that normally root in the ground but occasionally disperse to nearby trees (phorophytes). More consistent bark users, the true epiphytes, often occur nowhere else. Ecophysiologists recognize the varied and sometimes extraordinary structures and mechanisms that protect epiphytes against drought. No less noteworthy and certainly more unique to this vegetation are the adaptations for acquiring nutrient ions from sources other than earth soil.

Terrestrial and epiphytic bromeliads have proven exceptionally useful to study light relations and aspects of carbon and water economy associated with Crassulacean acid metabolism (CAM). Arboreal flora can also provide insights on broader phenomena in tropical forests, for example, why invertebrates are so diverse at certain locations and how intimate plant–animal mutualisms operate (e.g., myrmecotrophy). These plants supply extensive and incompletely inventoried resources for biota well beyond the usual pollinators, seed dispersers, and herbivores. Epiphytes influence processes in ecosystems such as mineral cycling and nutrient storage (Chapter 20). Arboreal flora have already commanded considerable attention from many

Copyright © 1995 by Academic Press, Inc.
All rights of reproduction in any form reserved.

kinds of biologists, but its potential to inform us about an even greater variety of subjects is underexploited.

II. General Characteristics, Paleobotanical History, and Geographic Range

Some of the requisites for regular occurrence on bark and associated aerial substrates are obvious (e.g., holdfast root systems and wind- or animal-carried propagules). Others, perhaps no less decisive for epiphytism and indeed sometimes obliging it, remain obscure (Table I). Most epiphytes can flourish in potting media, but they die on the ground in the forest even if they are dislodged as adults (Matelson *et al.*, 1993). Facultative forms succeed interchangeably as terrestrials or epiphytes, at least in part because rooting media where they occur are similarly moist or arid. Status as an obligate or facultative epiphyte often shifts within species that range across substantial humidity gradients (e.g., many Bromeliaceae).

Vascular epiphytism is largely restricted to the low latitudes and, within that region, arboreal flora reach greatest diversity and abundance at low- to midmontane elevations (Gentry and Dodson, 1987b). A few species, mostly exceptional members of fundamentally tropical clades, range into mild north and south temperate zones. *Epidendrum rigidum, Polypodium polypodioides,* and *Tillandsia usneoides* are the cold-hardiest of the North American epiphytes. Pteridophytes occur in higher latitudes along the Pacific coast. The most extensively colonized temperate forests are those of southeastern Australia, New Zealand, and Chile, where assortments of bromeliads, cacti, lily relatives, pteridophytes, and others grow protected from frost by nearby oceans. Epiphytes in nontropical families (e.g., Cornaceae, Ranunculaceae) at these locations show that cool, moist, and not just tropical climates foster transitions of flora from earth soil to tree crowns.

Most extant epiphytes are angiosperms; monocotyledons far outnumber the dicotyledons (Table II); leptosporangiate ferns constitute most of the remaining species. Fossils indicate that bark and related rooting media have supported vascular flora for more than 250 million years. For example, some Paleozoic forests supported arboreal cryptogams such as the coenopterid *Botryopteris forensis,* which anchored in the mantle of adventitious roots covering stems of the Pennsylvanian tree fern *Psaronius* (Rothwell, 1991). Many filmy ferns and *Tmesipteris* use trunks of extant tree ferns in much the same way (Fig. 1).

Even if epiphytes existed in proportions comparable to those present in some modern forests (up to 35% of the total flora; Gentry and Dodson,

Table I   Characteristics of Vascular Epiphytes[a]

1. Reproduction	Exclusively zoophilous, flowers
A. Pollination	tend to be showy, pollinators highly mobile (Benzing, 1990, Chapter 5)
B. Breeding systems	Little studied, although many orchids appear to be allogamous
C. Population structure	Little studied
D. Seed dispersal	Most families endozoochorous, most species anemochorous (because of the dominance of Orchidaceae)
E. Life history	Almost all iteroparous, long-lived perennials
2. Vegetative	
A. Foliage	Usually evergreen, often succulent, and xeromorphic generally
B. Habit	Woody (wet forests) to herbaceous (wet and dry forest)
C. Shoot architecture	Various
D. Roots	Adventitious, specialized for holdfast, often reduced
E. Special features[a]	Impounding shoots (e.g., Bromeliaceae) and root masses (e.g., ferns), velamentous roots and absorptive foliar trichomes to prolong contact with precipitation and canopy washes, often lack capacity to grow in earth soil
3. Mineral nutrition	
A. Mycorrhizas	Possibly significant in Orchidaceae and Ericaceae, probably relatively unimportant elsewhere compared to terrestrial flora
B. Myrmecotrophy[a]	Nearly exclusive to epiphytes
C. Carnivory	Underrepresented in aboreal flora
D. Saprotrophy[a]	Phytotelm and trash-basket types
E. Special features[a]	Tolerance for low pH (wet forest), effective nutrient scavengers (dry forests), frequent reliance on organic substrates for nutrient ions
4. Photosynthesis/water balance	
A. Photosynthetic pathways	CAM overrepresented, no typical C_4 types, much interesting detail probably remains underdescribed
B. Water economy	Often very high
C. Moisture requirements	Various
D. Other[a]	Much flexibility, e.g., facultative CAM, CAM–C_4 intermediates

[a] These characteristics distinguish arboreal from terrestrial flora more than the others.

1987a), an equivalent array of adaptive types was certainly lacking (Table III). Magnoliophyta accounts for most of the variety featured by modern arboreal flora, just as this division provides most of the substrates. Angio-

Table II   Taxonomic Distribution of Vascular Epiphytes[a]

Major group	Taxonomic category	Number of taxa containing epiphytes in each category	Percentage of taxa containing epiphytes in each category
All vascular plants	Classes	6	75
	Orders	44	45
	Families	84	19
	Genera	876	7
	Species	23,456	10
Ferns and allies	Classes	2	67
	Orders	5	50
	Families	13	34
	Genera	92	39
	Species	2593	29
Gymnosperms	Classes	2	67
	Orders	2	33
	Families	2	13
	Genera	2	3
	Species	5	<1
Angiosperms (dicots)	Subclasses	6	100
	Orders	28	44
	Families	52	16
	Genera	262	3
	Species	4251	3
Angiosperms (monocots)	Subclasses	4	80
	Orders	9	47
	Families	17	26
	Genera	520	21
	Species	16,608	31

[a] Following Kress (1986).

Table III   Functional/Ecological Categories of Vascular Epiphytes[a]

Scheme I. Categories based on relationship to the host

A. Autotrophs: plants supported mechanically by woody vegetation; no nutrients extracted from host vasculature
  1. Accidental
  2. Facultative
  3. Hemiepiphytic
    a. Primary
      i. Strangling
      ii. Nonstrangling
    b. Secondary
  4. "Truly" epiphytic (the "holoepiphytes" of A. F. W. Schimper)
B. Heterotrophs: plants subsisting on xylem contents and sometimes receiving a substantial part of their carbon supply from a host
  1. Parasitic (mistletoes)

Table III—*Continued*

Scheme II. Categories based on growth habit
A. Trees
B. Shrubs
C. Suffrutescent to herbaceous forms
  1. Tuberous
    a. Storage: woody and herbaceous
    b. Myrmecophytic: mostly herbaceous
  2. Broadly creeping: woody or herbaceous
  3. Narrowly creeping: mostly herbaceous
  4. Rosulate: herbaceous
  5. Root/leaf tangle: herbaceous
  6. Trash-basket: herbaceous

Scheme III. Categories based on water balance mechanisms
A. Poikilohydrous: many bryophytes and some lower vascular plants, mostly ferns and very few, angiosperms
B. Homoiohydrous
  1. Hygrophytes
  2. Mesophytes
  3. Xerophytes
    a. Drought-endurers
    b. Drought-avoiders
  4. Impounders

Scheme IV. Categories based on light requirements
A. Exposure types: largely restricted to sites in full or nearly full sun
B. Sun types: tolerant of medium shade
C. Shade-tolerant types: tolerant of deep shade

Scheme V. Categories based on substrates required
A. Relatively independent of rooting medium (obtain moisture and nutritive ions primarily from other sources)
  1. Mist and atmospheric forms with minimal attachment to bark
  2. Twig/bark inhabitants
  3. Species that create substitute soils (impounders) or attract ant colonies (ant-house epiphytes)
B. Tending to utilize a specific type of rooting medium for moisture and nutritive ions
  1. Humus-adapted
    a. Generalist types that root in shallow humus mats
    b. Deep humus types that penetrate knotholes or rotting wood
    c. Ant-nest garden and plant-catchment inhabitants (e.g., *Platycerium* nest endemics, *Utricularia humboltii* in tanks of *Brocchinia tatei,* numerous species comprising ant-nest garden flora)
  2. Mistletoes

[a] From Benzing (1990).

sperm forests emerged during the upper Cretaceous and families accounting for most of the canopy-based anthophytes (e.g., Araceae, Bromeliaceae, Orchidaceae) exhibit relatively advanced evolutionary status.

III. Evolution and Adaptive Types

Epiphytism has evolved repeatedly within Tracheophyta; arboreal taxa exist in more than 80 families (Kress, 1986). Some of the subsequent radiations were almost certainly magnified by capacities for epiphytism. Proliferations within Orchidaceae, the family responsible for over half of all the vascular epiphytes, are truly exceptional. Numerous large complexes of closely related, often co-occurring, mostly canopy-based species at least superficially exhibit indistinguishable substrate requirements (e.g., *Dendrobium, Epidendrum, Lepenthes, Pleurothallis*). In addition to Orchidaceae, Araceae and Bromeliaceae are exceptionally well disposed for canopy life. Comparable life histories and ecological propensities indicate similar arrays in these taxa (e.g., *Anthurium*, Araceae; *Tillandsia*, Bromeliaceae; and *Drymonia*, Gesneriaceae). As a group, the dicots are less diverse in canopy habitats. Most of the exceptions (e.g., Cactaceae, Gesneriaceae) share certain characteristics with their monocotyledonous counterparts: herbaceous habit, modular architecture, adventitious root systems, wind- or bird-carried seeds, tendencies for CAM, succulence beyond that required to store abundant malic acid, and other means to tolerate severe and unpredictable drought (Table I).

Several fundamentally woody angiospermous families deviate markedly from this stereotype and yet have colonized tree crowns (e.g., Ericaceae, Melastomataceae) in habitats where plentiful moisture permits extensive cambial activity. Tolerances for acidic, organic rooting media to match conditions prevailing in most suspended soils, perhaps promoted in some cases by mycorrhizal fungi (e.g., Ericaceae), may partially explain these seemingly improbable radiations. Few families are found in the driest forests where aridity combined with fragmented and temporary substrates severely challenge plant success (Benzing, 1981). Major exceptions are Bromeliaceae and Orchidaceae, families that possess suites of vegetative and reproductive characteristics, and scavenge and use scarce resources with unusual effectiveness. Generally massive and relatively immobile seeds, woody habits, costly wind pollination, and associated needs for high population densities probably helped maintain the almost exclusively terrestrial character of the gymnosperms. Certain otherwise successful (speciose) angiosperm families (e.g., Fabaceae, Poaceae) remain disproportionately earth-bound for reasons related to these and other phylogenetic constraints.

Ferns, a group second only to the orchids for occurrence (~35% of the species) in canopy habitats, have probably occupied this space in much the same way since the late Paleozoic. Tolerances for desiccation, including poikilohydry (ability to survive near-complete dehydration) in extreme cases, exceed that of most seed plants. Low rates of respiration, coupled with modest light compensation and saturation intensities, allow carbon gains in shade, where most of this free-sporing vegetation occurs (Raven, 1985). Profuse development of spongy masses of adventitious roots also provides benefits, as does the production of small propagules. The importance of dispersibility to the arboreal pteridophytes is probably illustrated by the much larger number of epiphytic homosporous (*Lycopodium sensu lato*) compared to heterosporous (*Selaginella*) lycopods.

IV. Categories of Vascular Epiphytes

Epiphytes are variously categorized according to the kinds of substrates used, mechanisms of water and nutrient balance, plant architecture, and other characteristics (Table III). Arrangements that emphasize how arboreal status is achieved reveal more about the peculiarities of epiphytism than do modes of pollination and seed dispersal. Discrimination based on the timing of mechanical dependence on a phorophyte during the life cycle also highlights important mechanisms.

Hemiepiphytes spend either early or later life wholly dependent on canopy media; at other growth stages, they exploit both earth and suspended soils. Primary hemiepiphytes (e.g., many aroids, Fig. 2) ascend trees as seedlings, losing contact with the ground later as older portions of vining shoots die, leaving only the younger, strictly arboreal roots functional. Secondary hemiepiphytes begin growth on bark and become anchored in the ground after roots reach the forest floor (Fig. 3). Robust, woody species sometimes kill supports by girdling and shading. Mistletoes, which are also arboreal plants in the strict sense, are neither close relatives of the true epiphytes nor do they use phorophytes in the same ways (Benzing, 1991, Chapter 6).

Epiphytes include a wide range of growth habits, from woody forms (including small trees) to herbs of various shapes, but mostly smaller sizes (Figs. 4, 8, and 9). Shoots tend to be compact; many species produce structures that impound debris (Fig. 5). Others entice colonies of plant-feeding ants by means of domatia fashioned from foliage or hollow stems (Figs. 14 and 15). Epiphytes native to dry forests or arid microsites at more humid locations (Fig. 6) deploy specialized organs or tissues (Fig. 18) suited to use transitory, flowing sources instead of more permanent, static supplies of moisture and essential ions. Stress tolerances reach exceptional levels in these epiphytes, which are the most fully air-exposed of all.

Photosynthetic pathways, which are usually reliable indicators of moisture supplies in native habitats, vary greatly among the epiphytes. CAM is widespread, present in all of its known versions, variable within individual plants and populations, and still poorly characterized for most arboreal flora (Benzing, 1990, Chapter 2). Patterns of gas exchange, diurnal

Figures 1–9   Types of vascular epiphytes distinguished by growth habits, types of substrates, and/or sources of nutrients and moisture. (1) *Tmesipteris tanensis,* growing in the mantle of adventitious roots produced by a tree fern in New Zealand: X0.20. (2) *Philodendron* sp., a primary hemiepiphyte; X0.10. Note that the oldest (lower) part of the stem is nearly rotted away. (3) *Ficus aurea,* a secondary hemiepiphyte growing on *Taxodium distichium* in South Florida. (4) *Catopsis berteroniana,* a reputed carnivorous phytotelm bromeliad in south Florida; X0.15.

fluctuations in citric and malic acids, stable carbon isotope and water-use data, and leaf anatomy indicate that epiphytes experience different degrees of stress and that they possess multiple mechanisms to use moisture in shared habitats. For example, drought-enduring, evergreen orchids some-times grow in the same trees with drought-deciduous types (Benzing *et al.*, 1982), although not necessarily on the same kinds of rooting media. The advantages of C_3 compared to CAM in several epiphytic bromeliads remain obscure (e.g., Griffiths *et al.*, 1986), as do the benefits of certain mixed syn-dromes (e.g., $CAM-C_4$ in some *Peperomia* spp.; Nishio and Ting, 1987). In addition to water and carbon economy, CAM may permit enhanced N-use efficiency and reduce vulnerability to photodamage. Some ferns (e.g., Winter *et al.*, 1986) and a large number of bromeliads (e.g., Martin *et al.*, 1985) exhibit uncommon shade tolerance for CAM plants, but the mecha-nisms responsible for these performances have been studied most intensely in terrestrial Bromeliaceae, particularly *Ananas* and *Bromelia* (e.g., Medina *et al.*, 1993). CAM, because of its often facultative nature and capacity to shift into the stress-mitigating CAM-idling mode, suits many epiphytes par-ticularly well. However, none of the necessarily extended studies has been conducted to determine how the versatility of this complex syndrome or any other carbon/water balance mechanism serving the epiphytes matches local climates and other environmental factors that determine moisture supplies and related plant demands.

V. Growing Conditions in Canopy Habitats

Conditions under which epiphytes grow are often presumed to be physi-cally stressful and the required rooting media are widely scattered and often ephemeral for such slow-growing plants. Actual data are few and the find-ings vary. Nutrient scarcities probably limit productivity more at some sites than others (e.g., Schlesinger and Marks, 1977; Benzing and Davidson, 1979). Some canopy washes contain N, P, and K at concentrations exceed-ing those in typical soils (e.g., Jordan and Golley, 1980). Nutrient-rich sol-ids, including plant debris and excrement, are available to many epiphytes. Suspended soils, while often less voluminous than those on the ground, can contain more concentrated critical ions (Benzing, 1990, Chapter 4).

---

(5) A "trash-basket" or "bird's nest" type *Asplenium* in Papua New Guinea. (6) *Camplocentrum micrantha*, "a shootless" twig orchid in southern Ecuador; X0.35. (7) *Vitaria lineata*, a humus epiphyte rooted in the debris accumulated in the persistent leaf bases of *Sebal palmetto* in South Florida; X0.20. (8) *Anthurium* sp., a "trash-basket" epiphyte in Amazonian Venezuela; X0.40. The arrow indicates intercepted litter. (9) *Tillandsia recurvata*, an "atmospheric" bromeliad in central Florida; X0.40.

Figures 10–17   Epiphyte biomass and vascular epiphytes with associated symbionts. (10) Epiphytic vegetation growing on trees left in place when pasture was created from humid forest near Rio de Janiero, Brazil. Elevated exposure probably increased epiphyte loads, a consequence with positive implications for the conservation of arboreal vegetation. Most of the biomass here is produced by representatives of the bromeliad genus *Neoregelia* and epiphytic cacti. (11) Flowering shoot of *Neoregelia cruenta,* the center of which is filled with water; X0.35. (12) *Guzmania* sp. illustrating how brightly ornamented foliage produces the phytotelma; X0.20. (13) *Brocchinia tatei* on Cerro Neblina in southern Venezuela. The arrow illustrates a

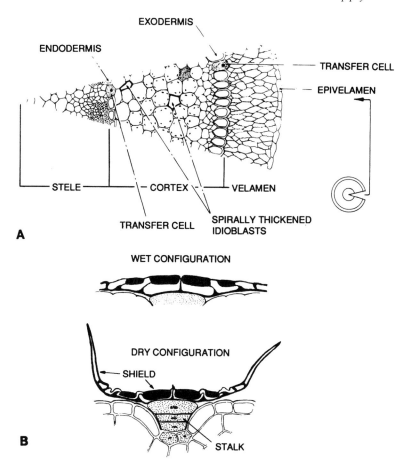

Figure 18 Absorptive organs specialized to promote the absorption of moisture and nutritive ions during transient contact with precipitation and canopy washes in forests and similar habitats (e.g., cliff faces). (A) A typical velamentous root on an epiphytic orchid; (B) the foliar trichome of an atmospheric bromeliad.

trap-bearing leaf of *Utricularia humboltii*, a carnivorous aquatic endemic to bromeliad shoots; X0.15. (14) *Hydnophytum moselyanum* in Papua, New Guinea. The domatium (the tuberous hypocotyle) has been cut open to expose the labyrinthine chambers routinely occupied by an ant colony; X0.25. (15) *Myrmecodium* sp. in Papua New Guinea, illustrating how ants occupying the ridged domatium have failed to deter folivores; X0.25. (16) An ant-nest garden in Ecuador constructed by *Camponotus femoratus* and possibly a parabiotic, second ant species with associated vascular plants representing at least seven species; X0.08. (17) Carton trail (right side) in Papua New Guinea constructed by *Iridomyrex cordatus* on the trunk of *Cocos nucifera*. Growing in that ant-provided medium are seedlings of the orchid *Dendrobium insigne;* X0.15. The trail to the left lacking orchids was produced by termites nesting in the same tree.

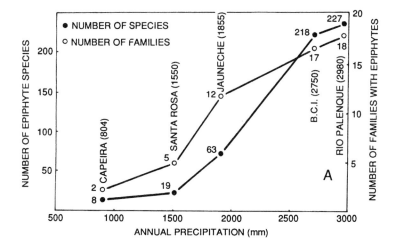

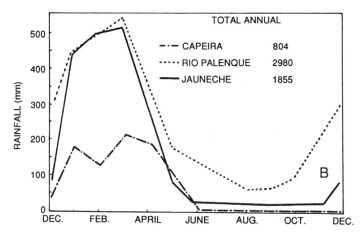

Figure 19   (A) The diversity of epiphyte families and species relative to annual rainfall at five locations and (B) the monthly distribution of rainfall at three of those sites. (From Gentry and Dodson, 1987b.)

Carnivory is underrepresented and ant-fed plants are almost exclusive to canopy habitats for reasons of plant resource economy according to Thompson (1981) and Givnish *et al.* (1984).

Plant distributions indicate that the most important regional environmental determinant of epiphyte success is less closely linked to the total annual rainfall than to rainfall distribution over the year (Gentry and Dodson, 1987a). Cool, continuously humid sites support the greatest diversity and usually the largest biomass of arboreal flora (Fig. 19). However, abun-

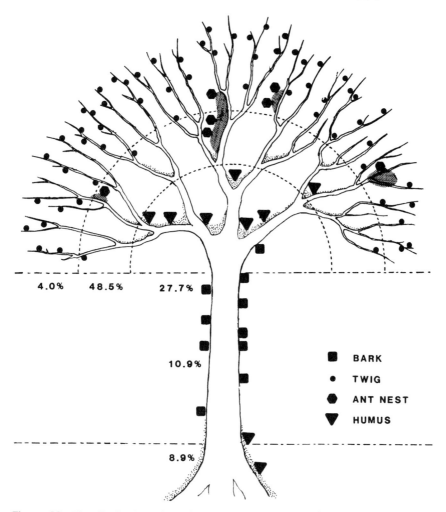

Figure 20 The distribution of epiphytes on a typical phorophyte in a humid forest. The percentage values are those recorded by Johansson (1974) for orchid species at a Nigerian site. The diverse epiphytes illustrated on this tree are distinguished by types of required substrates.

dance can be great despite low diversity (e.g., *Tillandsia* in Mesoamerican dry forests; Gardner, 1983). Segregations of populations within a single forest and within the crown of an individual tree follow additional, finer-scale gradients (Fig. 20). Local distributions also track the capacities of epiphytes to create substitutes for soil (Figs. 5, 7, 8, 16, and 17) and otherwise enhance the availability of transitory supplies of moisture (with special absorptive

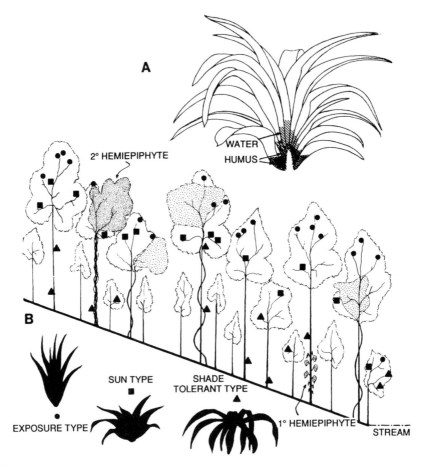

Figure 21   (A) Section of a shoot of a phytotelm bromeliad illustrating the locations of impounded moisture and humus in leaf axils. (B) The distribution of phytotelm bromeliads representing three exposure classes in humid forest in northern Trinidad and silhouettes to illustrate the corresponding differences in shoot structure. (From Pittendrigh, 1948.)

root and shoot tissues; Benzing and Pridgeon, 1983). Epiphytes nevertheless indicate climate more sensitively than co-occurring earth-rooted vegetation (Gentry and Dodson, 1987a).

Because terrestrial ancestors were so diverse, the structures and mechanisms underlying epiphytism differ greatly, sometimes within a single family. The Bromeliaceae is best understood in this respect, specifically how its members partition humid, primary forest more or less into three vertical life zones (Fig. 21). Pittendrigh (1948) first identified this kind of segregation in the mountains of Trinidad. Members of his "exposure" group occupied upper, well-illuminated perches. Distributions were broader in the more sparsely foliated, transparent crowns of trees in drier habitats elsewhere on the island. Exposure species were either "atmospheric" bromeliads (Fig. 9), succulent types heavily invested with absorbing, light-reflecting trichomes (Fig. 18), or less xeromorphic species equipped with shoots that impound moisture to create phytotelmata (Figs. 11 and 21). The "sun" group, most of which had sizable wider-spreading rosettes, inhabited intermediate sites (exposures) in dense montane forest. The most hydromorphic residents of all, the "shade-tolerant" populations, were closer to the ground. Their shallow impoundments (Fig. 21), constructed of lax foliage, were usually filled with moisture and tree debris. Versatility was greatest in these because full sun could be tolerated if humidity were sufficiently high (e.g., over a stream).

Subsequent studies were conducted to detect the underlying physiology responsible for Pittendrigh's three light-related categories (Benzing and Renfrow, 1971a,b; Griffiths and Smith, 1983; Smith *et al.*, 1985, 1986; Griffiths *et al.*, 1986; Lüttge *et al.*, 1986). Members of group three exhibit a wider tolerance for exposure than the upper or midcanopy specialists. Pittendrigh's proposal that the shade-tolerant species are cryptic heliophiles restricted to darker than optimal microsites by vulnerability to drought was only partially correct. They are sensitive to drought, but high light-use efficiencies (quantum yields) are possible at low photosynthetic photon flux densities (PPFD) (Benzing and Renfrow, 1971b), that is, these plants acclimate to deep shade. Thin-leaved forms equipped for C_3 photosynthesis (arguably the carbon fixation pathway best suited for deep shade with potentially low light compensation and saturation intensities) have high demands for moisture. Where vapor pressure deficits (VPD) around shoots are low and insolation is high, anthocyanins accumulate in the adaxial epidermis, perhaps to protect the underlying mesophyll from photoinhibition. Exposure and sun forms are often CAM types (as indicated by ^{13}C discrimination data) or, if equipped for C_3 photosynthesis, possess voluminous phytotelmata to match the associated high rates of transpiration imposed by substantial foliar conductances. Atmospheric types fail in darker sites because the dense, hygroscopic indumentum covering the foliage blocks light and impedes gas exchange while wet (Benzing and Renfrow, 1971a; Martin and Siedow, 1981).

Bromeliads, the most prolific contributors to suspended biomass in many Neotropical forests, partition aerial substrates somewhat less fastidiously than more root-dependent epiphytes, particularly orchids. Johansson (1974) identified five life zones by their resident orchids in West African trees, three depths in crowns plus one at the upper trunk and one at the lower trunk (Fig. 20). All five sectors had characteristic residents with some overlap. Percentages of the total local orchid flora (species) varied from 4% at the crown periphery to 48.5% at midcrown, where shade was moderate and moisture more abundant. Percentages for the crown center and upper and lower trunk zones were 7.7, 10.9, and 8.9, respectively. Presumably, survival on outermost twigs was challenged most by drought, whereas insufficient light limited growth deeper in the canopy. Gentry and Dodson (1987b) and Catling *et al.* (1986) noted similar apportionments of epiphytes at sites in Ecuador and Belize, respectively. Other studies (e.g., ter Steege and Cornelissen, 1989; Zimmerman and Olmsted, 1992) further confirmed nonrandom distributions and segregations of co-occurring epiphytes, but they have not substantially increased knowledge of the causes.

Factors responsible for structuring epiphyte assemblages within forests are complex and interactive. Especially puzzling are the twig orchids (Chase, 1987; Fig. 6), species that usually anchor only on the smallest branches even though similarly exposed, older bark (on thicker axes) occurs extensively in the same tree crowns (e.g., Catling *et al.*, 1986). Narrowly defined site requirements often mandate tolerances for coupled, but not necessarily equally influential growing conditions. Because suspended humic soils develop slowly (thickest layers are on older bark), plants relegated to this medium (probably by vulnerability to drought) must also tolerate the shade that usually prevails deep in the crowns of mature trees. Other arboreal flora (e.g., many evergreen orchids) occur exclusively on older but unembellished bark or root in knotholes or rotting branches (e.g., deciduous orchids). Arboreal, carton-manufacturing ants dictate rooting media (Figs. 16 and 17) and thus distributions for a broadly convergent group of nest-garden specialists representing many plant families (e.g., Davidson, 1988; Benzing and Clements, 1991). Fragrances and food associated with the seeds and fruits of this flora offer powerful incentives for myrmecochory, but near to complete fidelity to carton suggests that additional plant requirements exist for ant-provided substrates (Davidson, 1988). Other vectors influence the distributions of the nonmyrmecophytes, for example, frugivorous bats make *Billbergia zebrina* a regular knothole occupant in certain Brazilian forests.

VI. Host Specificity

Few workers have documented the restriction of an epiphyte to a specific species of phorophyte, and no mechanism has been demonstrated. Went (1940) considered tight pairings to be common and often grounded in chemical interactions. So predictable were the combinations of orchids and trees examined at Tjibodas, Java, that he could differentiate unidentified *Castanopsis* species by their associated arboreal flora. Most surveys have yielded contrary results (e.g., Johansson, 1974; Valdivia, 1977; Ackerman *et al.*, 1989) and relatively uninteresting alternatives explain at least some of the exceptions. For example, many of the putative, obligate epiphyte–tree pairings occur where too few kinds of trees grow to permit broader host ranges. Current wisdom generally agrees with the view that most epiphytes require conditions provided by many kinds of trees and that long histories of co-occurrence have no bearing on compatibility. Significant on this count are woody exotics (e.g., *Citrus* and *Coffea* in tropical America) that surpass many natives as substrates for indigenous epiphytes. More surprising than the lack of tight pairings is the absence of demonstrated axeny (epiphyte repulsion). Many trees grow relatively unencumbered (e.g., possibly allelopathic *Cinnamomum camphora*) while interspersed among others that support potentially damaging (see the following) loads of epiphytes. Whether their immunity is coincidental or the product of natural selection remains unclear.

Thorough phytosociological censuses, such as those conducted by Johansson (1974) and Valdivia (1977), should precede searches for the mechanisms responsible for putative host specificities. Surveys across a range of habitats may show that such epiphytes are less faithful to a single kind of phorophyte or compartment in the forest canopy than observations at a single or few sites would indicate. Many (perhaps most) trees are neither fully accommodating nor totally hostile to arboreal flora. Axeny may be conditional rather than absolute in the sense that even the most heavily infested phorophytes fail to support epiphytes on every type of surface to which seeds attach. Trees may become more or less exploitable as they age or the environment changes, particularly humidity and light and other factors that affect substrate characteristics. The availability of potential colonists is also important. *Rhizophora mangle* in Florida supports little or no canopy flora in regions where other phorophytes carry light to moderate loads of epiphytes. However, this same tree species accommodates at least a few bromeliads at other locations, and examination of these scattered anchorage sites could be revealing. *Rhizophora* may be colonized in the second instance simply because a denser seed rain reveals the few favorable attachment sites. Possibly only the limited areas represented by older or appropriately oriented bark, water courses, or colonies of lichens or bryophytes permit seedling establishment. Alternatively, the quality of the tree as a substrate may vary with conditions at the two sites.

Functional intimacy has little bearing on host–epiphyte fidelity. Exceptional one-on-one relationships involve both phloem-tapping mistletoes and orchids; close relatives in both instances exhibit much wider tolerances for substrates. Recurrent claims (e.g., Johansson, 1977) that the fungi that orchids need to germinate also help proscribe phorophytes for them remain untested. Bark stability, wettability, water-holding capacity, and surface texture probably determine in large measure the interception and fate of all attached epiphyte seeds. Allelopathy is another possibility that needs more detailed study (Went, 1940; Frei and Dodson, 1972). Effects could also operate in the other direction. Two *Tillandsia* species supposedly defoliate occupied trees by means of toxic secretions (Claver *et al.*, 1983), a

feat consistent with what are typically the demands of atmospheric bromeliads for high exposure.

VII. Succession

Reports of succession involving arboreal flora are limited to analyses of
presumed seral stages rather than of observed progressions. Lichens and
bryophytes apparently initiate protracted sequences of floristic change on
bark in certain Himalayan *Quercus* forest (Dudgeon, 1923). The first of
six recognizable stages begins with the arrival of crustose lichens on 3- to
4-year-old stems. Foliose and fruticose lichens follow after three to four
more seasons. Eventually ferns root in the resulting mat, to be joined later
by angiosperms. Van Oye (1924) recognized a briefer, three-stage sequence
in Java. Nonvascular plants appeared to condition sites for stress-resistant
ferns, followed by more demanding relatives and then by more vulnerable angiosperms. Bromeliad establishment, an event of substantial consequence in many tropical American forests, seems to be less dependent on
site preparation, in keeping with their roles as pioneers on arboreal substrates. Large Phytotelm species create new and unusually equable space for
a variety of dependent flora and substantial fauna.

Seral stages, where they can be said to exist, remain poorly characterized.
Catling and Lefkovich (1989) identified four epiphyte associations between
0.3 and 5 m above ground in a 2-ha plot of Guatemalan cloud forest. Composition ranged from two to five species per association and occurrence was
influenced by the age (stem thickness) of the substrate. Of two groups, one
early and one later in developing, the former contained fewer, but closely
related ferns and orchids. Members of the more diverse associations, which
included two unidentified *Tillandsia* spp., were larger and tended to have
seasonal rather than continuous flowering. More comprehensive knowledge of life histories and site requirements is needed to confirm roles as
pioneers and to identify patterns of community assembly. Long-term studies
should be initiated to follow succession in a variety of circumstances (e.g.,
kinds of trees, climates).

Elsewhere, an absence of characteristic patterns (e.g., Yeaton and Gladstone, 1982) suggests that early and subsequent colonists on arboreal substrates in dry forest are not strongly sorted by degrees of stress tolerance or
type of life history. The epiphytic flora of relatively seasonal forests (dominated by bromeliads and orchids) appear so robustly drought-hardy, grow
so slowly, and cover so little of what appears to be generally habitable space
in tree crowns that interactions or substrate modifications sufficient to assure succession seem unlikely. Ants responding to edible rewards or chemical lures associated with seeds influence patterns of epiphyte occurrence in

a more specialized context. Ant-nest garden composition varies with the identity of the responsible ants, the size and ingredients of their carton, and its exposure (Davidson, 1983). More shade-tolerant species sometimes replace initial occupants, perhaps with ant assistance.

VIII. Relationships with Fauna

Animals visit epiphytes for the same reasons that they visit other plants: to hide, feed, and coincidentally disperse pollen and seeds. Generalizations about the involvement of fauna in the reproductive biology of epiphytes would be premature. Only one report describes the genetic structures of populations (two *Tillandsia* species chosen to represent autogamous and outcrossing types; Soltis *et al.*, 1987). Traplining pollinators, usually insects, and floral mimicry may be overrepresented considering the prominence of Orchidaceae among the epiphytes (Dressler, 1981). Arboreal orchids deserve additional note as especially favorable subjects for population biologists (e.g., Calvo, 1993). Ramets are easily counted nondestructively, fruits typically result from single pollinations, and pollinia simplify paternity analysis.

Bats, birds, and insects pollinate epiphytic bromeliads; *Tillandsia* has received more study than all but a few orchid genera (e.g., Gardner, 1986). Its many floral syndromes, some differentiating closely related taxa, vary by corolla and bract color and shape, fragrance, timing of anthesis, and other characteristics. Many cultivated epiphytes are known for their ability or failure to set self-seed; documentation is scattered in the horticultural literature.

Determination of whether arboral flora and phorophytes tend to share or use different pollinators requires more knowledge and tests of what has become widely held conviction. Renner and Feil (1993), for example, present data that challenge the notion that dioecious, lowland tropical trees are pollinated by small, generalist bees. Instead, floral visitors included beetles, moths, and wasps, some of which are the same types of long-distance fliers that set fruit for many orchids and other large-flowered epiphytes (e.g., Ericaceae, Gesneriaceae; Benzing, 1990, Chapter 5). Epiphytes with close terrestrial relatives (e.g., species in *Anthurium* and various bromeliad and orchid genera) should be examined to see if they rely on the same vectors, possess the same breeding systems, and exhibit similarly structured populations. Zoochory among the epiphytes also remains little studied except for a few species (e.g., Benzing and Clements, 1991; Davidson and Epstein, 1989).

Too little is known about diseases and herbivory in canopy habitats (beyond that involving trees) to make comparisons with epiphytes. Better studied are other dependent biota, particularly the inhabitants of bromeliad phytotelmata and the ants that feed the myrmecotrophs and farm nest gardens (Huxley, 1980; Davidson, 1988; Benzing, 1991). However, epiphytes support far more canopy fauna by providing nesting sites and shelter in a world of abundant predators and climatic extremes. Only those taxa that create special opportunities to foster prolonged contact with fauna qualify for membership in five categories (not entirely exclusive): (1) ant-nest garden, (2) ant-fed, ant-house, (3) ant-guarded, (4) trash-basket (also called nest-forming or debris-collecting), and (5) phytotelm forms.

Ant-nest garden plants are most conspicuous in Neotropical forests, where the workers that tend them may dominate the local arboreal ant fauna (Wilson, 1987). Anyone who has collected ant-garden plants is aware that these myrmecophytes exist in intimate contact with some of the most aggressive predators that inhabit the canopies of American tropical lowland forests.

Ant-house epiphytes differ from ant-nest garden species by providing nesting cavities (Figs. 14 and 15) in modified stems (e.g., *Myrmecodia* spp.), individual leaves (e.g., *Dischidia* spp.), or whole shoots (e.g., *Tillandsia* spp.). Ant occupants are diverse and provide nutrients via excrement and nest debris, and may help deter herbivores. Which ant occupies the tuber of a specific *Myrmecodia* specimen (the best-studied genus of myrmecotrophs, e.g., Jebb, 1991) varies according to local circumstances. High-energy-demanding and aggressive ant taxa colonize only robust, well-exposed plants, whereas the more timid types usually occupy weaker-growing specimens in deeper shade. Experiments (e.g., Rickson, 1979) have demonstrated that a variety of substances can pass from ants into plants, but too little is known about the life histories of the ants or the nutrient budgets of the plants to quantify the benefits to either party. Although perhaps not as important to forest communities as ant-nest garden types, the myrmecotrophs offer opportunities to study aspects of ant and plant biology and mechanisms of coevolution.

Many epiphytes entice ant-guards with extrafloral nectar, the most thoroughly studied of this type being some ant-house orchids (e.g., Rico-Gray and Thein, 1989; Fisher *et al.*, 1990). In Panama, *Schomburgkia tibicinis* (Orchidaceae) either received protection or incurred significant damage, depending on which herbivores arrived and which ants were patrolling vulnerable plant organs at the time (Rico-Grey and Thein, 1989).

Ants frequently use nest sites in litter impounded by the masses of apogeotrophic roots (e.g., *Cryptopodium* spp.) or overlapping leaf bases (e.g., *Anthurium* spp., Fig. 8) produced by some epiphytes. To what extent (if any) these occupants enhance plant fitness remains undetermined, but some trash-basket types (*Platycerium* and *Crytopodium*, Fig. 5) produce extrafloral nectar.

The most extensively described biota associated with epiphytes is nurtured by phytotelm bromeliads (Figs. 13, 16, and 21). The cisternlike shoots of these plants have analogues in a few other families (e.g., Commelinaceae, Liliaceae), but nowhere else do they assume the variety of sizes, shapes, and colors that have encouraged much speculation (but generated little data) on functional significance. Findings from studies on simpler phytotelmata (e.g., the floral bracts of *Heliconia* spp.; Naeem, 1990) suggest that additional variables, such as the quantities and types of litter intercepted, also influence the structure of communities in bromeliad shoots.

Some of the more dependent and occasionally obligate vertebrates (e.g., salamanders; Wake, 1987) reported from bromeliads exhibit shapes well-suited to negotiate the narrow spaces between overlapping leaf bases. Tubular types also harbor frogs (e.g., some *Hyla, Eleuthrodactylus, Gastrotheca*), possible sources of plant nutrients, that use these shoots as refuges from predators and desiccation. Flat heads in some cases (e.g., *Hyla venulosa*) allow the animals to block shoot orifices, deterring all predators except those powerful enough (e.g., monkeys) to tear open the plants. Accounts of the invertebrates observed in bromeliad shoots are much more numerous (e.g., Frank and Lounibos, 1983).

From the botanical point of view, the impounding shoot, equipped in Bromeliaceae with absorbing trichomes, is the ultimate device for extracting nutrients from intercepted litter. This material represents a rich but uncertain source that mostly falls through the canopy (Nadkarni and Matelson, 1991). Provisioned with an extensive microflora that can include diazotrophs (Bermudes and Benzing, 1991), and harboring many soil detritivores, phytotelm bromeliads maintain an organic soil charged primarily with nutrients from phorophytes and the atmosphere. They may house a fauna sometimes more diverse and concentrated than that in adjacent terrestrial media (Paoletti *et al.,* 1991). Knowledge of how these multifaceted microcosms operate will reveal much about important support systems in the canopies of humid, Neotropical forests. A single bromeliad shoot can constitute a graded series of microhabitats ranging from fully aquatic ones at and near the center to increasingly humus-rich, drier compartments for air breathers in older, more peripheral leaf axils (Fig. 21).

IX. Influences on Phorophytes

Mistletoes parasitize trees, sometimes with sufficient impact to warrant status as pathogens. In contrast, epiphytes lack haustoria, although the belief persists that certain orchids have access to the vasculature of their hosts via interconnecting fungal hyphae (e.g., Johansson, 1977). However, no direct evidence indicates that the die-back reported for some heavily infested trees ("epiphytosis," see Ruinin, 1953) reflects direct nutrient loss to attached orchids. Another hypothesis has been offered to explain the diminished foliage of the putatively parasitic "shootless" orchids (e.g., Benzing and Ott, 1981, Fig. 6). Aspects of normal tree growth can account for the overrepresentation of epiphytes on dead and dying compared to healthy axes in colonized tree crowns (Benzing, 1979). Under certain conditions, arboreal flora may contribute to localized die-back and generally diminished tree vigor. Claver *et al.* (1983) attributed the rapid spread of *Tillandsia aeranthos* and *T. recurvata* at sites in Argentina to allelopathy, but did not characterize the responsible chemical agent. Weight alone may damage some heavily laden phorophytes, and the resulting gaps may contribute significantly to the maintenance of tree diversity in some forests (Strong, 1977). Shade produced by epiphytes can be another problem if cover is extensive (e.g., *Tillandsia usneoides*). Orchid roots reputedly girdle *Citrus* in some locations (Cook, 1926) and the fungi (e.g., *Rhizoctonia* spp.) inhabiting them are more virulent to some nonorchid hosts. These and other microflora associated with epiphytes may encourage disease in supporting trees. Another phenomenon, "nutritional piracy" (Benzing and Seemann, 1978), operates through intervention in nutrient cycling and under appropriate circumstances projects the illusion of parasitism. Studies are needed to determine the conditions under which epiphytes stress phorophytes enough to reduce their growth.

Arboreal vegetation is well-positioned by location, scavenging capacity (even for diverse pollutants, e.g., Schrimpff, 1981), and growth characteristics to deny phorophytes nutrients without acting as direct parasites or competitors. Epiphytes intercept wet and dry deposition from the atmosphere and extract ions from litter or canopy washes. All of these nutrients could otherwise be available for earlier use or reuse by the trees. Once incorporated into arboreal biomass, essential elements are sequestered (longer in slow-growing flora than in epiphytes of wetter sites), but in all cases they remain at least temporarily unavailable to phorophytes. Impact on fertile sites should be minimal, but where supplies are scarce, trees may experience nutritional stress, perhaps enough to explain some of the symptoms of epiphytosis.

Nutritional piracy certainly exists. Whether it merits recognition as a major force in forest communities is open to question, although circumstantial evidence suggests at least occasional importance. Many tropical forests are primarily rain-fed, their supporting soils being deep, heavily weathered, and accordingly impoverished. Moreover, several studies have demonstrated that substantial proportions of the nutrient capital contained in surveyed ecosystems were located in epiphyte biomass (e.g., Benzing and Seemann, 1978; Nadkarni, 1984).

X. Broader Impacts on Ecosystem Processes

Much is written in this volume about canopy involvement in the cycling of chemical elements in forest ecosystems (Chapter 20). Vascular and nonvascular epiphytic vegetation is depicted in the role of a "nutrient capacitor," intercepting inputs from the atmosphere and storing them for later release and subsequent use by other flora. Some of the nonarboreal flora, particularly trees with canopy roots, are well-positioned to draw on that reservoir (Nadkarni, 1981, 1984). Which of the many additional influences epiphytes may impose on ecosystems depends on the species and local circumstances. Benzing (1990, p. 100) compared flora anchored on a single trunk and the foliage of that tree in a Venezuelan seasonal forest as illustrative of larger systems. These co-occurring plants differed widely in their tolerances to drought and capacity for recovery following rainfall and in the ways they use resources. This example demonstrates that different photosynthetic pathways, phenological patterns, and tissue compositions assure that arboreal flora both store much nutrient capital and use it with different efficiencies than adjacent terrestrial vegetation. Biologically significant impacts require commensurate loads of epiphytes that can rival and even exceed the weight of foliage maintained by the supporting trees (e.g., Nadkarni, 1984; Tanner, 1980; Hofstede *et al.*, 1993, Fig. 10).

Though located in the productive portion of a plant (i.e., the green tissue comprising epiphyte biomass), nitrogen, for example, is deployed to capture energy at rates not necessarily equal to that of co-occurring trees. Nor will the energy harvested have the same influence in the ecosystem. Arboreal floras supported by high humidity in many low- to midmontane tropical forests are prolific and diverse, as are the resulting services and products for dependent biota. Such mixed, multilayered canopies use available nutrients and photons to create greater biological variety in more dimensions than similar systems without epiphytes. Conceivably, these more complex systems also harvest energy more efficiently (e.g., cal unit N/time) to achieve unusually high biological variety. Conversely, the fewer kinds of epiphytes that densely populate some drier forests might promote very different outcomes. There, scarce ions co-opted from trees to support the more modest photosynthetic outputs of CAM or poikilohydric (mostly nonvascular) epiphytes would reduce systemwide productivity and instantaneous mineral-use efficiency. Arboreal vegetation may also impose related consequences, for example, diminish an important support capacity by reducing forage for some keystone herbivore specialized to feed on the leaves of a suppressed phorophyte.

Disturbance imposes an additional layer of complexity across the entire range of forest types. Epiphytic vegetation in communities subjected to frequent perturbations would probably never achieve the sort of enhancement that Nadkarni (1981, 1984) envisions in her nutritional steady-state model. Epiphyte loads would be continuously regenerating at such sites and accumulating mineral ions, potentially at the expense of phorophytes, specifically by reducing their capacity for photosynthesis. These and many additional hypotheses must be tested at a variety of sites to determine how epiphytes influence the diverse and dynamic systems they inhabit.

XI. Conclusions

Epiphytes offer opportunities and challenges for biologists (Table IV). Much is known about the basic life processes of some of the more specialized arboreal flora, particularly orchids and bromeliads, but important pat-

Table IV   Research Questions on Epiphyte Ecology

Subject	Obvious	Questions remaining
1. Fidelity to canopy versus other substrates	Occurrence on trees ranges from accidental to obligate.	What factors differentiate canopy from terrestrial substrates for the obligate epiphyte? How has specialization for arboreal life compromised capacity to survive on the ground?
2. Requirements for specific types of arboreal substrates	Specific epiphytes typically colonize only subsets of the many types of substrates present in occupied tree crowns.	What plant characteristics determine microsite requirements for twig, bark, humus, ant-nest garden, etc., epiphytes?
3. Plant adjustments to the often transitory and relatively unpredictable supplies of moisture in forest canopies	Broadly occurring accommodations to drought (e.g., CAM, xeromorphy) are particularly well-developed among the epiphytes.	What is the nature of the moisture supply in forest canopies and how are mechanisms such as photosynthetic pathways, osmotic balance, and stomatal behavior fine-tuned to reduce risk and maximize effective use of available moisture?
4. Plant adjustments to the absence of mineral soil	A variety of organic substrates, including the products of mutualistic biota, serve in lieu of earth soil as sources of nutritive ions.	How is impounded litter processed for phytotelm epiphytes? How substantially do ant mutalists contribute to the nutrient budgets of associated epiphytes? How are the more oligotrophic epiphytes (e.g., atmospheric bromeliads) equipped to scavenge scarce ions and use them economically?

(continues)

Table IV—*Continued*

Subject	Obvious	Questions remaining
5. Impacts of arboreal ants	Some epiphytes require ants for dispersal and to provide rooting media.	How much arboreal flora beyond the obvious ant-nest garden and myrmecotrophic species are dependent on ants for dispersal, substrates, and defense?
6. Epiphytic vegetation as a resource for canopy fauna	Much arboreal fauna, particularly invertebrates, use epiphytes as resources.	What is the full extent of this dependence and what are the broader consequences of these dependencies for the forest community?
7. Epiphyte involvements in nutrient cycles	Nutritional piracy exists. Epiphyte biomass sometimes contains much of the nutrient capital present in a forest ecosystem.	To what degree and under what conditions does the presence of an epiphyte load have an impact on the nutritional status of a phorophyte?
8. Impacts on community productivity and patterns of resource use	Resources present in epiphyte biomass (e.g., N and P) at least sometimes yield photosynthetic returns at different rates than those of supporting soil-rooted vegetation.	How does the presence of substantial epiphyte biomass affect aggregate forest productivity and help determine overall resource-use efficiency?
9. Conservation	Because many epiphytes occupy narrow ranges (especially orchids), often in regions of rapid development, endangered status is correspondingly common.	What conservation strategies are likely to preserve the greatest diversity of epiphytes?
10. Indicators of habitat quality and global change	Some epiphytes possess characteristics that impart extraordinary utility as air quality monitors.	How can epiphytic vegetation be more effectively used to monitor changing conditions in the troposphere?
11. Succession	Presumed seral stages identified (e.g., Catling and Lefkovitch, 1989).	Do species displace one another on bark? If so, by what mechanisms?
12. Community organization	Species often co-occur in predictable assemblages (e.g., Catling and Lefkovitch, 1989), but often distribution and spacing among individuals are random (e.g., Yeaton and Gladstone, 1982).	Are the factors responsible for the distributions and combinations of species on bark primarily density-dependent or density-independent?

terns that require longer-term monitoring remain obscure. Numerous species constitute good systems to study pervasive plant phenomena such as xerophytism and the factors that force evolutionary radiations and mediate plant/animal mutualisms. Substantial natural history has also been recorded for the epiphytes. Far less studied and more important globally are the roles these plants play in ecosystems. We need to determine to what degree other biota are influenced by this vegetation and to what extent it promotes the high biodiversity typically found in the tropical humid forest. How might epiphytes help mold other characteristics of tropical forests, for example, their physical structure, patterns of nutrient capture and deployment, productivity, and responses to perturbation? Arboreal vegetation also represents a largely underutilized diagnostic system. Lugo and Scatena (1992) have called attention to the potential sensitivity of epiphytes to global change. Located at the interface between forest and atmosphere, and more dependent on airborne inputs than resources from soils, these plants mirror the status of important characteristics of the troposphere. Some bromeliads (and a variety of lichens) have provided information on air quality since early in this century (e.g., Wherry and Capen, 1928; MacIntire *et al.*, 1952; Schrimpff, 1981; Benzing *et al.*, 1992). It behooves us to learn more about epiphytic plants so that we can read the signals that this strategically situated and propitiously adapted vegetation offers on change in our increasingly threatened global support system.

XII. Summary and Directions for Future Research

Epiphytes constitute about 10% of all vascular flora and represent important components of a variety of tropical and subtropical ecosystems. These are diverse plants representing over 80 families and they illustrate a substantial array of structural, functional, and ecological types. Considerable data exist on the ecophysiology of certain taxa, particularly Bromeliceae and Orchidaceae. In contrast, except for some Neotropical bromeliads and orchids, little is known about reproductive biology, and for even fewer species the determinants of distributions in tree crowns or population structures.

Arboreal flora merit closer study for several reasons. As plants without access to terrestrial soil, they make good experimental subjects to study stress physiology and accommodations to unconventional sources of nutrients and discontinuous and unpredictable moisture supplies. Epiphytes also warrant more attention for the roles they play in forest dynamics, which include those affecting biodiversity, whole-community productivity, and mineral-use efficiency. Particularly promising is the opportunity this flora offers to gain information on important changes in the global atmosphere.

Until recently, difficult access and the absence of suitable equipment to measure gas exchange and other plant activities impeded research on epiphytes. Today, a variety of tree-climbing techniques and modified equipment are available to enter and work in the canopy. This improved access and newer, portable instruments have greatly improved opportunities to pursue the questions raised in this review.

References

Ackerman, J. D., Montalvo, A. M., and Vera, A. M. (1989). Epiphyte host specificity of *Encyclia krugii* a Puerto Rican endemic orchid. *Lindleyana* 4, 74–77.

Atwood, J. T. (1986). The size of the Orchidaceae and the systematic distribution of epiphytic orchids. *Selbyana* 9, 171–186.

Benzing, D. H. (1979). Alternative interpretations for the evidence that certain orchids and bromeliads act as shoot parasites. *Selbyana* 2, 135–144.

Benzing, D. H. (1981). Bark surfaces and the origin and maintenance of diversity among angiosperm epiphytes: An hypothesis. *Selbyana* 5, 248–255.

Benzing, D. H. (1990). "Vascular Epiphytes." Cambridge Univ. Press, Cambridge, UK.

Benzing, D. H. (1991). Myrmecotrophy: Origins, operations and importance. *In* "Anti–Plant Mutualism" (C. R. Huxley and D. F. Cutler, eds.), pp. 353–373. Oxford Univ. Press, Oxford.

Benzing, D. H., and Clements, M. (1991). Dispersal of the orchid *Dendrobium insigne* by the ant *Iridomyrmex cordatus* in Papua, New Guinea. *Biotropica* 23, 604–607.

Benzing, D. H., and Davidson, D. W. (1979). Oligotrophic *Tillandsia circinnata* Schlect. (Bromeliaceae): An assessment of its patterns of mineral allocation. *Am. J. Bot.* 66, 386–397.

Benzing, D. H., and Ott, D. (1981). Vegetative reduction in epiphytic Bromeliaceae and Orchidaceae: Its origin and significance. *Biotropica* 13, 121–140.

Benzing, D. H., and Pridgeon, A. (1983). Foliar trichomes of Pleurothallidinae (Orchidaceae): Functional significance. *Am. J. Bot.* 70, 173–180.

Benzing, D. H., and Renfrow, A. (1971a). Significance of the patterns of CO_2 exchange to the ecology and phylogeny of the Tillandsioideae (Bromeliaceae). *Bull. Torrey Bot. Club* 98, 322–327.

Benzing, D. H., and Renfrow, A. (1971b). The significance of photosynthetic efficiency to habitat preference and phylogeny among tillandsioid bromeliads. *Bot. Gar. (Chicago)* 132, 19–30.

Benzing, D. H., and Seemann, J. (1978). Nutritional piracy and host decline: A new perspective on the epiphyte–host relationship. *Selbyana* 2, 133–148.

Benzing, D. H., Bent, A., Moscow, D., Peterson, G., and Renfrow, A. (1982). Functional correlates of deciduousness in *Catasetum integerrimum* (Orchidaceae). *Selbyana* 7, 1–9.

Benzing, D. H., Arditti, J., Nyman, L. P., Temple, P. J., and Bennett, J. P. (1992). Effects of ozone and sulfur dioxide on four epiphytic bromeliads. *Environ. Exp. Bot.* 32, 25–32.

Bermudes, D., and Benzing, D. H. (1991). Nitrogen fixation in association with Ecuadorean bromeliads. *J. Trop. Ecol.* 7, 531–538.

Calvo, R. (1993). Evolutionary demography of orchids: Intensity and frequency of pollination and the cost of fruiting. *Ecology* 79, 1033–1042.

Catling, P. M., and Lefkovich, L. P. (1989). Associations of vascular epiphytes in a Guatemalan cloud forest. *Biotropica* 21, 35–40.

Catling, P. M., Brownwell, V. R., and Lefkovich, L. P. (1986). Epiphytic orchids in a Belizean grapefruit orchid: Distribution, colonization, and association. *Lindleyana* 1, 194–202.

Chase, M. W. (1987). Obligate twig epiphytes in the Oncidiinae and other Neotropical orchids. *Selbyana* 10, 24–30.

Claver, F. K., Alaniz, J. R., and Caldiz, D. O. (1983). *Tillandsia* spp.: Epiphytic weeds of trees and bushes. *For. Ecol. Manage.* 6, 367–372.

Cook, M. T. (1926). Epiphytic orchids, a serious pest on citrus trees. *J. Dep. Agric. P. R.* 10, 5–9.

Davidson, D. W. (1988). Ecological studies of Neotropical ant gardens. *Ecology* 69, 1138–1152.

Davidson, D. W., and Epstein, W. W. (1989). Epiphytic associations with ants. *In* "Vascular Plants as Epiphytes" (U. Lüttge, ed.), pp. 200–229. Springer-Verlag, Berlin.

Dressler, R. L. (1981). "The Orchids: Natural History and Classification." Harvard Univ. Press, Cambridge, MA.

Dudgeon, W. (1923). Succession of epiphytes in the *Quercus incana* forest at Landour, western Himalayas. *J. Indian Bot. Soc.* 3, 270–272.

Fisher, B. L., Sternberg, L. S. L., and Price, D. (1990). Variation in the use of orchid extrafloral nectar by ants. *Oecologia* 83, 263–266.

Frank, J. T., and Lounibos, L. P. (1983). "Phytotelmata." Plexus Publishing, Medford.

Frei, Sister J. K., and Dodson, C. H. (1972). The chemical effect of certain bark substrates on the germination and early growth of epiphytic orchids. *Bull. Torrey Bot. Club* 99 301–307.

Gardner, C. S. (1983). Tillandsias of gulf coastal Tamaulipas, Mexico. *J. Brom. Soc.* 33, 102–107.

Gardner, C. S. (1986). Inferences about pollination in *Tillandsia* (Bromeliaceae). *Selbyana* 9, 76–87.

Gentry, A. H., and Dodson, C. H. (1987a). Contribution of non-trees to species richness of a tropical rain forest. *Biotropica* 19, 149–156.

Gentry, A. H., and Dodson, C. H. (1987b). Diversity and biogeography of Neotropical vascular epiphytes. *Am. Mo. Bot. Gard.* 74, 205–233.

Givnish, T. J., Burkhardt, E. L., Happel, R., and Weintraub, J. (1984). Carnivory in the bromeliad *Brocchinia reducta,* with a cost/benefit model for general restriction of carnivorous plants to sunny, moist, nutrient-poor habitats. *Am. Nat.* 124, 479–497.

Griffiths, H., and Smith, J. A. C. (1983). Photosynthetic pathways in the Bromeliaceae of Trinidad: Relations between life-forms, habitat preference and the occurrence of CAM. *Geologia* 60, 176–184.

Griffiths, H., Lüttge, U., Stimmel, K. H., Crook, C. E., Griffiths, N. M., and Smith, J. A. C. (1986). Comparative ecophysiology of CAM and C_3 bromeliads. III. Environmental influences on CO_2 assimilation and transpiration. *Plant, Cell Environ.* 9, 385–393.

Hofstede, R. G. M., Wolf, G. H. D., and Benzing, D. H. (1993). Epiphytic biomass and nutrient status of a Colombian upper montane rain forest. *Selbyana* 14, 37–45.

Huxley, C. R. (1980). Symbiosis between ants and epiphytes. *Biol. Rev. Cambridge Philos. Soc.* 55, 321–340.

Jebb, M. (1991). Cavity structure and function in the tuberous epiphytic Rubiaceae. *In* "Ant–Plant Interactions" (C. R. Huxley and D. F. Cutter, eds.), pp. 274–389. Oxford Univ. Press, Oxford.

Johansson, D. R. (1974). Ecology of vascular epiphytes in western African rain forest. *Acta Phytogeogr. Suec.* 59, 1–136.

Johansson, D. R. (1977). Epiphytic orchids as parasites of their host trees. *Am. Orchid Soc. Bull.* 46, 705–707.

Jordan, C. F., and Golley, F. (1980). Nutrient scavenging of rainfall by the canopy of an Amazonian rain forest. *Biotropica* 12, 61–66.

Kress, W. J. (1986). A symposium: The biology of tropical epiphytes. *Selbyana* 9, 1–22.

Lugo, A. E., and Scatena, F. N. (1992). Epiphytes and climate change research in the Caribbean: A proposal. *Selbyana* 13, 123–130.

Lüttge, U., Stimmel, K. H., Smith, J. A. C., and Griffiths, H. (1986). Comparative ecophysiology of CAM and C_3 bromeliads. II. Field measurements of gas exchange of CAM bromeliads in the humid tropics. *Plant, Cell Environ.* 9, 377–383.

MacIntire, W. H., Hardin, L. J., and Hester, W. (1952). Measurement of atmospheric fluoride: Analysis of rain water and Spanish moss exposures. *Ind. Eng. Chem.* 44, 1365–1370.

Martin, C. E., and Siedow, J. N. (1981). Crassulacean acid metabolism in the epiphyte *Tillandsia usneoides* L. (Spanish Moss): Responses of CO_2 exchange to controlled environmental conditions. *Plant Physiol.* 68, 335–339.

Martin, C. E., McLeod, K. W., Eades, C. A., and Pitzer, A. F. (1985). Morphological and physiological responses to irradiance in the CAM epiphyte *Tillandsia usneoides* L. (Bromeliaceae). *Bot. Gaz. (Chicago)* 146, 489–494.

Matelson, T. J., Nadkarni, N. M., and Longino, J. T. (1993). Longevity of fallen epiphytes in a Neotropical montane forest. *Ecology* 74, 265–269.

Medina, E., Popp, E., Olivares, H., Janett, H. P., and Lüttge, U. (1993). Daily fluctuations in titratable acidity, content of organic acids (malate and citrate) and soluble sugars of varieties and wild relatives of *Ananas comosus* L. growing under natural conditions. *Plant, Cell Environ.* 16, 55–63.

Nadkarni, N. M. (1981). Canopy roots: Convergent evolution in rainforest nutrient cycles. *Science* 214, 1023–1024.

Nadkarni, N. M. (1984). Epiphyte biomass and nutrient capital of a neotropical elfin forest. *Biotropica* 16, 249–256.

Nadkarni, N. M., and Matelson, T. J. (1991). Fine litter dynamics within the tree canopy of a tropical cloud forest. *Ecology* 72, 2071–2082.

Naeem, S. (1990). Resource heterogeneity and community structure: A case study in *Heliconia imbricata* phytotelmata. *Oecologia* 84, 29–38.

Nishio, J. N., and Ting, I. P. (1987). Carbon flow and metabolic specialization in the tissue layers of the crassulacean acid metabolism plant *Peperomia camptotricha*. *Plant Physiol.* 84, 600–604.

Paoletti, M. H., Taylor, R. A. J., Stinner, B. R., Stinner, D. H., and Benzing, D. H. (1991). Diversity of soil fauna in the canopy and forest floor of a Venezuelan cloud forest. *J. Trop. Ecol.* 7, 373–384.

Pittendrigh, C. S. (1948). The bromeliad–*Anopheles*—malaria complex in Trinidad. I. The bromeliad flora. *Evolution (Lawrence, Kans.)* 2, 58–89.

Raven, J. A. (1985). Regulation of pH and generation of osmolarity in vascular plants: A cost–benefit analysis in relation to efficiency of use of energy, nitrogen and water. *New Phytol.* 101, 25–77.

Renner, S. S., and Feil, J. P. (1993). Pollination of tropical dioecious angiosperms. *Am. J. Bot.* 80, 1100–1107.

Rickson, F. R. (1979). Absorption of animal tissue breakdown products into a plant stem: The feeding of plants by ants. *Am. J. Bot.* 66, 87–90.

Rico-Gray, V., and Thein, L. B. (1989). Ant–mealybug interaction decreases reproductive fitness of *Schomburgkia tibicinis* (orchidaceae) in Mexico. *J. Trop. Ecol.* 5, 109–112.

Rothwell, G. W. (1991). *Botryopteris forensis* (Botryopteridaceae), a trunk epiphyte of the tree fern *Psaronius*. *Am. J. Bot.* 78, 782–788.

Ruinin, J. (1953). Epiphytosis, a second view on epiphytism. *Am. Bogor.* 2, 101–157.

Schlesinger, W. J., and Marks, P. L. (1977). Mineral cycling and the niche of Spanish moss, *Tillandsia usenoides* L. *Am. J. Bot.* 64, 1254–1262.

Schrimpff, E. (1981). Air pollution patterns in two cities of Colombia, S.A., according to trace substances content of an epiphyte (*Tillandsia recurrata* L.). *Water, Air, Soil Pollut.* 21, 279–315.

Smith, J. A. C., Griffiths, H., Bassett, J., and Griffiths, N. M. (1985). Day–night changes in leaf water relations of epiphytic bromeliads in the rain forest of Trinidad. *Oecologia* 67, 474–485.

Smith, J. A. C., Griffiths, H., Lüttge, U., Brook, C. E., Griffiths, N. M., and Stimmel, K. H. (1986).

Comparative ecophysiology of CAM and C_3 bromeliads. IV. Plant–water relations. *Plant, Cell Environ.* 9, 395–410.

Soltis, D. E., Gilmartin, A. J., Riesberg, L., and Gardner, S. (1987). Genetic variation in the epiphytes *Tillandsia ionantha* and *T. recuvata* (Bromeliaceae). *Am. J. Bot.* 74, 531–539.

Strong, D. R. (1977). Epiphyte loads, tree falls, and perennial forest disruption: A mechanism for maintaining higher tree species richness in the tropics without animals. *J. Biogeogr.* 4, 215–218.

Tanner, E. V. J. (1980). Studies on the biomass and productivity in a series of montane rain forests in Jamaica. *J. Ecol.* 68, 573–588.

ter Steege, H., and Cornelissen, J. H. C. (1989). Distribution and ecology of vascular epiphytes in lowland rain forest of Guyana. *Biotropica* 21, 331–339.

Thompson, J. M. (1981). Reversed animal–plant interactions: The evolution of insectivorous and ant-fed plants. *Biol. J. Linn Soc.* 16, 147–155.

Valdivia, P. E. (1977). Estudio botánico y ecológico de la región de Rió Uxpanapa, Veracruz. IV. Las epifitas. *Biotica* 2, 55–81.

Van Oye, P. (1924). Sur l'écologie des epiphytes de la surface des troncs d'arboles à Java. *Rev. Gén. Bot.* 36, 12–30, 68–83.

Wake, D. B. (1987). Adaptive radiation of salamanders in Middle American cloud forests. *Ann. Mo. Bot. Gard.* 74, 242–264.

Went, F. (1940). Soziologie der Epiphyten eines tropishen Urwaldes. *Ann. Jard. Bot. Buitenz.*

Wherry, E. T., and Capen, R. G. (1928). Mineral constituents of Spanish moss and ball moss. *Ecology* 9, 501–504.

Wilson, E. O. (1987). The arboreal ant fauna of Peruvian Amazon forest: A first assessment. *Biotropica* 19, 245–282.

Winter, K., Osmond, C. B., and Hubick, K. T. (1986). Crassulacean acid metabolism in the shade: Studies on the epiphytic *Pyrrosia longifolia* and other rain forest species from Australia. *Oecologia* 68, 224–230.

Yeaton, R. I., and Gladstone, D. E. (1982). The pattern of colonization of epiphytes on calabash trees (*Crescentia alata* HKB) in Guanacaste Province, Costa Rica. *Biotropica* 14, 137–140.

Zimmerman, J. K., and Olmstead, I. C. (1992). Host tree utilization by vascular epiphytes in a seasonally inundated forest (Tintal) in Mexico. *Biotropica* 24, 402–407.

12

The Ecology of Hemiepiphytes in Forest Canopies

Guadalupe Williams-Linera and Robert O. Lawton

I. Introduction

Epiphytic habits require a set of physiological, anatomical, and morphological adaptations to the rigors of the canopy environment. Hemiepiphytes have, at some time of their lives, an "umbilical" connection to the ground. Whether roots or stems, these connections buffer hemiepiphytes from problems of water and nutrient supply faced by strict epiphytes. Hemiepiphytes, as Holbrook (1991) noted, are thus the most "earthly" members of the epiphytic community.

In this review, we focus on three issues: (1) the differences between hemiepiphytes and both strict epiphytes and strict trees, (2) the role of hemiepiphytism and its umbilical connection to the ground in the evolution of epiphytic life-styles, and (3) the role of hemiepiphytes in vegetation dynamics within the forest canopy. First, we emphasize the intermediacy of hemiepiphytism and the role of metamorphic changes in hemiepiphytic life histories. Second, we suggest a possible evolutionary path from terrestrial to epiphytic habits through hemiepiphytic intermediates. Finally, we suggest that hemiepiphytes play a central role in the development of tropical forest (particularly cloud forest) canopy structure, both by competing with tree limbs in the colonization of specific canopy opportunities and by stabilizing mats of epiphytic organic soil. We emphasize woody hemiepiphytes and largely ignore the hemiepiphytic Araceae, to which Croat (1988) and Ray (1990, 1992) provide excellent introductions.

Copyright © 1995 by Academic Press, Inc.
All rights of reproduction in any form reserved.

A. Definitions

Hemiepiphytes are distinguished from epiphytes by their connections to the ground, and from lianas by shrubby or treelike crowns anchored to particular spots on their hosts. Hemiepiphytes begin their life cycle either as epiphytes and eventually send roots and/or shoots to the ground (primary hemiepiphytes) or as terrestrially established seedlings that secondarily become epiphytic by severing all connections with the ground (secondary hemiepiphytes) (Kress, 1986).

Hemiepiphytes comprise a group of great variety in growth form, in impact on their hosts, and even in their degree of dependence on hosts. They vary in a continuum of growth forms, from species that are erect and basically treelike to those that grow pendent from canopy tree limbs. In between are distinctly shrubby species, species with shelflike crowns extending from trunk or limb sides, and species that grow in scandent, clambering heaps. Their impacts on hosts range from lethal (e.g., strangler figs) to relatively benign (e.g., shrubby Ericaceae in Central American cloud forests). Some hemiepiphytes are obligate and are never (or only rarely) found on the ground. Others are facultative and grow in the canopy or on the ground. Occasionally, plants that are ordinarily terrestrial (e.g., the palm *Geonoma surtuba* in the cloud forests of Costa Rica) grow as "accidental" epiphytes (*sensu* Kress, 1986).

B. Historical and Cultural Importance

The ability of hemiepiphytes to colonize cliffs and rock outcrops (Schimper, 1888; Richards, 1952; Benzing, 1990) extends to human structures. Archeological ruins throughout the tropics are being damaged (and in some cases held together) by roots of hemiepiphytes. Old buildings in many towns in the tropics are decorated by strangler and nonstrangler hemiepiphytes (G. Williams-Linera and R. O. Lawton, personal observation). *Ficus* species, for example, are a common sight on buildings in Singapore (Wee, 1992).

Hemiepiphytes have been important symbolic elements for many ancient cultures. In the Mayan cosmogony, for instance, the strangler fig—the tree that develops from the heavenly to the terrestrial plane—represents the immortal soul that descends to a mortal body and kills it. Cora and Huichol peoples of Mexico also venerate the stranglers as symbols of life. They believe that the soul that meditated about its existence on earth undertakes a descent to the infra-world, and eventually to the "Great Tree" (Lenz, 1984). In India and Southeast Asia, stranglers have mythic stature; the Buddha found enlightenment while meditating under *Ficus religiosa*.

On a more mundane level, Mayan and Aztec texts were written on *amate*, a paper made from the bark of figs, including stranglers. *Choco* (a nonstrangler hemiepiphyte, *Oreopanax capitatus;* Araliaceae) is used in Mexico to prepare *tamal de choco,* a special dish made of corn and wrapped in *choco* leaves, which is eaten especially during the celebration of All Souls' Day. Throughout Central America the berries of some epiphytic species of *Vaccinium* (Ericaceae) and the petals of *Blakea* spp. (Melastomaceae) are prized snack foods.

II. Distribution and Origin

A. Taxonomic Distribution

The phylogenetic distribution of hemiepiphytes suggests that this life-history pattern has evolved independently a number of times. Why one particular lineage developed canopy dependence while another did not is unknown (Putz and Holbrook, 1986; Gentry and Dodson, 1987; Benzing, 1989). Twenty-five families and approximately 59 genera contain hemiepiphytes (Table 1), with at least 823 species of primary hemiepiphytes and 649 species of secondary hemiepiphytes (Richards, 1952; Kress, 1986; Putz and Holbrook, 1986; ter Steege and Cornelissen, 1989; Faber-Langendoen and Gentry, 1991). This is probably an underestimate, due largely to collecting biases. Hemiepiphytic species are most easily collected when they grow on logs or stumps in pastures or along roadbanks, which may obscure their hemiepiphytic nature. Furthermore, hemiepiphytes may be relatively undercollected because of the difficulty of gathering fertile material of large plants from the forest canopy, particularly in montane rain forests, where hemiepiphytic growth is most luxuriant and diverse.

The description of the 38 fig species native to Costa Rica by Burger (1977) illustrates the difficulty of determining how many of these are hemiepiphytes or stranglers. First, Moraceae are generally underrepresented in collections (Burger, 1977); six *Ficus* species were known from three or fewer collections, and two species were described as common, but were rarely collected. Nineteen species are presented as freestanding trees, but at least one of these (*Ficus crassiuscula*) grows as a strangler. Ten species are found as stranglers, freestanding trees, or hemiepiphytes, as circumstances (and presumably life-history stage) dictate, and nine species are presented as hemiepiphytes or freestanding trees. This confusion probably reflects both our ignorance about the plants and the genuine lability in life histories and growth forms.

B. Abundance in Different Vegetation Types

Although some hemiepiphytes are found in wet temperate forests (Oliver, 1930; Sharp, 1957), they are mainly tropical. They range from sea level to about 2500 m and occur in many different vegetation types. Although quantitative data are lacking, hemiepiphytes increase in density and diver-

Table I   Families and Genera That Contain Hemiepiphytes[a]

SECONDARY HEMIEPIPHYTES
Monocotyledonae
1. Araceae
    Amydrium Schott, 4/4 Malaysia
    Anthurium Schott, 200/550 Neotropics
    Caladiopsis Engl., 2/2 South America
    Epipremnum Schott, 15/15 Indomalaya
    Monstera Adans, 24/25 Neotropics
    Pedicellarum Hotta, 1/1 Borneo
    Philodendron Schott, 133/275 Neotropics
    Porphyrospatha Engl., 3/3 Neotropics
    Pothos L., 25/75 Indomalaya and Pacific
    Rhaphidophora Hassk., 100/100 Indomalaya and Pacific
    Syngonium Schott, 18/25 Neotropics
2. Cyclanthaceae
    Asplundia Harling, 20/82 Neotropics
    Carludovica Ruiz & Pav., 1/3 Central America
    Ludovia Brongn., 2/2 South America
    Sphaeradenia Harling, 7/38 Neotropics
    Thoracocarpus Harling, 1/1 South America

Dicotyledonae
3. Marcgraviaceae
    Caracasia Szyszyl., 2/2 Venezuela
    Marcgravia L., 50/55 Neotropics
    Norantea Aubl., 20/35 Neotropics
    Souroubea Aubl., 20/25 Neotropics
    Ruyschia Jacq., 2/10 Neotropics

PRIMARY HEMIEPIPHYTES
4. Araliaceae
    Didymopanax Decne. & Planch, Neotropics
    Oreopanax Decne. & Planch., Neotropics
    Pentapanax Seem., 2/15 Java to Formosa
    Polyscias J.R.&G. Forst, 5/80 Malaya to New Zealand
    Schefflera J.R.&G. Forst, 60/200 Pantropics
    Sciadophyllum P. Br., 5/30 South America and West Indies
    Tupidanthus Hook.f.& Thoms., 1/1 Indomalaya
5. Bignoniaceae
    Schlegelia[b]
6. Burseraceae
    Bursera, 1 Costa Rica
7. Celastraceae
    Euonymus L., 2/175 Himalayas
8. Clusiaceae
    Clusia L., 85/145 Africa, Madagascar, Neotropics
    Clusiella Planch. & Triana, 3/7 South America
    Havetiopsis Planch. & Triana, 3/7 South America
    Odematopus Planch. & Triana, 1/10 South America
    Quapoya Aubl., 1/3 South America
    Renggeria Meisn., 1/3 Brazil

(continues)

Table I—*Continued*

PRIMARY HEMIEPIPHYTES, continued
9. Cunoniaceae
    Ackama A. Cunn., 1/3 New Zealand
    Weinmannia L., 3/170 New Zealand and Neotropics
10. Dulongiaceae
    Phyllonoma Willd. ex Schult., 1/8 Neotropics
11. Ericaceae
    Cavendishia (2 spp.),[c] Neotropics
    Gonocalyx, Neotropics
    Disterigma, South America
    Sphyrospermum, South America
12. Euphorbiaceae
    Schradera (2 spp.)[c]
13. Gesneriaceae
    Drymonia (2 sp.),[c] Central America
14. Griseliniaceae
    Griselinia Forst.f., 3/6 New Zealand and Chile
15. Melastomataceae
    Blakea P. Br., 60/70 Neotropics
    Topobea Aubl., 20/50 Neotropics
16. Moraceae
    Coussapoa Aubl., 20/45 Neotropics
    Ficus L., 500/800 Pantropics
17. Myrsinaceae
    Grammadenia Benth., 6/15 Neotropics
18. Myrtaceae
    Metrosideros Banks ex Gaertn., 3/60 New Zealand
19. Potaliaceae
    Fagraea Thunb., 20/35 Malaysia-Pacific
20. Rubiaceae
    Posoqueria Aubl., 1/15 Neotropics
    Cosmibuena Ruiz & Pav., Neotropics
21. Rutaceae
    Zanthoxylum, Central America
22. Saxifragaceae
    Hydrangea, Neotropics
23. Solanaceae
    Markea,[b] Neotropics
24. Violaceae
    Melicitus
25. Winteraceae
    Drimys

[a]Number of species/number of total species in the genus. After Madison (1977) and Putz and Holbrook (1986).
[b]From ter Steege and Cornelissen (1989).
[c]From Faber-Langendoen and Gentry (1991).

sity as conditions get wetter, and appear more luxuriant in premontane, lower montane, and montane rain forests (*sensu* Holdridge, 1967). Hemi-epiphytes are particularly prominent in midmontane cloud forests of the tropical trade-wind zone, where the forest canopy is consistently bathed in orographic clouds.

1. Cloud Forest Hemiepiphytism   Hemiepiphytes are very abundant in lower and midmontane cloud forests (Beard, 1946, 1949; Richards, 1952; van Steenis, 1972; Rzedowski, 1978; Kelly, 1985; Gomez, 1989), but accounts are largely qualitative. Nonetheless, the increase in hemiepiphyte abundance in cloud forests is one of the most striking physiognomic changes in tropical vegetation as elevation increases. The abundance of epiphytes and hemi-epiphytes declines, however, at elevations above 2000–2500 m (Gentry, 1988), apparently in concert with the decline in precipitation and cloud cover.

Quantitative data on the contributions of hemiepiphytes to cloud forest vegetation are few. Although no exhaustive studies of hemiepiphytes exist for the Monteverde Cloud Forest Reserve of the central Cordillera de Ti-larán of Costa Rica, the density of hemiepiphytic trees has been cursorily examined (R. O. Lawton, unpublished data) in a series of 10 × 50-m plots in the very wet windward cloud forests (*sensu* Lawton and Dryer, 1980). The species examined [*Clusia* spp. (Clusiaseae), *Didymopanax pittieri* and *Oreopanax nubigenum* (Araliaceae), *Weinmannia pinnata* (Cunoniaceae), and *Hillia valerii* (Rubiaceae)] are all facultative primary hemiepiphytes in this setting. At maturity, these reach stem diameters >10 cm and heights of 4 m or more above the point of primary rooting or initial establishment. Individuals of these species were tallied if they were taller than 1 m from their point of rooting.

In sheltered forest in ravines and on leeward slopes, about 70% of the trees >30 cm dbh were occupied by at least one individual taller than 1 m from the hemiepiphytic tree species above. About 13% of the sampled host trees supported three or more of these hemiepiphytes. On windward slopes, about 85% of the trees >30 cm dbh held at least one of these hemi-epiphytes and 25% had three or more. In the dwarfed, or elfin, forest on very windy ridgecrests, 55% of the trees >30 cm dbh had these hemiepi-phytes and 11% had three or more. Smaller individuals of these hemiepi-phytes were not tallied in this study, but are presumably even more abundant. These results underestimate hemiepiphyte abundance because they omit smaller species [e.g., *Gonocalyx, Cavendishia* and *Vaccinium* spp. (Ericaceae), *Drymonia* spp. (Gesneriaceae), *Hydrangea peruviana* (Saxifraga-ceae), *Grammadenia* sp. (Myrsinaceae), the pendent *Begonia estrellensis*] and the large strangler (*Ficus crassiuscula*), all of which are abundant and conspicuous components of the vegetation. Hemiepiphytes occupy virtually all medium to large trees in this forest.

In this same forest, canopy structure was sampled as part of an investigation of light penetration (Lawton, 1990 and unpublished data). Hemiepiphytes contributed to 9.3% of the area in the uppermost canopy foliage layer. The strangler *Ficus crassiuscula* contributed 2%.

Other forests in similar settings are similar. In a Jamaican lower montane rain forest, the largest components of the biomass of epiphytes are hemiepiphytes of the genus *Clusia* (two species); other hemiepiphytes includes species of *Blakea, Ficus,* and *Oreopanax* (Kelly, 1985). The same genera grow in the montane forests of the El Triunfo Biosphere Reserve, Chiapas, Mexico (Long and Heath, 1991). Hemiepiphytes contribute substantially to the structure of cloud forest canopies.

2. Hemiepiphytes in Lowland Tropical Forests   Hemiepiphytes are important components of some lowland tropical forests. In the lowland wet forests of La Selva Biological Station, Costa Rica, Hammel (1990) found that plants growing or climbing on others account for about one-third of the flora, and that about 25 species (roughly 10% of the epiphytic species) are stranglers (*Ficus* spp. and *Clusia* spp.) or hemiepiphytes. In very wet lowland sites such as the Choco of Colombia, hemiepiphytes (including six species of *Topobea,* four of *Clusia,* two of *Cavendishia, Marcgravia,* and *Schradera,* and one of *Drymonia, Norantea,* and *Ficus*) are conspicuous components of the vegetation (Gentry, 1986b). Hemiepiphytes in such situations seem to be ecological replacements for lianas, a feature in which such sites resemble montane ones (Faber-Langendoen and Gentry, 1991).

In most lowland forests, hemiepiphytes occur on ~10% of the trees (Putz and Holbrook, 1986). In the moist forests of Barro Colorado Island, Panama, 9.8% of the trees were occupied by hemiepiphytes of *Clusia* (1 sp.), *Cosmibuena* (1 sp.), *Coussapoa* (2 spp.), *Ficus* (12 spp.), *Havetiopsis* (1 sp.), *Oreopanax* (1 sp.), *Souroubea* (1 sp.), and *Topobea* (1 sp.) (Todzia, 1986). In the evergreen forest of San Carlos, Venezuela, 13% of the trees supported *Clusia* spp. and *Ficus* spp. (Putz, 1983). In dry forests, some species of *Ficus* can be relatively abundant. In the floodplain forest on the Mana Pools, Zimbabwe, 12.6% of the trees supported three strangler species of *Ficus* (Guy, 1977). In very dry tropical forests, however, hemiepiphytes are not present even where vines and lianas are abundant (Lot *et al.,* 1987; R. O. Lawton and G. Williams-Linera, personal observations).

Variation in the tempo or pattern of forest dynamics may influence hemiepiphyte abundance. Lellinger (1989) noted that epiphytic ferns are conspicuously less abundant and diverse in old second-growth forest than in primary forest stands. This may simply reflect limited time for epiphytic colonization on young trees. This pattern may also result either from differences in the physical structure (and consequently microclimate) of old- versus second-growth forests or from differences in characteristics of shade-intolerant and shade-tolerant tree species.

Palm savannas are special cases for hemiepiphytes. In Los Llanos, Vene-zuela, 41% of the palms (*Copernicia tectorum*) in a seasonally flooded sa-vanna are inhabited by *Ficus pertusa* and *F. trigonata;* the density of strangler figs reaches 141 ha[-1] (Putz and Holbrook, 1989). In the savannas in Vera-cruz, along the Gulf of Mexico, *Sabal mexicana* similarly serves as host to the strangler *Ficus tecolutensis.* On the Pacific Coast of Mexico, *Sabal rosei* forms monospecific stands in which most palms host a strangler fig (Ryder, 1978; G. Williams-Linera, personal observation). Strangling figs are also wide-spread on palms in India (Davis, 1970) and can be pests in African oil palm plantations (Vanderyst, 1922).

3. Microhabitat Variation and Local Distribution   Within a single region, mi-croenvironmental conditions that vary at different spatial scales may deter-mine the uneven distribution of hemiepiphytes, for example, *Oreopanax capitatus* is significantly more abundant at forest edges than in the forest interior of a Mexican lower montane forest in Veracruz (Williams-Linera, 1992). In Monteverde, Costa Rica, *Ficus tuerckheimii* is abundant in cove for-ests but almost absent from contiguous leeward cloud forests 0.1 – 1 km dis-tant (Lawton and Dryer, 1980). In the very wet lower montane rain forests at Monteverde, *Ficus crassiuscula* is a common strangler in the taller forest of lee slopes and ravine bottoms but is essentially absent from wind-exposed montane thickets and elfin forest on the windward slopes and ridgecrests less than 100 m away (Daniels and Lawton, 1991). The extent to which such patterns of spatial distribution are the result of partitioning epiphytic op-portunities is unknown.

C. Facultative Hemiepiphytism and Evolution of the Hemiepiphytic Habit

There are advantages to starting life as an epiphyte. The most conspicu-ous is that the forest canopy is generally better lit than the forest understory (Schimper, 1888; Richards, 1952; Lawton, 1990). Not all sites in the canopy are sunny, however, and there are shade-tolerant as well as shade-intolerant epiphytes (Went, 1940; Benzing, 1990). Epiphytes may also avoid flooding, fire damage, and browsing by terrestrial animals (Putz and Holbrook, 1986).

These advantages probably first accrued by accidents of dispersal. Acci-dental epiphytes possess no characters that are obviously specific for canopy life (Oliver, 1930; Sharp, 1957; Benzing, 1990). Rather, their occurrence is related to the high precipitation and humidity that allow the germination of seeds in favorably moist epiphytic sites. In wet temperate forests, acci-dental epiphytes [e.g., *Rhododendron maximum* (Ericaceae) and *Rubus cana-densis* (Rosaceae) on *Betula alleghaniensis* in the Smoky Mountains of east-ern North America] may occupy such wet "safe sites" in knotholes, near crotches, in crevices of bark, and even in the moss on the upper surfaces of large branches (Sharp, 1957). In tropical cloud forests, such sites are much more common. In the wind-exposed lower montane rain forests in Monteverde, the seedlings, saplings, and poles of most tree species are occasionally found as accidental epiphytes (Lawton and Putz, 1988). Members of 37 terrestrial plant species occur as accidental epiphytes in a southern Ecuadorian cloud forest (Bøgh, 1992). Accidental epiphytes such as these are probably similar to the plants from which modern facultative and obligate hemiepiphytes evolved.

Accidental and facultative epiphytes face an evolutionary dilemma: terrestrial and arboreal environments differ. Two differences are particularly critical: drought frequency and what Raup (1957) termed the "stability of the site." To use the relatively well-lit environment of the forest canopy, epiphytes must both hold onto their hosts and acquire water from an environment where it is scarce, relative to the ground. These functions are generally a matter for roots, although many strict epiphytes (as opposed to hemiepiphytes) have striking foliar and stem adaptations for water acquisition, storage, and management. Genetically based variation in root system structure exists (Zobel, 1975), given the phenotypic plasticity of growth allocation involved in root system development (Levitt, 1980). Selection among the accidentally or facultatively epiphytic individuals of a population should favor those with root systems that most effectively and efficiently attach themselves, and those with root systems that can most readily establish an umbilical root connection to the soil as their growth outstrips the water supplies available in their hosts' crowns.

The abundance of accidental and facultative epiphytes makes tropical montane and lower montane rain forests important foci for the evolution of epiphytism. This arises in large part because the distinction between terrestrial and epiphytic rooting becomes blurred in such forests. Sheathing mats and pendent masses of bryophytes grow on limbs, trunks, fallen logs, and prop-root tangles, accumulating an organic soil by their decomposition (e.g., Nadkarni, 1981, 1984; Nadkarni and Matelson, 1991). The arboreal and terrestrial habitats may be quite similar, particularly at the spatial and temporal scales of seedlings, and so reduce the mechanical and water acquisition problems associated with epiphytism.

The biogeography of cloud forests also encourages adaptive radiation in epiphytes. Cloud forests, which are dependent on orographic or convective channeling of air up mountainsides, have inherently patchy distributions. The genetic isolation thus imposed should foster local adaptation. Further, the accidents of "island" colonization could provide different taxa with access to the same adaptive zone in different cloud forests. Gentry and Dodson (1987) suggested that these factors produced an "explosive speciation" in the foothill areas of the northern Andes and southern Central America (Gentry, 1982, 1986a,b).

1. A Case History: The Facultative Hemiepiphyte Didymopanax pittieri   The shade-intolerant tree *Didymopanax pittieri* (Araliaceae) occupies lower montane and montane rain forests in Costa Rica and western Panama (Lawton, 1983). As a facultative hemiepiphyte in the very wet forests of the Monteverde Cloud Forest Reserve, it leads an intriguing split existence. In the taller cloud forest of sheltered slopes and ravines (the Windward Cloud Forest of Lawton and Dryer, 1980), *D. pittieri* is found only as a hemiepiphyte perched in the crowns of the 20- to 30-m-tall canopy trees at a density of two to three reproductive hemiepiphytic adults (trees larger than 10 or 15 cm in trunk diameter) per hectare. Seedlings and saplings are difficult to detect from the ground. This hemiepiphytic portion of the population is widespread, covering about 10 km^2 in the central Cordillera de Tilarán. In the dwarfed elfin forests of wind-exposed ridgecrests, however, *D. pittieri* is the dominant shade-intolerant terrestrially rooted tree and one of the most abundant trees overall. There, the terrestrial portion of the population occurs at a density of 40–50 trees ha^{-1}.

Hemiepiphytic *D. pittieri* are abundant even in the elfin forest, particularly in smaller size classes. Saplings are found commonly as hemiepiphytes and grow on nurse logs in treefall gaps. Seedlings, however, are very rare near the ground. They are sparsely scattered in the bryophyte and humus mats on nurse logs, limbs of fallen trees, and on tilted trunks of understory trees within gaps. Climbing into the canopy reveals that seedlings are quite abundant as epiphytes and clearly adapted for epiphytic establishment. Soon after germination within the bryophyte mat, a tuberous swelling develops at the base of the stem, apparently a water storage adaptation for canopy drought (Putz and Holbrook, 1986).

In the elfin forest, the hemiepiphytic and terrestrial portions of the population are demographically coupled. Saplings on light-gap nurse logs have in the lower part of the stem the stout stems and very short internodes that are characteristic of wind-exposed, small hemiepiphytic individuals and the twigs of adult terrestrially established trees. In light gaps, which are more protected from the wind, the saplings produce long and thin internodes. The terrestrial portion of the *D. pittieri* population generally starts by epiphytic germination and seedling establishment. Some trees may grow to maturity as hemiepiphytes, whereas others may survive host collapse, continue growing on the nurse log, and eventually become terrestrial adults.

What are the evolutionary dynamics within this species? Is *D. pittieri* a hemiepiphyte for which the relatively gentle fall of elfin forest trees allows juveniles to grow to terrestrial adulthood? If *D. pittieri* is a hemiepiphyte, we have the paradox that it reaches larger sizes, greater density, and is more fecund as a terrestrial adult. Is this a terrestrial lineage that has evolved characteristics allowing it to use epiphytic juvenile establishment as a regenerative strategy? If so, it has lost terrestrial seedlings. Each case presents possible adaptive transitions between hemiepiphytic and terrestrial existence.

2. Another Case History: The Strangler Ficus crassiuscula   The canopy tree *Ficus crassiuscula* (subgenus Pharmacosycea) is restricted to tall lower montane rain forest in the Monteverde Cloud Forest Reserve (Windward Cloud Forest of Lawton and Dryer, 1980). It has evolved the strangling habit independently of the better-known stranglers of the subgenus Urostigma (Lawton, 1986). Although there is some confusion about its nearest relatives (DeWolf, 1960; Burger, 1977; Ramirez, 1988), it is derived from a lineage of terrestrially established trees (Lawton, 1986, 1989). *Ficus crassiuscula* exhibits host, as well as habitat, preferences (Daniels and Lawton, 1991) and has a complex, metamorphic life history (Daniels and Lawton, 1993).

Seeds of *F. crassiuscula* germinate in relatively well-lit bryophyte mats, usually in the forest canopy. The seedling grows into a viny sapling as the initial stem elongates and branches sparsely to form a sprawling tangle, consisting of many (2–20 or more) primary stems generally less than 0.5 cm in diameter. The stems grow over and through existing epiphytic vegetation and are at least partially appressed to the epiphytic mat of bryophytes and soil-like organic debris. The stems root freely where nodes contact the substrate. Individual viny saplings can cover several square meters and even sprawl from one host to another. Any stem of a viny sapling may undergo a transition to an erect trunk with secondary growth, but usually only one becomes treelike and sends a root down the host trunk to the ground. Eventually, both crown and root enlarge, and the host is both overtopped and encased in roots (Daniels and Lawton, 1993). Although the demographic data are lacking, the process must take on the order of a century.

There are many presumed advantages of the strangling habit (e.g., early access to canopy light conditions, reduced investment in mechanical support, rapid maturation, and avoidance of falling debris, terrestrial browsers, and soil-based pathogens; Richards, 1952; Dobzhansky and Murca-Pires, 1954; Whitmore, 1975; Corner, 1976; Putz and Holbrook, 1986), but the evolutionary path to strangling is not clear. *Ficus crassiuscula* has a terrestrial lineage, which probably (like so many other species) produced accidental epiphytes in the very wet cloud forest habitat. In growing large, they probably fell from their hosts, having lost their grip or overburdened their perch. This is a problem in apparently well-adapted modern stranglers (G. Williams-Linera and R. O. Lawton, personal observations). It is easy to imagine that in the ancestral *F. crassiuscula* population, individuals of lax growth form early in life might sprawl about to end in more secure spots, like major crotches, and thus be more likely to survive to grow large and reproduce.

The genus *Ficus,* with its diversity of growth forms, is evolutionarily labile with respect to development (Corner, 1976). The metamorphosis from viny sapling to a hemiepiphyte with an erect trunk allows *Ficus crassiuscula* to "forage" for suitable sites to create a trunk. A heterochronic retention of the viny growth form with respect to the onset of sexual maturity would produce a derived viny species, as Corner (1976) has suggested for some of the viny Old World figs.

III. Ecology

A. Physiological and Morphological Adaptation to Life in the Canopy

The arboreal environment places strong restrictions on the size, activities, and life span of its inhabitants. Resource acquisition in plants is influenced by the area available for exploitation, and for epiphytes this area is constrained. The position of epiphyte crowns depends on the placement of supports. Roots can only run over the host bark or extend into the air. Furthermore, growth itself poses problems. The epiphyte that becomes too heavy or binds too tightly runs the risk of killing its limb and falling to the forest floor. Life span is limited in a similar manner; epiphytes can live no longer than their hosts. Trees must generally grow large to be good hosts, so epiphytes must commonly live between host accession to the forest canopy and host death. Hemiepiphytism is a way of life by which these restrictions can be loosened or avoided in part. Hemiepiphytes adjust to problems of size and scale largely by metamorphosis, a notion implicit in the work of Putz and Holbrook (1986, 1989) and explicitly elaborated in Ray's (1990, 1992) analyses of growth and foraging behavior of climbing Araceae.

1. Photosynthesis, Water, and Nutrient Relations   Hemiepiphytes share many physiological and morphological adaptations to the canopy setting with true epiphytes. They both must deal with a limited water supply, and so commonly have somewhat xeromorphic foliage, with thick cuticles, sunken stomata, and a water-storing hypodermis (Putz and Holbrook, 1986). Some succulent-leaved *Clusia* species exhibit CAM photosynthesis (Tinoco and Vazquez-Yanes, 1983; Ting *et al.,* 1985, 1987; Schmitt *et al.,* 1988). *Clusia minor* has facultative control of its photosynthetic pathway; it uses C_3 photosynthesis in the rainy season and CAM in the dry season (Ting *et al.,* 1987; Borland *et al.,* 1992). None of the following exhibit CAM: *Ficus* spp. (neither epiphytic nor terrestrially rooted individuals of the species), *Blakea grandifolia* (Melastomaceae), *Coussapoa villosa* (Rubiaceae) (Ting *et al.,* 1987), or *Oreopanax capitatus* (Araliaceae) (G. Williams-Linera, unpublished data).

The primary advantage to hemiepiphytism is the root connection to the ground, which seems to be more important for water supply than for nutrient acquisition (Putz and Holbrook, 1986, 1989, unpublished data). Hemiepiphytes in their epiphytic youth generally do not appear to suffer nutrient deficiencies, and epiphytic humus is often nutrient-rich (Putz and Holbrook, 1986, 1989). In cloud forests, epiphytic humus is comparable to the organic horizon of the terrestrial soil (Nadkarni, 1981, 1984). Indeed, apogeotropic roots from terrestrial trees may leave the soil and ascend trunks, apparently following nutrient concentration gradients (Putz and Holbrook, 1986, 1989; Sanford, 1987), and in very wet forests adventitious canopy roots may exploit the organic soil on the stems that sprouted them (Nadkarni, 1981). *Ficus pertusa* and *F. trigonata* on *Copernicia* palms in Venezuela have vesicular-arbuscular mycorrhizae on terrestrial roots, but not on epiphytic ones (Putz and Holbrook, 1989), which suggests that the carbohydrate cost of the symbiosis is not worth the nutrient gain. Putz and Holbrook (1989) note that this could be a colonization problem, but it is more likely that it is simply not cost-effective to maintain the symbiosis in a nutrient-rich medium.

Root connections to soil water supplies may have a set of morphological and physiological consequences. Surprisingly, the root xylem of epiphytic and descending roots of *Ficus pertusa* and *F. trigonata* differ neither in vessel density nor vessel diameter (Putz and Holbrook, 1989), which suggests no anatomical adjustment to increased water flow. Five species of hemiepiphytic figs have higher stomatal densities in the terrestrially rooted phase, which implies greater transpiration capacity, but *Oreopanax pittieri* and *Clusia minor* do not show similar increases in stomatal density (Putz and Holbrook, 1986).

Physiological and structural attributes of nine species of hemiepiphytes in a palm savanna, a mid-elevation, and a high-elevation cloud forest have been studied by Holbrook and Putz (1992; F. Putz, personal communication). Epiphytic individuals have lower stomatal conductances, higher leaf water potentials, higher osmotic potentials at full saturation, and more elastic cell walls compared with conspecific trees. Specific leaf areas ($cm^2 \ g^{-1}$) of epiphytic *F. pertusa* and *F. trigonata* were higher than in terrestrially established trees (Putz and Holbrook, 1989), and leaf life spans were shorter for epiphytes than for trees (F. Putz, G. Romano, and N. Holbrook, unpublished data).

The extent to which water availability influences hemiepiphyte phenology is unclear. Strangler figs in the semideciduous lowland forest on Barro Colorado Island, Panama, initiate fruiting and produce new leaf flushes year-round. Nevertheless, mean dates of fruit initiation and leaf flush of all species studied fell within the 4-month dry season (Windsor *et al.*, 1989). In contrast, leaf production of *Oreopanax capitatus* in a Mexican lower montane forest (G. Williams-Linera, unpublished data) and *F. pertusa* and *F. trigonata* in the Venezuelan llanos (F. Putz, G. Romano, and N. Holbrook, unpublished data) was concentrated during the months with higher precipitation. *Oreopanax capitatus* showed a flowering peak in August–September, but fruiting was not confined to any particular season. Fig flowering phenology showed no seasonality, as was expected owing to the dependence on particular pollinators.

2. Roots of Attachment and Travel   Metamorphosis in hemiepiphytes is most marked in their root systems. Three functions and associated adaptive morphologies are distinguished: foraging, attachment, and travel. Foraging for and acquiring nutrients and water are "normal" root functions. Attaching the shoot to the substrate is also a normal root function, but it may become unusually complex for hemiepiphytes (Schimper, 1888). Extended root travel through the air is unique to hemiepiphytes.

Juvenile hemiepiphytes, like small true epiphytes, may require no special modifications of the roots for attachment to limbs or trunks. The accumulation of organic humus and a binding mat of roots may produce a rooting medium sufficiently like a terrestrial soil (particularly in cloud forests; Nadkarni, 1981, 1984; Nadkarni and Matelson, 1991) to render unnecessary any root modification for support. However, hemiepiphytes with roots to the ground are not limited in size by the water- and nutrient-holding capacity of their arboreal perch, and as they grow large, they may encounter mechanical difficulties. Hemiepiphytes are essentially limb-mimics and must bind themselves to the host as effectively as a limb of similar size, but without the advantage of wood continuity.

The root frameworks by which hemiepiphytes hold on are mechanically sophisticated structures, both anatomically and morphologically. Tension wood develops in the aerial roots of *Ficus benjamina* as they connect to the ground, which is documented by anatomical examination of secondary cell wall structure (Zimmermann *et al.*, 1968). Such tension wood development is widespread in hemiepiphyte attachment frameworks, as can be demonstrated by sawing apart elements of the frameworks and observing the shortening typical of tension woods (R. O. Lawton, personal observation of *Clusia* spp., *Oreopanax* spp., *Blakea* spp., and *Didymopanax pittieri*). Root grafts, by which anastomosing webworks of roots are created, form anatomically in *Ficus globosa* by fusion of root hairs, followed by compression of the cortices of the roots in contact. Rays at the periphery of the contact zone produce meristematically active parenchyma, which fuses the roots and develops an active cambium (Rao, 1966).

Engineering analyses by Mattheck and co-workers suggest that the attachment network of root cables and trusses has shapes optimized to eliminate or minimize notch stresses at joints and branches (Mattheck, 1991; Mattheck and Burkhardt, 1990). Because stresses focus at notches, this shape optimization allows plants to maintain strength and reliability while reducing carbohydrate expenditure. The sophistication of hemiepiphyte root frameworks is such that Mattheck and Burkhardt (1990) have used them as structural models in engineering simulations that follow an "adaptive growth" algorithm to maintain constant surface stress on the mechanical elements (Kubler, 1987). Mattheck and Burkhardt (1990) have developed computer-aided optimization procedures to design structures of remarkable durability, which comes as no surprise to anyone who has depended on hemiepiphyte roots to climb tropical trees. The specialization of hemiepiphyte root attachment frameworks suggests a history of strong selection for staying with the host.

The canopy environment is, in most circumstances, a difficult place from which to get a root to the ground. Hemiepiphytes appear most likely to get established in relatively moist humus accumulations in crotches and knotholes, or on major limbs. Roots growing toward the ground must go through inclement territory. Roots are generally designed to absorb and transport water, but the anatomical attributes and surface : volume ratio appropriate for water acquisition are inappropriate for roots descending through the air, or over arid zones of the host trunk. The freely descending aerial roots characteristic of hemiepiphytic Clusiaceae and Araceae, and some of the hemiepiphytic species of *Ficus,* resemble stems more than roots (Putz and Holbrook, 1986). In *Ficus benghalensis,* they are thicker than ordinary roots, lack root hairs, and possess a large pith, a thicker cortex and pericycle, and a well-developed periderm with chloroplasts and numerous lenticels (Kapil and Rustagi, 1966). The apices of aerial roots of some species of hemiepiphytes are covered with a gelatinous slime, as described for *Hedyosmum arborescens* by Gill (1969). These attributes are best interpreted as reducing water loss and dehydration of the traveling root.

B. Population Structure and Dynamics

Our understanding of hemiepiphyte population biology is dismal. The broad outlines of hemiepiphyte life history are best known, but the recent discovery of two-stage dispersal systems in the strangler *Ficus microcarpa* (Kaufmann *et al.,* 1991) and the recognition of the importance of metamorphic transitions in both secondary and primary hemiepiphytes (Daniels and Lawton, 1991, 1993; Ray, 1992) suggest that broader and more detailed studies of hemiepiphyte life histories will provide us with important insights. More appalling is our lack of information on hemiepiphyte population structure and demography, as our ability to assess the conservation challenges faced by this distinctive and structurally important component of tropical forest canopies is lacking.

1. Life Cycles

a. Seed Dispersal and Establishment Sites   Hemiepiphyte seeds face the difficulty of reaching safe sites for germination. All plants require safe sites for water, nutrients, opportunities to establish partnerships with symbionts, and havens from predators, parasites, and disease. Hemiepiphyte safe sites are also restricted to the world of boles, limbs, and crotches—a skimpy target, and one that some seed dispersers cannot reach. Because hemiepiphytes become a growing burden to their hosts and may lower host fitness, selection may occur for host attributes that hinder hemiepiphyte establishment and growth.

Seeds of hemiepiphytes are mainly dispersed by animals, particularly birds and bats, but there are some conspicuous and successful wind-dispersed hemiepiphytes (e.g., *Metrosideros* spp. in the Australasian region and *Cosmibuena* spp. in the neotropics) (Putz and Holbrook, 1986). All produce many small seeds rather than a few large ones. Large-seeded taxa well represented in cloud forests (e.g., Lauraceae and Neotropical Myrtaceae) are notably absent from the hemiepiphyte flora. Dispersal has been little studied, particularly at the community level. Do hemiepiphytes, as a group, differ from other plants in the same community in their dispersal mechanisms and characteristics? Are hemiepiphytes dispersed in similar ways in different biogeographic realms, or in different habitats within a region?

Seed dispersal from hemiepiphytes seems to depend on plant population structure and fruit production. For a *Clusia* species (Mesquita, 1992) and the mistletoe *Phoradendron robustissimum* (Sargent, 1993), seed removal increases with fruit production of individual plants, and with the distance to conspecific neighbors. The salient feature of frugivory on *Ficus pertusa* in Monteverde seems to be variation among individuals—in plant attributes (e.g., fruit size, time of fruiting, and crop size) and in the kinds of frugivores that visit, visitation rates, the timing and length of the visits, and feeding rates (Bronstein and Hoffmann, 1987).

Clear and effective adaptations to the problems of sticking seeds to the limb on which they are deposited exist (e.g., in mistletoes; Docters van Leeuwen, 1954), but how widespread such adaptations are is not known. Some hemiepiphytic figs (in both subgenera Urostigma and Pharmacosycea) produce seeds that are defecated with a sticky, hygroscopic, viscid coat (King, 1888; Bessey, 1908; Ramirez, 1976, 1977).

Another adaptation to the dispersal problems of hemiepiphytes is the multiphase dispersal system, which is more common than we currently realize. *Ficus microcarpa* in south Florida, for example, has a two-stage dispersal system in which small seeds are ingested and evacuated in groups by vertebrates, and then secondarily moved by ants (Kaufmann *et al.*, 1991). Because the seeds are small and provide few reserves, the seedlings must become established quickly after germination. Ants that forage in the canopy remove seeds from bare branches and carry them to humus pockets, where their nests are located. The effect is to focus the seed rain on sites that are most favorable to germination and establishment. In the lowland forests of Manu, Peru, 202 of 203 *Ficus paraensis* examined by Davidson

(1988) were growing in association with epiphytic ant gardens. The ant *Camponotus femoratus,* which is instrumental in the establishment of epiphytic ant gardens in this setting, quickly carries to its nest seeds of the hemiepiphytic *Ficus paraensis,* but not seeds of a terrestrial *Ficus* species.

b. Germination and Establishment   The bulk of what we know of hemiepiphyte seed germination comes from a series of experimental studies on species of *Ficus.* Both light and moisture are important. Darkness inhibits germination of *Ficus aurea* (Bessey, 1908), *F. pertusa,* and *F. tuerckheimii* (Titus *et al.,* 1990), but not *Oreopanax capitatus* (G. Williams-Linera, unpublished data). Germination of *F. tuerckheimii* (but not *F. pertusa*) is enhanced in permanently moist Petri dishes compared to well-watered potting soils (Titus *et al.,* 1990), and watering the soil under *F. religiosa* results in substantial germination (Galil and Meiri, 1981). Passage through frugivores has no impact on germination of *F. pertusa* or of *F. tuerckheimii,* but significantly improves both percentage germination and speed of germination of *F. benghalensis* (Midya and Brahmachary, 1991). R. L. Brahmachary (personal communication) suggests that passage through birds is necessary to destroy germination inhibitors in the pulp. In contrast, Ramirez (1976) found that ingestion by birds and bats did not alter seed viability, nor did it relieve the inhibition of germination by the viscid coat, which was deactivated by soil bacteria. In *F. crassiuscula,* gut passage seems necessary for expansion of the viscid seed coating (Talley and R. O. Lawton, unpublished data). The potentially dislodging effect of throughfall and stemflow would seem to select for rapid germination in hemiepiphytes (unless two-stage dispersal is in force), but we have little evidence. *Clusia* species in the lower montane rain forests of Monteverde germinate immediately; clusters of seedlings can be found in fresh bird droppings.

Germination and establishment in the field are difficult to follow, and most evidence is anecdotal. For example, Schimper (1888) noted that *Clusia rosea* almost always becomes established at the base of large epiphytic bromeliads, the leaking tanks of which might create moist sites. Direct evidence of the benefits of epiphytic establishment was recently provided by Mesquita (19920, who reported that seeds of *Clusia grandiflora* germinating in the canopy were more likely to survive their first year and the survivors grew more than seeds germinating in the understory or on the forest floor.

Substantial accumulations of epiphytic humus seem to be of importance in early establishment of hemiepiphytes (Putz and Holbrook, 1986; Laman, 1993). The common establishment of hemiepiphytes in crotches (Bessey, 1908; Kelly, 1985; Todzia, 1986) and in palm leaf axils (Vanderyst, 1922; Troth, 1979; Putz and Holbrook, 1989) is probably related to humus accumulation. Laman (1993) noted that *Ficus stupenda* produces a large number of seeds, and fruits are taken by a diverse group of frugivores, but adults occupied only a small proportion of the available large substrate of dipterocarps that seem to be its preferred hosts. Experimental seed plantings revealed that germination depended largely on the availability of a humus substrate, but humus is no guarantee of success. Most seemingly suitable sites were unoccupied by fig seedlings, perhaps because growth and survival—<2% of seedlings survived a year—depended on the availability of light as well as a humus soil. Fig seed densities in pockets of epiphytic humus can be quite high (10–20/ml of epiphytic humus from *Ficus tuerckheimii;* Titus *et al.,* 1990), so attack by fungi or insects (e.g., Slater, 1972) should also be examined.

Laman (1993) suggested that the coexistence of large groups of ecologically similar hemiepiphytic figs is mediated in the dipterocarp forests of Southeast Asia by the low levels of competition for space that result from failures in early establishment. Accumulations of epiphytic humus are not necessarily competitively benign places; Vance and Nadkarni (1992) found fine root densities in epiphytic humus in crotches at Monteverde to be 20% greater than the terrestrial root mat.

Although the upper reaches of forests are generally better lit than the areas near the ground (e.g., Torquebiau, 1988; Lawton, 1990), canopy disturbance may be important in providing high-light environments for the establishment of shade-intolerant hemiepiphytes. In the tropical rain forests along the Ivory Coast, about 80% of the hemiepiphytic *Ficus* individuals occurred in conjunction with damaged sites in the crowns of their hosts (Michaloud and Michaloud-Pelletier, 1987).

c. Vegetative Propagation and Fusion   The plasticity of growth that is so marked in hemiepiphytes contributes to interesting life-history phenomena. There is no quantitative information on the extent of vegetation propagation in the wild, but many hemiepiphytes are readily transplanted and propagated vegetatively in horticultural settings (Putz and Holbrook, 1986). In the cloud forests of Monteverde, some hemiepiphytes (e.g., *Weinmannia* spp., *Hillia valerii, Oreopanax* spp., and *Blakea* supp.) are able to survive serious damage, including fragmentation, and can thus propagate vegetatively. Hemiepiphytes with scrambling or sprawling growth forms seem particularly prone to develop several separate webs of attachment roots. The extent to which this type of vegetative spread is responsible for clusters of conspecific hemiepiphyte "individuals" within a host is unknown.

The recognition that genetic individuals within species of strangling figs can fuse (Thompson *et al.,* 1991) raises the botanically strange specter of the functional "individual" being a social entity. The extent of physiological integration involved in such circumstances needs investigation, but the potential mechanical benefits of constructing a common trunk are clear. No single small individual could metabolically afford to construct a tall self-supporting trunk, but several together could. In very wet cloud forests, some "trees" prove to be amalgams of hemiepiphytes with roots wrapped about each other in a caricature of a trunk, reflecting the existence of a host long dead and decomposed. Such fusion and amalgamation raise the prospect that stranglers might be analogous to pack predators like wolves. Are hosts more likely to be overgrown and killed by a group than by single genets?

2. Demography We are almost entirely ignorant of hemiepiphyte demography as these plants are difficult to approach, much less measure. For a few species, there is some information on population density, and somewhat less on population size or life-history stages (e.g., Todzia, 1986; Putz and Holbrook, 1989; Daniels and Lawton, 1991). In montane cloud forest, topographic influences on microclimate and forest structure may impose a metapopulation structure on hemiepiphytes (e.g., Lawton, 1983; Daniels and Lawton, 1991), but the demographic and evolutionary consequences are unexplored. These consequences are likely to be most striking in the case of facultative hemiepiphytes, such as *Didymopanax pittieri* at Monteverde.

The central problem in hemiepiphyte population biology is measuring growth rates, or at least the rates of transition among life-history stages. Hemiepiphytes are generally too remote from the ground to easily measure stem diameter, and it taxes ingenuity to pick a meaningful place to measure stem growth on strangling figs, or on the welter of stems in scrambling shrubs. Ground-based techniques using transits mounted at marked permanent stations to measure the angles subtended by stems or crowns is a potential solution to the problem. The unavoidable measurement errors will dictate long periods between remeasurements to ascertain growth rates. A life-history stage-based demography faces similar problems in obtaining stage-to-stage transition rates.

C. Community Ecology

Hemiepiphytes contribute to biogeochemical cycling, increase the diversity of structure and biological interaction in forests, and influence patterns and rates of disturbance. These influences increase as conditions become wetter, because the abundance and diversity of hemiepiphytes increase. They are greatest in tropical lower montane and montane rain forests.

1. Biogeochemical Cycling Hemiepiphytes contribute directly to nutrient cycling via litterfall and uptake in rough proportion to their contribution to the forest canopy, but may also serve as epiphytic keystones by providing the most durable and extensive components of the weft of epiphytic roots. Free-climbing in trees suggests that these roots supply most of the structural integrity of the mats of epiphytic organic matter that is characteristic of diverse cloud forests (R. O. Lawton, personal observation). Consequently, the amount of epiphytic organic matter that accumulates may depend on the patterns of hemiepiphyte establishment. Because hemiepiphyte establishment is likely to be influenced by the presence of accumulated epiphytic organic matter, this establishes a potential positive feedback, eventually arrested by disturbance in the canopy community.

2. The Hemiepiphytic Role in Canopy Structure   Hemiepiphyte community structure, particularly its dynamic aspects, is largely unexplored. Biological interactions influence hemiepiphyte growth and distribution, and disturbance to forest canopies profoundly influences the canopy environment and may determine critical features of the opportunities and hazards faced by hemiepiphytes.

   a. Host Specificity   Hemiepiphytes may grow on many types of hosts, but their distribution is not random. Individual hemiepiphyte species have preferences among local habitats (Daniels and Lawton, 1991; Williams-Linera, 1992), among host species within habitats (Todzia, 1986; Daniels and Lawton; 1991), among trees of different sizes (Leighton and Leighton, 1983; Daniels and Lawton, 1991), and among different places on trees (Todzia, 1986; Michaloud and Michaloud-Pelletier, 1987; Daniels and Lawton, 1991). Different hemiepiphytes within a forest may differ in their establishment preferences (Todzia, 1986; Putz and Holbrook, 1986).

   Palms seem to provide especially favorable opportunities for hemiepiphytes, particularly in alternately dry and flooded savannah environments (Vanderyst, 1922; Troth, 1979; Putz and Holbrook, 1986, 1989). This may be because palms are favored roosts of dispersers, particularly bats (Guy, 1977; Morrison, 1978; August, 1981), but nutrient relations are also important. Putz and Holbrook (1986, 1989) showed that the leaf bases of the palm *Copernicia tectorum,* the favored host in Venezuela of the two strangler fig species *F. pertusa* and *F. trigonata,* hold a nutrient-rich epiphytic humus thoroughly permeated by fig roots. Species of hemiepiphyte may, however, differ in the extent to which they inhabit palms. In the old-growth tropical moist forest on Barro Colorado Island, Panama, the *Scheelea zonensis* palms hosted 20% of the hemiepiphytic *Ficus,* but no individuals of *Clusia odorata, Coussapoa panamensis,* or *Souroubea sympetala* (Todzia, 1986). Tree ferns also seem "easy targets" in some places; in a Jamaican montane forest, Newton and Healey (1989) found that the facultative hemiepiphyte *Clethra occidentalis* established epiphytically only on the tree fern *Cyathea pubescens.*

   In wetter and more complex tropical forests, however, host specificity is less apparent. In general, hemiepiphytes occur more often in large trees. In Borneo, 64% of hemiepiphytic *Ficus* had host trees >60 cm dbh, but these comprised only 17% of the trees >20 cm dbh (Leighton and Leighton, 1983). In forests of the Ivory Coast, mean diameter of host trees of *Ficus* was 72 cm (Michaloud and Michaloud-Pelletier, 1987). At La Selva, hemiepiphytes were found only on trees >45 cm dbh (Clark and Clark, 1990). In a Mexican cloud forest, hemiepiphytes establish only on individuals >30 cm dbh (Williams-Linera, 1992).

This pattern raises a question: Do hemiepiphytes occupy big trees because they are intrinsically better places to get established and grow to maturity, or are big trees just bigger targets for random establishment? In the case of the strangler *Ficus crassiuscula* in lower montane rain forest at Monteverde (Daniels and Lawton, 1991), host abundance (considering all forest trees as potential hosts) was assessed in two ways: the number of host stems per host size class and the host trunk surface area as a proportion of that of the whole forest. Using the first measure, *F. crassiuscula* was five times more abundant than expected on hosts >80 cm dbh, twice as abundant on hosts 40–60 cm dbh, and one-third as abundant on hosts 10–20 cm dbh. This seems to be only a matter of bigger targets. When host abundance is calculated as the contribution of a host size class to the total trunk surface of the forest, then *F. crassiuscula* occupies trees in proportion to the surface they present.

The question of whether some host species are occupied more than others presents similar problems. Some examples are clear; 58% of *Hura crepitans* host hemiepiphytes, but only 1% of *Quararibea asterolepis* do (Todzia, 1986). Even though mature *H. crepitans* are larger trees with stout trunks, and *Q. asterolepis* trees are more slender, this pattern reflects real differences in the likelihood of hemiepiphyte establishment and growth to maturity on these species. The strangler *Ficus crassiuscula* at Monteverde was three to five times as abundant on *Guarea* spp., and half as abundant on *Persea schiedeana,* as expected from these trees' contributions to the total trunk surface of the forest (Daniels and Lawton, 1991). These results raise questions on the patterns of host occupation in the hemiepiphyte community, and on the mechanisms governing such host preferences. Do all hemiepiphyte species prefer the same host species, or is there partitioning of the tree guild by hemiepiphyte species? How are these host preferences determined?

Some sites in the forest canopy appear better for hemiepiphytes than others, although the matter has been difficult to address quantitatively. Went (1940) provided the classic plant sociological analysis of epiphytic communities in the montane forests of Java, using binoculars to establish the existence of epiphytic associations related to host species, substrate (humus accumulations versus bare bark), and position on the host. The approach is seldom used for hemiepiphytes; there have been few attempts to reduce patterns of distribution to life-history processes, and fewer to explore underlying physiological mechanisms. In the tropical moist forest of Barro Colorado Island, some hemiepiphytes clearly differ from others in the sites they occupy on their hosts (Todzia, 1986). *Coussapoa panamensis,* and the bat- and bird-dispersed strangler figs, grow mainly on the upper half of host trunks, whereas *Clusia odorata* grows mainly on crotches or primary branches. Examination of the sites occupied by life-history stages or size classes may provide insights into the origins of such patterns (Daniels and Lawton, 1991). Viny saplings, the youngest life-history stage, of *Ficus crassiuscula* at Monteverde are found mainly low on the sides of tree trunks, and less commonly in crotches, on limbs, and on stump tops (Daniels and Lawton, 1991). In comparison, juveniles that have established an erect trunk are half as abundant on trunk sides and twice as abundant on crotches and on stump tops. Adult *F. crassiuscula* are four times as abundant as juveniles on crotches, and commensurately less abundant on trunk sides and the tops of stumps. Assuming that the adults grew from viny saplings that were distributed as the current population of juveniles, successful growth to maturity for these hemiepiphytes is more likely on crotches than elsewhere.

3. Disturbance in the Canopy   Hemiepiphytes influence the rates and patterns of canopy disruption in two ways: by burdening their hosts to the point of host collapse and by losing their grip and falling on their own. The extent of this influence increases with the luxuriance of hemiepiphytic growth. Although quantitative data are almost completely lacking, the processes are basically clear and related to the better understood relationships of lianas and their hosts.

Let us enter the web of relationships at the point of hemiepiphyte colonization of a host. Because increased light availability in the canopy seems to be the main advantage of the hemiepiphytic life-style, the likelihood of hemiepiphyte establishment and growth should be higher on hosts that offer well-lit spots than on those that do not. hence, snags or trees with damaged crowns, or trees adjacent to edges of treefall gaps, streams, roads, or agricultural land, should be more susceptible to establishment of hemiepiphytes. Williams-Linera (1992) has reported this edge effect for *Oreopanax capitatus* in Mexican montane forest, and Michaloud and Michaloud-Pelletier (1987) pointed out that strangling figs in the forests of the Ivory Coast are often established where limbs have broken from their hosts.

Increased hemiepiphyte growth might lead to host deterioration. First, crown competition may occur between hemiepiphyte and host. Some hemiepiphytes are more shade-tolerant than their hosts and grow in the crown interior. Others (e.g., *Coussapoa panamensis*) grow laterally to position their crown at the edge of the host's, and others (e.g., *Clusia odorata*) produce a compact crown that may minimize host shading (Todzia, 1986). Many strangling figs simply spread their crown above that of their host. In all but the first case, host crown space is usurped. Second, as hemiepiphytes grow, root competition with the host should increase. Third, as the epiphytic root mat extends, more epiphytic humus accumulates, other epiphytes colonize, and the web of attachment roots binds the host more tightly, which may constrict host transport and growth. Fourth, this epiphytic accumulation increases not only the static load on the host but also the dynamic wind stress by increasing the drag of the ensemble supported by the host. These problems must influence allocation within the host, and in particular demand increased allocation to mechanical support to maintain a constant safety margin. As a result of these processes, hemiepiphytes should decrease host growth and increase the likelihood of damage to their host's crown and of host collapse (Strong, 1977).

This scenario suggests that hemiepiphytes as mechanical parasites could be modeled as a disease, and the matter of host–hemiepiphyte interaction could be explored epidemiologically. Traditional epidemiological tools such as logistic regression might be used to examine risk factors (both quantitative and qualitative) associated with the likelihood of hemiepiphyte occupation of hosts. Questions to be addressed include: Is the likelihood of hemiepiphyte occupation influenced by the specific identity of the host, by host size, by prior damage to the host, by presence or abundance of other epiphytic vegetation, by distance to an edge, or by the canopy exposure of the host? A multivariate approach allows the assessment of confounding and interaction among potential risk factors (Hosmer and Lemeshow, 1989). Multivariate logistic regression might disentangle the roles of host age, size, canopy status, and damage behind the common observation that hemiepiphytes are abundant in emergents with damaged crowns.

Hemiepiphytes also fall out of their hosts. Sometimes this is due to the death of a supporting limb, but sometimes it involves failure of the webwork of hemiepiphyte attachment roots. Hemiepiphytes are at times partially dislodged; a hemiepiphyte's collapse may be arrested by a provident limb, or major roots may hold, while the hemiepiphyte dangles. The disruption in the epiphytic community may initiate epiphytic gap-phase regeneration in the forest canopy. Because these events are difficult to assess from below, canopy exploration (particularly low-level aerial examination) of cloud forests will provide insights into patterns of regeneration and gap-phase seral dynamics in epiphytic communities.

Epiphytic and terrestrial matters are difficult to distinguish in some habitats, notably the elfin forests of wind-exposed cloud forest sites in the tradewind zone (Lawton and Putz, 1988). In this setting, where facultative hemiepiphytes are prominent (even dominant) members of the terrestrially rooted forest, trees often topple slowly in the tangle and can take years to become prostrate. Meanwhile their hemiepiphytes clamber about, struggling to reorient (Lawton, 1990). Epiphytic seedling establishment may be a viable strategy for a gap colonist (as described earlier for *Didymopanax pittieri*), and hemiepiphytic shrubs that survive host collapse may competitively exclude more traditional colonists from some gaps.

IV. Conservation Biology

Hemiepiphytes are important in the conservation of the cloud forest biodiversity owing to their contribution to the structure and composition of the vegetation. They may also play crucial conservation roles in other habitats. Figs are "pivotal" or "keystone" mutualists for many animal species in tropical forests (August, 1981; Leighton and Leighton, 1983; Terborgh, 1983; McKey, 1989). As a result of their asynchronous fruiting both among trees and among genets of a single "tree" (Thompson *et al.*, 1991), figs may be a fairly constant source of food, in contrast to distinctly seasonal fruit-bearing species.

In some areas, hemiepiphytic figs are among the most important isolated trees in pastures derived from tropical rain forest. These trees attract frugivores that play a major role in seed dispersal, and the conditions under their canopy facilitate the establishment of native rain forest tree species (Guevara *et al.*, 1992). Neotropical figs may be important in reforesting areas of degraded pasture in the tropics. Figs (especially species of stranglers) can be planted as "living fence posts" (or forests) merely by opening a hole in the soil and inserting a recently cut branch during the wet season. Planting *Ficus* species for shade, moisture, and wildlife could be a useful first step in renewing these lands (Windsor *et al.*, 1989).

Because large (hence old) trees seem to be the favored host of most hemiepiphytes, carefully planned conservation of old-growth forest will be important to avoid the extinction of many hemiepiphytes. The association between figs and the larger, commercially most valuable timber trees in Southeast Asia suggests that low-intensity selective logging, even if only of trees >50 cm dbh as prescribed in Indonesia, could drastically alter the densities of fig trees (Leighton and Leighton, 1983). In any tropical forest, the disappearance of fig trees that results from deforestation or profound alteration of the natural forest may bring about serious local consequences for the survival of important webs of ecological interactions.

V. Future Research Needs

Almost any work on hemiepiphytes will be rewarding. Systematic work on the hemiepiphytic flora of cloud forests and very wet lowland forests will undoubtedly reveal a great deal of very local endemism. Owing to their growth plasticity, hemiepiphytes may prove to be useful models for physiological examination of changes in resource allocation in response to water relations, nutrient availability, and light. The matter of host preference should be pursued at a mechanistic level. Because hemiepiphytes may have detrimental impacts on their hosts, the prospect of allelopathic interactions between host bark and hemiepiphyte seeds, seedlings, and roots should be explored. The population biology of hemiepiphytes needs serious study, particularly for those species of conservation interest, such as the strangler figs. Larger community issues, such as the role that hemiepiphytes play in forest dynamics, also need more detailed assessment.

References

August, P. V. (1981). Fig fruit consumption and seed dispersal by *Artibeus jamaicensis* in the llanos of Venezuela. *Biotropica* 13, 70–76.

Beard, J. S. (1946). The natural vegetation of Trinidad. *Oxford For. Mem.* 20.

Beard, J. S. (1949). The natural vegetation of the Windward and Leeward Islands. *Oxford For. Mem.* 21.

Benzing, D. H. (1989). The evolution of epiphytism. *In* "Vascular Plants as Epiphytes" (U. Lüttge, ed.), pp. 15–41. Springer-Verlag, Berlin.

Benzing, D. H. (1990). "Vascular Epiphytes." Cambridge Univ. Press, Cambridge, UK.

Bessey, E. A. (1908). The Florida strangling figs. *Annu. Rep. Mo. Bot. Gard.* 19, 25–34.

Bøgh, A. (1992). Composition and distribution of vascular epiphytes in a montane forest in southern Ecuador. *Selbyana* 13, 25–34.

Borland, A. M., Griffiths, H., Maxwell, C., Broadmeadow, M., Griffiths, N. M., and Barnes, J. D. (1992). On the ecophysiology of the Clusiaceae in Trinidad: Expression of CAM in *Clusia minor* L. during the transition from wet to dry season and characterization of three endemic species. *New Phytol.* 122, 349–357.

Bronstein, J. L., and Hoffmann, K. L. (1987). Spatial and temporal variation in frugivory at a neotropical fig, *Ficus pertusa*. *Oikos* 49, 261–268.

Burger, W. (1977). Moraceae. *Fieldiana, Bot.* 40, 94–215.

Clark, D. B., and Clark, D. H. (1990). Distribution and effects on tree growth of lianas and woody hemiepiphytes in a Costa Rican tropical wet forest. *J. Trop. Ecol.* 6, 321–331.

Corner, E. J. H. (1976). The climbing species of *Ficus*: Derivation and evolution. *Philos. Trans. R. Soc. London, Ser. B* 273, 359–386.

Croat, T. B. (1988). Ecology and life forms of Araceae. *Aroideana* 11, 4–55.

Daniels, J. D., and Lawton, R. O. (1991). Habitat and host preferences of *Ficus crassiuscula*, a Neotropical strangling fig of the lower-montane rain forest. *J. Ecol.* 79, 129–141.

Daniels, J. D., and Lawton, R. O. (1993). Natural history of *Ficus crassiuscula*. *Selbyana* 14, 59–63.

Davidson, D. W. (1988). Ecological studies of Neotropical ant gardens. *Ecology* 69, 1138–1152.

Davis, T. A. (1970). Epiphytes that strangulate palms. *Principes* 14, 10–25.

DeWolf, G. P. (1960). *Ficus* (Moraceae). Flora of Panama. *Ann. Mo. Bot. Gard.* 47, 146–265.

Dobzhansky, T., and Murca-Pires, B. J. (1954). Strangler trees. *Sci. Am.* 190, 78–80.

Docters van Leeuwen, W. M. (1954). On the biology of some Javanese Loranthaceae and the role birds play in their life histories. *Beaufortia* 41, 105–206.

Faber-Langendoen, D., and Gentry, A. H. (1991). The structure and diversity of rain forests at Bajo Calima, Chocó Region, western Colombia. *Biotropica* 23, 2–11.

Galil, J., and Meiri, L. (1981). Drupelet germination in *Ficus religiosa* L. *Isr. J. Bot.* 30, 41–47.

Gentry, A. H. (1982). Neotropical floristic diversity: Phytogeographical connections between Central and South America, Pleistocene climatic fluctuations, or an accident of the Andean orogeny? *Ann. Mo. Bot. Gard.* 69, 557–593.

Gentry, A. H. (1986a). Endemism in tropical vs. temperate plant communities. *In* "Conservation Biology: The Science of Scarcity and Diversity" (M. Soulé, ed.), pp. 153–181. Sinauer Press, Sunderland, MA.

Gentry, A. H. (1986b). Species richness and floristic composition of Choco region plant communities. *Caldasia* 15, 71–91.

Gentry, A. H. (1988). Changes in plant community diversity and floristic composition on environmental and geographical gradients. *Ann. Mo. Bot. Gard.* 75, 1–34.

Gentry, A. H., and Dodson, C. D. (1987). Diversity and biogeography of Neotropical vascular epiphytes. *Ann. Mo. Bot. Gard.* 74, 205–233.

Gill, A. M. (1969). The ecology of an elfin forest in Puerto Rico. 6. Aerial roots. *J. Arnold Arbor., Harv. Univ.* 50, 197–209.

Gomez P., L. D. (1989). "Vegetación de Costa Rica." Editorial Universidad Estatal a Distancia, San José, Costa Rica.

Guevara, S., Meave, J., Moreno-Casasola, P., and Laborde, J. (1992). Floristic composition and structure of vegetation under isolated trees in Neotropical pastures. *J. Veg. Sci.* 3, 655–664.

Guy, P. R. (1977). Notes on the host species of epiphytic figs (*Ficus* spp.) on the flood-plain of the Mana Pools Game Reserve, Rhodesia. *Kirkia* 10, 559–562.

Hammel, B. (1990). Flora of La Selva. *In* "Four Neotropical Rainforests" (A. H. Gentry, ed.). Yale Univ. Press, New Haven, CT.

Holbrook, N. M. (1991). Small plants in high places: The conservation and biology of epiphytes. *Trends Ecol. Evol.* 6, 314–315.

Holbrook, N. M., and Putz, F. E. (1992). The water balance of hemiepiphytes: A model system for studies of drought adaptation? *Selbyana* 13, 154.

Holdridge, L. R. (1967). "Life Zone Ecology." Tropical Science Center, San José, Costa Rica.

Hosmer, D. W., and Lemeshow, S. L. (1989). "Applied Logistic Regression." Wiley, New York.

Kapil, R. N., and Rustagi, P. N. (1966). Anatomy of the aerial and terrestrial roots of *Ficus benghalensis* L. *Phytomorphology* 16, 382–386.

Kaufmann, S., McKey, D. B., Hossaert-McKey, M., and Horvitz, C. (1991). Adaptations for a two-phase seed dispersal system involving vertebrates and ants in a hemiepiphytic fig (*Ficus microcarpa:* Moraceae). *Am. J. Bot.* 78, 971–977.

Kelly, D. L. (1985). Epiphytes and climbers of a Jamaican rain forest: Vertical distribution, life forms and life histories. *J. Biogeogr.* 12, 223–241.

King, G. (1888). The species of *Ficus* of the Indo-Malayan and Chinese countries. *Ann. R. Bot. Gard.* (*Calcutta*) 1, 1–184.

Kress, W. J. (1986). The systematic distribution of vascular epiphytes: An update. *Selbyana* 9, 2–22.

Kubler, H. (1987). Growth stresses in trees and related growth properties. *For. Abstr.* 48, 131–189.

Laman, T. G. (1993). Seedling establishment of the hemiepiphyte *Ficus stupenda* in the Bornean rain forest canopy. *Bull. Ecol. Soc. Am.* 74, Suppl., 321.

Lawton, R. O. (1983). *Didymopanax pittieri. In* "Costa Rican Natural History" (D. H. Janzen, ed.), pp. 233–234. Univ. of Chicago Press, Chicago.

Lawton, R. O. (1986). The evolution of strangling by *Ficus crassiuscula. Brenesia* 25/26, 273–278.

Lawton, R. O. (1989). More on strangling by *Ficus crassiuscula* Warb. Ex Standley: A reply to Ramirez. *Brenesia* 32, 119–120.

Lawton, R. O. (1990). Canopy gaps and light penetration into a wind-exposed tropical lower montane rain forest. *Can. J. For. Res.* 20, 659–667.

Lawton, R. O., and Dryer, V. (1980). The vegetation of the Monteverde cloud forest reserve. *Brenesia* 18, 101–116.

Lawton, R. O., and Putz, F. E. (1988). Natural disturbance and gap-phase regeneration in a wind-exposed tropical cloud forest. *Ecology* 69, 764–777.

Leighton, M., and Leighton, D. R. (1983). Vertebrate responses to fruiting seasonality within a Bornean rain forest. *In* "Tropical Rain Forest Ecology and Management" (S. L. Sutton, T. C. Whitmore, and A. C. Chadwick, eds.), pp. 181–196. Blackwell, Oxford.

Lellinger, D. B. (1989). The ferns and fern-allies of Costa Rica, Panama, and the Choco (Part 1: Psilotaceae through Dicksoniaceae). *Pteridologia* 2A.

Lenz, H. (1984). "Cosas del Papel en Mesoamerica." Editorial Libros de Mexico, Mexico.

Levitt, J. (1980). "Responses of Plants to Environmental Stresses," 2nd ed. Academic Press, London.

Long, A., and Heath, M. (1991). Flora of the El Triunfo Biosphere Reserve, Chiapas, Mexico: A preliminary floristic inventory and the plant communities of polygon I. *An. Inst. Biol., Univ. Nac. Auton. Mex.* 62, 133–172.

Lot, J. L., Bullock, S. H., and Solis-Magallanes, A. (1987). Floristic diversity and structure of upland and arroyo forests of coastal Jalisco. *Biotropica* 19, 228–235.

Madison, M. (1977). Vascular epiphytes: Their systematic occurrence and salient features. *Selbyana* 2, 1–13.

Mattheck, C. (1991). "Trees: The Mechanical Design." Springer-Verlag, Berlin.

Mattheck, C., and Burkhardt, S. (1990). A new method of structural shape optimization based on biological growth. *Int. J. Fatigue* 12, 185–190.

McKey, D. (1989). Population biology of figs: Applications for conservation. *Experientia* 45, 661–673.

Mesquita, R. C. G. (1992). Factors influencing selection of reproductive traits in a hemiepiphyte: *Clusia grandiflora. Selbyana* 13, 159.

Michaloud, G., and Michaloud-Pelletier, D. (1987). *Ficus* hemiepiphytes (Moraceae) et arbres supports. *Biotropica* 19, 125–136.

Midya, S., and Brahmachary, R. L. (1991). The effect of birds upon germination of banyan (*Ficus bengalensis*) seeds. *J. Trop. Ecol.* 7, 537–538.

Morrison, D. W. (1978). Foraging ecology and energetics of the frugivorous bat *Artibeus jamaicensis. Ecology* 59, 716–723.

Nadkarni, N. M. (1981). Canopy roots: Convergent evolution in rainforest nutrient cycles. *Science* 214, 1023–1024.

Nadkarni, N. M. (1984). Epiphyte biomass and nutrient capital of a neotropical elfin forest. *Biotropica* 16, 249–256.

Nadkarni, N. M., and Matelson, T. (1991). Fine litter dynamics within the tree canopy of a tropical cloud forest. *Ecology* 72, 2071–2082.

Newton, A. C., and Healey, J. R. (1989). Establishment of *Clethra occidentalis* on stems of the tree-fern *Cyathea pubescens* in a Jamaican montane rain forest. *J. Trop. Ecol.* 5, 441–445.

Oliver, W. R. B. (1930). New Zealand epiphytes. *J. Ecol.* 18, 1–50.

Putz, F. E. (1983). Liana biomass and leaf area of a 'Tierra Firme' forest in the Rio Negro basin, Venezuela. *Biotropica* 15, 185–189.

Putz, F. E., and Holbrook, N. M. (1986). Notes on the natural history of hemiepiphytes. *Selbyana* 9, 61–69.

Putz, F. E., and Holbrook, N. M. (1989). Strangler fig rooting habitats and nutrient relations in the llanos of Venezuela. *Am. J. Bot.* 76, 781–788.

Ramirez B., W. (1976). Germination of seeds of New World Urostigma (*Ficus*) and of *Morus rubra* L. (Moraceae). *Rev. Biol. Trop.* 24, 1–6.

Ramirez B., W. (1977). Evolution of the strangling habit in *Ficus* L. subgenus Urostigma (Moraceae). *Brenesia* 12/13, 11–19.

Ramirez B., W. (1988). A reply to Lawton's paper on *Ficus crassiuscula* Warb. as a strangler. *Brenesia* 29, 115–116.

Rao, A. N. (1966). Developmental anatomy of natural root grafts in *Ficus globosa. Aust. J. Bot.* 14, 269–276.

Raup, H. M. (1957). Vegetational adjustment to the instability of the site. *Proc. Pap. Tech. Meet., Int. Union Conserv. Nat. Nat. Res., 6th,* Edinburgh, pp. 36–48.

Ray, T. S. (1990). Metamorphosis in the Araceae. *Am. J. Bot.* 77, 159–169.

Ray, T. S. (1992). Foraging behaviour in tropical herbaceous climbers (Araceae). *J. Ecol.* 80, 189–203.

Richards, P. W. (1952). "The Tropical Rain Forest." Cambridge Univ. Press, Cambridge, UK.

Ryder, V. P. (1978). Fly and drive for Mexican palms. *Principes* 22, 110–111.

Rzedowski, J. (1978). "Vegetación de Mexico." Editorial Limusa, Mexico City.

Sanford, R. L. (1987). Apogeotropic roots in an Amazon rain forest. *Science* 235, 1062–1064.

Sargent, S. (1993). Effects of two sizes of plant neighborhoods on avian seed dispersal in a Neotropical mistletoe. *Bull. Ecol. Soc. Am.* 72, Suppl., 425.

Schimper, A. F. W. (1888). "Die epiphytische Vegetation Amerikas," Bot. Mitt. Tropen II. Fischer, Jena.

Schmitt, A. K., Lee, H. S. J., and Lüttge, U. (1988). The response of the C_3–CAM tree, *Clusia rosea,* to light and water stress. *J. Exp. Bot.* 39, 1581–1590.

Sharp, A. J. (1957). Vascular epiphytes in the Great Smoky Mountains. *Ecology* 38, 654–655.

Slater, J. A. (1972). Lygaeid bug (Hemiptera: Lygaeidae) as seed predators of figs. *Biotropica* 4, 145–151.

Strong, D. R. (1977). Epiphyte loads, treefalls, and perennial forest disruptions: A mechanism for maintaining higher tree species diversity in the tropics without animals. *J. Biogeogr.* 4, 215–218.

Terborgh, J. (1983). "Five Neotropical Primates: A Study in Comparative Ecology." Princeton Univ. Press, Princeton, NJ.

ter Steege, H. L., and Cornelissen, J. H. C. (1989). Distribution and ecology of vascular epiphytes in lowland rainforest of Guyana. *Biotropica* 21, 331–339.

Thompson, J. D., Herre, E. A., Hamrick, J. L., and Stone, J. L. (1991). Genetic mosaics in strangler fig trees: Implications for tropical conservation. *Science* 254, 1214–1216.

Ting, I. P., Lord, E. M., Sternberg, L. da S. L., and De Niro, M. J. (1985). Crassulacean acid metabolism in the strangler *Clusia rosea* Jacq. *Science* 229, 969–971.

Ting, I. P., Hann, J., Holbrook, N. M., Putz, F. E., Sternberg, L. da S. L., Price, D., and Goldstein, G. (1987). Photosynthesis in hemiepiphytic species of *Clusia* and *Ficus. Oecologia* 74, 339–346.

Tinoco, C., and Vazquez-Yanes, C. (1983). Especies CAM en la selva húmeda tropical de Los Tuxtlas, Veracruz. *Bol. Soc. Bot. Mex.* 45, 150–153.

Titus, J. H., Holbrook, N. M., and Putz, F. E. (1990). Seed germination and seedling distribution of *Ficus pertusa* and *F. tuerckheimii:* Are strangler figs autotoxic? *Biotropica* 22, 425–428.

Todzia, C. (1986). Growth habits, host tree species, and density of hemiepiphytes on Barro Colorado Island, Panama. *Biotropica* 18, 22–27.

Torquebiau, E. F. (1988). Photosynthetically active radiation environment, patch dynamics and architecture in a tropical rainforest in Sumatra. *Aust. J. Plant Physiol.* 15, 327–342.

Troth, R. G. (1979). Vegetational types on a ranch in the central llanos of Venezuela. *In* "Vertebrate Ecology in the Northern Neotropics" (J. F. Eisenberg, ed.), pp. 17–30. Smithsonian Institution Press, Washington, DC.

Vance, E. D., and Nadkarni, N. M. (1992). Root biomass distribution in a moist tropical forest. *Plant Soil* 142, 31–39.

Vanderyst, R. P. H. (1922). Nouvelle contribution à l'étude de *Ficus* epiphitiques sur l'*Elaeis. Rev. Zool. Bot. Afr., Suppl.* 10, 65–74.

van Steenis, C. G. G. J. (1972). "The Mountain Flora of Java." E. J. Brill, Leiden.

Wee, Y. C. (1992). The occurrence of *Ficus* spp. on high-rise buildings in Singapore. *Int. Biodeterior. Biodegr.* 29, 53–59.

Went, F. W. (1940). Soziologie der Epiphyten eines tropischen Urwaldes. *Ann. Jard. Bot. Buitenz.* 50, 1–98.

Whitmore, T. C. (1975). "Tropical Rain Forests of the Far East." Clarendon Press, Oxford.

Williams-Linera, G. (1992). Distribution of the hemiepiphyte *Oreopanax capitatus* at the edge and interior of a Mexican lower montane forest. *Selbyana* 13, 35–38.

Windsor, D. M., Morrison, D. W., Estribi, M. A., and De Leon, B. (1989). Phenology of fruit and leaf production by 'strangler' figs on Barro Colorado Island, Panama. *Experientia* 45, 647–653.

Zimmermann, M. H., Wardrop, A. B., and Tomlinson, P. B. (1968). Tension wood in aerial roots of *Ficus benjamina* L. *Wood Sci. Technol.* 2, 95–104.

Zobel, R. W. (1975). The genetics of root development. *In* "The Development and Function of Roots" (J. G. Torrey and D. T. Clarkson, eds.), pp. 261–275. Academic Press, London.

13

Ecology and Population Biology of Mistletoes

Nick Reid, Mark Stafford Smith, and Zhaogui Yan

I. Introduction

To laypeople, mistletoes are plants that you kiss under at Christmas. Biologists use the term "mistletoe" to define both a life-form (or growth habit) and a taxonomic category. Mistletoes are shrubby angiosperms in the Santalales that parasitize the stems of other perennial plants. Almost all mistletoes are aerial stem-parasites, so "mistletoe" is synonymous with the habit of shrubby epiphytic parasites (Kuijt, 1990). Canopy dependency and parasitism make mistletoes an ecologically distinctive group of plants, as well as producing problems in many parts of the world in terms of timber loss and mortality of desirable trees and shrubs (Hawksworth, 1983). About 1400 species of mistletoe occur in forests, woodlands, and shrublands on every continent (except Antarctica), with most species in the tropics. The Loranthaceae (ca. 950 spp.) and Viscaceae (ca. 365 spp.) contain most species; the remainder belong to the Misodendraceae, Eremolepidaceae, and Santalaceae (*Phacellaria* and *Henslowia*) (Atsatt, 1983; Kuijt, 1990).

The objectives of this chapter are: (1) to summarize the distinctive features of the mistletoe life cycle; (2) to review recent advances in ecological research on mistletoes; (3) to collate information about the life-history strategies of several Australian mistletoe species in different environments; and (4) to pose hypotheses about the population biology and evolutionary ecology of mistletoes that should be addressed.

Copyright © 1995 by Academic Press, Inc.
All rights of reproduction in any form reserved.

II. The Parasitic Mistletoe Life-Form

Ecologically important differences occur between mistletoes and other flowering plants in fruit and seed anatomy, seed dispersal, seedling development, and the process of and anatomical structures associated with host infection. In the Loranthaceae and Viscaceae, the seed consists of an embryo and endosperm, but lacks a testa. The fruit contains the seed surrounded by the pericarp, which consists of succulent tissues and an external rindlike epicarp. The succulent fruit pulp (mesocarp) contains nutrients attractive to avian dispersers and the endocarp is usually viscous in order to cement the seed to a host stem. Most aerial stem-parasitic mistletoes are dispersed by frugivorous birds that ingest the fruits or diaspores and defecate or regurgitate the seeds onto host stems. Two genera of Viscaceae, the dwarf mistletoes *Arceuthobium* and *Korthalsella,* differ from bird-dispersed mistletoes. *Arceuthobium* has explosive fruits; the pericarp abscises violently from the pedicel upon maturation and propels the diaspore over horizontal distances of up to 40 m in tall forest (Hawksworth, 1961). In *Korthalsella,* the diaspore is also released from the pericarp spontaneously but not explosively (Stevenson, 1935; D. Nickrent, personal communication). Both genera are subsequently dispersed by birds and mammals over longer distances because the small viscous seeds stick to plumage or pelage (Barlow, 1983; Nicholls *et al.,* 1984).

Mistletoes have a brief free-living phase between germination and infection after which the seedling is dependent on the host. Mistletoe seeds are short-lived and germinate immediately upon release from the fruit given appropriate external conditions (Lamont, 1983). The seeds of many species germinate in dry air, and embryo elongation is enhanced by light. Warm-climate species such as *Amyema preissii* have optimum temperature ranges of $10-25°C$ for germination and $25-30°C$ for embryo elongation (Lamont, 1982a). Cool-temperate species germinate at temperatures between $0°C$ and $25-30°C$, with optimum embryo elongation at $15-20°C$ (Lamont, 1983). The naked seedling is susceptible to desiccation at high temperatures. *Amyema preissii* seedlings grown at $40°C$ had half the water content of plants grown at $7°C$ after 10 days, and seedling growth declined rapidly between 30 and $35°C$ as a result (Lamont, 1982a). Seedlings and developing fruits are also susceptible to subzero temperatures (Baranyay and Smith, 1974; Hudler and French, 1976; Clay *et al.,* 1985; Yan, 1993a).

Once the radicle has made contact with the branch, its tip thickens to form a club-shaped holdfast. Mistletoe tissue from the holdfast penetrates the host bark and cortex via a combination of enzymatic digestion and mechanical pressure (Thoday, 1951) and grows by means of a terminal primary meristem (Sallé, 1983). The holdfast and associated penetration structures are collectively called the haustorium. Upon reaching the host vascular cambium, further growth of the mistletoe endophyte is assumed by an intercalary meristem that allows correlated growth between host and mistletoe. An apoplastic continuum is quickly established by direct xylem-to-xylem contact, and the mistletoe xylem is subsequently embedded in the host wood during annual periods of coordinated growth. The haustorium often grows large and woody with age. This physiological connection between host and parasite physically anchors the mistletoe in the host wood and subtends the parasite's aerial shoots.

III. Recent Research on the Ecology of Mistletoes

A. Methods of Canopy Access

Like other denizens of tree canopies, forest and woodland mistletoes present terrestrial researchers who wish to study the plants *in situ* with a major logistic problem, that of access to the canopy and its organisms. In recent Australian studies, this problem has been circumvented in a number of ways. Mistletoes in tall shrubs or small trees in low woodland are sampled by ladder, from vehicle rooftops, or by climbing. In woodland and open forest, sufficient numbers of individuals sometimes occur close to the ground for observation and manipulative experiments. Such mistletoes do not appear to differ in any important respect (except position in the canopy) from plants higher in the canopy. In disturbed woodland and forest, regenerating stands of saplings infected by mistletoes offer accessible study populations. For studies of dispersion and population dynamics in woodland and open forest with low leaf area indices, we have estimated mistletoe sizes and successfully mapped their position from the ground, using binoculars and sketches of mistletoe position in relation to tree architecture. For studies in which samples of mistletoe tissue are required, pole-pruners, slingshots, and shotguns are part of the researcher's arsenal (D. Nickrent, personal communication). Hydraulic bucket-hoists ("cherrypickers") are also useful for making detailed measurements or removing entire plants high in the canopy.

B. Ecophysiology

Mistletoes are often divided into autotrophic and heterotrophic groups according to the presumed mode of carbon acquisition. Dwarf mistletoes are considered to be primarily heterotrophic because their foliage contains little chlorophyll, they tap both the xylem and phloem of the host, and a substantial proportion of their organic carbon is absorbed directly from the host phloem (Hull and Leonard, 1964; Knutson, 1983). Other mistletoes (e.g., Loranthaceae and most Viscaceae) have high foliar concentrations of chlorophyll and tap only the host xylem stream. The latter group were therefore assumed to absorb water and mineral nutrients but to be auto-trophic. However, studies have shown that these "xylem-tapping" mistle-toes are also partly heterotrophic. Room (1971) detected labeled carbon fixed by host foliage in the leaves of *Tapinanthus bangwensis*. Raven (1983) calculated that about 20% of the carbon in xylem-tapping parasites may be derived from the host in the form of organic nitrogen compounds dissolved in the xylem sap. Marshall and Ehleringer (1990) estimated that 62% of the carbon in *Phoradendron juniperinum* was derived from its juniper host in the form of organic acids dissolved in host xylem water. Pate *et al.* (1991b) esti-mated that 24% of the carbon requirements for dry matter accumulation in *Amyema linophyllum* was met by intake of xylem-borne organic com-pounds from *Casuarina* hosts. Schulze *et al.* (1991) calculated the degree of carbon heterotrophy of *Tapinanthus oleifolius* on a range of hosts to be 47–67%.

The resources and physiological processes that limit mistletoe primary production have also been the subject of recent ecophysiological interest. Xylem-tapping mistletoes transpire prodigious amounts of water, transpira-tion rates being 1.5 to 10 times higher than those of their hosts (Fisher, 1983; Ullmann *et al.*, 1985). Because mistletoes have relatively low rates of carbon assimilation and therefore very low water-use efficiencies (Ehleringer *et al.*, 1985, 1986), autotrophic carbon assimilation is unlikely to be the primary function of profligate water use. High transpiration rates may be a means of acquiring nitrogen from the host xylem stream, as nitrogen is regarded as the macronutrient most limiting to mistletoe growth (Schulze *et al.*, 1984; Schulze and Ehleringer, 1984; Ehleringer *et al.*, 1985, 1986). Alternatively, high transpiration rates might provide the means for maximizing hetero-trophic carbon gain, thereby lessening the need for host-derived nitrogen, because photosynthesis is particularly demanding of nitrogen (Stewart and Press, 1990). Access to heterotrophic carbon could reduce the need for nitrogen, which could be important in nitrogen-limited environments. A composite carbon:nitrogen (C:N) balance hypothesis has been suggested by Marshall and Ehleringer (1990): mistletoes maintain a low water-use ef-ficiency because they assimilate host carbon and therefore do not require an equivalent autotrophic carbon assimilation capacity to achieve the same C:N tissue ratio as their hosts. This hypothesis could be tested if the C:N ratio in the host xylem stream could be manipulated experimentally.

C. Mechanisms of Host Specificity and Infectivity

Mistletoes vary in host specificity, that is, the identity and frequency of the host species parasitized by a mistletoe population. Host-generalist mis-tletoes parasitize most or all potential host species to which they are dis-persed. In Australia, mistletoes that occur in rain forest (e.g., *Amylotheca dictyophleba, Amyema queenslandicum,* and *Benthamina alyxifolia*) or have extensive geographical ranges (e.g., *Dendrophthoe falcata*) have low host specificity and infect numerous, taxonomically diverse hosts (Barlow, 1981). In contrast, host specialists parasitize only a restricted subset of species, (as few as one) of the potential host species encountered. Australian examples include mistletoes of open forest and woodland, such as *Amyema lucasii* on *Flindersia maculosa, Muellerina bidwillii* on *Callitris* spp., and *A. gaudichaudii* on *Melaleuca decora.* Similar patterns of host specialization and generalization have been observed elsewhere. The host range of the dwarf mistletoes varies from extremely specific (*Arceuthobium apachecum* and *A. blumeri* are only known to parasitize *Pinus strobiformis*) to dozens of host species (e.g., *A. globosum;* Hawksworth and Wiens, 1972).

Patterns of host specificity are generated over evolutionary time as a result of mistletoe populations adapting to potentially suitable host populations that are frequently and predictably encountered (Atsatt, 1983). In habitats with low tree species dominance, such as humid tropical forest, high host specificity is selectively disadvantageous because seed dispersal is unlikely to be sufficiently targeted to guarantee delivery of a high proportion of seeds to a particular host species. Rather, the ability for a mistletoe to infect and grow on a wide range of hosts is likely to be crucial to mistletoe persistence. In open habitats dominated by one species or a small number of congeners, selection will favor physiological adaptations for growth on the predominant tree species or genus, possibly rendering the mistletoe population less fit to grow on other hosts (Barlow, 1981).

A number of proximate (ecological) factors may account for observed patterns of host infection. First, mistletoes must maintain a lower water potential than the host in order to obtain host xylem water and solutes. Accordingly, mistletoes cannot survive on tree and shrub species that maintain very low water potentials (Lamont, 1982b; Yan, 1993b). Second, the behavior of avian dispersers may be responsible for patterns of host specificity if they consistently bias their movements toward certain types and sizes of trees and shrubs (Lamont, 1982b, 1985). Third, incompatibility mechanisms may prevent mistletoes from establishing or reproducing on some hosts. The mechanism may be a chemical or physical barrier to infection or postinfection development in the host, and it may be present prior to seed dispersal or induced by the presence of the seed or haustorial penetration (Hariri *et al.,* 1991). A number of types of host incompatibility response have been described, although the precise mechanisms are not always known:

1. Initial haustorial penetration may provoke a wound periderm response in the host cortex that isolates the invading endophyte from the host's vascular cambium (Atsatt, 1983; Hoffmann *et al.,* 1986; Yan, 1993c). In the case of poplar (*Populus*) cultivars infected by *Viscum album*, both

preexisting and induced defenses help block the passage of the mistletoe endophyte (Hariri *et al.,* 1991). Resistant poplars (*P. nigra*) have high concentrations of polyphenols (flavonoids) in the periderm, cortical parenchyma, and cortical fibers compared to susceptible cultivars. In response to haustorial penetration, resistant cultivars develop several layers of inner periderm and secrete more flavonoids, which chemically and physically block further growth of the endophyte.

2. The mistletoe endophyte may successfully penetrate the cortex and make contact with the host xylem but subsequently languish there and die (Hoffmann *et al.,* 1986; Yan, 1993c).

3. Incompatibilities may arise after successful establishment but prior to the mistletoe attaining reproductive maturity. Large individuals of *Amyema pendulum* are only rarely found on *Eucalyptus blakelyi* in the field (Z. Yan and N. Reid, unpublished data, 1995). However, a high proportion of *A. pendulum* seedlings establish and grow on *E. blakelyi* hosts for up to 2.5 years and attain canopy diameters of up to 10 cm prior to death.

Little is known about the biochemistry of mistletoe–host recognition and incompatibility responses (Knutson, 1983), but host chemical cues appear to be necessary for holdfast development in *Phoradendron juniperinum* and perhaps *P. macrophyllum* (Dawson and Ehleringer, 1991; Lichter and Berry, 1991). There is also the intriguing observation of Hoffmann *et al.* (1986) that *Kageneckia oblonga* hosts bear only one large mistletoe of *Tristerix tetrandrus.* Experimental inoculations demonstrated that *T. tetrandrus* seedlings establish only on uninfected hosts, suggesting that mistletoe infection induces host resistance to subsequent infection.

Another intriguing feature is the phenomenon of "host exclusion," the tendency for sympatric mistletoe populations to partition available hosts between them and infect mutually exclusive subsets. Although sympatric mistletoe populations often infect the same host populations (and individuals), Hawksworth and Wiens (1972) described instances involving dwarf mistletoes where a host population is not infected by secondary parasites if the host's primary parasite is present in the area. Conversely, if the principal parasite is absent, the chances of parasitism by other dwarf mistletoes are high. A possible explanation is that the historical presence of the primary parasite selects for higher levels of resistance to dwarf mistletoe infection in local host populations, rendering the hosts less susceptible to secondary (less compatible) parasite species. There is mounting evidence of infraspecific variation in host susceptibility to mistletoe infection (Scharpf, 1987) and for local adaptation of mistletoe populations to primary hosts (May, 1971; Atsatt, 1983; Clay *et al.,* 1985). However, there is no evidence of coevolutionary responses to mistletoe infection in host populations, nor is it clear over what geographical and temporal scales such evolutionary "arms races" between hosts and mistletoes might operate.

D. Haustorial Structure

Mistletoes exhibit a variety of haustorial types (Hamilton and Barlow, 1963; Kuijt, 1969). The simplest is a primary haustorium that forms a ball-like attachment to the host, sometimes with the production of sinkers around the margins of the ball-like union or the longitudinal development of the haustorium along the branch. Many species develop secondary haustoria. These may develop from external rootlike runners that grow from the primary haustorium above the bark and along the host's branches, periodically sending sinkers (secondary haustoria) into the host tissue and giving rise to aerial shoots. Alternatively, strands of mistletoe tissue ramify beneath the bark in the host cortex. Ramifying cortical strands radiate out from the primary haustorium, periodically subtending sinkers into the host wood and may give rise to aerial shoots that erupt through the host bark.

Dwarf mistletoes produce different types of ramifying cortical haustorium. Typically, dwarf mistletoe infection leads to the production of dense masses of distorted host branches, called "witches' brooms," near the site of infection. Two types of witches' brooms are formed (Hawksworth and Wiens, 1972): (1) systemic types, in which the growth of the endophyte keeps pace with growth of the infected branch, and the dwarf mistletoe shoots are scattered along the host branches, often concentrated at nodes; and (2) non-systemic types, in which the dwarf mistletoe shoots remain concentrated near the original site of infection. In a few host–dwarf mistletoe combinations, witches' brooms are not formed at all.

The variety of haustorial types presumably conveys selective advantages in different environments. In humid tropical forest, host-generalist mistletoes often have external runners (Hamilton and Barlow, 1963; Kuijt, 1969) that infect multiple hosts in interlocking canopies. Secondary haustoria are also advantageous on long-lived hosts and in ecologically stable habitats, enabling individual mistletoes to persist after the primary attachment and infected stem are senesced or abscised and to relocate within the canopy to optimize light interception. Ramifying cortical haustoria are advantageous in arid habitats, as the mistletoe can develop maximum absorbing contact with the host wood and be protected from desiccation by the host bark (Hamilton and Barlow, 1963). On short-lived hosts or in disturbance-prone environments, elaborate haustorial structures and secondary attachments are unlikely to convey greater fitness than a simple ball-like union.

E. Techniques for Aging Mistletoes

Accurate information about plant population biology and dynamics is difficult to obtain without being able to age the individuals in a population. In host species with identifiable annual growth rings, the age of a mistletoe can be determined destructively by counting the maximum number of annual host rings that bury the mistletoe haustorium in the host wood (Dawson *et al.*, 1990). However, many host species do not exhibit annual growth rings. Plant size is therefore sometimes used as a surrogate for age in plant population studies. Reid and Lange (1988) argued that the maximum diameter of the host branch proximal to the haustorium (and hypertrophied section of host branch) was proportional to mistletoe age, because seedlings of most species establish on young host branches. Host branch diameter has the advantage of increasing monotonically through time, unlike mistletoe canopy dimensions, number of branches, and other variables that regress in senescent individuals. However, the utility of host branch diameter as an age index suffers from its sensitivity to the overall growth rate of the host individual (some hosts grow faster than others) and position of the branch in the canopy (some branches grow faster than others).

Fortunately, many mistletoe species have an annual growing season and exhibit systematic vegetative growth and branching patterns. Some of the morphological growth patterns are identifiable as annual growth increments, thus providing vegetative means of aging the oldest aerial shoot on a plant. In the case of *Phoradendron juniperinum,* the number of bifurcate branching points on the longest stem is linearly related to mistletoe age (Dawson *et al.*, 1990). Consistent internodal patterns in annual shoot growth may also be useful in aging the longest vegetative shoot in some mistletoe species (G. Orshan, personal communication).

F. Mutualisms Involving Mistletoes and Birds

Mistletoe dispersal systems involving birds often display an unusual degree of mutualistic specificity (Reid, 1991). Birds that eat mainly mistletoe fruit are responsible for most of the mistletoe dispersal in many parts of the world. Avian specialization on mistletoes has independently evolved in unrelated taxa on different continents: tinkerbirds (*Pogoniulus*) in Africa; euphonias (*Euphonia*), tyrant flycatchers (*Zimmerius*), and cotingids (*Zaratornis stresemanni, Phibalura flavirostris,* and *Idopleura pipra*) in the neotropics; flowerpeckers (*Dicaeum*) in India and Southeast Asia; and the painted honeyeater (*Grantiella picta*) in Australia.

Two attributes related to the aerial stem-parasitic habit underlie mistletoe interactions with avian frugivores (Reid, 1991): (1) the requirement for targeted seed dispersal to young compatible host stems and (2) the viscidity of the seeds (for adherence to stems). Specialized dispersers are generally small birds and have a variety of anatomical and behavioral adaptations to a diet of viscous mistletoe fruit. Dicaeids and euphonias, for example, have highly reduced gastrointestinal tracts and both wipe their cloacas on the perch to be rid of sticky fecal seed strings. The factors responsible for initial coevolution of specialized mistletoe-dispersing birds and mistletoes may have been (1) the requirement of small birds to use perches to discard regurgitated or defecated seeds, (2) the preference of small birds to perch on small stems, and (3) the concomitant selective advantage that this afforded mistletoes in terms of directed seed dispersal to safe sites. Under these conditions, natural selection would have favored mistletoes with fruiting displays that differentially targeted small dispersers (Reid, 1991). Over time, selection would also have favored birds with specialized gastrointestinal tracts to efficiently process large quantities of viscous fruit. The rapid passage time of mistletoe fruits and concomitant gentle treatment of naked mistletoe seeds in the gut of specialized dispersers may be a secondary selective pressure driving bird–mistletoe coevolution (Murphy *et al.*, 1993).

Many Loranthaceae are pollinated by nectar-feeding birds. Old World sunbirds (Nectarini), Oriental flowerpeckers (Dicaeini) and white-eyes (*Zosterops*), Australasian honeyeaters (Meliphagidae), and Neotropical hummingbirds (Trochilidae) are the most frequent avian pollinators (Davidar, 1984; Docters van Leeuwen, 1954; Gill and Wolf, 1975; Reid, 1986). The specificity of the mutualistic interactions between nectar-feeding birds and mistletoes is low in some regions. In Australia, a mistletoe's flowers are likely to be visited by a variety of honeyeater species that also visit a taxonomically diverse range of other flowers in the community (Bernhardt, 1983; Reid, 1986). In the Oriental region, however, a greater degree of specificity between avian pollinators and mistletoes has evolved, involving the mistletoes' dispersers. At higher elevations in the Nilgiris of southern India, several mistletoe species (*Taxillus recurvus, T. cuneatus,* and *Dendrophthoe neelgherrensis*) have evolved relatively small, dull-colored flowers that explosively open when visited by nectar-feeding birds (Davidar, 1983, 1984). Nectar is contained in the mature bud but is not secreted after anthesis. Fruiting in these species is delayed until just prior to the following year's flowering so that fruiting merges into flowering. This unusual suite of floral and phenological characteristics has evidently evolved in response to the pollinator service of flowerpeckers such as *Dicaeum unicolor,* which also disperses the mistletoes' seeds. In the same region, another set of mistletoes is dispersed by the flowerpecker but has a more typical suite of ornithophilous floral and phenological characters. These species have large reddish flowers that open unaided, secrete nectar after anthesis, and are pollinated principally by sunbirds. Davidar (1983) interpreted the fruitlike attributes of the flowers of the flowerpecker-pollinated mistletoes as a case of facultative mimicry, inducing the flowerpecker to perform both a pollinator and a dispersal service.

Such remarkable specialization may have arisen for several reasons. Many species of Loranthaceae co-occur in the region, so specialization on quite different avian pollinator types may reflect selection to minimize interspecific competition for pollinators or to minimize the deleterious effects of heterospecific pollen flow between species that flower simultaneously (Davidar, 1984). Some Loranthaceae in the region are pollinated by insects rather than birds, so such selective pressures may have been strong in the past. The selection may have been reinforced by the scarcity of more conventional avian pollinators in the region. At present, the only common sunbird in the region migrates seasonally to lower altitudes. Assuming that the flowerpecker dispersal role predates the pollinator role in these mistletoes, selection presumably favored plants that could "co-opt" the services of faithful avian dispersers already in attendance to double as pollinators.

Although the specificity of mistletoe–pollinator interactions may be low in Australia, mistletoe flowering is important to communities of avian pollinators in Australian woodlands and forests. On the Northern Tablelands of New South Wales, the winter–spring flowering of *Amyema pendulum* and the summer–autumn flowering of *A. miquelii* provide keystone nectar resources for the honeyeater community (G. Barrett, personal communication). Several species of honeyeater move between different eucalypt-dominated habitats of the two mistletoes in autumn and spring in response to annual flowering. In *Acacia papyrocarpa* woodland in arid South Australia, *Amyema quandang* is the primary nectar source for honeyeaters for more than six months of the year (Reid, 1990). The fruit crop of the same mistletoe sustains the dominant honeyeater, *Acanthagenys rufogularis*, for the remainder of the year. These observations, and the abundance of mistletoes in Australian woodlands generally (Barlow, 1986), suggest that mistletoes are commonly keystone nectar resources for honeyeaters in floristically depauperate communities.

G. Effects of Mistletoes on Host Populations

Mistletoes damage tree and shrub hosts in forests, plantations, orchards, parks, gardens, and rural landscapes in many parts of the world (Hawksworth, 1983). Accordingly, information is accumulating about the level of damage exacted by mistletoes on host populations. Light infection (i.e., a low ratio of mistletoe to host foliage) generally has no measurable effect on host vigor. Severe mistletoe infection, however, affects hosts in a variety of ways, including reduction in height, diameter, foliage growth, and reproductive output; predisposition to attack by secondary agents such as insects and decay fungi; and premature mortality (Hawksworth, 1983). Most data are available for dwarf mistletoes that parasitize commercially important conifers in North American forests. Severe infection of *Pinus ponderosa* by *Arceuthobium vaginatum* depresses mean radial growth over a 5-year period by 52% in dominant trees (Hawksworth, 1961). Severe infection of *Abies grandis* by *Arceuthobium abietinum* reduces the merchantable volume increment (averaged over 25 years) by 45% in trees with live-crown ratios greater than 80% (Filip, 1984). In both *P. ponderosa* and *Abies grandis,* the effect of severe infection is less pronounced in subdominant and suppressed trees. Infection of *P. ponderosa* by *Arceuthobium vaginatum* and *Pseudotsuga menziesii* by *A. douglasii* increases host mortality over a 10-year period from 1% in uninfected trees to 38 and 45%, respectively, in the most severely infected hosts (Hawksworth and Shaw, 1984).

Tree age may influence the severity of the impacts of dwarf mistletoe. The reduction in radial growth of old-growth (>300 years) *Pinus contorta* as a result of *A. americanum* infection was 8–20% less than that in 80-year-old trees of equivalent infection status (Hawksworth *et al.*, 1992). The physiological basis for this difference is uncertain, but trees in old-growth stands may be more tolerant of dwarf mistletoe as a result of (1) the mortality of the most susceptible hosts at earlier stages, leaving a residual of more resistant trees, or (2) self-thinning in old-growth stands, leaving remaining trees relatively free of intraspecific competition and therefore less susceptible to dwarf mistletoe impacts.

Less information is available on the impact of xylem-tapping mistletoes on hosts. *Amyema miquelii* increases mortality of and reduces radial stem growth in farm eucalypts in rural Australia (Reid and Yan, 1995). The percentage reduction in radial increment is approximately equivalent to the percentage infestation level (Nicolson, 1955; Reid *et al.*, 1994), that is, the reduction in host growth is equivalent to the replacement of the host canopy by mistletoe foliage.

H. Management of Overabundant Mistletoe Populations

Mistletoe populations that damage host trees to a significant degree are almost invariably natural components of the forest and woodland ecosystems in which they achieve pest status. In the case of garden, orchard, and plantation damage, they occur naturally in the surrounding districts. Eradication is therefore not an appropriate objective and the designation of mistletoes as "pests" is valid only when populations affect specified management goals, such as timber production, the aesthetics of a recreation area, or the maintenance of remnant farm trees (Wicker and Hawksworth, 1988). Because all mistletoes are obligate parasites and at least some species are host-specific and have relatively low rates of growth and spread, strategies for controlling mistletoe populations at acceptable levels may be integrated with programs for overall ecosystem management. Separate control programs are likely to be undesirable on economic grounds, and preventive management strategies (such as planting resistant tree germplasm) will be more efficient and effective than reactive (curative) tactics (such as repeated pruning of host trees).

The development of preventive management strategies for mistletoes requires an understanding of the ecosystem's dynamics, the ecology of the mistletoe population, and its interactions with other ecosystem components. In rural Australia, concern about the increase in *Amyema miquelii* populations in eucalypt woodlands since 1900 has coincided with widespread fire suppression, a reduction in the density of natural predators such as common brushtail possums (*Trichosurus vulpecula*), tree clearance and habitat fragmentation, and grazing-induced suppression of natural tree regeneration. All of these are likely to contribute to an imbalance between mistletoes and remaining farm trees (Reid and Yan, 1995). Preventive strategies to avoid or tolerate moderate levels of mistletoe infestation should emphasize long-term planning for farm tree management, including regeneration of host and nonhost species, revegetation works, encouragement of natural predators, and, in appropriate areas, prescribed burning.

IV. Evolutionary Ecology of Mistletoes

Life-history strategies of plants differ widely in response to the environments and selective pressures that have molded their evolution. Significant influences on the ecology and evolution of mistletoes are short-term climate (e.g., temperature and soil moisture, the latter affecting the supply of resources provided by the host) and damaging agents (e.g., fire and predators). Over evolutionary time, the same factors are important selective pressures that forge each mistletoe's life-history strategy. Host attributes are also a key selection pressure in mistletoe evolution.

It is important to distinguish between ecological and evolutionary responses to environmental stimuli. The ecological effects of a severe drought on a mistletoe may be retarded growth and eventual death. The ultimate effect of recurrent drought, however, might be to select for high growth rates so that the mistletoe population can grow rapidly and expand in good seasons between droughts (a "boom-and-bust" strategy). Conversely, drought might select for a conservative growth strategy and persistence mechanisms, so that individuals have a high probability of surviving severe droughts. Host strategies are likely to influence mistletoe strategies. A short-lived host population that adopts a boom-and-bust strategy in an arid environment constrains its parasite population to a similar strategy. Long-lived xerophytic hosts that invest heavily in persistence mechanisms and grow slowly, on the other hand, limit their parasite population to evolving a conservative strategy, because the apoplastic continuum between host and parasite is maintained by coordinated growth.

In the following sections, we explore the proposition that mistletoe life histories can be explained in terms of key environmental influences such as fire, drought, and host attributes. We hypothesize that in successional or disturbance-prone habitats, in environments where temperature and soil moisture conditions for plant growth are excellent, and on fast-growing hosts, mistletoe growth should be rapid. This allows sexual reproduction to be initiated from as early an age as possible. Conversely, in stable (rarely disturbed) habitats, under conditions difficult for plant growth, and on slow-growing hosts, mistletoe growth should be slow. More mistletoe resources should be invested in vegetative propagation and mechanisms for persistence, such as secondary haustoria.

To examine these hypotheses, we compare aspects of the population biology of several Australian Loranthaceae. The host–mistletoe combinations and the environments in which they occur are given in Table I. Most of our comparisons are between different species of mistletoe and their hosts, but we expect the hypotheses to hold for species of mistletoe growing in widely different environments or on widely differing host species.

A. Haustorium

The mistletoe species in Table I differ in haustorial structure in predictable ways. Mistletoes that infect short-lived hosts (*Amyema preissii*) or occur in fire-prone environments (*A. miquelii*) have haustoria that are basically ball-like. These species do not invest in complex structures that facilitate vegetative propagation and persistence after senescence of the primary attachment because of the likelihood of early mortality. *Amyema quandang*, on the other hand, infects slow-growing, long-lived host trees in arid environments where fire is rare or infrequent. The Middleback and Atartinga populations of the species have an elaborate ramifying cortical haustorium that facilitates persistence in severe drought, permits vegetative propagation after the senescence of the primary attachment (and perhaps after burning), and allows the plant to grow along infected stems over time.

Lysiana exocarpi has a simple ball-like haustorium. The species is host-generalist and parasitizes phreatophytic hosts in addition to xerophytes. A simple haustorial structure might facilitate or indeed be a prerequisite for infecting a taxonomically diverse range of hosts.

In these mistletoe species, haustorial type is correlated with the likelihood of survival of the host branch distal to the mistletoe. On small branches, the endophytic tissues of mistletoes with a ball-like haustorium rapidly girdle the stem and cut off the distal portion of the host branch from the vascular stream. Consequently, most *A. miquelii, A. preissii,* and *L. exocarpi* plants rapidly occupy a distal position on the host branch. Progressive infestation by these species can lead to marked foliage loss and decline of the host (Reid *et al.,* 1994). The ramifying cortical haustorium of arid-zone populations of *A. quandang,* on the other hand, does not tend to girdle the host branch. Host foliage generally flourishes distally to large mistletoes, and infestations of these mistletoes are consequently benign and rarely lead to host deaths (Reid and Lange, 1988).

B. Growth

Southern and central Australian mistletoes vary greatly in growth rates (Fig. 1). *Amyema quandang* parasitizes long-lived, slow-growing acacia trees and is itself slow-growing. In the same habitats in central Australia, *A. preissii*

Table I Study Sites, Host–Mistletoe Combinations, and Host Attributes of Selected Woodland Mistletoes in Southern, Eastern and Central Australia[a]

Site	Mean annual rainfall (mm)	Fire interval (years)[a]	Mistletoe	Haustorial type	Host	Host attributes
Brookfield, South Australia	260	5–20	Amyema preissii	Ball-like	Acacia nyssophylla	Shrub to 2 m; growth rate and max. age unknown
			Lysiana exocarpi	Ball-like	Heterodendrum oleifolium	Small tree to 5 m; slow growth rate; max. age 500 yr (Purdie, 1969)
				Ball-like	Myoporum platycarpum	Tree to 8 m; moderately slow growth rate; max. age unknown
Alice Springs, Northern Territory	280	5–10	Amyema preissii	Ball-like	Acacia victoriae	Multistemmed shrub to 9 m; rapid growth rate; max. age 15–25 yr (M. Stafford Smith and N. Reid, unpublished data)
Middleback, South Australia	210	Rare	Amyema quandang	Diffuse with ramifying cortical strands and secondary sinkers and shoots	Acacia papyrocarpa	Spreading tree to 7 m; very slow-growing; max. age 250 yr (Lange and Purdie, 1976)
			Lysiana exocarpi	Ball-like	Myoporum platycarpum, Heterodendrum oleifolium, Eremophila spp., Exocarpos aphyllus	Slow to moderately slow growers (see above; N. Reid, unpublished data)

Location			Amyema species	Haustorium	Host	Description
Atartinga, Northern Territory	305	Rare	Amyema quandang	Diffuse with ramifying cortical strands and secondary sinkers and shoots	Acacia georginae	Spreading tree to 8 m; slow-growing; max. age 200 yr (M. Stafford Smith, unpublished data)
Armidale, New South Wales	760	1–5	Amyema pendulum	?	Eucalyptus caliginosa, E. laeviphinea, E. viminalis	Tree to 25 m, moderately slow growth rate for E. caliginosa and E. laeviphinea (radial increment 0.2 cm/yr; J. Duggin, personal communication); moderate growth rate for E. viminalis (radial increment 1–2 cm/yr; J. B. Williams, personal communication); max. age 300 yr[b]
		1–5	Amyema miquelii	Ball-like with secondary sinkers and shoots at edge of primary haustorium	Eucalyptus blakelyi and E. melliodora	Trees to 20 m, moderate growth rate (radial increment 0.8–1 cm/yr; Reid et al., 1994); max. age 300 yr[b]

[a]Data from Walker (1981), J. Duggin and R. D. B. Whalley (personal communication), and the authors (unpublished data).
[b]Maximum ages from J. Duggin and J. B. Williams (personal communication).

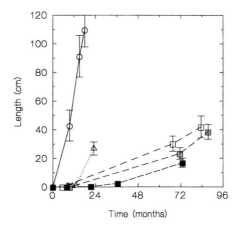

Figure 1   Seedling growth rates of *Amyema* spp. Data are the mean (± 1 s.e.) maximum canopy diameter for an Alice Springs population of *Amyema preissii* (deployed in May 1992, ○) and three cohorts of *A. quandang:* Middleback (March 1983, ■), Middleback (July–August 1983, □), and Atartinga (June 1986, ■). For *A. miquelii* (Armidale, October 1990, △), data are the mean (± 1 s.e.) length of the longest shoot.

parasitizes fast-growing acacia shrubs and grows rapidly. In temperate eastern Australia, *A. miquelii* parasitizes woodland eucalypts. Under benign growing conditions and in habitats subject to recurrent fire, *A. miquelii* grows considerably faster than *A. quandang* but not as fast as *A. preissii*. The host-generalist *Lysiana exocarpi* also has an intermediate growth rate that varies between host species. It grows more rapidly on an acacia shrub (of uncertain growth rate) than on slow-growing xerophytic trees in Brookfield, South Australia (Fig. 2). It also grows more slowly than *A. preissii*, although it parasitizes the same host (*Acacia nyssophylla*).

C. Mortality and Herbivory

Host death and host branch atrophy are frequent causes of mortality of mature mistletoes. The haustoria of old mistletoes become convoluted as a result of the hypertrophy of the host wood and ultimately the atrophy of the vascular connections between host and mistletoe. Secondary haustoria developed at some distance from the primary attachment enable mistletoes to escape the atrophy of senescent vascular connections.

Fire is an important mortality factor, whether or not host trees are killed. In the coniferous forests of North America, frequent forest fires were thought to have maintained dwarf mistletoes in low abundance because the regeneration of host populations was more rapid than dwarf mistletoe reinvasion (Alexander and Hawksworth, 1975). In most Australian eucalypt forests and woodlands, eucalypts survive wildfires and resprout epicormically,

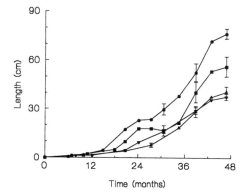

Figure 2 Seedling growth rates of *Amyema preissii* on *Acacia nyssophylla* (●) and *Lysiana exocarpi* on three different hosts at Brookfield. Data are the mean (± 1 s.e.) length of the longest shoot of mistletoes deployed as seeds in June 1985. *Lysiana* hosts were *A. nyssophylla* (■), *Myoporum platycarpum* (▲), and *Heterodendrum oleifolium* (▼).

whereas *Amyema miquelii* and *A. pendulum* are killed by hot fires (Cleland, 1940; P. Kelly and N. Reid, unpublished data, 1995; Reid and Yan, 1995). Mature mistletoes may also be susceptible to severe freezes. At Kingstown, New South Wales, an unusually severe freeze of −9 to −13°C in different parts of the district killed much of the *A. miquelii* on eucalypt hosts in 1965 and reduced the abundance of the species in the district for several years.

Severe drought may increase mortality in susceptible mistletoe species even if the host survives. *Lysiana exocarpi* suffered increased mortality during a severe drought in arid South Australia, unlike the slow-growing *Amyema quandang* (Reid and Lange, 1988). In large individuals of the latter species, some large haustorial branches died, but smaller branches emanating from different parts of the haustorium survived. Selective branch death had the effect of reducing transpiration losses. Large branches are also more expensive to maintain in terms of the hydraulic resistance to water flow (Dawson *et al.*, 1990).

In contrast to the preceding factors, severe predation is not lethal to species of *Amyema*. *Amyema miquelii* and *A. pendulum* have dormant buds beneath the bark that resprout after removal of foliage. This allows individuals of all sizes to survive even complete defoliation (P. Kelly and N. Reid, unpublished data, 1995). Most individuals of *A. miquelii* also survive branch-pruning as far back as the haustorium by resprouting from epicormic buds. The ability to survive severe levels of foliage and branch loss may have been selected for by high rates of herbivory and browsing. The foliage of Australian Loranthaceae is preferred to host eucalypt foliage by arboreal mammals (Choate *et al.*, 1987; Porter, 1990), and mammalian browsing is thought to have been the primary selective force for the evolution of host crypsis in

Australian mistletoes (Barlow and Wiens, 1977; Barlow, 1981, 1986). Mistletoes that have evolved in the absence of sustained browsing pressure by arboreal mammals may not be as resilient as Australian Loranthaceae. In New Zealand, the remaining five extant species of Loranthaceae are threatened by the browsing pressure of introduced common brushtail possums (Wilson, 1984; Ogle, 1987; Norton, 1991).

D. Fecundity

Age at first reproduction is negatively correlated with the intrinsic rate of population increase. We hypothesize that mistletoes that parasitize fast-growing hosts and that are themselves fast-growing will reproduce at an earlier age than slower-growing mistletoes. *Amyema preissii* is a fast grower (Figs. 1 and 2) and the most vigorous individuals first flowered in their third year at Brookfield (Yan, 1993b). *Amyema quandang* is slow-growing and the first of 71 seedlings deployed on *Acacia papyrocarpa* at Middleback flowered in the fifth year. However, most (53%) flowered for the first time in the seventh year. At Atartinga, *A. quandang* reproduction on *Acacia georginae* was even slower. The first seedling to flower was in its eighth year. Our experimental cohorts of *A. miquelii* near Armidale are now in their fourth year and have yet to flower (Z. Yan and N. Reid, unpublished data, 1995).

E. Recruitment

Australian Loranthaceae reproduce annually. They flower and fruit once per year and fruiting seasons may span many months (Pate *et al.*, 1991a; Reid, 1990; Reid and Yan, 1995; Yan, 1990). Recruitment may occur semicontinuously (at least annually) or episodically (i.e., once every few years). Evidence of both patterns of recruitment is seen in the size-class distributions of different mistletoe species. In arid South Australia, recruitment of *Amyema quandang* is continuous, as indicated by a preponderance of seedlings and young plants; small plants of *A. quandang* are most abundant and larger size classes are progressively less abundant (Reid and Lange, 1988). The mistletoe fruits year-round and maintains permanent populations of dispersers (Reid, 1990). The woodlands in which it occurs rarely burn. The ramifying cortical haustorium and conservative growth strategy of *A. quandang* enable the species to withstand severe droughts and frosts. Cryptic fruits that ripen slowly throughout the year deter heavy losses to visually orienting seed predators such as parrots and emus. The net result is continuous recruitment into the population.

Episodic recruitment in mistletoe populations is evident in temperate eucalypt woodlands in eastern Australia. In three different stands near Armidale, New South Wales, the size-class distributions of *Amyema miquelii* and *A. pendulum* are bell-shaped (Fig. 3), with a preponderance of intermediate size classes. Small plants are rare, suggesting that years of high recruitment

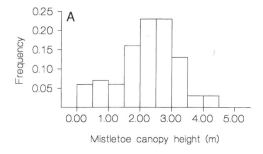

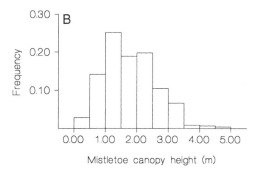

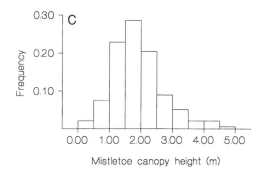

Figure 3   Size-class distributions of mistletoe canopy height of three populations of *Amyema* spp. near Armidale in 1989–1990: (A, B) two populations of *A. miquelii* separated by about 1.5 km ($n = 100$ and 708, respectively); (C) *A. pendulum* ($n = 539$).

occur infrequently. Although fruiting intensity varies somewhat between years (Reid and Yan, 1995), the variation is unlikely to account for the episodic nature of recruitment suggested in Fig. 3. More likely hypotheses are that dispersers are present in high densities only in some years, or that climatic conditions and predator effects in combination prevent significant recruitment in most years. *Amyema miquelii* seedlings are susceptible to frost and drought (Z. Yan and N. Reid, unpublished data, 1995; J. Nicholson, personal communication), and pre- and postdispersal predators are also prevalent; an insect parasitoid annually infects a large percentage of the developing fruit crop of *A. miquelii,* and rosella parrots (*Platycercus* spp.) eliminate entire cohorts of *Amyema* seedlings in their free-living phase (Z. Yan and N. Reid, unpublished data, 1995).

Episodic recruitment has also been reported in *Phoradendron juniperinum* parasitizing *Juniperinus osteosperma* (Dawson *et al.,* 1990). A greater than expected number of individuals was found in the 5- to 7-year age class, which coincided with years of above-average summer precipitation. Dawson *et al.* (1990) suggested that recruitment is likely to be highest in years with favorable water status or in trees in favored sites.

F. Mistletoe Life Histories

An increasingly clear picture of the population biology of the Australian Loranthaceae is emerging. Host attributes, climatically extreme events, fire, and predators appear to have been important selective factors in their evolution. The integrated set of morphological and life-history attributes that respond to selection includes haustorial structure, growth rate, age at first reproduction, shoot regenerative ability from dormant buds, reproductive phenology, and fruit display.

V. Future Directions for Mistletoe Research

We have summarized and synthesized several areas of mistletoe ecology and population biology. There are many loose threads, and students who wish to pursue them will advance knowledge of these interesting plants. Certain hypotheses and issues deserve special priority in mistletoe research, which are outlined in this section.

Mistletoe seeds are short-lived (Lamont, 1983). The lack of a persistent seed bank leads to the potential for rapid genetic selection, particularly after density-independent events such as wildfire or hurricanes that wipe out most or all individuals. Such catastrophes may result in genetic bottlenecks. After recolonization by founder events, genetic drift may result in rapid genetic changes in the new mistletoe population. The contagious patterns of distribution in mistletoe populations that result from disperser behavior (Reid and Lange, 1988; Reid, 1989) presumably reinforce the drift due to founder effects among local populations. Because biotic interactions with hosts, dispersers, and herbivores are of fundamental importance in mistletoe populations, the potential for rapid genetic selection may be reflected in unusually specific and labile coevolutionary relationships with hosts, mutualists, predators, and hyperparasites. The extent to which the adaptive shifts are one-sided on the part of the mistletoe or reciprocated by interacting organisms will depend on the genetic lability of the latter. Isozyme analysis has proved useful in understanding infraspecific and interspecific genetic relationships in dwarf mistletoes in relation to host range (Nickrent and Stell, 1990; Nickrent and Butler, 1990, 1991). Genetic analysis of mistletoe populations and their hosts, mutualists, and predators could reveal the spatial and temporal scales of and limits to coadaptation among higher plants, vertebrates, and insects. A benefit of this line of inquiry may be a better understanding of the genetic lability of host resistance to mistletoe infection. Knowledge of intra- and interpopulational differences in host resistance to mistletoe infection will assist with the development of strategies for managing pest mistletoes.

Mistletoes are also proving to be useful test systems for plant population biology studies. The facility with which mistletoe seeds can be monitored and manipulated in natural systems (Clay *et al.*, 1985; May, 1971; Reid, 1989; Yan, 1993a,c) and the discovery of nondestructive means of accurately aging plants (Dawson *et al.*, 1990; Dawson and Ehleringer, 1991) are likely to provide fundamental insights into how plants interact with and adapt to their environment.

Current information needs relating specifically to the population biology and management of mistletoes include:

1. *Seed shadows of bird-dispersed mistletoes.* New seedlings in mistletoe populations can be readily monitored. Better information is needed on how dispersers move in and about populations of fruiting mistletoe if the spatial patterns of mistletoe population increase are to be predicted over a range of host dispersions and densities. This information is needed to help develop mistletoe control strategies based on disinfection or removal of infested trees and stands by mechanical means or prescribed burning.

2. *Host attributes that influence mistletoe establishment and persistence.* Although many pests and pathogens selectively attack weaker hosts, the reverse situation may apply to obligate parasites such as mistletoes. Hawksworth (1961) proposed that establishment and persistence may be greatest on vigorous hosts because such individuals afford the greatest quantity of resources to invading mistletoe seedlings. (Paradoxically, such hosts will suffer the greatest impact in terms of growth reduction from severe mistletoe infection.) Gehring and Whitham (1992) proposed the alternative hypothesis, based on the assumption that host vigor and nutrition are important to host resistance to infection. They argued that stressed trees are least able to resist mistletoe attack and therefore will be more heavily parasitized than unstressed trees. Surprisingly, such fundamental hypotheses concerning the relationship between host nutritional and water status and mistletoe

infectivity have received little attention. Other host attributes that are likely to affect the chances of mistletoe establishment and persistence, such as patterns of host branch retention, factors influencing host branch abscission, position of the host in the stand (in terms of root competition with neighbors and light interception), and position of the mistletoe in the canopy relative to shading, light interception, and receipt of xylem sap, also merit study.

3. *Influence of predation on mistletoe demography.* Some pest mistletoe populations, such as *Amyema miquelii* in eastern Australia, sustain high levels of leaf, flower, and fruit predation by insect grazers, parasitoids, and parrots. The importance of this degree of predation is uncertain. Heavy levels of predation could potentially limit recruitment. Alternatively, recruitment and mortality in these populations may be regulated by climatic events or disperser abundance, and predation may be redundant in terms of demographic impact. The question is relevant in a managerial context because predator abundance can be more easily manipulated than disperser abundance or climate.

Acknowledgments

We are grateful to Dan Nickrent for helpful comments on a draft of this chapter, and to John Williams, John Duggin, and Wal Whalley for information about eucalypt growth rates and longevities. The research was partly funded by the Australian Research Council (large grant AO8930825), the University of New England Internal Research Grants Scheme, and the UNE Newholme Funds Allocation Scheme.

References

Alexander, M. E., and Hawksworth, F. G. (1975). Wildland fires and dwarf mistletoes: A literature review of ecology and prescribed burning. *USDA For. Serv. Gen. Tech. Rep. RM* RM-14.

Atsatt, P. R. (1983). Host parasite interaction in higher plants. *Encycl. Plant Physiol., New Ser.* 12C, 259–275.

Baranyay, J. A., and Smith, R. B. (1974). Low temperature damage to dwarf mistletoe fruit. *Can. J. For. Res.* 4, 361–365.

Barlow, B. A. (1981). The loranthaceous mistletoes in Australia. *In* "Ecological Biogeography of Australia" (A. Keast, ed.), pp. 556–574. Dr. W. Junk Publ., The Hague, The Netherlands.

Barlow, B. A. (1983). Biogeography of Loranthaceae and Viscaceae. *In* "The Biology of Mistletoes" (M. Calder and P. Bernhardt, eds.), pp. 19–46. Academic Press, Sydney.

Barlow, B. A. (1986). Mistletoes. *In* "The Ecology of the Forests and Woodlands of South Australia" (H. R. Wallace, ed.), pp. 137–143. Government Printer, South Australia.

Barlow, B. A., and Wiens, D. (1977). Host parasite resemblance in Australian mistletoes: The case for cryptic mimicry. *Evolution (Lawrence, Kans.)* 31, 69–84.

Bernhardt, P. (1983). The floral biology of *Amyema* in south-eastern Australia. *In* "The Biology of Mistletoes" (M. Calder and P. Bernhardt, eds.), pp. 87–100. Academic Press, Sydney.

Choate, J. H., Andrews, R. A., and Barlow, B. A. (1987). Herbivory and cryptic mimicry in Australian Loranthaceae. *In* "Parasitic Flowering Plants" (H. C. Weber and W. Forstreuter, eds.), pp. 127–135. Phillips Universität, Marburg.

Clay, K., Dement, D., and Rejmanek, M. (1985). Experimental evidence for host races in mistletoe (*Phoradendron tomentosum*). *Am. J. Bot.* 72, 1225–1231.

Cleland, J. B. (1940). Rejuvenation of vegetation after the bushfires in the National Park, South Australia. *South Aust. Nat.* 20, 43–48.

Davidar, P. (1983). Similarity between flowers and fruits in some flowerpecker pollinated mistletoes. *Biotropica* 15, 32–37.

Davidar, P. (1984). Ecological interactions between mistletoes and their avian pollinators in south India. *J. Bombay Nat. Hist. Soc.* 82, 45–60.

Dawson, T. E., and Ehleringer, J. R. (1991). Ecological correlates of seed mass variation in *Phoradendron juniperinum*, a xylem-tapping mistletoe. *Oecologia* 85, 332–342.

Dawson, T. E., King, E. J., and Ehleringer, J. R. (1990). Age structure of *Phoradendron juniperinum* (Viscaceae), a xylem-tapping mistletoe: Inferences from a non-destructive morphological index of age. *Am. J. Bot.* 77, 573–583.

Docters van Leeuwen, W. M. (1954). On the biology of some Javanese Loranthaceae and the role birds play in their life-histories. *Beaufortia* 4, 103–204.

Ehleringer, J. R., Schulze, E.-D., Ziegler, H., Lange, O. L., Farquhar, G. D., and Cowan, I. R. (1985). Xylem tapping mistletoes: Water or nutrient parasites? *Science* 227, 1479–1481.

Ehleringer, J. R., Cook, C. S., and Tieszen, L. L. (1986). Comparative water use and nitrogen relationships in a mistletoe and its host. *Oecologia* 68, 279–284.

Filip, G. M. (1984). Dwarf mistletoe and *Cytospora* canker decrease grand fir growth in central Oregon. *For. Sci.* 30, 1071–1079.

Fisher, J. T. (1983). Water relations of mistletoes and their hosts. *In* "The Biology of Mistletoes" (M. Calder and P. Bernhardt, eds.), pp. 161–183. Academic Press, Sydney.

Gehring, C. A., and Whitham, T. G. (1992). Reduced mycorrhizae on *Juniperus monosperma* with mistletoe: The influence of environmental stress and tree gender on a plant parasite and a plant–fungal mutualism. *Oecologia* 89, 298–303.

Gill, F. B., and Wolf, L. L. (1975). Foraging strategies and energetics of east African sunbirds at mistletoe flowers. *Am. Nat.* 109, 491–510.

Hamilton, S. G., and Barlow, B. A. (1963). Studies in Australian Loranthaceae. II. Attachment structures and their interrelationships. *Proc. Linn. Soc. N.S.W.* 88, 74–90.

Hariri, E. B., Sallé, G., and Andary, C. (1991). Involvement of flavonoids in the resistance of two poplar cultivars to mistletoe (*Viscum album* L.). *Protoplasma* 162, 20–26.

Hawksworth, F. G. (1961). Dwarf mistletoe of ponderosa pine in the Southwest. *USDA For. Serv. Tech. Bull.* 1246.

Hawksworth, F. G. (1983). Mistletoes as forest parasites. *In* "The Biology of Mistletoes" (M. Calder and P. Bernhardt, eds.), pp. 317–333. Academic Press, Sydney.

Hawksworth, F. G., and Shaw, C. C. (1984). Damage and loss caused by dwarf mistletoe in coniferous forests of western North America. *In* "Plant Diseases: Infection, Damage, and Loss" (R. K. S. Wood and G. J. Jellis, eds.), pp. 285–297. Blackwell, Oxford.

Hawksworth, F. G., and Wiens, D. (1972). Biology and classification of dwarf mistletoes (*Arceuthobium*). *USDA For. Serv., Agric, Handb.* 401.

Hawksworth, F. G., Moir, W. H., and Janssen, J. C. (1992). Effects of dwarf mistletoe in old-growth lodgepole pine stands at Fraser Experimental Forest, Colorado. *In* "Old-Growth Forests in the Southwest and Rocky Mountain Regions" (M. R. Kaufmann, W. H. Moir, and R. L. Bassett, Tech. Coords.), Gen. Tech. Rep. RM-213, pp. 60–65. USDA For. Serv., Fort Collins, CO.

Hoffmann, A. J., Fuentes, E. R., Cortes, I., Liberona, F., and Costa, V. (1986). *Tristerix tetrandrus* (Loranthaceae) and its host-plants in the Chilean matorral: Patterns and mechanisms. *Oecologia* 69, 202–206.

Hudler, G., and French, D. W. (1976). Dispersal and survival of seed of eastern dwarf mistletoe. *Can. J. For. Res.* 6, 335–340.

Hull, R. J., and Leonard, O. A. (1964). Physiological aspects of parasitism in mistletoes (*Arceuthobium* and *Phoradendron*). I. The carbohydrate nutrition of mistletoe. *Plant Physiol.* 39, 996–1007.

Knutson, D. M. (1983). Physiology of mistletoe parasitism and disease responses in the host. *In* "The Biology of Mistletoes" (M. Calder and P. Bernhardt, eds.), pp. 295–316. Academic Press, Sydney.

Kuijt, J. (1969). "The Biology of Parasitic Flowering Plants." Univ. of California Press, Berkeley.

Kuijt, J. (1990). Correlations in the germination pattern of santalacean and other mistletoes. *In* "The Plant Diversity of Malesia" (P. Baas, K. Kalkman, and R. Geesink, eds.), pp. 63–72. Kluwer Academic Publishers, Dordrecht, The Netherlands.

Lamont, B. (1982a). Gas content of berries of the Australian mistletoe *Amyema preissii* and the effect of maturity, viscin, temperature and carbon dioxide on germination. *J. Exp. Bot.* 33, 790–798.

Lamont, B. (1982b). Host range and germination requirements of some South African mistletoes. *S. Afr. J. Sci.* 78, 41–42.

Lamont, B. (1983). Germination of mistletoes. *In* "The Biology of Mistletoes" (M. Calder and P. Bernhardt, eds.), pp. 129–143. Academic Press, Sydney.

Lamont, B. (1985). Host distribution, potassium content, water relations and control of two co-occurring mistletoe species. *J. R. Soc. West. Aust.* 68, 21–25.

Lange, R. T., and Purdie, R. (1976). Western myall (*Acacia sowdenii*), its survival prospects and management needs. *Aust. Rangel. J.* 1, 64–69.

Lichter, J. M., and Berry, A. M. (1991). Establishment of the mistletoe *Phoradendron macrophyllum:* Phenology of early stages and host compatibility studies. *Bot. Gaz. (Chicago)* 152, 468–475.

Marshall, J. D., and Ehleringer, J. R. (1990). Are xylem-tapping mistletoes partially heterotrophic? *Oecologia* 84, 244–248.

May, D. S. (1971). The role of population differentiation in experimental infection of *Prosopis* by *Phoradendron*. *Am. J. Bot.* 58, 921–931.

Murphy, S. R., Reid, N., Yan, Z., and Venables, W. N. (1993). Differential passage time of mistletoe fruits through the gut of honeyeaters and flowerpeckers: Effects on seedling establishment. *Oecologia* 93, 171–176.

Nicholls, T. H., Hawksworth, F. G., and Merrill, L. M. (1984). Animal vectors of dwarf mistletoe with special reference to *Arceuthobium americanum*. *In* "Biology of Dwarf Mistletoes: Proceedings of the Symposium" (F. G. Hawksworth and R. F. Scharpf, Tech. Coords.), Gen. Tech. Rep. RM-111, pp. 102–110. USDA For. Serv., Ft Collins, CO.

Nicholson, D. I. (1955). "The Effect of 2,4-D Injections and of Mistletoe on the Growth of *Eucalyptus polyanthemos,*" For. Timber Bur. Leafl. No. 69. Government Printing Office, Canberra.

Nickrent, D. L., and Butler, T. L. (1990). Allozymic relationships of *Arceuthobium campylopodum* and allies in California. *Biochem. Syst. Ecol.* 18, 253–265.

Nickrent, D. L., and Butler, T. L. (1991). Genetic relationships in *Arceuthobium monticola* and *A. siskiyouense* (Viscaceae): New dwarf mistletoe species from California and Oregon. *Biochem. Syst. Ecol.* 19, 305–313.

Nickrent, D. L., and Stell, A. L. (1990). Electrophoretic evidence for genetic differentiation in two host races of hemlock dwarf mistletoe (*Arceuthobium tsugense*). *Biochem. Syst. Ecol.* 18, 267–280.

Norton, D. A. (1991). *Trilepidea adamsii:* An obituary for a species. *Conserv. Biol.* 5, 52–57.

Ogle, C. C. (1987). The incidence and conservation of animal and plant species in remnants of native vegetation within New Zealand. *In* "Nature Conservation: The Role of Remnants of Native Vegetation" (D. A. Saunders, G. W. Arnold, A. A. Burbidge, and A. J. M. Hopkins, eds.), pp. 79–87. Surrey Beatty, Chipping Norton, N.S.W.

Pate, J. S., True, K. C., and Kuo, J. (1991a). Partitioning of dry matter and mineral nutrients during a reproductive cycle of the mistletoe *Amyema linophyllum* (Fenzl.) Tieghem parasitizing *Casuarina obesa* Miq. *J. Exp. Bot.* 42, 427–439.

Pate, J. S., True, K. C., and Rasins, E. (1991b). Xylem transport and storage of amino acids by S.W. Australian mistletoes and their hosts. *J. Exp. Bot.* 42, 441–451.

Porter, A. (1990). The relationship between possums and mistletoes with particular reference to feeding behavior. B. Nat. Res. Thesis, University of New England, Armidale, N.S.W.

Purdie, R. W. (1969). The population structures of selected arid zone tree species. B.Sc. (Hons.) Thesis, University of Adelaide, Adelaide S. A.

Raven, J. A. (1983). Phytophages of xylem and phloem: A comparison of animal and plant sapfeeders. *Adv. Ecol. Res.* 13, 135–234.

Reid, N. (1986). Pollination and seed dispersal of mistletoes (Loranthaceae) by birds in southern Australia. *In* "The Dynamic Partnership: Birds and Plants in Southern Australia" (H. A. Ford and D. C. Paton, eds.), pp. 64–84. Government Printer, South Australia.

Reid, N. (1989). Dispersal of mistletoes by honeyeaters and flowerpeckers: The components of seed dispersal quality. *Ecology* 70, 137–145.

Reid, N. (1990). Mutualistic interdependence between mistletoes (*Amyema quandang*), and spiny-cheeked honeyeaters and mistletoebirds in an arid woodland. *Aust. J. Ecol.* 15, 175–190.

Reid, N. (1991). Coevolution of mistletoes and frugivorous birds? *Aust. J. Ecol.* 16, 457–469.

Reid, N., and Lange, R. T. (1988). Host specificity, dispersion and persistence through drought of two arid zone mistletoes. *Aust. J. Bot.* 36, 299–313.

Reid, N., and Yan, Z. (1995). Mistletoe biology, pathology and management. *In* "Eucalypt Diseases" (G. A. Kile, K. Old, and P. J. Keane, eds.). CSIRO, Melbourne (in press).

Reid, N., Yan, Z., and Fittler, J. (1994). Impact of mistletoes (*Amyema miguelii*) on host (*Eucalyptus blakelyi*, and *E. melliodora*) survival and growth in eastern Australia. *For. Ecol. Manag.* 70, 55–65.

Room, P. M. (1971). Some physiological aspects of the relationship between cocoa, *Theobroma cacao*, and the mistletoe *Tapinanthus bangwensis* (Engl. and K. Krause). *Ann. Bot. (London)* [N.S.] 35, 169–174.

Sallé, G. (1983). Germination and establishment of *Viscum album* L. *In* "The Biology of Mistletoes" (M. Calder and P. Bernhardt, eds.), pp. 145–159. Academic Press, Sydney.

Scharpf, R. F. (1987). Resistance of Jeffrey pine to dwarf mistletoe. *In* "Parasitic Flowering Plants" (H. C. Weber and W. Forstreuter, eds.), pp. 745–753. Phillips Universität, Marburg.

Schulze, E.-D., and Ehleringer, J. R. (1984). The effect of nitrogen supply on growth and wateruse efficiency of xylem-tapping mistletoes. *Planta* 162, 268–275.

Schulze, E.-D., Turner, N. C., and Glatzel, G. (1984). Carbon, water and nutrient relations of two mistletoes and their hosts: A hypothesis. *Plant, Cell Environ.* 7, 293–299.

Schulze, E.-D., Lange, O. L., Ziegler, H., and Gebauer, G. (1991). Carbon and nitrogen isotope ratios of mistletoes growing on nitrogen and non-nitrogen fixing hosts and on CAM plants in the Namib Desert confirm partial heterotrophy. *Oecologia* 88, 457–462.

Stevenson, G. B. (1935). The life history of the New Zealand species of the parasitic genus *Korthalsella*. *Trans. Proc. R. Soc. N. Z.* 64, 175–190.

Stewart, G. R., and Press, M. C. (1990). The physiology and biochemistry of parasitic angiosperms. *Annu. Rev. Plant Physiol. Mol. Biol.* 41, 127–151.

Thoday, D. (1951). The haustorial system of *Viscum album*. *J. Exp. Bot.* 2, 1–19.

Ullmann, I., Lange, O. L., Ziegler, H., Ehleringer, J., Schulze, E.-D., and Cowan, I. R. (1985). Diurnal courses of leaf conductance and transpiration of mistletoes and their hosts in central Australia. *Oecologia* 67, 577–587.

Walker, J. (1981). Fuel dynamics in Australian vegetation. *In* "Fire and the Australian Biota"

(A. M. Gill, R. H. Groves, and I. R. Noble, eds.), pp. 101–127. Australian Academy of Science, Canberra.

Wicker, E. F., and Hawksworth, F. G. (1988). Relationships of dwarf mistletoes and intermediate stand cultural practices in the northern Rockies. *USDA For. Serv., Gen. Tech. Rep. INT* INT-243, Ogden.

Wilson, P. R. (1984). The effect of possums on mistletoe on Mt Misery, Nelson Lakes National Park. *In* "Protection and Parks: Essays in the Preservation of Natural Values in Protected Areas" (P. R. Dingall, ed.), pp. 53–60. Department of Lands & Survey, Wellington.

Yan, Z. (1990). Host specificity of two mistletoe species, *Amyema preissii* and *Lysiana exocarpi*, in a semi-arid environment. Ph.D. Thesis, Flinders University, Adelaide.

Yan, Z. (1993a). Germination and seedling development of two mistletoes, *Amyema preissii* and *Lysiana exocarpi:* Host specificity and mistletoe–host compatibility. *Aust. J. Ecol.* 18, 419–429.

Yan, Z. (1993b). Low water potentials as a deterrent to mistletoes. *Haustorium* 27, 2.

Yan, Z. (1993c). Resistance to haustorial development in two mistletoes growing on their host and non-host species. *Int. J. Plant Sci.* 154, 386–394.

14

Vines in Treetops: Consequences of Mechanical Dependence

Francis E. Putz

. . . appearing very far away in the deceptive light, the big trees of the forest, lashed together with manifold bonds by a mass of tangled creepers, looked down at the growing young life at their feet with the sombre resignation of giants that had lost faith in their strength. And in the midst of them the merciless creepers clung to the big trunks in cable-like coils, leaped from tree to tree, hung in thorny festoons from the lower boughs, and, sending slender tendrils on high to seek out the smallest branches, carried death to their victims in an exulting riot of silent destruction.
—*Joseph Conrad, "Almayer's Folly" (1895, p. 134)*

I. Introduction

The dependence of vines on trees, other vines, boulders, or brick walls for mechanical support saves them wood, to be sure, but it also profoundly influences many other aspects of their biology. Vine stems, for example, are slender but have anatomical characteristics that provide flexible strength (Putz and Holbrook, 1991), promote recovery from mechanical damage (Fisher and Ewers, 1991), and allow rapid xylem and phloem transport (e.g., Ewers and Fisher, 1991). Freed from the need to prop up their shoots, vine roots tend to be thinner, more far-reaching, and better at extracting soil resources than the plants on which they depend for support (Dillenburg *et al.*, 1993a,b; F. Putz, unpublished data). In some forests, the sea-

Copyright © 1995 by Academic Press, Inc.
All rights of reproduction in any form reserved.

sonal rhythms of leaf flush, flowering, and fruiting of vines are different from those of trees (Putz and Windsor, 1987; Opler *et al.*, 1991).

Rather than describe the peculiarities of vines that were discussed in a recent volume edited by Putz and Mooney (1991), here I contemplate the biology of vines at the canopy level, a little-explored portion of our planet but one that provides fertile ground for investigation. Do not look to this chapter, however, for definitive answers to questions of vine–tree interactions at the top of the forest. I lack detailed data or particularly revealing insights and can only hope to stimulate thought and experiment. Let readers also be aware that I climb to the treetops reluctantly, have not yet shared the pleasure of canopy cranes or rafts, and as a vehicle for canopy access generally prefer the rocking chair on my porch, looking down on wild grapevines (*Vitis rotundifolia*) draped over pignut hickory (*Carya glabra*) and live oak (*Quercus virginiana*) saplings.

II. The Climbing Challenge

Little may be known about what vines do when they reach the canopy, but to get there, they face a major problem; lack of suitable trellises keeps most vine shoots from reaching the treetops. Trellis suitability is a function of the diameter of its struts (tree trunks and branches) and distances between struts, as well as of the climbing mechanism of the vines and their capacity to span gaps between trellises (review by Putz and Holbrook, 1991). Tendril-climbers with flexible leaders may be nimble but are relegated to twiggy realms, whereas robust twiners can ascend stems up to 20–30 cm in diameter. Differences in trellis requirements may explain the prevalence of tendril-climbers in short-statured or young forest and the abundance of twiners in taller or more mature stands (Hegarty and Caballé, 1991). Vines that climb with the aid of adventitious roots or adhesive tendrils (or both) can climb bigger trees or even the sides of buildings; spanning trellis gaps, however, is not their forte. The influence of bark texture on trellis quality (from the vine perspective) is not clear; bark exfoliation may result in vine dislodgement, but surface roughness may benefit climbers. We will revisit the issue of the ability of vines to span gaps, but note here that vine species differ greatly in their gap-spanning capacities (Putz, 1984).

III. Vine Height Distributions

Most vines are relatively light-demanding, or at least they proliferate vegetatively, flower, and fruit only in well-lit realms. It is not clear, however, why some species flourish beneath the upper leaves of their supporting trees whereas others cover their hosts with veritable carpets of foliage. In the case of species that climb with the aid of adventitious roots and/or adhesive tendrils, proliferation below the uppermost tree twigs is understandable on mechanical grounds. Most vines of this sort do not often grow on small twigs and lack the capacity to span all but the narrowest interbranch gaps. *Hedera helix* (English ivy) and *Toxicodendron radicans* (poison ivy) are root climbers familiar to many north temperate botanists; several prominent *Piper* spp. climb in this fashion in Australasia (Hegarty, 1991). For other types of climbers, failure to reach up to full sun may be due to physiological preferences, more subtle biomechanical impediments, or both.

The uppermost leaves in a forest are exposed to full sun, extreme temperatures, severe vapor pressure deficits, and high wind speeds (Chapter 3). Vine species that attain canopy status by growing up through the forest understory or up the edges of small treefall gaps may not be able to withstand the rigors of the uppermost canopy. Although many vines display marked change in leaf size and shape as they grow taller or older (Lee and Richards, 1991) and also presumably have well-developed physiological plasticity (Castellanos, 1991), few species may have the capacity to produce leaves that function efficiently in the understory and also flourish in the upper canopy. It may be argued that many canopy tree species accomplish this degree of physiological acclimation. Most tree leaves, however, are at least partly shaded by other leaves on the same individual, whereas vines on the uppermost canopy grow horizontally and thus are subjected to little self-shading.

Completely carpeting the top surface of a tree may seem to be the ultimate expression of vine power, but displaying leaves just below the uppermost tree leaves may be more advantageous. Few plants use full sun efficiently (e.g., Kozlowski *et al.*, 1991); some vines may avoid excessive heat, high photon flux densities, and extreme vapor pressure deficits by leaving the very top of the forest to the trees.

Although the canopy is the focus of this chapter, not all vines are treetop-bound. Some species flourish in forest understories, albeit usually not in the gloomiest reaches (Castellanos *et al.*, 1992). Other vines are difficult to distinguish from rhizomatous or stoloniferous plants, and climb only infrequently (Hegarty and Caballé, 1991).

IV. Tree Crown Exploitation by Vines

Many vines that attain canopy status grow from one tree to another. For example, in the Malaysian state of Sarawak, the average canopy vine grew on 1.4 trees (maximum = 4; Putz and Chai, 1987); on Barro Colorado Island in Panama, the average was very similar (1.6) but the range was huge (maximum = 49; Putz, 1984). The Panamanian vine that connected the crowns of 49 trees was a member of the pantropical genus *Entada* (Leguminosae); Caballé (1980) described a different species of *Entada* in Gabon that connected the crown of 13 trees. The number of intercrown passages made by a canopy vine is a function of numerous factors, but canopy structure and vine stem biomechanics seem preeminent. I think that the former is generally the more consequential.

Most vines can span lateral gaps between trees but this capacity varies greatly. Species with stiff leader shoots on which leaf expansion is delayed until the shoot is mechanically anchored (Raciborski, 1900) may span intercrown gaps of up to several meters (Putz, 1984); often, however, a horizontal gap exceeding 2 m precludes intercrown growth. A vine in the canopy of a tree isolated by more than a few meters need not be entirely stranded if lower-statured trees are arrayed in such a way that they form bridges between trees in the upper canopy. I am not aware of reports of vines in the canopy producing the specialized horizontal "foraging" shoots that extend for great distances through the understories of many tropical forests without branching or expanding normal leaves (Peñalosa, 1982; Courdurier, 1992); high light conditions in the upper canopy may preclude production of such structures.

Intercrown growth is advantageous for vines. By providing larger areas over which to display foliage, flowers, and fruits, vines benefit from having access to more than one tree crown. The occurrence of loosely hanging vines between fairly widely spaced trees is evidence that a vine growing in more than one tree need not fall upon the demise of any of its hosts. Finally, for vines that flourish only in upper canopy positions, being able to grow between crowns is a hedge against being overtopped by other trees.

Caballé (1993) suggested that there is a correlation between the tendency to proliferate widely in the canopy and anatomical characteristics of liana stems. According to his observations, liana species that frequently grow between tree crowns generally have stems with multiple cambia or otherwise divided xylem cylinders. That anomalous secondary growth ("cambial variants" *sensu* Carlquist, 1991) and intercrown growth are more common among Neotropical than African lianas led Caballé to suggest that light in Neotropical forests may be more limited and more heterogeneously distributed than in Africa.

Forests in different parts of the world and stands within the same forest vary in canopy height, tree growth rates, and the distances between neighboring crowns. If securing a trellis of dimensions appropriate for canopy ascent is a major problem for most forest vines, then tree crowns in short forests would be more accessible than in otherwise similar but taller forests. Fast-growing trees may outpace vines in upward ascent and tend to remain vine-free, especially if they self-prune (review by Hegarty, 1991). Once they have attained the canopy, intercrown vine growth would likely be limited by the distances between adjacent crowns. These distances are primarily influenced by the density of canopy trees. Even in forests fully stocked with trees, however, intercrown spacing varies strikingly. This variation in what have been called "crown shyness gaps" is due to factors such as differences in shade tolerance (Ng, 1977) and tree flexibility (Putz *et al.,* 1984). Light-demanding trees and flexible trees on which wind-induced swaying results in mechanical pruning of marginal branches are expected to be spatially isolated from neighbors. Crown shyness gaps also occur between branch systems within individual tree crowns, thereby reducing the likelihood of a single vine carpeting an entire tree.

Trees with crowns composed of separated "crownlets" may avoid becoming completely inundated by vines. Although each crownlet may become covered or otherwise thoroughly infested by vines, if the intercrownlet gaps are sufficiently large, the same vine is unlikely to invade the entire crown. The cauliflowerlike appearance of the crowns of many Dipterocarpaceae and the even more widely separated crownlets of some Araliaceae spring to mind as examples of this intriguing although unverified phenomenon.

Considering the diversity of tree crown geometries (Chapter 2) and the various mechanisms vines employ to climb, carpeting the tops of some tree species may be mechanically impossible for some vines. Climbing palms that do not branch above the ground, for example, have almost no capacity for lateral expansion upon reaching the canopy (Putz, 1990). Mechanical limitations of canopy-carpet formation may likewise affect twining vines, albeit less so. Twiners require structures about which to twine, generally either tree branches or other vines. Where canopy tree branches are vertical or nearly so, even if each branch supports a twiner, the result would not be carpetlike. Only canopy trees with horizontal branches or large pinnately compound leaves, and trees that already have horizontal vine superstructures, may be prone to carpeting by twiners. Branching patterns of trees in the Myristicaceae and Annonaceae, for example, may increase their susceptibilities to vine carpeting; the long compound leaves of many trees in the Meliaceae may have the same effect.

From my porch in Florida, I can gaze down upon a good example of differential vine susceptibility. There are apparently marked differences in the capacity of grapevines to carpet the *Quercus virginiana* (live oak) sapling and the nearby *Carya glabra* (hickory) of the same height. Both are vine-infested, but whereas few of the pinnately compound hickory leaves are visible under the grape foliage, the oak has several vine-free upright shoots piercing an otherwise complete net of vines. The nearby presence of a 3-m-tall *Vaccinium arboreum* (sparkleberry) with small leaves and vertical branches completely smothered by a grape, however, is evidence that these characteristics do not vine resistance make. I thus add rapid vertical growth to small leaves and vertical branches as potential vine carpet-avoiding characteristics. These observations are consistent with the suggestion of Beekman (1981) that closely spaced, slow-growing, and damaged trees are particularly susceptible to becoming covered by vines.

Once vines have spanned the gap between trees, they must secure their position against the discordant swaying of their hosts. This capacity may vary with the type of "prehensile apparatus" (Baillaud, 1962) they possess. The connections that tendril climbers make between trees would be tenuous at best (at least initially) and are likely to be broken when the connected trees sway in opposite directions. Because coiled twiner stems might slide or uncoil rather than break during discordant tree movements, twiners may avoid breakage and dislodgement until their position is made secure by vegetative proliferation. Scrambling vines (i.e., those that sprawl over their host trees with no discrete attachment mechanism) might be best suited for intercrown growth because they do not securely attach to anything. In Central America, *Acacia hayesii* displays the supreme development of scrambling from tree to tree. Its thorny branches grow to substantial diameters and weave a messy vine carpet heavy enough to crush supporting trees, thereby opening even larger canopy gaps in which this species flourishes. Scramblers, however, are rarely high climbers.

V. Vine Effects on Tree Crowns

Scaling to the treetops may represent success for vines but it spells doom (or at least gloom) for trees. Vines often inflict mechanical damage on the twigs and branches to which they cling. They weigh down and break branches and entire trees, increase the likelihood of trees being pulled down or otherwise damaged by neighbors, shade their hosts, and compete effectively for water and nutrients (review by Hegarty, 1991). Smith (1973) proposed that by binding trees together, vines may increase the structural integrity of the canopy. Although this "vines as guy wires" hypothesis seems plausible, most evidence is to the contrary. Finally, although the amounts of herbivore damage between vine-infected and vine-free trees have not yet been compared (Lowman, 1994), folivorous mammals such as sloths are known to disproportionately frequent trees with vines (Montgomery and Sunquist, 1978).

The degree to which canopy vines are deleterious to their host trees probably varies with type of climber but also with whether vine foliage is displayed above or beneath that of the supporting tree. Subcanopy vine proliferation, for example, reduces both the shade cast and the mechanical torque effect that canopy-top vines inflict on their hosts. Vines that remain slender (e.g., climbing palms) would also be less damaging than more robust climbers such as the *Acacia hayesii* discussed earlier.

VI. Vine Tangles near the Ground

Those of us who spend much of our time near the ground most often encounter high densities of vines where trees have fallen or been felled. These pedestrian vine tangles are often deeper than those that develop in the canopy, perhaps because the ground provides more mechanical support than the treetops. Furthermore, fallen vine stems, their sprouts, and vine seedling colonists all contribute to making some of these tangles formidable for both humans and trees; horizontal and vertical passage are impeded and often painful.

How trees surmount vine tangles (and they eventually do) is still something of a mystery. Some vine tangles may lose their vigor in the shade of trees that are protected by vine-snipping ants or that are otherwise able to avoid and shed vines and thereby pierce and overtop the tangle. Neotropical trees of the genus *Cecropia* (Moraceae), which are protected by species of *Azteca* ants, often perform this feat of vine domination (Schupp, 1986). In Asia, *Macaranga* (Euphorbiaceae) trees protected from vines by *Crematogaster* ants often do the same. Trees with especially large compound leaves or phyllomorphic (i.e., leaflike) branches (e.g., *Anthocephalus,* Rubiaceae) are also frequent vine carpet-piercers, perhaps because they shed their vine loads upon leaf or branch abscission. Once overtopped and weakened, vine tangles may be open to invasion by slower-growing, more shade-tolerant, and less vine-proof trees. A less biologically innovative but equally effective mechanism by which arboreal dominance over vines is retained is through the lateral expansion of the crowns of trees adjacent to but not greatly afflicted by vine tangles.

In the absence of vine-proof trees, such as *Cecropia* and *Macaranga,* large vine tangles can persist for decades; eventually, however, they are surmounted by trees. The process by which trees grow up through and overtop vine tangles is often very slow. I know of some tangles that have persisted, approximately unchanged, for more than two decades. Driven by negative geotropism and fueled by the little light that filters through the vine carpet and soil resources that are not usurped by vine roots, trees find their way up through the vines, only to be grabbed once again by a vine tendril or twining stem. Some of these temporarily successful tree shoots achieve height growth that is not completely lost when they are once again smothered by vines. By this iterative process, the vine carpet is slowly elevated above the ground. Occasionally, a tree will completely escape the clutches of vines, rapidly grow taller, and cast shade on the vine tangle, thereby reducing its vigor. Because vines are generally less fire resistant than some tree species (F. E. Putz, personal observation), ground fires may facilitate tree dominance of vine tangles. In general, however, once a vine tangle develops either at ground level or in the canopy, trees are likely to suffer for a very long time.

Figure 1   Vine carpets in (A) a natural treefall gap in Venezuela, and (B) a logging area in Malaysian Borneo.

Given the tremendous proliferation of vines after logging and other forms of forest abuse, the question of how trees surmount vine tangles is of more than academic concern. Vine tangles often start along roads through forests subjected to over-logging. Their capacity for vegetative expansion is favored where establishment from seed is precluded by compacted or otherwise degraded soils. From a strong roadside foothold, the scene over time appears like a vine carpet being unrolled over the surrounding forest (Fig. 1). Protecting forest from indiscriminate harvesting is the best solution for this problem, but prescriptions for tree regeneration in vine tangles also need to be developed.

VII. Vine versus Vine

When vines of different species encounter each other in the top of a tree, which one will attain ascendancy depends on both relative extension growth rates and climbing mechanisms. Darwin (1875) observed that climbing with the aid of tendrils requires less biomass per unit growth in length than twining. Thin-stemmed vines may benefit from height growth efficiency as well as from their reliance on more pliant, smaller-diameter, but more numerous structures for support (e.g., petioles or rachises of other vines). To a vine, using another vine as a support presents few mechanical

Figure 1—*Continued*

constraints; vine stems provide continuous trellises and are generally narrow enough to be grasped by even the most diminutive tendril-climber.

I am not aware of studies on intervine encounters of a competitive kind, but vines often facilitate one another's canopy access. Large remnant trees in logged forests in Malaysia, for example, are often first climbed by vines with adventitious roots and adhesive tendrils (e.g., *Piper* spp.). These colonizing vines are used as ladderlike trellises for twining species (e.g., *Merremia*). Finally, tendril-climbers (e.g., *Passiflora* spp.) ascend the outside of the mass of other types of vines (Pinard and Putz, 1994). The same sort of facilitation is obvious in Rock Creek Park, the remnant forest in Washington, D.C., where grapes (*Vitis* spp.) and bittersweet vines (*Celastrus* spp.) grasp onto root-climbing poison and English ivy on their way to the canopy (Putz, 1994).

VIII. Canopy Vines and Canopy Creatures

Legions of canopy vertebrates and invertebrates undoubtedly depend on canopy vines for food, shelter, escape from predators, and access to resources. Frugivores (e.g., Charles-Dominique *et al.,* 1981; Malcolm, 1991; Langtimm, 1992) and folivores (Montgomery and Sunquist, 1978) are known to make use of canopy vines for intercrown passage; other trophic groups can be added to this list without qualms but also apparently without data (Emmons and Gentry, 1983). In light of silvicultural prescriptions that recommend reducing vine populations in forests managed for timber production (review by Putz, 1991), more detailed studies on the roles of vines in the lives of animals are needed. Compromises must be made between conservation-minded but timber production-oriented forest managers and pure preservationists. Questions for which answers are needed include: Are there some vine species that, like figs, should be left unscathed? Is controlling some vine species not justifiable on economic grounds because they are not very injurious to trees? How many vines are "enough" to maintain viable wildlife populations?

IX. Future Research Possibilities

Much of what is not yet known about vines in the canopy could be learned from upstairs porches, childrens' treehouses, step ladders, or even from the ground. The challenge of canopy research is not so much physical as it is a test of our powers of imagination, observation, and deduction. Giant cranes, canopy rafts, trams, and walkways will all facilitate this research, but more modest equipment such as rocking chairs on upstairs porches will also bear fruit. Place vines of different sorts on a backyard shrub and see what happens; replicate this simple experiment ten or twenty times with different types of vines. Watch squirrels, marmosets, caterpillars, pangolins, or lemurs, and record the frequency with which they use vines for moving about the canopy. Make model trees with different architectures out of sticks and wires, place them in vine tangles, and see which ones are most rapidly surmounted by vines. Where canopy access is possible, treetop experiments of the same nature would also be interesting, as would studies requiring more sophisticated equipment. For example, one could gently move vines that typically grow beneath the upper layer of tree leaves into full sunlight and determine whether their photosynthetic rates decline or if they suffer thermal damage. Testing the hypothesis that these vines do not grow on the very top of trees because they are biomechanically unable to do so could be tested by installation of some simple treetop trellises. Vine succession is also ripe for research that could be conducted either from the ground or in the treetops.

A great deal can be learned about canopy biology if people stop thinking like terrestrial mammals and renounce arboricentrism by seeing the forest, not just the trees. Vine biology will also be promoted by replacing the evil reputation of merciless creepers wreaking silent havoc on defenseless trees (e.g., Conrad, 1895) with an image of gentle vine cascades in leafy cloudlands (Hudson, 1904).

X. Summary

Sacrificing mechanical rigidity for rapid extension growth results in vines being susceptible to trellis deficiency. Although the upper canopy is characterized by an abundance of small branches that serve well as trellis struts, gaps between tree crowns often preclude intercrown vine growth. Within individual tree crowns, the interplay between tree branching patterns, vine climbing mechanisms, and vine leaf light and temperature requirements may all influence whether vines grow over or among tree branches. Knowledge of canopy-level processes is accumulating, but much remains to be learned about the proximal and evolutionary effects of vines on trees and trees on vines.

References

Baillaud, L. (1962). Mouvements autonomes des tiges, vrilles et autres organes à l'exception des organes volubiles. *Handb. Pflanzenphysiol.* 17, 562–634.

Beekman, F. (1981). "Structural and Dynamic Aspects of the Occurrence and Development of Lianas in the Tropical Rain Forest." Department of Forestry, Agricultural University, Wageningen.

Caballé, G. (1980). Charactères de croissance et déterminisme chorologique de la liane *Entada gigas* (L.) Fawcett & Rendle (Leguminosae–Mimosoideae) en forêt dense du Gabon. *Adansonia* [2] 20, 309–320.

Caballé, G. (1993). Liana structure, function, and selection: A comparative study of xylem cylinders of tropical rainforest species in Africa and America. *Bot. J. Linn. Soc.* 113, 41–60.

Carlquist, S. (1991). Anatomy of vine and liana stems: A review and synthesis. *In* "The Biology of Vines" (F. E. Putz and H. A. Mooney, eds.), pp. 53–71. Cambridge Univ. Press, Cambridge, UK.

Castellanos, A. E. (1991). Photosynthesis and gas exchange of vines. *In* "The Biology of Vines" (F. E. Putz and H. A. Mooney, eds.), pp. 181–204. Cambridge Univ. Press, Cambridge, UK.

Castellanos, A. E., Duran, R., Guzmán, S., Briones, O., and Feria, M. (1992). Three-dimensional space utilization of lianas: A methodology. *Biotropica* 24, 396–401.

Charles-Dominique, P., Atramentowitz, M., Charles-Dominique, M., Gérard, H., Hladik, A., Hladik, C. M., and Prévost, M. F. (1981). Les mammifères frugivores arboricoles nocturnes d'une forêt guyanaise: inter-relations plantes-animaux. *Rev. Ecol. (Terre Vie)* 35, 341–435.

Conrad, J. (1895). "Almayer's Folly." Fisher Unwin, London.

Courdurier, T. (1992). Sur la place des lianes dans la forêt guyanaise. Thèse de Doctorat, Université de Montpellier 2 Sciences et Techniques du Languedoc.

Darwin, C. (1875). "The Movements and Habits of Climbing Plants." Murray, London.

Dillenburg, L. R., Whigham, D. F., Teramura, A. H., and Forseth, I. N. (1993a). Effects of below- and aboveground competition from the vines *Lonicera japonica* and *Parthenocissus quinquefolia* on the growth of the tree host *Liquidambar styraciflua*. *Oecologia* 93, 48–54.

Dillenburg, L. R., Whigham, D. F., Teramura, A. H., and Forseth, I. N. (1993b). Effects of vine competition on availability of light, water, and nitrogen to a tree host (*Liquidambar styraciflua*). *Am. J. Bot.* 80, 244–252.

Emmons, L. H., and Gentry, A. H. (1983). Tropical forest structure and the distribution of gliding and prehensile-tailed vertebrates. *Am. Nat.* 121, 513–524.

Ewers, F. W., and Fisher, J. B. (1991). Why vines have narrow stems: Histological trends in *Bauhinia* (Fabaceae). *Oecologia* 88, 233–237.

Ewers, F. W., Fisher, J. B., and Fichtner, K. (1991). Water flux and xylem structure in vines. *In* "The Biology of Vines" (F. E. Putz and H. A. Mooney, eds.), pp. 127–160. Cambridge Univ. Press, Cambridge, UK.

Fisher, J. B., and Ewers, F. W. (1991). Structural responses to stem injury in vines. *In* "The Biology of Vines" (F. E. Putz and H. A. Mooney, eds.), pp. 99–124. Cambridge Univ. Press, Cambridge, UK.

Hegarty, E. E. (1991). Vine–host interactions. *In* "The Biology of Vines" (F. E. Putz and H. A. Mooney, eds.), pp. 357–375. Cambridge Univ. Press, Cambridge, UK.

Hegarty, E. E., and Caballé, G. (1991). Distribution and abundance of vines in forest communities. *In* "The Biology of Vines" (F. E. Putz and H. A. Mooney, eds.), pp. 313–335. Cambridge Univ. Press, Cambridge, UK.

Hudson, W. H. (1904). "Green Mansions: A Romance of the Tropical Forests." Duckworth & Company, London.

Kozlowski, T. T., Kramer, P. J., and Pallardy, S. G. (1991). "The Physiological Ecology of Woody Plants." Academic Press, San Diego.

Langtimm, C. A. (1992). Specialization for vertical habitats within a cloud forest community of mice. Ph.D. Dissertation, University of Florida, Gainesville.

Lee, D. W., and Richards, J. H. (1991). Heteroblastic development in vines. *In* "The Biology of Vines" (F. E. Putz and H. A. Mooney, eds.), pp. 205–243. Cambridge Univ. Press, Cambridge, UK.

Lowman, M. (1994). In preparation.

Malcolm, J. R. (1991). The small mammals of Amazonian forest fragments. Ph.D. Dissertation, University of Florida, Gainesville.

Montgomery, G. G., and Sunquist, M. E. (1978). Habitat selection and use by two-toed and three-toed sloths. *In* "The Ecology of Arboreal Folivores" (G. G. Montgomery, ed.), pp. 329–359. Smithsonian Institution Press, Washington, DC.

Ng, F. S. P. (1977). Shyness in trees. *Nat. Malays.* 2, 34–37.

Opler, P. A., Baker, H. G., and Frankie, G. W. (1991). Seasonality of climbers: A review and example from Costa Rican dry forest. *In* "The Biology of Vines" (F. E. Putz and H. A. Mooney, eds.), pp. 377–391. Cambridge Univ. Press, Cambridge, UK.

Peñalosa, J. (1982). Morphological specialization and attachment success in two twining lianas. *Am. J. Bot.* 69, 1043–1045.

Pinard, M. A., and Putz, F. E. (1994). Vine infestation of remnant trees in logged forest in Sabah. *J. Trop. For. Sci.* 6, 302–309.

Putz, F. E. (1984). The natural history of lianas on Barro Colorado Island, Panama. *Ecology* 65, 1713–1724.

Putz, F. E. (1990). Liana stem diameter growth rates on Barro Colorado Island, Panama. *Biotropica* 22, 103–104.

Putz, F. E. (1991). Silvicultural effects of vines. *In* "The Biology of Vines" (F. E. Putz and H. A. Mooney, eds.), pp. 493–501. Cambridge Univ. Press, Cambridge, UK.

Putz, F. E. (1994). Relay ascension of big trees by vines in Rock Creek Park, District of Columbia. *Castanea* (in press).

Putz, F. E., and Chai, P. (1987). Ecological studies of lianas in Lambir National Park, Sarawak, Malaysia. *J. Ecol.* 75, 523–532.

Putz, F. E., and Holbrook, N. M. (1991). Biomechanical studies of vines. *In* "The Biology of Vines" (F. E. Putz and H. A. Mooney, eds.), pp. 73–97. Cambridge Univ. Press, Cambridge, UK.

Putz, F. E., and Mooney, H. A., eds. (1991). "The Biology of Vines." Cambridge Univ. Press, Cambridge, UK.

Putz, F. E., and Windsor, D. M. (1987). Liana phenology on Barro Colorado Island, Panama. *Biotropica* 19, 334–341.

Putz, F. E., Parker, G. G., and Archibald, R. M. (1984). Mechanical abrasion and intercrown spacing. *Am. Midl. Nat.* 112, 24–28.

Raciborski, M. (1900). Über die Vorläuferspitze. *Flora (Jena)* 87, 1–25.

Schupp, E. W. (1986). *Azteca* protection of *Cecropia:* Ant occupation benefits juvenile trees. *Oecologia* 70, 379–385.

Smith, A. P. (1973). Stratification of temperate and tropical forest. *Am. Nat.* 107, 671–683.

15

Life on the Forest Phylloplane: Hairs, Little Houses, and Myriad Mites

David Evans Walter and Dennis J. O'Dowd

I. Introduction

The discovery that forest canopies are inhabited by an amazing array of organisms has galvanized interest in biodiversity and unleashed fears that increasing deforestation is accelerating rates of extinction. Yet, when we gaze up from the forest floor into the canopy of a rain forest tree, the diversity and abundance of life are rarely apparent. Lurking among the leaves, branches, and epiphytes are few vertebrates; these are relatively small and often well-camouflaged. Even smaller are the invertebrates, and among the smallest of these are the mites.

Many studies of canopy invertebrates have specifically targeted insects (Basset, 1991, Table 1; Chapter 5), which are thought to be responsible for most canopy biodiversity and abundance (Erwin, 1983; Stork, 1988). Other studies have attempted to account for all canopy arthropods, including mites. These studies leave the impression that mites are minor players in the forest canopy. Most surveys list few mites outside of the crown humus (Nadkarni and Longino, 1990), and these often represent less than 1% of the total arthropods collected (Table I).

Free-living mites are among the least studied of animals, and little is known about them in the forest canopy (World Conservation Monitoring Centre, 1992). Robert May (1988) noted that "our ignorance of tropical mites . . . is at least as great as the ignorance about beetles and other arthropods." What are mites? Are they abundant or diverse in forest canopies?

Copyright © 1995 by Academic Press, Inc.
All rights of reproduction in any form reserved.

Table I Some Studies of Canopy Arthropods Based on Chemical Knockdown Techniques with the Number of Mites (Acari) and Total Number of Arthropods Reported

Site/forest type	Sampling method	Acari	Total arthropods	Percent-age Acari	Source
Tropical					
Cameroons, lowland rain forest	Branch fumigation	25	2271	1.1	Basset et al. (1992)
Borneo, lowland rain forest	Canopy fogging	151	23,874	0.6	Stork (1991)
Indonesia, lowland rain forest	Canopy fogging	?	?	3.8	Stork (1988)
Australia, lowland rain forest	Canopy fogging	380	9967	3.8	Kitching et al. (1993)
Subtropical					
Australia, montane rain forest	Restricted fogging; traps	0	51,600	0	Basset (1991)
Australia, montane rain forest	Canopy fogging	1503	22,984	6.5	Kitching et al. (1993)
Temperate					
Southeast U.S.A., deciduous forest	Branch fumigation	3	1120	0.3	Costa and Crossley (1991)
Australia, eucalypt woodland	Canopy fogging; branch fumigation	93	20,527	0.5	Majer and Recher (1988)
Australia, various forests	Branch fumigation	0	24,283	0	Woinarski (1984)
Australia, cool rain forest	Canopy fogging	807	5360	15.0	Kitching et al. (1993)
Australia, cool rain forest	Canopy fogging	8646	27,600	31.3	Yen and Lillywhite (1990)
South Africa, various trees	Canopy spraying	>10	6847	?	Southwood et al. (1982)
England, deciduous trees	Canopy spraying	>23	34,997	?	Southwood et al. (1982)
Japan, cedar plantation	Tree fumigation	22,077	99,469	22.2	Hijii (1986)

What factors affect their distribution at a scale appropriate to such minute animals? Do mites harm or benefit their "hosts"? Over the last few years, our research has provided some answers to these questions and one firm conclusion. Insects may be the most diverse of canopy arthropods, but they are not the most abundant. That distinction belongs to the mites.

II. The Stage

Although most researchers have given canopy mites short shrift, we have required of them (and of ourselves) a far more severe penance. The penalty was simple, but tedious—putting vegetation under microscopes and collecting mites. Our published studies (Walter, 1992; Walter and O'Dowd, 1992b; Walter and Behan-Pelletier, 1993; Walter *et al.*, 1993, 1994) have been narrowly focused and matter of fact, as journals generally require. However, these conventions have not prevented us from recording the larger drama played out in the rain forest canopy. In this chapter we review methods before introducing the cast and presenting the play itself.

A. The Forest Canopy

To date we have removed mites from 177 species of rain forest trees, shrubs, and lianas in eastern Australia [see Christophel and Hyland (1993), Costermans (1981), and Stanley and Ross (1983, 1986, 1989), for authorities of names of rain forest plants cited in text]. "Rain forest" is defined as fire-protected or fire-resistant forests with closed canopies composed of broad-leaved, evergreen vegetation not dominated by species of *Eucalyptus* or *Acacia* (Gell and Mercer, 1992), or understorys of rain forest vegetation in sclerophyll forests. The rain forests sampled grew from southern Tasmania (43°S) to northern Queensland (16°S) and transversed three climatic zones—temperate, subtropical, and tropical—and, two elevational zones—lowland (including littoral) and montane (Figure 1).

Most of our samples came from crowns easily accessible by hand or by extendable pole-pruner, that is, the lower canopy (1–5 m above the ground) of trees and lianas and mixed levels in saplings and shrubs. Stretching the pole-pruner to its limit, slopes, and ingenuity often allowed us to reach to 8 m above ground. A shanghai (slingshot) and lead weight were used to loop a monofilament line over branches up to 35 m above ground on Mt. Lewis, Queensland. The monofilament line was then used to draw up a stout rope, and the ensnared branch was pulled down (Hyland, 1972). Additional upper canopy (6–37 m) samples of leaves and stems (< 1 cm diameter) of large forest trees and lianas were obtained through the access provided by canopy walkways and towers located at Dorrigo National Park in New South Wales, and at O'Reilly's adjacent to Lamington National Park,

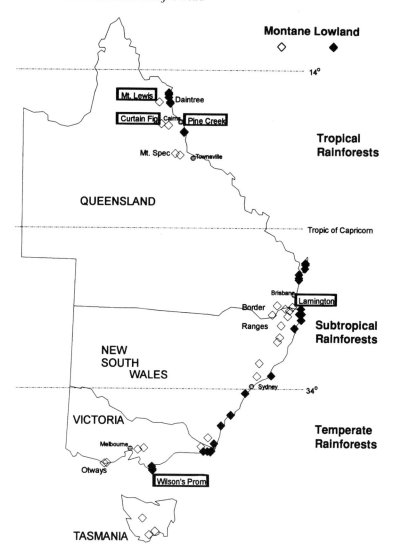

Figure 1   Map of eastern Australian showing areas with rain forests sampled for foliar mites. Primary research sites are boxed.

Gambubal State Forest near Emu Vale, Pine Creek near Gordonvale, and Curtain Fig near Atherton in Queensland. To determine if mite abundances varied between canopy levels, we compared mean values for lower and upper crowns of *Acmena smithii, Elaeocarpus reticulatus,* and *Pomaderris aspera* at Wilson's Promontory National Park in Victoria; *Caldcluvia panicu-*

losa at Dorrigo, *Sloanea woollsii* at Gambubal, *Austrosteenia glabrista, Parsonsia fulva, Randia benthamiana,* and *Synoum glandulosum* at O'Reilly's; and *Argyrodendron peralatum* at Curtain Fig.

B. The Leaf and Stem

Leaves and small segments of stem were used to sample mite populations. Most foliar mites are less than 0.5 mm in length, and even small leaves represent a very large sample of mite habitat. Individual trees, shrubs, and lianas were selected for ease of access, or by a haphazard or a random technique along transects. From each tree 1–15 individual shoots (i.e., the terminal growth of a small stem) were haphazardly chosen, and 1–10 leaves were subsampled from each shoot. Shoots/leaves were placed in plastic bags and kept in an ice chest or refrigerator until processed, usually within 48 hr. Leaves (> 13,000 total) were processed individually by scanning the abaxial and then the adaxial surface under a dissecting microscope (× 20–80), and then dissecting any hiding places with a scalpel. Stems were similarly scanned, and cracks and crevices dissected.

C. The Phylloplane

Mites are not always easily seen on the leaf surface. Many young rain forest leaves have smooth, simple surfaces (Fig. 2A) that shed rain readily and allow easy counting and collecting (using a fine brush) of foliar mites. However, as these leaves age, silk webbing (from spiders, psocids, caterpillars, and mites), casts, skins of arthropods, damage from insect feeding (pits, mines), galls, lichens, liverworts (Fig. 2D), and other debris accumulate, producing a complex habitat within which mites are neither easy to see nor to collect. Other leaves are inherently complex with strong, raised veins and dense layers of hairs (Fig. 2C). Mites on complex surfaces are captured only after laborious probing and dissection of leaf surface structures with a scalpel. Numbers of active galls and predatory mites and inquilines within galls were noted, but no attempt was made to count the numerous eriophyoid mites within them (although this would greatly increase estimates of the number of mites per leaf). Species identifications were based on mites cleared in Nesbitt's solution and mounted in Hoyer's medium on glass slides (Krantz, 1978).

Leaves of many woody dicots have domatia (Figs. 2E and 2H) and extrafloral nectaries (Koptur, 1992). The influence of domatia on mite numbers was tested by using a bitumen paint to block access to domatia (Walter and O'Dowd, 1992b) on leaves of *Randia benthamiana* at O'Reilly's just after the spring flush of leaves. Five months later, the leaves with blocked domatia and a set of control leaves (domatia left open, but dabs of paint placed below the domatia) were collected and mite abundances were compared.

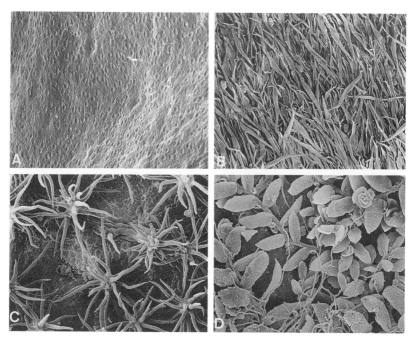

Figure 2   Leaf surfaces that are smooth and featureless (**A**: lilly pilly, *Acmena smithii* ×180) or covered with appressed hairs (**B**: musk daisy, *Olearia argophylla*, ×130) have few hiding places for mites. Leaves with dense layers of hairs (**C**: hazel pomaderris, *Pomaderris aspera*, ×250) or covered with epiphylls (**D**: upper surface of leaf of *Geissois benthamii* covered with Lejeuneaceae liverworts, ×95) have many hiding places and many mites. Some plants produce specialized structures in their vein axils (domatia) that act as shelters for mites (**E**: tuft domatium on leaf of *Synoum glandulosum*, ×90; **F**: pocket domatium on leaf of *Elaeocarpus obovatus*, ×60; **G**: pit domatium on leaf of *Timonius timon*, ×280; **H**: pit domatium transected to show oribatid mite, ×330).

We tested the effect of nectaries on predatory mites in the laboratory at room temperature and 70% relative humidity using a garden shrub (*Viburnum tinus* L.) that has a pair of nectaries on the lower leaf margin. Shoots with four leaves were cut from the shrub and the stems inserted into individual vials of water (a bouquet). Bouquets were randomly assigned to two treatments: nectaries excised with a scalpel or nectaries retained, but a wedge of tissue above the nectary excised. Four adult female phytoseiid mites (*Typhlodromus occidentalis* Nesbitt) and seven adult females of the spider mite *Tetranychus urticae* Koch (to provide prey) were added to each bouquet. After 10 days, bouquets were scanned under a microscope and all mites were counted.

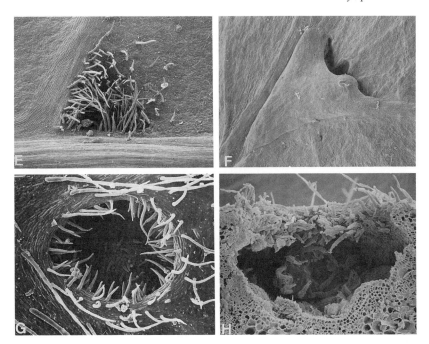

Figure 2—*Continued*

III. The Cast

Mites (phylum Chelicerata, subclass Acari) are minute arthropods (mostly less than a millimeter) with chelicerae (pincer- or styletlike mouthparts), six-legged larvae, eight-legged nymphs and adults, and a propensity for infesting every terrestrial, freshwater, and intertidal to deep benthic marine habitat. Two major lineages are recognized, the orders Parasitiformes and Acariformes, both with numerous species inhabiting forest canopies throughout the world. The vertebrates and many of the insects that inhabit forest canopies have diverse faunas of parasitic, commensal, and mutualistic mites, but our chapter is limited to those mites that live on plants, especially those that inhabit the phylloplane (leaf surface) of trees, shrubs, and lianas.

IV. The Roles

What do canopy mites do? We used our own observations, the extensive literature on arboreal mites in agroecosystems (see reviews in Jeppson *et al.*,

1975; Hoy, 1982; Hoy *et al.*, 1983; Helle and Sabelis, 1985), and information on related taxa that inhabit soils (Moore *et al.*, 1988; Dindal, 1990; Walter *et al.*, 1988, 1991) to assign arboreal mites to three feeding guilds: herbivores, predators, and scavengers/microbivores, and to characterize their dispersal behaviors and generation times.

A. Herbivores

We consider herbivorous mites to be those parasitic on living higher plants; mites feeding on lichens, fungi, and algae are treated as scavengers and microbivores. Most herbivorous species use cheliceral stylets to puncture plant cells and extract the cell contents, and belong to the order Acariformes, primarily to two superfamilies. The Tetranychoidea includes the Tetranychidae (called spider mites, because of the dense silken webbing produced by some species), major agricultural pests capable of defoliating and killing plants; the Tenuipalpidae (false spider mites, Fig. 3A), also important agricultural pests; and the Tuckerellidae. Eriophyoidea includes most of the mite species capable of causing galling and deformation of leaves, buds, and fruits (Keifer *et al.*, 1982), and numerous species that wander the leaf surface (leaf vagrants) and whose feeding causes a silvery to rust-colored damage. Most woody dicots, and many monocots and ferns, are attacked by eriophyoid mites, and most of these mites appear to be species- or genus-specific parasites. A number of genera of Tarsonemidae feed on higher plants, whereas others feed on ferns (Lindquist, 1985). Species in one genus of Stigmaeidae (*Eustigmaeus*) feed on mosses (Gerson, 1972), and other mites are often associated with mosses and liverworts (Gerson, 1987).

Feeding by large populations of herbivorous mites may cause rapid decline in resource quality (including defoliation) and attract large numbers of predators. Perhaps because of these ever-present dangers, most herbivorous mites readily "cast their fates to the winds" by using air currents to disperse. Many spider mites "balloon" on strands of silk. Generation times tend to be 2–4 weeks at 25–30°C.

Two families of parasitiform mites also have numerous species that have evolved a specialized plant-parasitic life-style. In the neotropics, a very diverse assemblage of ascid mites in the genera *Rhinoseius, Proctolaelaps,* and *Lasioseius* live in flowers pollinated by hummingbirds, use the birds for phoretic transport between plants, and feed on nectar (and probably on pollen) within the flowers (Colwell, 1985; Heyneman *et al.*, 1991). We have recently discovered that ameroseiid mites in the genera *Hattena* and *Afrocypholaelaps* inhabit flowers in Australia. The species that we have observed readily feed on nectar and pollen and can reach densities of several hundred per flower. *Hattena* mites are associated with honeyeaters (Domrow, 1979) and *Afrocypholaelaps* mites with bees (Eickwort, 1990).

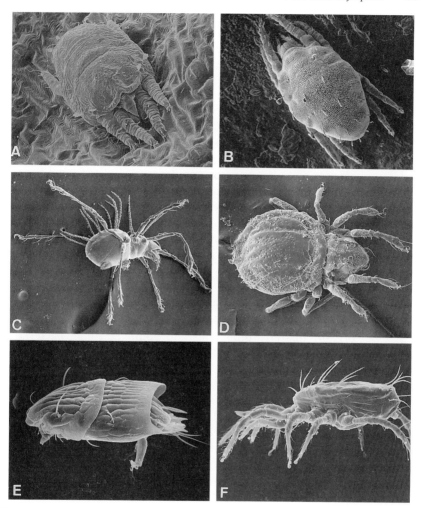

Figure 3   Some canopy mites are plant parasites (**A:** false spider mite, Tenuipalpidae, ×850), but most are potentially beneficial to the plant because they feed on microbes growing on the leaf surface (**B:** tydeid mite, ×750; **C:** *Tuparazetes* sp., ×200, and **D:** *Adhaesozetes polyphyllos,* both oribatid mites, ×380; **E:** *Daidalotarsonemus* sp., a tarsonemid mite, ×880), or feed on plant-parasites (**F:** *Phytoseius woolwichensis* Schicha).

B. Predators

Predators are those mites that capture and eat other mites, small insects (e.g., aphids, scale insects, and thrips) and their eggs, nematodes, and other small invertebrates. Two prey capture strategies are evident: active search and ambush (Walter and Kaplan, 1991). Most parasitiform mites found on leaves are active and voracious predators, and often have rapid developmental rates (a week or less at 25–30°C) leading to generation times shorter than those of their prey. Many of these predators also subsist on plant products (pollen, nectar, cell contents) or microbes when prey are scarce (Tanigoshi, 1982; McMurtry, 1992). Phytoseiidae (Fig. 3F) are reared commercially and are used to suppress outbreaks of phytophagous mites in agricultural systems. In tropical rain forests, some of the diversity of phytoseiid mites is replaced by close relatives in the Ascidae. Predatory acariform mites include the actively searching Stigmaeidae, Eupallopsellidae, Cunaxidae (Cunaxoidinae), Bdellidae, Anystidae, Adamystidae, Erythraeidae, and Trombidiidae; ambush predators are found in the Cheyletidae and Cunaxidae (Cunaxinae). Dispersal by arboreal predatory mites may entail drifting on air currents, phoresy (hitching rides on insects and vertebrates), and hiking. Not all predators of mites are mites. Some thrips and the tiny larvae of insect species in at least three families (Coniopterygidae, Coccinellidae, and Cecidomyiidae) feed on mites and are commonly found on leaves in Australian rain forests.

C. Scavengers

Scavengers are mites that graze on the leaf or stem microbiota or scavenge dead organic matter. Epiphyllous and epiphytic fungi, bacteria, lichens, and algae appear to be the primary foods, but pollen and decaying plant matter are also eaten. Scavengers are primarily members of the Acariformes and are usually the most abundant phylloplane mites. Included in the suborder Prostigmata are the Tydeidae (Fig. 3B), Tarsonemidae (Fig. 3E), and Eupodidae. Most of these mites appear to be algivores, fungivores, and pollen-feeders, but at least some species are facultative predators. Chelicerae are stylet-like and are used to puncture foods and ingest fluids. Generation times tend to be 2–4 weeks at 25–30°C.

Numerous arboreal species (in dozens of families) in the suborders Astigmata and Oribatida (Aoki, 1973; Spain and Harrison, 1968; Walter and Behan-Pelletier, 1993) use chelate-dentate chelicerae to bite off particles of fungi, algae, lichens, and detritus; and hence have gut contents that can be used to infer feeding behavior. Lichens often are attacked by oribatid mites, especially by the immature stages that may burrow into the thallus (Gerson and Seaward, 1977). Adult oribatid mites (Figs. 3C and 3D) are heavily armored, slow-moving, and often relatively large (>0.5 mm). Metabolic rates tend to be low, resulting in relatively long generation times (>4 weeks at 25–30°C), and many species may have no more than one or two generations a year (Norton, 1993). Little is known about how oribatid mites disperse, but astigmatid mites usually have a deutonymphal dispersal morph (hypopus) that attaches to insects or other arthropods (Krantz, 1978).

V. The Play

A. Act I. Abundance of Canopy Mites

We found that mites were extraordinarily abundant in rain forest canopies. All the 177 species of woody plants sampled harbored mites. At the four subtropical rain forest sites in Table II, upper canopies (>5 m) of 16 tree and liana species averaged 11 mites per leaf and 3 mites for every 5 cm of stem examined. Most (68%) of the mites collected from leaves were scavengers, especially oribatid mites (37% of total). Predators (17%) and herbivores (16%) were almost equally abundant. Similar results were obtained from 18 species of trees growing at the three tropical rain forest sites in north Queensland and from the 4 tree species in a warm-temperate rain forest at Chinaman Creek in southernmost Victoria (Table III). The three tropical rain forests averaged 8 mites per leaf and the warm-temperate rain forest 16 mites per leaf. Scavengers and predators were the mites most likely to be encountered, but herbivorous mites were occasionally very abundant.

We found no significant difference in mite abundance between canopy levels (mean = 14.9 per leaf ≤5 m versus 20.4 >5 m; paired *t*-test $P = 0.465$), using mean values for 10 of the plant species from Tables II and III. These included warm-temperate, subtropical, and tropical rain forest plant species. On some trees, lower crown mite densities were highest, but in others this was reversed with no apparent pattern, suggesting that density differences were not directly related to height. Nor did Kitching *et al.* (1993), one of the few studies to seriously consider mites in forest canopies, find much difference between lower (2–6 m) and upper crown mite abundances.

How many mites live in these rain forest canopies? We can approximate an answer to this question by looking at a 10-m-high, 13-cm-dbh native gardenia (*Randia benthamiana*) growing beside Mick's Tower at O'Reilly's. This tree had 8 vertical meters of crown spread across 17 large branches. Each branch carried 44 ± 2 ($n = 3$) shoots; each shoot supported 52 ± 9 ($n = 5$) leaves. Rounding down the branch and shoot means gave us a conservative estimate ($17 \times 40 \times 50$) of 34,000 leaves in the crown. Mite densities were determined by sampling 3 leaves from each of 3–5 shoots taken at 2, 5, 6, 8, and 10 m above the ground. Over all heights, the gardenia averaged 11 ± 1 mites per leaf ($n = 50$), a value near the mean for leaves in the upper crown samples in Table II. Multiplying up ($11 \times 34,000$) suggested that 374,000 mites were scurrying about the leaves of this one small tree. This did not include mites on stems and trunk, or in crown humus (Nadkarni & Longino, 1990); nor is this a particularly high number of mites per leaf for *R. benthamiana* at Lamington. Samples from 9 m in the canopy of another gardenia tree taken in September 1991, February 1992, and

Table II Mite Populations on Upper-Canopy Tree or Liana Leaves at Four Subtropical Rain Forest Sites in Eastern Australia [a]

Site/species	Leaf surface	Domatia	Scavengers	Predators	Herbivores	Total mites/leaf	Range
Mick's Tower (March 10, 1993)							
Parsonsia fulva (L)	Hairy	Pocket	74.7	12.5	20.0	107.2 ± 21.6	20–443
Austrosteenisia glabrista (L) [b]	Flat hairs	None	13.2	4.5	0	17.8 ± 2.2	6–32
Synoum glandulosum (T) [b]	Smooth	Pit/tuft	21.0	6.2	0	27.3 ± 2.7	3–59
Randia benthamiana (T) [b]	Smooth	Pit/tuft	11.5	2.1	0.1	13.6 ± 1.7	0–36
Caldcluvia paniculosa (T) [b]	Smooth	Tuft	8.7	2.3	0	10.9 ± 2.7	0–38
Orites excelsa (T)	Smooth	None	0.3	0.5	0.4	1.2 ± 0.3	0–11
Euodia micrococca (T) [b]	Smooth	None	0.1	0.3	0.1	0.4 ± 0.1	0–4
Piper novae-hollandiae (L) [b]	Smooth	None	0.2	0.1	0.1	0.4 ± 0.1	0–2
Geissois benthamii (T) [b]	Smooth	None	0.1	0	0	0.4 ± 0.1	0–1
Stenocarpus salignus (T) [b]	Smooth	None	0.1	0	0	0.4 ± 0.1	0–2
O'Reilly's (March 11, 1993)							
Parsonsia fulva (L)	Hairy	Pocket	24.2	7.0	15.6	46.8 ± 16.6	16–106
Caldcluvia paniculosa (T)	Smooth	Tuft	15.7	1.3	1.3	18.2 ± 3.6	0–51
Argyrodendron actinophyllum (T)	Smooth	Pocket	4.6	2.2	4.3	11.1 ± 2.7	0–58
Orites excelsa (T)	Smooth	None	3.4	0.4	1.2	5.0 ± 0.9	0–14
Pseudoweinmannia lachnocarpa (T) [b]	Smooth	None	0.5	0.3	3.7	4.4 ± 0.9	0–34
Doryphora sassafras (T)	Smooth	None	1.7	0.1	0.2	2.0 ± 0.9	0–40
Ficus watkinsiana (T)	Smooth	None	0.1	0	0	0.1 ± 0.1	0–1
Gambubal (September 23, 1991)							
Sloanea woolsii (T)	Smooth	Tuft	11.4	2.9	0	14.3 ± 1.8	1–32
Dorrigo (March 12, 1993)							
Caldcluvia paniculosa (T)	Smooth	Tuft	57.4	23.6	1.2	82.2 ± 13.7	32–139
Ehretia acuminata (T) [b]	Smooth	Pocket	1.2	0.4	0	1.6 ± 0.4	0–7
Doryphora sassafras (T) [b]	Smooth	None	0.2	0.1	0	0.2 ± 0.1	0–2
Average leaf			7.2 ± 0.8	1.8 ± 0.2	1.7 ± 0.5	10.7 ± 1.2	
Per centimeter of stem			0.5 ± 0.2	0.1 ± 0.03	0.01 ± 0.01	0.6 ± 0.2	

[a] A leaf of an upper-canopy (6–30 m) tree (T) or liana (L) harbored up to 443 mites. Mites that scavenge for microbes and detritus on the leaf surface (scavengers) were most abundant. Mites that prey on other mites and small insects (predators) were as abundant as mites feeding on leaf cells (herbivores), but the latter were more patchily distributed. Numbers represent mean (±standard error) per leaf or cm of stem. A total of 582 leaves and 529 cm of stem yielding 6533 mites was sampled. Within each site, plants are listed in approximately decreasing order of structural complexity of leaf surfaces, ranging from those with many hiding places among hairs and in domatia, to those where domatia provide the only mite-sized space, to those with smooth, glossy surfaces without refugia.

[b] Per centimeter of stem.

November 1993 averaged, respectively, 52 ± 10 ($n = 5$), 17 ± 2 ($n = 15$), and 39 ± 8 ($n = 10$) mites per leaf. And a 2-m-high sapling with 343 leaves growing under the gardenia at Mick's Tower averaged 11 ± 1 ($n = 8$) mites per leaf in November 1993 (or about 3800 foliar mites on this small sapling).

High densities of foliar mites are not unusual in relatively undisturbed Australian rain forests. At the Curtain Fig Tower near Atherton in north Queensland, a brown tamarind (*Diploglottis diphyllostegia*) 26 m tall carried about 4000 large, pinnately compound leaves in its canopy. Each leaf (4–8 leaflets) averaged about 95 mites (Table III), suggesting a population of foliar mites (380,000) similar to that of the gardenia at Mick's Tower. A cluster of three scentless rosewood (*Synoum glandulosum*) trees (12–15 m high, 9–37 cm dbh) adjacent to Mick's Tower at Lamington averaged 28 ± 3 mites per leaf. One of these trees carried about 16,500 pinnately compound leaves, and therefore a population of nearly a half million phylloplane mites. Even trees like the red silky oak (*Stenocarpus salignus*) in Table II, which averaged only about 1 mite for every 10 smooth, glossy leaves examined, must harbor many thousands of mites living and feeding in its canopy.

B. Act II. Diversity of Canopy Mites

During the last three years, we conducted a study of the foliar mite fauna on blueberry ash (*Elaeocarpus reticulatus*) growing in the warm-temperate rain forest along Chinaman Creek on Wilson's Promontory, Victoria. During that time we examined 481 leaves from 36 different ash trees. On average, each leaf was inhabited by 7.5 ± 0.6 mites, and in total we found 25 species of foliar mites (Table IV). Four species of herbivorous mites and a *Tarsonemus* species (predatory on eriophyoid mites) were found only on blueberry ash and also occurred on this tree at other sites in Victoria and New South Wales. The remaining predatory and scavenging mites were not restricted to blueberry ash, but used leaves of several tree species growing at Chinaman's Creek.

If most foliar mites are tree-generalists, then their site diversity may not be high. However, at Chinaman Creek we also sampled a total of 430 leaves from four other tree species (*Acmena smithii, Bedfordia arborescens, Olearia argophylla,* and *Pomaderris aspera*) and collected 26 additional species of foliar mites (Table IV). As on blueberry ash, the herbivorous mites and a second *Tarsonemus* species (assumed to be a scavenger) were restricted to single tree species, but most scavengers and predators were found on two or more tree species (Walter, 1992; Walter and Behan-Pelletier, 1993). Therefore, even within a temperate forest with most species diversity represented by mites that are not host-specific, doubling the number of leaves examined doubled the number of species collected.

Table III Mite Populations on Tree Leaves at Three Tropical Rain Forest Sites in North Queensland[a]

Site/forest type/species	Leaf surface	Domatia	Scavengers	Predators	Herbivores	Total mites/leaf	Range
Mt. Lewis, montane tropical (16°32'S, 145°15'E)							
Elaeocarpus largiflorens	Hairy	Pocket	15.8	1.6	4.6	22.1 ± 2.5	1–81
Elaeocarpus ferruginiflorus	Hairy	Pocket	2.7	0.2	2.8	5.7 ± 0.8	0–27
Aceratium ferrugineum	Hairy	None	24.3	3.8	5.1	33.2 ± 4.2	6–166
Elaeocarpus foveolatus	Flat hairs	Pocket	1.1	0.1	0.1	1.3 ± 0.3	0–3
Sloanea langii	Smooth	Tuft	4.5	1.4	2.9	8.9 ± 1.4	1–35
Elaeocarpus sp. (RFK 864)	Smooth	Pocket	0.4	0.1	5.7	6.2 ± 2.6	0–86
Elaeocarpus sp. (RFK 714)	Smooth	Pocket	5.3	0.5	0	5.8 ± 1.3	0–48
Elaeocarpus elliffii	Smooth	Pocket	0.2	0.3	0.02	0.5 ± 0.1	0–6
Aceratium doggrellii	Flat hairs	None	1.4	0.1	0	1.5 ± 0.2	2–17
Total			4.1 ± 0.4	0.7 ± 0.1	1.5 ± 0.3	6.3 ± 0.6	
Curtain Fig Tower, montane tropical (17°16'S, 145°34'E)							
Diploglottis diphyllostegia	Hairy	None	16.3	32.1	47.0	95.4 ± 12.2	27–160
Toona australis	± Smooth	Pocket	18.8	8.8	1.2	28.8 ± 5.6	14–46
Argyrodendron peralatum	Flat scales	None	12.1	1.8	0	13.9 ± 2.0	0–33
Aleurites moluccana	Smooth	None	1.2	1.8	0	3.0 ± 0.9	0–5
Total			12.8 ± 1.6	12.4 ± 2.5	15.5 ± 4.2	40.7 ± 7.0	

Pine Creek Tower, lowland tropical (16°59'S, 145°50'E)						
Sarcopteryx stipata	Hairy	Pocket	4.6	5.2	78.2	88 ± 29.5	27–177
Beilschmiedia bancroftii	Smooth	None	0.1	0.1	1.5	1.6 ± 0.4	0–1
Polyalthia sp.	Smooth	None	0.6	0.4	0	1.1 ± 0.5	0–17
Flindersia pimenteliana	Smooth	None	0.8	0.3	0	1.1 ± 0.5	0–7
Acacia aulacocarpa	Smooth	None	0.1	0.1	0	0.2 ± 0.2	0–8
Total			0.5 ± 0.1	0.3 ± 0.1	2.9 ± 1.4	3.6 ± 1.5	
Chinaman Creek, lowland warm-temperature (38°55'S, 146°23'E)						
Pomaderris aspera	Hairy	None	19.4	3.5	7.4	30.4 ± 4.3	2–172
Elaeocarpus reticulatus	Smooth	Pocket	2.5	1.1	19.3	23.4 ± 4.2	0–105
Olearia argophylla	Flat Hairs	None	5.2	1.3	0.1	6.7 ± 0.9	0–26
Acmena smithii	Smooth	None	1.7	0.5	0.8	3.0 ± 0.9	0–32
Total			7.2 ± 0.8	1.6 ± 0.2	6.9 ± 1.3	15.9 ± 1.8	

[a] Leaves in the canopies of rain forest trees at Mt. Lewis (2–35 m), Curtain Fig Tower (18–37 m), and Pine Creek Tower (8–30 m) harbored up to 177 mites, and similar numbers were found in lower canopies (1–3 m) of a warm-temperate rain forest in southern Victoria. Abundances are presented as mean (± standard error) mites per leaf, with feeding guilds and plant order as in Table II. A total of 4029 mites were collected from 638 leaves at Mt. Lewis on June 4, 1992, 1870 mites from 46 leaves at Curtain Fig on April 24, 1993, 591 mites from 160 leaves at Pine Creek on April 23, 1993, and 3172 mites from 200 leaves at Chinaman Creek on November 18, 1990.

Table IV   Numbers of Mite Species Collected from Leaves of *Elaeocarpus reticulatus* and Four Other Warm-Temperate Rain Forest Tree Species at Chinaman Creek, Wilson's Promontory, Victoria (November 1990 to February 1993), a Subtropical Rain Forest at Lamington, Queensland (early February 1993), and a Montane Tropical Rain Forest at Mt. Lewis, Queensland (June 4, 1992)

	Chinaman Creek		Lamington	Mt. Lewis
	E. reticulatus (481 leaves)	5 species (911 leaves)	14 species (581 leaves)	9 species (638 leaves)
Predators				
Adamystidae	0	1	0	1
Anystidae	0	0	0	1
Ascidae	0	0	0	4
Bdellidae	0	1	1	1
Cheyletidae	1	2	2	0
Cunaxidae	1	1	3	3
Eupallopsellidae	0	0	1	0
Phytoseiidae	3	7	7	11
Stigmaeidae	2	3	2	2
Tarsonemidae	1	1	0	0
Total predator species	8	16	16	23
Scavengers				
Astigmata	1	2	2	2
Eupodidae	0	1	1	1
Oribatida	4	12	12	12
Tarsonemidae	3	4	6	10
Tydeidae	5	5	5	5
Total scavenger species	13	24	26	30
Herbivores				
Eriophyoidea	2	7	11	14
Tetranychidae	1	1	4	1
Tenuipalpidae	1	4	2	2
Total herbivore species	4	11	17	17
Total foliar mite species	25	51	59	70

Samples from a subtropical (Lamington) and a tropical (Mt. Lewis) rain forest (Table IV), although much more limited in time and in leaf area than those from the warm-temperate rain forest at Chinaman Creek, both contained more species of mites. There was some overlap of species among sites; one oribatid mite (*Adhaesozetes polyphyllos* Walter & Behan-Pelletier) and one phytoseiid mite (*Typhlodromus dachanti* Collyer) occurred at all three sites. The apparent increase in diversity from temperate to tropical sites in Table IV was supported by a more in-depth analysis of predatory parasitiform mites (Phytoseiidae, Ascidae) across all our sites (Fig. 1). Mean number of families, genera, and species was higher in rain forests in north Queensland than in those farther south (Walter *et al.*, 1994).

C. Act III. Effects of Leaf Structure

Although the average leaf in the canopy of the subtropical rain forest at Lamington had >10 mites, this value varied over 1000-fold across plant species (Table II). Large differences in mite abundance were observed even when leaves of different species were interspersed in the canopy. At the top of Mick's Tower, for example, furry silkpod (*Parsonsia fulva*), a liana with densely hairy leaves and pocketlike domatia, was supported by the crown of a mountain silky oak (*Orites excelsa*), a tree with smooth, simple leaves. Furry silkpod averaged 127 times as many mites per leaf as those of mountain silky oak at 14 m (127 versus 1), and 35 times as many at 17 m (48 versus 1.5). From a mite's perspective, leaves with the entire undersurface covered in a dense forest of raised hairs (Figure 2C) present a vast array of mite-sized habitat space. Leaves with flat, closely appressed hairs (Fig. 2B) or those with a smooth, glossy surface (Fig. 2A) give mites room to run, but nowhere to hide.

One type of leaf surface structure that can occur on leaves that are otherwise smooth is the domatium (Figs. 2E and 2F). Even though domatia typically comprise <1% of leaf surface area, they are very important to foliar mites as shelters and nurseries (Walter and O'Dowd, 1992a,b). We tested the effect of domatia on mites by experimentally blocking access to domatia on leaves of a native gardenia. All leaves were intermingled on branches 9 m above the ground at Lamington, and leaves with open domatia had dabs of bitumen paint applied below the domatia (control for effects of paint). Five months after domatia were closed off with bitumen paint, total mite numbers were seven-fold lower than on leaves with open domatia (Fig. 4). Domatia, including the few missed in the blocked treatment, housed 257 of the 291 mites collected at the end of the experiment. Both predators and scavengers were dramatically reduced when domatia were blocked, and this effect was consistent across all mite taxa present (horizontal rules in Fig. 4).

Leaves of many rain forest trees and vines have extrafloral nectaries with carbohydrate-rich secretions (Koptur, 1992). Most previous research on extrafloral nectaries emphasized their use by ants (Koptur, 1992), but predatory and scavenging mites also drink nectar (McMurtry and Scriven, 1965; Bakker and Klein, 1992; Pemberton, 1993). We tested the effect of nectaries on predatory mites in the laboratory using a garden shrub (*Viburnum tinus* L.) that has a pair of nectaries on the lower leaf margin. After 10 days, leaves with nectaries had seven times (1.6 ± 0.5) as many phytoseiid mites and eight times (3.5 ± 1.0) as many total predatory mites (including naturally occurring Stigmaeidae and Cunaxidae) as leaves with nectaries excised

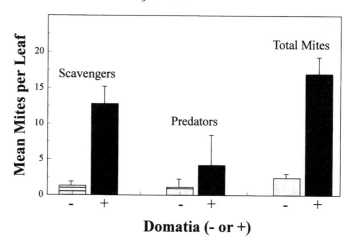

Domatia (- or +)

Figure 4 Leaf structural features such as domatia are primary determinants of the abundance of phylloplane mites. Five months after randomly selected leaves at 9 m in the canopy of a native gardenia (*Randia benthaminana*) had their domatia closed off with a bitumen paint (−), total mite numbers were much lower (P < 0.0001, $F_{1,28}$ = 62.9, ANOVA log-transformed data) than on control leaves (+, domatia open, daubs of paint below domatia). Both scavengers (from base to top segments = Oribatida, Tydeidae, and others) and predators (segments = Phytoseiidae, Stigmaeidae, and others) were strongly affected. No herbivores were present on treated leaves.

($F_{1,13}$ = 8.8, P = 0.011). This suggests that nectaries also are used by predatory mites, and that rain forest plants with extrafloral nectaries will have more of these predators on their leaves.

VI. Epilogue

Forest canopies are full of mites. Every forest we sampled, from cool-temperate rain forests in southern Tasmania to the wet tropics of north Queensland, contained high densities of foliar mites. Across a similar range of Australian rain forests, Kitching *et al.* (1993) also found mites in canopies, but at relatively low abundance. If we found small trees with hundreds of thousands of foliar mites in their canopies (let alone flower, stem, trunk, and epiphyte inhabitants), why have so few mites been reported in studies of canopy arthropods (Table I)? For example, black booyong has been intensively studied by Basset (1991) on Mt. Glorious in southeastern Queensland, and although he reported a collection of over 50,000 arthropods, none were mites (Table I). Yet, at O'Reilly's we examined 25 leaves (each with 5–9 palmate leaflets) of black booyong (*Argyrodendron actinophyllum*)

(Table II). Only two leaves lacked mites, and the average (11 mites/leaf) was very close to the average for all the species sampled in Table II.

We suspect that three factors are primarily responsible for the plethora of relatively mite-free data sets in canopy arthropod studies. First, most mites are minute (<1 mm), translucent to pale brown in color, and may go unnoticed in samples. We expect that most mites reported in other studies are conspicuous (e.g., large, active, red Prostigmata and the larger Oribatida). Second, some workers explicitly exclude minute arthropods from their analyses (e.g., Woinarski and Cullen, 1984); others may have implicitly followed similar exclusion criteria (e.g., Kitching *et al.*, 1993, p. 189 reported that Basset, 1991, excluded Acari and Collembola). Finally, chemical knockdown methods may not be efficient for arthropods as small as mites. Although chemical fogging produces a fine mist that rises through the canopy (Stork, 1988), droplets must contact animals to be effective. Most of the mites we collected from the leaves of black booyong were found within domatia. There were 9—84 of these pocketlike structures on each leaf, and each had to be dissected open with a scalpel to find the mites. It seems unlikely that pesticidal fogs easily penetrate these structures, the dense mat of hairs present on some leaves, or crevices in bark. Even if mites are intoxicated within these refugia, it seems unlikely that most will then stumble out and fall many meters onto a drop cloth or into a funnel.

Phylloplane mite abundances appear to be far greater than those of insects, but mite diversity may not be as great. Because of the time involved in collecting mites from leaves, we sampled only a minute fraction of the leaves in a canopy (e.g., 0.15% of the leaves on the native gardenia at Mick's Tower). In contrast, chemical fogging studies usually sample the insects from large sections of canopy, including leaves, stems, trunks, and epiphytes. Erwin (1983) collected 1080 species of adult beetles from four forest types in the Amazon Basin. Of these, the vast majority (795 species) were herbivores. At our three rain forest sites (Table IV), we identified only 45 species of herbivorous mites. However, the numbers of predatory and scavenging mite species clearly rival those of functionally similar beetles. Erwin (1983) reported 145 species of predatory beetles and observed that 83% of beetle species were present in only one of the four forest types he sampled; therefore, 30 to 55 predatory beetle species could be expected at a site. Our much more restricted samples ranged between 16 and 23 species of predatory mites at a site. Similarly, each of Erwin's sites appeared to harbor 27 to 49 species of scavenging and fungivorous beetles, whereas our more restricted samples ranged between 24 and 30 species of such mites.

During tropical downpours, the need of mites for shelter is apparent. Protection from the elements afforded by leaf hairs and domatia is undoubtedly a factor in the correlation between mite numbers and leaf surface structure that is obvious in Tables II and III. Smooth, rain-shedding leaves may also shed mites, and in the intense sunlight after the storm, desiccation would be more severe on a smooth leaf than on one with a dense tomentum or cavelike refugia (Grostal and O'Dowd, 1994). Mites fall prey to larger arthropods or even small vertebrates, such as mite-eating frogs (Simon and Toft, 1991). Safe havens from these predators are provided by the mite-sized spaces between hairs and the portals of domatia.

Although individual mites are minute, large numbers of scavenging and predatory mites roam the surface of rain forest leaves, and their numbers increase with increasing leaf surface structural complexity. By consuming fungal pathogens, sooty molds, and mildews, scavengers may reduce plant disease and increase photosynthetic area by removing epiphyll cover. Predatory mites, by attacking herbivorous mites and small insects, may decrease leaf damage. If these behaviors increase plant fitness, then natural selection may even shape structures that retain scavenging and predatory mites on leaves (O'Dowd *et al.*, 1991).

The high number of rain forest plant species that have independently evolved leaf domatia (O'Dowd and Willson, 1991) suggests that mites are important in the ecology of rain forests. Short on evidence but long on imagination, Lundstroem (1887) interpreted one of the most obscure of leaf structures—the domatium—in just this way. He argued that they are the basis of a protective mutualism between plants and mites. Later studies (O'Dowd and Willson, 1989; Pemberton and Turner, 1989; Willson, 1991; Walter and Denmark, 1991) set the mite–domatia association on a firmer quantitative basis. Furthermore, comparisons and experiments showed that domatia elevate the numbers of scavenging and predatory mites on leaves (Walter and O'Dowd, 1992a,b). But without evidence of any effect on plant fitness, the long-standing controversy remains. Are domatia billets for a standing army of mite bodyguards or are they an elaborate "whim of plant structure" (van Steenis, 1976) used as convenient squats by hordes of mites short on housing?

VII. Future Directions

Most woody plants that humankind finds useful are attacked by herbivorous mites capable of reducing fruit set and seed production, and sometimes of killing plants (Jeppson *et al.*, 1975). When pesticide applications disrupt populations of predators, herbivorous mite populations often reach levels that cause serious economic loss (Helle and Sabelis, 1985). If one could remove all domatia from all leaves of a sufficient number of replicate rain forest trees, we expect that similar herbivorous mite outbreaks and fitness effects on plants could eventually be demonstrated. However, we also expect that even the most industrious student of canopy mites would find this an impossible task. Given these difficulties, what questions might be more amenable to study?

A. What Factors Limit the Populations of Plant-Parasitic Mites in the Rain Forest Canopy?

At our sites in undisturbed forests, leaves with herbivorous mites are rare and leaves with predatory mites are frequently and consistently encountered. Based on their effectiveness against plant-parasitic mites in agricultural systems, it seems likely that predatory mites are a major check on the population growth of herbivorous mites in rain forests. But what factors influence the success of predatory mites? Do predatory mites have home ranges, that is, do they tend to remain on attractive leaves, or is there constant movement and turnover? Do structures like domatia and extrafloral nectaries tend to draw mites away from the "highways" of the stems and petioles onto the leaf surface and increase the chance of encountering an herbivore? Are leaves without mite lures protected from herbivorous mites by especially effective chemical defenses (and hence a potential source of new acaricides), or do they rely on especially active predators that can regularly make the journey from refugia on the bark?

B. Who Eats Whom in the Rain Forest Canopy?

If leaf domatia function as a type of constitutive defense against herbivorous mites by increasing the probability that predatory mites are resident on leaves and ready to attack herbivorous mites before their populations reach damaging levels, then the predatory mites inhabiting domatia must be able to find alternative sources of food. Many predatory mites do have very broad diets, but some species are specialized predators of herbivorous mites (Tanigoshi, 1982; McMurtry, 1992). How narrow or broad are the diets of the predatory mites that inhabit leaf domatia? Does the more diverse tropical rain forest fauna include more specialized predatory mites than temperate forests?

Although herbivorous mites are patchily distributed and relatively rare in our samples, scavenging mites, especially oribatid and tydeid mites, are very abundant. Adult oribatid mites are heavily armored; we have yet to see one of these mites being eaten by a predator, but immature oribatids are less well protected and possibly more vulnerable. In contrast, soft-bodied tydeid mites can be important alternative prey for predators in agricultural systems when herbivore numbers are low (Knop and Hoy, 1983). Our observations on Australian species indicate that tydeid mites are readily eaten by ambush predators (Cunaxinae, Cheyletidae), but infrequently consumed by more active predators (Phytoseiidae, Stigmaeidae). Perhaps high reproductive output, very broad diets, including the eggs and the active stages of the smaller herbivorous mites (Hussein and Perring, 1986), and the rapid speed at which tydeid mites move allow these mites to maintain high densities in the face of strong predation pressure? Do these high populations of scavengers help support the high densities of predators and indirectly benefit the plants on which they live?

C. Are Scavenging Mites Good for Plants?

Mites that scavenge on epiphyllous fungi or eat plant parasites were the most abundant and diverse phylloplane mites in our samples. In the domatia-blocking experiment, 52% of the mites inhabiting leaves of native gardenia were oribatid mites. Even though oribatid mites are usually considered soil inhabitants (Dindal, 1990), we found as many as 94 oribatid mites on a single leaf in the upper canopy in the subtropical rain forest at Lamington; an average of four oribatid mites lived on each leaf. The guts of these mites are filled with fungal spores and hyphae, confirming that they graze on fungi growing on the phylloplane. What are the effects of large populations of scavengers on the leaf microflora?

Do mites that scavenge leaf detritus and graze on leaf microbes directly or indirectly affect plant fitness? Scavenging mites may reduce the rate at which epiphylls capture leaves, but many tropical rain forest trees retain their leaves for many years. Older leaves tend to be completely covered in growths of lichens, fungi, and liverworts; it is difficult to understand what benefit they may be providing the plant. These old, epiphyll-encrusted leaves also support herds of scavenging mites. By grazing on foliar microbes and excreting on the leaf, mites may free nutrients sequestered by epiphylls, make them available for foliar absorption, and provide rain forest trees with the equivalent of slow-release fertilization.

D. How Important Are Mites to Canopy Biodiversity?

Insects attack plants in numerous ways, chewing, sucking, and burrowing through most plant tissues. This diversity in feeding ecology perhaps contributes to the very high diversity of herbivorous insects. In contrast, herbivorous mites seem to be limited to stabbing cells and sucking out their contents. Because most herbivorous mites appear to be species- or genus-specific parasites (Collyer, 1973; Jeppson *et al.*, 1975; Keifer *et al.*, 1982; Krantz, 1978), it is not unreasonable to assume that most rain forest tree species support different species of herbivorous mites, and we have found as many as six species of herbivorous mites on the leaves of a single tree. Still, herbivorous beetle diversity (Erwin, 1983) alone would seem to dwarf that of leaf-feeding mites.

However, foliar mite species diversity is not equivalent to canopy mite species diversity. For example, one of our students, V. Barnes, found that ornate false spider mites (Tuckerellidae) are abundant on musk daisy, feeding in shallow crevices on small stems. During three years of sampling only the leaves of the same musky daisy trees, we never encountered ornate false spider mites. If mites—predators, scavengers, and herbivores—finely partition the canopy habitat, then they may be the most diverse as well as most abundant canopy arthropods.

VIII. Summary

Mites have generally been ignored in studies of canopy arthropods, however, in Australian rain forests even small trees have hundreds of thousands of mites living on their leaves, and countless others inhabit flowers, stems, trunks, hanging humus, epiphytes, and the other, larger canopy animals. Most arboreal acarine abundance is represented by scavenging microbivores and by predators, and species diversity in these groups rivals that of similar feeding guilds in the more diverse insect orders (e.g., beetles). However, herbivorous mite species diversity is relatively low and distributions are patchy, presumably because of the large populations of predators. Diversity in all three mite feeding guilds is higher in tropical than in temperate rain forests.

Leaf structures, especially domatia, tomenta, and extrafloral nectaries, are associated with elevated abundances of mites that may be beneficial to the plants they inhabit. Removal of either domatia or extrafloral nectaries results in decreased abundances of predatory mites, and presumably reduced protection against attack by plant-parasitic mites. Mites that scavenge and feed on the leaf microflora also respond to leaf structures, suggesting that there may be a functional relationship between these tiny grazers and rain forest plants.

Acknowledgments

Peter O'Reilly knew that one's life wouldn't be complete without a visit to a rain forest canopy, and he built a pathway to a fine one. We thank him for his foresight, and for his kindness and help during our research. We thank the members of the Tropical Research Centre, CSIRO, Atherton, for their help, and especially thank Bruce Grey, a man with a wealth of knowledge about rain forest trees, and a dead shot with a shanghai. Andrew Beattie, Rob Colwell, Mike Cullen, Bruce Halliday, Mark Harvey, Jerry Krantz, Evert Lindquist, and Roy Norton all made useful suggestions on earlier drafts of this manuscript; Meg Lowman faxed fearlessly at an unresponsive lab until it all came together. Our research was often aided by courteous and knowledgeable rangers and foresters throughout Eastern Australia, and we thank them and the Department of Parks, Wildlife and Heritage, Tasmania, the Department of Conservation and Environment, Victoria, the National Parks & Wildlife Service and Forestry Commission of New

South Wales, the Queensland Forestry Commission, and the Queensland National Parks and Wildlife Service for their collection permits. This research was supported in part by the Australian Research Council.

References

Aoki, J. (1973). Soil mites (oribatids) climbing trees. *Proc. Int. Cong. Acarol. 3rd,* Prague, *1971,* pp. 59–65.

Bakker, F. M., and Klein, M. E. (1992). Transtrophic interactions in cassava. *Exp. Appl. Acarol.* 14, 293–311.

Basset, Y. (1991). The taxonomic composition of the arthropod fauna associated with an Australian rain forest tree. *Aust. J. Zool.* 39, 171–190.

Basset, Y., Aberlenc, H., and Delvare, G. (1992). Abundance and stratification of foliage arthropods in a lowland rain forest of Cameroon. *Ecol. Entomol.* 17, 310–318.

Christophel, D. C., and Hyland, B. P. M. (1993). "Leaf Atlas of Australian Tropical Rain Forest Trees." CSIRO Publications, East Melbourne.

Collyer, E. (1973). New species of the genus *Tenuipalpus* (Acari: Tenuipalpidae) from New Zealand, with a key to the world fauna. *N. Z. J. Sci.* 16, 915–955.

Colwell, R. K. (1985). Community biology and sexual selection: Lesions from hummingbird flower mites. *In* "Community Ecology" (J. Diamond and T. J. Case, eds.), pp. 406–424. Harper & Row, New York.

Costa, J. T., III, and Crossley, D. A. (1991). Diel patterns of canopy arthropods associated with three tree species. *Environ. Entomol.* 20, 1542–1548.

Costermans, L. (1981). "Native Trees and Shrubs of South-Eastern Australia." Weldon Publishing, Sydney.

Dindal, D. L., ed. (1990). "Soil Biology Guide." John Wiley, New York.

Domrow, R. (1979). Ascid and ameroseiid mites phoretic on Australian mammals and birds. *Rec. West. Aust. Mus.* 8, 97–116.

Eickwort, G. C. (1990). Mites: An Overview. *In* "Honey Bee Pests, Predators, and Diseases" (R. A. Morse and R. Nowogrodzki, eds.), 2nd ed., pp. 188–199. Cornell Univ. Press, Ithaca, NY.

Erwin, T. L. (1983). Beetles and other insects of tropical forest canopies at Manaus, Brazil, sampled by insecticidal fogging. *In* "Tropical Rain Forests: Ecology and Management" (S. L. Sutton, T. C. Whitmore, and A. C. Chadwick, eds.), pp. 59–75. Blackwell, Oxford.

Gell, P., and Mercer, D. (1992). "Victoria's Rain Forests: Perspectives on Definition, Classification and Management," Monash Publ. Geogr., No. 41.

Gerson, U. (1972). Mites of the genus *Ledermuelleria* (Prostigmata: Stigmaeidae) associated with mosses in Canada. *Acarologia* 13, 319–343.

Gerson, U. (1987). Mites which feed on mosses. *Symp. Biol. Hung.* 35, 721–724.

Gerson, U., and Seaward, M. R. D. (1977). Lichen-invertebrate associations. *In* "Lichen Ecology" (M.R.D. Seaward, ed.), pp. 69–119. Academic Press, London.

Grostal, P., and O'Dowd, D. J. (1994). Plants, mites, and mutualism: Leaf domatia and the abundance and reproduction of mites on *Viburnum tinus* (Caprifoliaceae). *Oecologia* 97, 308–315.

Helle, W., and Sabelis, M. W. (1985). "Spider Mites, Their Biology, Natural Enemies, and Control." Elsevier, New York.

Heyneman, A. J., Colwell, R. K., Saeem, S., Dobkin, D. S., and Bernard, H. (1991). Host Plant Discrimination: Experiments with Hummingbird Flower Mites. *In* "Plant–Animal Interactions: Evolutionary Ecology in Tropical and Temperate Regions" (P. Price, T. M. Lewinsohn, G. W. Fernandes, and W. W. Benson, eds.), pp. 455–485. Wiley, New York.

Hijii, N. (1986). Density, biomass, and guild structure of arboreal arthropods as related to their inhabited tree size in a *Cryptomeria japonica* plantation. *Ecol. Res.* 1, 97–118.

Hoy, M. A., ed. (1982). "Recent Advances in Knowledge of the Phytoseiidae," Publ. No. 3284, University of California, Division of Agricultural Sciences, Berkeley.

Hoy, M. A., Cunningham, G. L., and Knutson, L., eds. (1983). "Biological Control of Pests by Mites," Spec. Publ. No. 3304. University of California, Division of Agriculture and Natural Resources, Berkeley.

Hussein, N., and Perring, T. M. (1986). Feeding habits of the Tydeidae with evidence of *Homeopronematus anconai* (Acari: Tydeidae) predation on *Aculops lycopersici* (Acari: Eriophyidae). *Int. J. Acarol.* 12, 215–221.

Hyland, B. (1972). A technique for collecting botanical specimens in rain forests. *Flora Males. Bull.* 26, 20.

Jeppson, L. R., Keifer, H. H., and Baker, E. W. (1975). "Mites Injurious to Economic Plants." Univ. of California Press, Berkeley.

Keifer, H. H., Baker, E. W., Kono, T., Delfinado, M., and Styer, W. E. (1982). An illustrated guide to plant abnormalities caused by eriophyid mites in North America. *U.S. Dep. Agric., Agric. Hand.* 573.

Kitching, R. L., Bergelson, J. M., Lowman, M. D., McIntyre, S., and Carruthers, G. (1993). The biodiversity of arthropods from Australian raniforest canopies: General introduction, methods, sites and ordinal results. *Aust. J. Ecol.* 18, 181–191.

Knop, N. F., and Hoy, M. A. (1983). Biology of a tydeid mite, *Homeopronematus anconai* (n. comb.) (Acari: Tydeidae), important in San Joaquin Valley vineyards. *Hilgardia* 51, 1–30.

Koptur, S. (1992). Extrafloral nectary-mediated interactions between insects and plants. *In* "Insect-Plant Interactions" (E. Bernays, ed.), Vol. IV, pp. 81–129. CRC Press, Boca Raton, FL.

Krantz, G. W. (1978). "A Manual of Acarology." Oregon State University Book Stores, Corvallis.

Lindquist, E. E. (1985). The world genera of Tarsonemidae. *Mem. Entomol. Soc. Can.* 136, 1–517.

Lundströem, A. N. (1887). Planzenbiologische Studien. II. Die Anpassungen der Planzen an Thiere. *Nova Acta Regiae Soc. Sci. Ups.* [3] Series 3, 13, 1–87.

Majer, J. D., and Recher, H. F. (1988). Invertebrate communities on Western Australian eucalypts: A comparison of branch clipping and chemical knockdown procedures. *Aust. J. Ecol.* 13, 269–278.

May, R. M. (1988). How many species are there on earth? *Science* 241, 1441–1449.

McMurtry, J. A. (1992). Dynamics and potential impact of 'generalist' phytoseiids in agroecosystems and possibilities for establishment of exotic species. *Exp. Appl. Acarol.* 14, 371–382.

McMurtry, J. A., and Scriven, G. T. (1965). Life-history studies of *Amblyseius limonicus*, with comparative observations on *Amblyseius hibisci* (Acarina: Phytoseiidae). *Ann. Entomol. Soc. Amer.* 58, 106–111.

Moore, J. C., Walter, D. E., and Hunt, W. H. (1988). Arthropod regulation of micro- and mesobiota in below-ground detrital food webs. *Annu. Rev. Entomol.* 33, 419–439.

Nadkarni, N. M., and Longino, J. T. (1990). Invertebrates in canopy and ground organic matter in a Neotropical montane forest, Costa Rica. *Biotropica* 22, 286–289.

Norton, R. A. (1993). Evolutionary aspects of oribatid mite life histories and consequences for the origin of the Astigmata. *In* "Mites: Ecological and Evolutionary Analysis of Life-History Patterns" (M. A. Houck, ed.), pp. 99–135. Chapman & Hall, New York.

O'Dowd, D. J., and Willson, M. F. (1989). Leaf domatia and mites on Australian plants: Ecological and evolutionary implications. *Biol. J. Linn. Soc.* 37, 191–236.

O'Dowd, D. J., and Willson, M. F. (1991). Associations between mites and leaf domatia. *Trends Ecol. Evol.* 6, 179–182.

O'Dowd, D. J., Brew, C. R., Christophel, D. C., and Norton, R. A. (1991). Mite-plant associations from the Eocene of southern Australia. *Science* 252, 99–101.

Pemberton, R. W. (1993). Observations of extrafloral nectar feeding by predaceous and fungivorous mites. *Proc. Entomol. Soc. Wash.* 95, 642–643.

Pemberton, R. W., and Turner, C. E. (1989). Occurrence of predatory and fungivorous mites in leaf domatia. *Am. J. Bot.* 76, 105–112.

Simon, M. P., and Toft, C. A. (1991). Diet specialization in small vertebrates: Mite-eating in frogs. *Oikos* 61, 263–278.

Southwood, T. R. E., Moran, V. C., and Kennedy, C. E. J. (1982). The richness, abundance and biomass of the arthropod communities on trees. *J. Anim. Ecol.* 51, 635–649.

Spain, A. V., and Harrison, R. A. (1968). Some aspects of the ecology of arboreal Cryptostigmata (Acari) in New Zealand with special reference to the species associated with *Olearia colensoi* Hoof. f. *N. Z. J. Sci.* 11, 452–458.

Stanley, T. D., and Ross, E. M. (1983). "Flora of South-Eastern Queensland," Vol. I. Queensland Department of Primary Industries, Brisbane.

Stanley, T. D., and Ross, E. M. (1986). "Flora of South-Eastern Queensland," Vol. II. Queensland Department of Primary Industries, Brisbane.

Stanley, T. D., and Ross, E. M. (1989). "Flora of South-Eastern Queensland," Vol. III. Queensland Department of Primary Industries, Brisbane.

Stork, N. E. (1988). Insect diversity: Facts, fiction and speculation. *Biol. J. Linn. Soc.* 35, 321–337.

Stork, N. E. (1991). The composition of the arthropod fauna of Bornean lowland rain forest trees. *J. Trop. Ecol.* 7, 161–180.

Tanigoshi, L. K. (1982). Advances in knowledge of the Phytoseiidae. *In* "Recent Advances in Knowledge of the Phytoseiidae" (M.A. Hoy, ed.), Publ. No. 3284, pp. 1–22. Univ. of California, Division of Agricultural Sciences, Berkeley.

van Steenis, C. G. G. J. (1976). Autonomous evolution in plants. Differences in plant and animal evolution. *Gard. Bull. Singapore* 29, 103–126.

Walter, D. E. (1992). Leaf surface structure and the distribution of *Phytoseius* mites (Acarina: Phytoseiidae) in south-east Australian forests. *Aust. J. Zool.* 40, 593–603.

Walter, D. E., and Behan-Pelletier, V. (1993). Systematics and ecology of *Adhaesozetes polyphyllos,* sp. nov. (Acari: Oribatida: Licneremaeoidea) leaf-inhabiting mites from Australian rain forests. *Can. J. Zool.* 71, 1024–1040.

Walter, D. E., and Denmark, H. A. (1991). Use of leaf domatia on wild grape (*Vitis munsoniana*) by arthropods in central Florida. *Fla. Entomol.* 74, 440–446.

Walter, D. E., and Kaplan, D. T. (1991). Observations on *Coleoscirus simplex* (Acarina: Prostigmata), a predatory mite that colonizes greenhouse cultures of rootknot nematode (*Meloidogyne* spp.), and a review of feeding behavior in the Cunaxidae. *Exper. Appl. Acarol.* 12, 47–59.

Walter, D. E., and O'Dowd, D. J. (1992a). Leaves with domatia have more mites. *Ecology* 73, 1514–1518.

Walter, D. E., and O'Dowd, D. J. (1992b). Leaf morphology and predators: Effect of leaf domatia on the abundance of predatory mites (Acari : Phytoseiidae). *Environ. Entomol.* 21, 478–484.

Walter, D. E., Hunt, H. W., and Elliott, E. T. (1988). Guilds or functional groups? An analysis of predatory arthropods from a shortgrass prairie soil. *Pedobiologia* 31, 247–260.

Walter, D. E., Kaplan, D. T., and Permar, T. A. (1991). Missing links: A review of methods used to estimate trophic links in soil food webs. *Agric. Ecosyst. Environ.* 34, 399–405.

Walter, D. E., Halliday, R. B., and Lindquist, E. E. (1993). A review of the genus *Asca* (Acarina: Ascidae) in Australia, with descriptions of three new leaf-inhabiting species. *Invertebr. Taxon.* 7, 1327–1347.

Walter, D. E., O'Dowd, D. J., and Barnes, V. (1994). The forgotten anthropods: Foliar mites in the forest canopy. *Mem. Queensl. Mus.* 36, 221–226.

Willson, M.F. (1991). Foliar mites in the eastern deciduous forest. *Am. Midl. Nat.* 126, 111–117.

Woinarski, J. C. Z. (1984). Ecology of pardalotes in south-eastern Australia. Ph.D. Thesis, Monash University, Clayton, Australia.

Woinarski, J. C. Z., and Cullen, J. M. (1984). Distribution of invertebrates on foliage in forests of south-eastern Australia. *Aust. J. Ecol.* 9, 201–232.

World Conservation Monitoring Centre (1992). "Global Biodiversity: Status of the Earth's Living Resources." Chapman & Hall, London.

Yen, A. L., and Lillywhite, P. K. (1990). Preliminary report into the invertebrates collected from the canopy of Tasmanian rain forest trees, with special reference to the myrtle beech. Unpublished report to the Tasmanian Department of Parks, Wildlife and Heritage, Hobart.

16

Nonvascular Epiphytes in Forest Canopies: Worldwide Distribution, Abundance, and Ecological Roles

Fred M. Rhoades

I. Introduction

Nonvascular (cryptogamic) epiphytes are categorized into three groups: lichens, bryophytes, and free-living algae. These organisms are poikilohydric, as they depend on an atmospheric supply of water and inorganic nutrients from precipitation, dew, or fog interception. Free-living algae are occasionally dominant in very moist canopy habitats but are not dealt with in this review. Although lichens and bryophytes are very different kinds of organisms (lichens are symbiotic fungi and algae, and bryophytes are plants), they occupy many similar habitats and often are studied together. They are able to survive long periods of drought and inactivity, but revive to become photosynthetically active within minutes of being rehydrated. They particularly thrive in areas that alternate between wet and dry conditions, such as in the forest canopies of coastal temperate zones, boreal taiga, and montane cloud forests throughout the world. Although these organisms often make up a minor part of the biomass of forests, they may play critical roles in water relations, mineral cycling, and providing habitats for other organisms. Henceforth the general term "canopy cryptogams" will be used for both sets of organisms.

Canopy cryptogams have been the focus of little research, except for a few studies in northwestern North America (Pike *et al.*, 1972, 1975, 1977; Rhoades, 1977, 1978, 1981, 1983; Nadkarni, 1981, 1983, 1984a; McCune, 1990), northern Appalachian Mountains (Lang *et al.*, 1980), and tropi-

Copyright © 1995 by Academic Press, Inc.
All rights of reproduction in any form reserved.

cal America (Nadkarni, 1983, 1984b, 1986; Nadkarni and Matelson, 1991, 1992). Otherwise, canopy species are included in studies of "epiphytes," and this literature often includes the lower trunk habitat. Some general references that discuss epiphytic lichens include Barkman (1958), Ahti (1977), Carroll (1980), Kappen (1988), Seaward (1988), Sipman and Harris (1989), Galloway (1991a), Bates and Farmer (1992), and Boucher and Stone (1992). Some general references that discuss epiphytic bryophytes include Barkman (1958), Carroll (1980), Slack (1982, 1988), Smith (1982), Pócs (1982), Scott (1982), Richards (1984), Ramsay *et al.* 1987, Schuster (1988), Gradstein and Pócs (1989), Frahm (1990b,d), Longton (1992), and Bates and Farmer (1992).

I first review aspects of the biology of lichens and bryophytes for noncryptogamic biologists. The sections on population biology elaborate on aspects of reproductive biology. Hale (1983), Hawksworth and Hill (1984), and Lawrey (1984) provide details on the lichen biology; Schuster (1984) and Schofield (1985) give background on bryophyte biology.

A. Background on Lichens

Rather than a single, taxonomically distinct category, lichens are symbiotic "ways of life" shared by many taxonomically unrelated groups of fungi. A fungus (the mycobiont) is "lichenized" if it forms a physical and physiological association (a thallus) with a photosynthetic partner (the photobiont). Evolution of the lichenized habit has occurred independently many times and exists in at least 18 orders of fungi and among a few fungallike protists (Hawksworth, 1988a,b). Lichens take their names from the mycobiont, which in most cases are ascomycetes (about half the known species of ascomycetes are lichenized). Approximately 25,000 species of lichens have been described.

Photobionts are usually members of the Chlorophyta, but Cyanobacteria occur in many lichens and genera with both types of photobionts are not uncommon. Photobiont species may be specific to a single mycobiont, one photobiont may be found with a number of different mycobionts in different lichens, or a single mycobiont may form the same lichen with different photobionts (Tucker *et al.*, 1991; Ahmadjian, 1993). In some situations, a single mycobiont forms distinctly different thalli depending on whether the dominant photobiont is a green alga or a cyanobacterium, a factor that depends on the environment.

Generally, the mycobiont gives a lichen its overall form and provides the bulk of the thallus biomass, outer protective layer(s), and a looser, inner layer that functions in physical absorption and storage. Usually, the photobiont is restricted to a layer just below the protective covering provided by the mycobiont. For the most common lichen–algal genus, *Trebouxia,* species probably do not survive in the free-living condition (Ahmadjian, 1988).

Reproduction by lichens is commonly asexual, by fragmentation or with specialized propagules that include all associates. Many mycobionts also produce sexual spores. The relative role of sexual spores of lichen mycobionts in dispersal (requiring resynthesis of the lichen thallus by association with the proper photobiont) is unclear for most species, but is variable among different species of lichens. Shaw (1992) reviewed some of the intriguing aspects of the reproductive capacities of lichens.

Because the lichens represent diverse evolutionary groups, the standard growth forms are somewhat arbitrary, but are useful to describe functional groups in canopy habitats. "Foliose" refers to leaflike, usually with dorsoventral arrangements of mycobiont and photobionts. "Fruticose" refers to thalli without distinctive dorsoventral arrangements, usually relatively narrow in cross section and of pendulous or shrubby habits. "Crustose" refers to thalli firmly cemented to a substrate, without a lower surface. Combinations of form and intergradations between these types are common. During (1992b) discussed lichen growth form in relationship to ecological strategy.

Physiologically, the mycobiont is dependent on carbohydrates obtained from the photobiont. Both the mycobiont and photobiont may provide the other associate(s) with metabolic products (lichen acids, vitamins, fixed nitrogen, proteins) that the other(s) cannot provide themselves, and that aid the collective life of the thallus. Lichenized mycobionts produce elaborate mixtures of complex organic compounds. These compounds likely function in three major defensive roles: antimicrobial, allelopathic, and antiherbivoral (Lawrey, 1986, 1993). Mineralization of substrate and light-screening roles are often mentioned, but are not well supported by experimental studies (Lawrey, 1986). Whole thalli leak carbohydrates and a variety of organic compounds (Cooper and Carroll, 1978; Carroll, 1980).

According to the International Code of Botanical Nomenclature, lichen species are given the name of their mycobiont; photobiont names are subsidiary. Morphology (sexual structures, asexual structures, and vegetative surface characters) and thallus chemistry are important species characters. It is impossible to know if all the morphological species that taxonomists delimit function as classical biological species.

B. Background on Bryophytes

The division (phylum) Bryophyta includes plants that lack true vascular tissue and vascularized organs (roots, stems, and leaves). The division is divided into three classes: Musci (mosses), Hepaticae (liverworts), and the Anthocerotae (hornworts). Although they exhibit similarities in morphology and life cycle, these three classes may not be closely related evolutionarily (Schofield, 1985; Shaw, 1985, 1992). Bryophytes in canopy habitats include the mosses (with about 10,000 species worldwide) and leafy liverworts (leafy hepatics or scale mosses, with about 7200 species worldwide). Thallose (strap- or fan-shaped) liverworts and hornworts are largely terrestrial and, as epiphytes, are usually restricted to moist, lower trunks and dead wood, occasionally becoming abundant on branches in tropical montane forests.

Bryophytes alternate between two phases. Dominant, photosynthetic gametophytes produce haploid eggs and sperm that combine sexually with those of other "parents" within very short distances (a few meters at most). Many gametophytes reproduce asexually and disperse over short to moderate distances by various types of buds (gemmae), by indeterminate growth and branching, or through thallus fragmentation. Fertilized eggs grow into sporophytes that are partly parasitic on the gametophytes. Sporophytes disperse aerial spores from terminal capsules that are disseminated by wind and ultimately germinate to form gametophytes.

Gametophytes absorb water rapidly and have adaptations that help conserve water. Leaves are generally one cell thick and lack the water-resistant coverings of vascular plants. They may be folded, deeply divided, or produce sack like outgrowths to aid in water retention. Leaf cells of some mosses have bulging or roughened cell walls (mamillae and papillae) that slow evaporation. On trunks and leaves, bryophytes may have closely attached, creeping growth forms. Moss stems often are covered with mats of rhizoids that absorb and transmit water. Many species form tight cushions or spherical balls that expose less surface area, hydrate relatively easily, and remain hydrated after precipitation has stopped. Some tropical leafy hepatics produce mucilage, often copiously, from slime papillae at shoot apices. This may protect them from drying, help gemmae adhere, or help attract invertebrate dispersers (Thiers, 1988). Bryophytes have evolved physiological adaptations that allow them to survive periods of complete dehydration. This capacity is well developed in the bryophytes found on small twigs in areas that experience frequent hydration and drying, particularly in tropical cloud forest canopies, where they predominate (Norris, 1990). Schofield (1981), During (1992b), and During and Van Tooren (1987) discussed bryophyte growth forms in relationship to ecological strategy and habitats.

Frey *et al.* (1990) discussed the relationship of bryophyte growth forms to altitude and vegetation type in epiphytic bryophytes of North Borneo. Three possible adaptations of water-conducting and water-storing mechanisms were proposed: draining off surplus water, storage of water in dry season (the principal adaptation), and condensation of water vapor (allowing for rapid absorption and improving efficiency by which surplus water drips off). In physiological studies, Hosokawa and Kubota (1957) showed that the osmotic ability to withstand drying increased with height among different species of mosses in canopies of Japanese beech trees.

Sporophytes produce spores by meiosis and efficiently disperse these. When present, peristome teeth of mosses are hygroscopic. Liverwort capsules may contain hygroscopic hairs. These structures flex slowly with changes in humidity, aiding spore dispersal. Sporophytes of many moss species adapted to the periodically xeric nature of epiphytic habitats possess common sets of characters (Vitt, 1981): short, straight, stout setae; erect, oblong to ovate, sometimes strongly ribbed, capsules; and a reduced or absent peristome.

Bryophyte spores can potentially be distributed over wide ranges, at least within latitudinal regions. How far bryophyte spores travel and remain viable is the subject of debate and experimentation. Van Zanten and Gradstein (1987, 1988) compared the characteristics of spores and sporelings of Neotropical liverworts to the species' geographical ranges. They have greater longevity than previously thought, but shorter than those of mosses. The spores of transoceanic liverwort species (in contrast to endemics) have good resistance to drought and frost, but UV light is lethal to most species. They concluded that in the tropics, aerial long-distance transport can occur by wet air currents (e.g., cyclonic tropical storms) at high altitude and by dry air currents at low altitude, but long-distance transport at jet-stream altitudes is unlikely.

C. Techniques of Access and Analysis

Two technical problems have plagued canopy biologists: difficulty of access to the canopy and difficulty of sampling structurally diverse environments. Most studies of epiphytic cryptogams have been limited to trunk surfaces below 2 m or have depended on litterfall analysis. Neither technique accurately describes canopy communities. Although study of processes can be done in a laboratory (Frahm, 1987a,b, 1990c), some questions can only be answered by *in situ* work and should also be carried out to confirm laboratory studies (Hosokawa and Odani, 1947; Hosokawa and Kubota, 1957; Hosokawa *et al.*, 1964). Thus, direct canopy access is of prime importance to understand canopy cryptogams.

Access has improved considerably with the development of rope-climbing techniques in the northwestern United States (Pike *et al.*, 1972; Denison *et al.*, 1972) and the extension of those techniques into tropical canopies (Verhoeven and Lee, 1976; Perry, 1984; Nadkarni, 1988; ter Steege and Cornelissen, 1988). Although other techniques have facilitated access [aerial walkways (Mull and Liat, 1970), the floating raft suspended on tropical canopies (Hallé, 1990), canopy crane (Parker *et al.*, 1992)], none has been used to study cryptogams. In many areas of the world, access to canopy environments is no longer a limiting factor, and our knowledge of the players and their roles should develop more rapidly (Lowman and Moffett, 1993; Chapter 1).

Nimis (1991) reviewed the historical development of schools of vegetation sampling as they pertain to lichen studies. Slack (1984) discussed the need for bryologists trained in suitable techniques of ecological analysis. Pike (1971) used a wide variety of sampling and analytic methods that should be included in any survey of epiphyte communities. Coddington *et al.* (1991) discussed sampling methods in tropical canopies from a zoological perspective, but their conclusions hold for any group of organisms. Methods need to be rapid, because time for study in canopy environments may be limited; reliable, because workers need to apply them in diverse areas to generate comparable data; and simple and cheap, because the problem of extinction is most severe in developing tropical countries where the scientific and museum infrastructure is often rudimentary. McCune and Lesica, 1992 contrasted work carried out by systematists and ecologists: collections by systematists, which may adequately represent an area's species richness, may be intractable statistically; sampling procedures designed to answer ecological questions usually provide for statistical analysis, but often poorly represent the total flora. Questions of biological diversity may end up in political or legal arenas (e.g., lichen and bryophyte diversity in Spotted Owl habitat in the old-growth conifer forests of the northwestern United States; Thomas *et al.*, 1993) and data must be as statistically "tight" as possible.

Obtaining accurate, statistically sound data on epiphytes involves careful planning. Pike (1981) and Boucher and Stone (1992) discussed ways to subsample a complex substrate to give reliable estimates. These methods have been used to describe cryptogamic communities in old-growth conifer forests of the Pacific northwestern United States (Pike *et al.*, 1972; Denison *et al.*, 1972; Rhoades, 1978, 1981, 1983) and the physical characteristics of this habitat (Pike *et al.*, 1977; Massman, 1982). McCune (1990, 1993) and McCune and Lesica (1992) reviewed the problems involved in rapid sampling for floristic and ecological information and tested a variety of techniques in temperate coniferous forests. Hoffmann (1987) compared the phytosociological "relevé" method with a plotless method of assessing community associations. John (1992) used a plotless, point-sampling scheme to study the interactions between corticolous lichens that might be modified for canopy habitats.

II. Characteristics of Canopy Habitats from a Cryptogamic Perspective

Canopy cryptogams grow on the bark surfaces of trunks, large branches, twigs, leaf surfaces, layers of suspended soil, and on some animals that live within the zone where branches and foliage occur (Fig. 1) (Brodo, 1973; Pócs, 1982; Richards, 1984). Epiphyllic (foliicolous) habitats are found only in regions where the humidity is continually high and evergreen leaves and

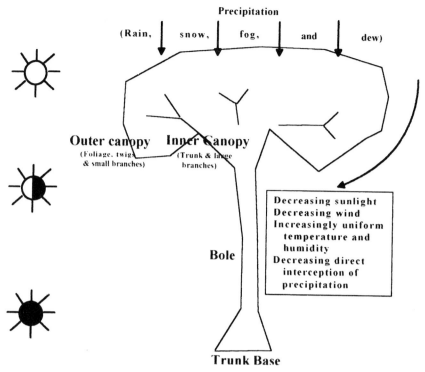

Figure 1   Canopy habitats seen from a cryptogamic perspective. The canopy does not include the trunk base nor branch-free lower bole. Adapted from Kürschner, 1990a, with permission.

needles provide surfaces for several years' growth, typically in lower-canopy and understory habitats of tropical and subtropical regions, and occasionally in temperate regions (Vitt *et al.,* 1973). Pócs (1982) and Richards (1984) reviewed the tropical "epiphyllae" and they are not considered here. A few canopy cryptogams are found on epizoic habitats (on the bodies of living beetles) that are canopy residents (Richards, 1984; Seaward, 1988; Gradstein and Pócs, 1989). These cryptogamic species are normal epiphytes or epiphylls and may exist as living, attached thalli or nonattached fragments. None is known to be restricted to the animal host.

Epiphytic cryptogams "see" the environment differently than terrestrial or epiphytic vascular plants (McCune and Antos, 1982; Norris, 1990). Because cryptogams are less protected by complex tissues and organs, they are particularly sensitive to changes in microenvironments. Light intensity, relative humidity, temperature, wind, and the specific chemical and physical nature of the surfaces play roles in habitat definition. The three-dimensional arrangement of branches, foliage, and other organisms in forest canopies influences these characteristics. The phorophyte surfaces and biota in the outer canopy intercept radiation and water, retard air currents, and lead to gradients of light, temperature, and humidity (Fig. 1). These factors are further influenced by forest density, geographical location, and structural and chemical features specific to the phorophyte species.

In a forest, each phorophyte and each part of the canopy may support a different set of cryptogams. Some cryptogams are obligately epiphytic, whereas others are facultatively epiphytic, inhabiting both the forest floors and soillike environments on the upper, larger branch surfaces in the canopy (Smith, 1982). Although some species show narrow phorophyte ranges, most cryptogams are relatively general in their host demands and occupy habitats on many phorophyte species, as long as the microclimatological characteristics meet a species' needs (Trynoski and Glime, 1982; Beever, 1984; Schmitt and Slack, 1990; Wolf, 1994). It is difficult to separate studies of epiphytes on lower tree boles and tree bases from those on trunk surfaces in the canopy per se. The following section includes trunk studies when they apply to true canopy habitat, or where few studies in the true canopy are available.

A. Bark Characteristics

Bark chemistry, particularly pH, is an important factor that determines species presence. Trees with basic bark (e.g., *Acer, Fraxinus, Populus*) appeal to one set of lichen species, and trees with acidic bark (e.g., most conifers, *Quercus, Fagus*) appeal to another. The pH of bark on bryophyte-loaded phorophytes in Malaysia is phorophyte-dependent but not correlated with bark structure (Frahm, 1990a). Most values fall within the acidic range of pH 4 to 6, as do bark acidity values in montane rain forest of Colombia (Wolf, 1993b). Bark and moisture were the two most important variables affecting the abundance of nitrogen-fixing lichens in the southern Appalachian Mountains (Becker, 1980).

Trees with flaky bark rarely carry large epiphyte loads, as this surface is ephemeral and large epiphytes tend to pull the bark off. Kenkel and Bradfield (1981) used ordination methods to study bryophyte communities on lower boles of conifers in British Columbia and found that the compositional variation is determined in part by bark type. This pattern was true for bryophytes in Malaysia (Frahm, 1990a). The relative richness of canopy flora on Douglas fir versus coastal redwood in a northern California forest may partly be due to differences in bark texture and sloughing (Sillett and Sillett, 1992).

B. Moisture

The quantity and fluctuation of available water are critical factors in defining cryptogamic habitats (Brodo, 1973; Richards, 1984; Larson, 1987; Thiers, 1988; Proctor, 1990). Proper conditions may occur periodically throughout the year or be restricted to one or more seasons (Pitkin, 1975). The extent of saturation varies, but is usually at least 75%. Whereas green algal photobionts can make use of hydration by water vapor alone for positive assimilatory gain, lichens with cyanobacterial photobionts must be hydrated by liquid water to achieve positive assimilatory status (Lange *et al.*, 1993). Mosses higher in trees tend to have higher osmotic values, hence offering increased resistance to desiccation. Excess moisture can impede growth of both bryophytes (Hosakawa and Kubota, 1957) and lichens (Redon and Lange, 1983b).

Heavy cryptogamic epiphyte loads can greatly influence the hydrological cycle in forest canopies by intercepting large amounts of water in the form of fog and/or by retaining water and maintaining high humidity, long after precipitation (including cloud interception) has ceased (Redon and Lange, 1983b; Veneklaas *et al.*, 1990). The high humidities required to keep epiphytic bryophytes saturated are maintained by the epiphytes themselves, as shown in the study of an elevational gradient in Malaysia (Frahm, 1990a).

Pócs (1980, 1982) compared values for biomass and water retention capacity of bryophytes of epiphytes in two forests in Tanzania (see the following). He proposed that excess moisture beyond 200 mm per month in tropical communities has little effect on the structure of the phorophyte population, but correlates with increases in biomass of cryptogamic epiphytes in the outer branches.

C. Light

If moisture levels are adequate and temperatures are not too high (the latter causes increased evaporation and raises the compensation points for photosynthesis), light levels are a critical factor in determining cryptogamic habitat. Frahm (1990a) discussed the complex relationships between light intensity, temperature, and moisture in determining whether conditions are favorable for bryophyte growth in the tropics. Canopy epiphyte loads and phorophyte distribution may cut light levels below the canopy to only 2% of daylight. At lower elevations, where temperatures are higher, the intensity in lower/inner canopy areas may be far too low to support growth because of the higher compensation values required at higher temperatures.

In three types of Australian rain forest, light interception was relatively constant throughout the year (Lowman, 1986). This was attributed to the complex, three-dimensional and multistory nature of the rain forest canopy. In the temperate zone, lichens are found in areas of higher light intensity than bryophytes, so lichens predominate in the outer canopy and bryophytes predominate in the inner and lower canopy. The relative richness of canopy flora on Douglas fir versus coastal redwood in California forest may be due in part to differences in shading due to crown morphology (Sillett and Sillett, 1992).

D. Temperature

Increases in temperature raise the compensation point for photosynthesis, often above that sustainable by the light intensities present. For species growing close to bark substrates in well-lit, sun-facing habitats, thallus temperatures may be quite different from the surrounding air. Coxson *et al.* (1984) examined three lichen species with different growth forms on *Pinus* branches and trunks in Alberta, Canada. Closely appressed *Hypogymnia* thalli growing on branches were up to 12° C higher than ambient air temperature and up to 28° C higher on lower trunks where there was little convective cooling. Pendulous *Usnea* was only 4° C higher, whereas *Letharia,* with an intermediate, tufted growth form, was intermediate, 9° C above ambient temperature.

E. Other Factors

Other aspects that determine epiphyte loads are stand age and closure (McCune, 1993; Rose, 1976), the general hydrological properties of the total mass of surrounding epiphytes (Veneklaas *et al.*, 1990), competitive pressure from other species (Trynoski and Glime, 1982), and the presence of human influences such as atmospheric pollution, general urbanization, and forest management practices (Lesica *et al.,* 1991; Rose, 1992).

An overriding factor that modifies the effects of all others is the role of chance in establishment of new thalli. Even on neighboring branches exposed to the same general meteorological conditions, communities of epiphytes tend to be variable. Epiphytic cryptogam community heterogeneity was first noted by Schuster (1957) for bryophytes on *Thuja* trees and confirmed by Barkman (1958) for lichens. Wolf (1993a,b,c) found similar extreme heterogeneity of epiphytes in Colombian montane forests; among 600 relevés taken, the same combination of species was never observed twice. Schuster (1982) has proposed that chance arrival and recruitment at newly available sites for colonization plays a major role in generating this heterogeneity.

F. Interactions

Many studies have investigated which of the interacting aspects of the canopy environment are most important. Most conclude that a combination of these aspects determines which and how much of each species is present (Hale, 1952; Barkman, 1958; Brodo, 1973; Grough, 1975; Kallio and Kärenlampi, 1975; Slack, 1976, 1977; Smith, 1982; Trynoski and Glime, 1982; Eversman *et al.*, 1987; Frahm, 1987a,b; Schmitt and Slack, 1990; Canters *et al.,* 1991; Bates, 1992a; Pórto, 1992).

Multivariate statistical approaches, such as those used by Bryant *et al.* (1973), Oksanen (1988), and Bates (1992a), may help clarify the interactions among all of these characteristics. Bates (1992a) found that height on the tree, bark slope, and bark roughness are the primary factors related to floristic patterns of lichen and bryophyte species on lower trunks of *Quercus* and *Fraxinus* in western Scotland. Altitude and height in the tree are the most important environmental factors interacting to cause floristic patterns among epiphytic bryophytes in upper montane forests of Colombia (Wolf, 1994). In Oksanen's (1988) ordination analysis, over 97% of the variation in the data could not be related to complex combinations of different environmental factors.' This supports the important function of chance encounters of new individuals in newly available habitat.

III. Floras

A. Regional

Only a few studies have focused on worldwide biogeography of bryophytes (Schofield, 1992) or lichens (Hawksworth and Hill, 1984), and then without specific reference to or distinction of canopy habitats. Hawksworth and Ahti (1990) provided the most recent bibliography of worldwide lichen floras; Green and Harrington (1988, 1989) provided the most recent bibliographies of worldwide bryophyte floras. Floristic inventories in the earliest studies tend to be simple lists of collections, with later studies focusing more on community structure, productivity, and ecological roles. For many regions where inventories of macrolichens and bryophytes have been compiled, inventories of crustose lichens are lacking or incomplete. This is unfortunate because crustose lichens appear to exhibit greater tree-to-tree specificity than other cryptogams in many forests (Ahti, 1977) and are the dominant cryptogamic form in outer canopies, particularly in tropical rain forests (Sipman and Harris, 1989). Following are brief reviews by large, geographical region. Table I lists references for each region.

1. Northern Boreal and Temperate Forests Considerable work on canopy cryptogams has been done in this region. The region is subdivided into several floristically distinct subregions that blend into each other on their margins. Although some floristic elements of the northern floras exist in the canopies of the Southern Hemisphere forests, the phorophytes and balance of lichens and bryophytes are different.

The canopies of boreal forests may be broadly continuous and uniform or, in areas close to the northern and higher-altitudinal limits, disrupted and uneven. There is no comparable floristic zone in the Southern Hemisphere, although some form of subalpine forest exists almost continuously along the mountain chains that run the length of the American continents, and less continuously in southern Asia and eastern Africa. Because boreal and subalpine forests are relatively uniform in phorophyte composition

Table I Sources of Floristic Surveys Including Canopy Cryptogams by Geographical Region

General region	Specific region	Sources
Northern boreal and temperate	Boreal-subalpine, Northern Hemisphere	Ahti (1977); Biazrov (1971); Brodo and Hawksworth (1977); Davis (1964); Esseen (1985); Eversman (1990); Eversman et al. (1987); Gough (1975); Halonen et al. (1991); La Roi and Stringer (1976); Nimis (1985); Schuster (1957); Wein and Speer (1975); Wetmore (1967); Lang et al. (1980)
	Intermountain, temperate North America	Eversman (1982); Lesica et al. (1991); McCune (1982); McCune and Antos (1982); Rosentreter (1993); Schofield (1992); Bird and Marsh (1973)
	Pacific moist coastal, northwestern North America	Boucher and Nash (1990); Coleman et al. (1956); Nadkarni (1981, 1983, 1984a); Pike (1971); Pike et al. (1972, 1975, 1977); Rhoades (1984, 1983); Sillett and Sillett (1992); Stone (1989)
	Eastern North America, Europe, Central Asia	Barkman (1958); Becker (1980); Becker et al. (1977); Degelius (1964, 1978); Dey (1976); Gowan and Brodo (1988); Gilbert (1984); Hale (1952); Hosokawa and Odani (1947); Novichkova-Ivanova (1983); Rose (1988); Schofield (1992); Schuster (1957); Sjögren (1961); Slack (1977); Stubbs (1989); Wilmanns (1962); Makryi (1985)
South Australia, Tasmania, New Zealand, temperate South America, associated islands	Southern Australasia	Catcheside (1982); Galloway (1987, 1988a, 1991b, 1991c, 1992); Kantvilas (1988, 1990); Kantvilas and James (1987); Ochi (1982); Ramsay et al. (1987); Rundel et al. (1979); Schuster (1982); Vitt (1991); Vitt and Ramsay (1985)
	Temperate South America, including cloud forests	Galloway (1991c, 1992); Jørgensen (1977, 1979); Redon and Lange (1983a,b); Schuster (1982); Stevens (1987)

364

Southern Africa	Almborn (1988); Brusse (1991); Schofield (1992)
General reviews	Aptroot (1991); Arvidsson (1991a,b); Buck and Thiers (1989); Fulford (1951); Galloway (1991a,b); Gradstein (1992); Gradstein and Pócs (1989); Krog (1991); Marcelli (1991); Nishida (1989); Pócs (1982); Richards (1984, 1988); Sipman (1991); Sipman and Harris (1989); Smith (1991); Stevens (1991); Tucker *et al.* (1991); Wolseley (1991)
Neotropical	Arvidsson (1991a); Cornelissen and ter Steege (1989); Frahm (1987a,b, 1990d); Galloway and Arvidsson (1990); Gradstein *et al.* (1990); Hofstede *et al.* (1993); Marcelli (1991); Nadkarni and Matelson (1991, 1992); Pórto (1992); Richards (1954); Russell and Miller (1977); Schuster (1982); Sipman (1989, 1991, 1992); ter Steege and Cornelissen (1988); Van Reenen and Gradstein (1984); Wolf (1993a–d, 1994)
Paleotropical	Arvidsson (1991b); Frahm (1990a,b,d); Hyvönen *et al.* (1987); Krog (1987, 1991); Kürschner (1990a,b); Piipo *et al.* (1987); Pócs (1980, 1982); Ramsay *et al.* (1987); Schuster (1982); Stevens (1987, 1991); Thaithong (1984); Van Zanten and Pócs (1981); Vitt and Ramsay (1985); Wolseley (1991)
Cloud forests and elevational transects	Delgadillo (1979); Delgadillo and Cárdenas (1989); Frahm (1987a,b, 1990a–d); Frey *et al.* (1990); Gradstein *et al.* (1989); Krog (1987); Kürschner (1990a,b); Pórto (1992); Sipman (1989); Van Leerdam *et al.* (1990); Van Reenen and Gradstein (1984); Wolf (1993a–d, 1994)

and shorter in stature, the canopies of these forests are more easily studied than those of continental and coastal forests farther south and at lower altitude in the temperate zone. There have been few canopy studies on cryptogams in these areas and those have been in conjunction with general inventories or studies of the ecology of other organisms (e.g., animal consumption).

The dry, coniferous forests of the intermountain regions in temperate North America have been examined floristically for cryptogams for many years. This zone overlaps with the boreal subalpine zones, until the lower-elevation forests disappear with decreasing precipitation and increasing temperature in the Rocky Mountains of south-central United States. To the west, the continental forests are divided by the northern Sierra/Cascade ranges from the much more mesic forests of the Pacific Coast of the northwestern United States. In Europe, this forest type is largely subsumed by the boreal forest type. Similar regions in Asia are less well studied, and the results of studies are difficult to obtain. In the Southern Hemisphere, a climatically similar region occurs in parts of Australia, with different floristic elements (see Section III,A,2).

In both the boreal-subalpine and intermountain zones, epiphytic bryophytes are usually restricted to lower trunks, and only a few mosses (*Orthotrichum, Ulota*) and leafy liverworts (*Frullania*) are found in the canopies. The knowledge of the lichen flora, with the exception of the crustose species, is fairly complete for Europe and North America, but less so for northern Asia. Macrolichens are common and include species in the Usneaceae (*Letharia, Evernia,* and *Usnea*), Alectoriaceae (*Alectoria* and *Bryoria*), Sphaerophoraceae (*Sphaerophorus*), Parmeliaceae (*Parmelia sensu lato, Platismatia, Parmeliopsis,* and *Tuckermannopsis*), and Hypogymniaceae (*Hypogymnia*). In regions where the canopies are closed and the humidity is high (areas immediately along lakes and rivers and in upper montane cloud forest regions), the abundance of lichens can be particularly high.

A special forest region extends from southeastern Alaska south along the Pacific Coast of Canada and the United States into northern California. It is bounded on the east by the Coast Mountains of British Columbia and the Cascade Mountains and Sierra Nevada in the United States. Wintertime rainfall is considerable, temperatures are mild, and, to the south, summer drought increases. These forests have greater diversity of coniferous phorophytes and, because the individual trees are so large, considerably greater substrate diversity than the coniferous forests to the east (Pike *et al.,* 1977). In the Southern Hemisphere, similar climatic regions include broadly similar cryptogamic communities in the canopies of very different phorophytes (see Section III,A,2).

The canopy communities of the wettest forests are dominated by the same groups found in the boreal forests, with the addition of many nitrogen-fixing lichens in the Peltigeraceae (*Peltigera* and *Nephroma*), Pannariaceae (*Pannaria* and *Parmeliella*), and Lobariaceae (*Lobaria, Pseudocyphellaria,* and *Sticta*). Bryophytes are dominant on old, lower branches close to the trunk, with mosses in the Hypnaceae (*Hypnum*), Brachytheciaceae (*Isothecium* and *Eurhynchium*), Dicranaceae (*Dicranum*), Leucodontaceae (*Antitrichia*), and Neckeraceae (*Metaneckera* and *Neckera*). In redwood forest canopies, to the south of this region, there is a considerable difference in canopy floras between Douglas fir and coastal redwood in California (Sillett and Sillett, 1992). Similar floristic differences are also observed between Douglas fir and hemlock and the "cedars" (*Thuja plicata* and *Chamaecyparis* spp.) to the north of this region. Rhoades (1981) studied an unusually rich subalpine fir forest at low elevations that is floristically similar to the canopies of neighboring low-altitude, old-growth conifer forests.

In this region are areas dominated by deciduous trees, including moist lowland terraces with big-leaf maple (Nadkarni, 1981, 1983, 1984a), dry-valley oaks in Oregon (Pike, 1971; Stone, 1989), and the coastal woodlands with large populations of *Ramalina menziesii* in southern California (Boucher and Nash, 1990).

In North America east of the Great Plains, mixed conifer–hardwood forests grade into subalpine and boreal coniferous forests at higher elevations and to the north, respectively. These conifer–hardwood forests are richer in phorophyte species, particularly hardwoods, than the northern boreal forests. Consequently, the canopy cryptogam floras are richer in species, although not much more in genera. Similar forests occur in western Europe. A number of studies in these areas include mention of canopy species, but few are directed specifically at this habitat (Table I).

2. South Australia, Tasmania, New Zealand, Temperate South America, and Intervening Islands   Particularly rich in these regions are the canopy lichen floras in fog belts and temperate "rain forests" on the windward sides of mountain regions, such as those in southern New Zealand and Tasmania. The genera of these floras are similar to those along the Northwest Coast of North America, with some additions of genera and changes in the relative numbers of species. In the fog belt of northern, coastal Chile, fruticose lichens in the Alectoriaceae (*Oropogon*), Usneaceae (*Usnea,* and *Everniopsis*), and Ramalinaceae (*Ramalina*) are predominant.

In the temperate rain forest areas of Australia, Tasmania, and New Zealand, the rich lichen flora contains species from the Hypogymniaceae (*Hypogymnia* and *Menegazzia*), Parmeliaceae (*Parmelia, Pannoparmelia,* and *Cetraria*), Sphaerophoraceae (*Sphaerophorus*), and the nitrogen-fixing families Lobariaceae (*Pseudocyphellaria* and *Sticta*), Collemataceae (*Collema, Leptogium, Rammalodium,* and *Physma*), and Pannariaceae (*Pannaria, Parmeliella, Leioderma, Degelia, Psoroma,* and *Psoromidium*).

Bryophytes in the temperate coastal rain forests of eastern Australia and New Zealand share some floristic elements with both the tropical rain forests to the north (particularly the large moss genus *Macromitrium*) and other temperate remnants of Gondwanaland in the Southern Hemisphere. Important moss genera in this area are *Dicnemon, Ephemerum, Leptostomum, Meterorium, Papillaria, Macrocoma, Macromitrium, Orthotrichum,* and *Zygodon.* The hepatic flora of this region is particularly rich and unusual (liverworts are much more common here than in the Northern Hemisphere), with 21 families with one or more endemic genera.

Cryptogams are less abundant in the open, drier forests of southern and southwest Australia. The moss floras of that region have little in common with the more tropical floras to the north and east, but show some phytogeographical affinity to floras elsewhere in the south temperate regions or, disjunctively, with floras in the north temperate zone (Catcheside, 1982; Ochi, 1982).

3. Southern Africa   The most unusual canopy lichen flora of the temperate zone in the Southern Hemisphere occurs in a few residual dry forests in southern South Africa. The canopy flora is poorly known but includes monotypic genera and species and fewer of the nitrogen-fixing species than in comparable forests discussed previously. Species endemism is also high for bryophytes and the flora has a greater affinity to dry areas farther north in Africa rather than to the floras of the Neoantarctic and Australian kingdoms.

4. Tropics   Cryptogams of tropical forest canopies are the least well known in the world. Because of increased interest in tropical biodiversity, recent reviews of cryptogams have focused on these regions (Tables I and II). Forests of tropical regions can be divided into three general forest types: (a) lowland wet tropical forests with abundant rainfall resulting in forests with dense, closed canopies; (b) lowland dry forests (savannahs) with open or discontinuous canopies; and (c) upland cloud forests in areas of cooler temperatures and higher humidity, resulting in forests of relatively short stature and abundant, diverse cryptogamic floras. Most work in this region has concentrated on the lowland floras of the moist tropics or documented the differences in the floras with altitude. Not as much attention has been paid to the less abundant canopy inhabitants in the savannah areas, where open tree canopies mostly offer homes to crustose and closely appressed, foliose lichens.

In the lowland wet tropics, canopy habitats may be particularly rich in crustose species. The dominant lichens are species in the families Arthoniaceae, Graphidaceae, Phyllopsoraceae, Thelotremataceae, and Trypetheliaceae, which predominate in the outer canopy. The richness of foliose and fruticose species increases with altitude. Species in the Collemataceae, Pan-

Table II Species Richness of Epiphytic Cryptogams in Worldwide Forest Types [a]

Location / forest type	Latitude (°N)	Number of trees sampled	Mosses	Liverworts	Total bryophytes	Macrolichens	Source
Guyana; dry evergreen *Eperua* spp.	5?	11	28	53	81	33	Cornelissen and Ter Steege (1989)
French Guyana; mixed lowland rain forest	5	4	43	61	104	21	Montfoort and Ek (1990)
Colombia; montane rain forest, 1500 m	5	4	22	36	58	49	Wolf (1993c)
Colombia; montane rain forest, 2550 m	5	4	33	102	135	51	Wolf (1993c)
Colombia; montane rain forest, 3510 m	5	4	19	63	82	37	Wolf (1993c)
Guyana; mixed lowland rain forest	7?	5	28	60	88	19	Cornelissen and Ter Steege (1989)
Oregon, United States; low, mixed coniferous forest	44	11	11	6	17	37	Pike et al. (1975); upper trunk and branches of overstory
Wisconsin, United States; mixed conifers and hardwoods	46	Many	14	3	17	29	Hale (1952); species collected from 15 m and higher
Montana, United States; old-growth *Abies*	48	5	4	1	5	34	Lesica et al. (1991); lower canopy stratum only
Montana, United States; managed, second-growth *Abies*	48	5	1	0	1	37	Lesica et al. (1991); lower canopy stratum only
Washington, United States; low elevation fir forest on lava flow	49	Many	8	5	13	53	Rhoades (1981); plus additional data
Sweden; deciduous forest	56	Many	78	17	95	—	Sjögren (1961)

[a] Within the tropical areas, sampling included the canopy and, in the Guianas, epiphyllous species. Modified from Table 1 in Wolf (1993c).

nariaceae, Usneaceae, and Coccocarpiaceae are dominant around 1000 m, and members of the Lobariaceae and Parmeliaceae predominate above 1000 m. The dominant moss families in the lowland tropics are the families Calymperaceae, Hookeriaceae, Hypnaceae, Orthotrichaceae, and Sematophyllaceae. Leafy liverworts are an important element in canopy habitats in the tropics and the Southern Hemisphere and are probably the dominant bryophyte in the lowland forests. Species in the family Lejeuneaceae are particularly well represented, along with species in the Lepidoziaceae, Plagiochilaceae, and Frullaniaceae. The moss genus *Macromitrium* is distributed widely throughout the Australasian region and is an important canopy epiphyte in wet tropical forests of northern Australia and drier forests everywhere in the region.

5. Moss or Cloud Forests   Extremely large epiphyte biomass has been recorded in forest canopies exposed to periodic moisture from cloud condensation in montane localities throughout the world (Table III and Fig. 2). In temperate regions in the Northern and Southern Hemispheres, such epiphyte communities are dominated by lichens, particularly the genera *Alectoria, Bryoria, Usnea,* and *Hypogymnia*. In subtropical to tropical regions, these communities are usually dominated by bryophytes, with the leafy liverworts most abundant. In the tropics, many floras have been documented in the context of altitudinal transects (Table I). In general, foliose and (known) crustose lichens peak in numbers of taxa at midaltitudes and then drop with altitude, whereas fruticose and bryophyte taxa increase in number with altitude and diminish with decreasing size and numbers of phorophytes. Liverworts are important elements in forests at higher elevation, often representing 60–70% of the total cover.

B. Endemism and the Evolution of Cryptogamic Floras

Distributions of lichens and bryophytes in forest canopies depend on many interacting factors: evolution and distribution of suitable phorophyte hosts, distribution of suitable climates for both phorophytes and their epiphytes (both present and former distribution of climates), sources and dispersal mechanisms of the organisms, and rates of evolution within each taxonomic group of organisms. Van Zanten and Pócs (1981) reviewed data on evolutionary dispersal of bryophytes and discussed the relative importance of plate tectonics, step-by-step dispersal, and long-range dispersal as important mechanisms. Evolution of canopy habitats in the tropics has provided for rapid species evolution among the leafy liverworts as new surfaces were made available and terrestrial habitats were eliminated by heavy litter layers and rapid decomposition (Schuster, 1988). Terrestrial habitats in the lowland tropics are limited by the same factors for mosses (Richards, 1988) and lichens (Sipman and Harris, 1989). Galloway (1987, 1991b) dis-

Table III Epiphytic Biomass Estimates from All Sources[a]

Group	Study area	Latitude	Forest type	Elevation (m)	kg (dry weight) ha⁻¹		Source
Epiphytic, nitrogen-fixing lichens	North Carolina piedmont	35° N ?	80–100 yr oak–hickory forest	270	BG[b]	0.127?	Becker et al. (1977)
Epiphytic, nitrogen-fixing lichens	Southern Appalachian Mountains	35° N	Gray beech and oak–chestnut forests of wet, high-elevation beech gaps	500?	BG	8.4	Becker (1980)
Lichens (mostly Ramalina menziesii and other G) and bryophytes	California	36° N	Blue oak	490	Lichen Bryo	754 36	Boucher and Nash (1990)
Bryophytes (dominated by liverworts, Frullania atrata, Hubertus junipe- roideus, and Metzgeria leptoneura, and moss, Phyllogonium fulgens)	Guadeloupe	16° N	Cloud forest domi- nated by 3 to 4 m Clusia mangle	1330	Upper canopy Lower canopy	10,126 2210	Coxson (1991)
Epiphytes: vascular, bryophytes, and lichens	Papua, New Guinea	6° S	Montane rain forest in Marafunga basin	2500	Total epiphytes = 4300 (including 3400 pure epiphytes + half the epiphyte-trapped soil that was said to be half epiphyte material) Vascular, dicots ≅ 2279 nondicot ≅ 1011 lichens + bryophytes ≅ 1011		Edwards and Grubb (1977)
Epiphytes, mostly G lichens (Alectoria and Bryoria)	British Columbia, Wells Gray Park	52° N	Douglas fir, spruce, fir, pine in various mixtures	792 1128 1829	Site 1 G Site 2 G Site 3 G Site 4 G BG = 0 Bryo = n.s.[c]	838 283 756 3291	Edwards et al. (1960)

continues

Table III—*Continued*

Group	Study area	Latitude	Forest type	Elevation (m)	kg (dry weight) ha⁻¹	Source
Epiphytic macrolichens	Sweden	62° N	*Picea abies* forest	450 (mean of 2 sites)	1737?	Esseen (1985)
Epiphytic macrolichens	Colombia	5° N	Montane rain forest	2700	6.9	Forman (1975)
Epiphytic bryophytes	Tropical, North Borneo, Mt. Kinabalu, BRYOTROP project	6° N	Tropical lowland to tropical subalpine, along an altitudinal gradient	3400	1320	Frahm (1990a)
				3100	125	
				3000	200	
				2700	170	
				2500	200	
				2100	440	
				1900	18.2	
				1700	24	
				1500	16.2	
				1300	1.8	
				1100	2.4	
				900	0	
				700	0.06	
				500	0	
				20	1	
Epiphytic bryophytes	Eastern slope of Andes, Peru	6° S	Various		g m⁻² tree surface	Values read from Frahm (1990d) Fig. 1. Because data are given per tree surface, not included in Fig. 2
				3200	160	
				3000	100	
				2800	55	
				2600	80	
				2300	105	
				2100	45	
				1900	75	
				1700	80	
				1500	40	
				1300	25	
				1100	55	
				900	1	
				700	1	
				500	10	
				300	1	

Epiphyte description	Location	Latitude	Forest type	Elevation		Biomass (kg ha⁻¹)	Reference
Epiphytes; cryptogams not distinguished	Darien, Panama	9° N	Premontane tropical forest	500?		1400	Golley et al. (1971), in Nadkarni (1984a, 1986)
Mainly bryophytes (Herbertus, Leptoscyphus, Plagiochila, Lepicolea, and Porotrichodendron) and suspended soil (> 60%)	Colombia	5° N	Montane tree line, rain forest with Weinmannia	3700		44,000 / 30,800 — If 30% of the former total is noncryptogam material in the suspended soil	Hofstede et al. (1993)
Epiphytic bryophytes on boles up to 2 m height	Transvaal	25° S ?	Trees	1000?		34 g m⁻² of tree regrowth 7.6, 15.45, 27.5% after 1, 2, and 3 yr, respectively; biomass estimate for total study area avg. 340 kg ha⁻¹	Jacobsen (1978), in Frahm (1990d)
Lower boles; results may also include terrestrial spp.	Southern Venezuela	1–5° N ?	Amazonian Caatinga (heath) forest	100		1–19 kg ha⁻¹ (avg. 7)	Klinge and Herrera (1983), in Frahm (1990d)
Epiphytic lichens on boles and branches	Mt. Moosilauke, New Hampshire	48° N	Temperate montane forest, dominant Abies balsamea	1220 m NE face / 1295 m E face / 1250 m W face	44 yr / 52 / 78 / 25 / 35 / 61 / 67 / 22 / 31 / 79	923 / 591 / 1603 / 133 / 125 / 1283 / 635 / 204 / 120 / 676	Lang et al. (1980)
Cryptogamic epiphytes	Three forest stands of different ages in Northwest U.S., Cascade Range (Oregon, Washington)	44° N / 46° N	Pseudotsuga–Tsuga forests / 400+ yr / 145 yr / 95 yr	915 / 500 / 550	Lichen / 1870 / 870 / 950	Bryo / 780 / 10 / 180	McCune (1993)
Epiphytes	Northwest U.S. coastal lowlands, Olympic Peninsula, Washington	46° N / 48° N	Pure stands of Acer macrophyllum	180 & 195		6870 — Ferns 2% / Club mosses 6% / Bryophytes 74% / Org. matter 18% / Bryo = 5084	Nadkarni (1981, 1983, 1984a, 1986)

373

continues

Table III—Continued

Group	Study area	Latitude	Forest type	Elevation (m)	kg (dry weight) ha⁻¹			Source
Epiphytes	Monteverde, Costa Rica	10° N	Cloud forest	1500	Total = 4820 22% byro = 964 (rest org. matter and vascular plants)			Nadkarni (1983), in Nadkarni (1984a,b, 1986)
Epiphytes (all G lichens on twigs and branches < 20 yr)	Oregon, Willamette Valley	44° N	Scrub oak woodland	85	G & BG 1775 (mostly G) Bryo 0			Pike (1971)
Epiphytic lichens	Northeast U.S. (New Hampshire)	44° N	Northern hardwood forest, dominant *Acer saccharum*	300?	100			Pike (1978)
Epiphytes	Northwest U.S. (H. J. Andrews Forest, Oregon)	44° N	Old-growth coniferous forest, dominant *Pseudotsuga menziesii*	500	BG 650 G 250 Bryo 350			Pike *et al.* (1977)
Epiphytic bryophytes on trunks and canopy microhabitats	Uluguru Mountains, Tanzania	7° S	Submontane evergreen rain forest, with understory of tree ferns	1415	1773 Total for all epiphytes = 2130			Pócs (1980, 1982)
Epiphytic bryophytes	Bondwa Peak, Uluguru Mountains, Tanzania	7° S	Mossy elfin forest 4- to 6-m-tall trees with 1- to 2.5-m half-woody Acanthaceae	2120	Total on branches, trunk and roots = 10,300 Total for all epiphytes = 13,650			Pócs (1980, 1982)
Epiphytic lichens and bryophytes	Northwest U.S. (Mt. Baker, Washington)	49° N	Low-elevation open coniferous forest on lava flow, dominant *Abies lasiocarpa*	Site I 660 Site II 550		I 846 581 419	II 1600 479 4520	Rhoades (1981)
Epiphytic bryophytes	Temperate, Belgium?	51° N ?	Oak	100?	335			Schnock (1972), in Pócs (1982)
Epiphytes, mostly G lichens (*Bryoria, Tuckermannopsis, Parmelia, Evernia, Usnea*)	Black Lake, northern Saskatchewan	59° N	Spruce, pine	300	G 1460 avg. Spruce 1078 Pine 1846 Bryo ? 0			Scotter (1962) Values adjusted down 10% to approx. oven-dry weight
Epiphytic mosses on trunks to 1.5 m	Síkfőkút, near Eger, northern Hungary	48° N	Turkey oak (*Quercus petrae*)-sessile oak (*Q. cerris*)	300	41.4			Simon (1974)

374

Epiphyte description	Location	Latitude	Forest / tree	Elevation (m)	Biomass / production	Type[b]	Reference
Epiphytic lichens (*Alectoria sarmentosa* and 3 *Bryoria* spp.)	Oulanka, northern Finland	68° N	*Pinus sylvestris*	300?	150 ?, assuming the 15 kg ha^{-1} yr^{-1} given is 10% of standing crop		Sulkava and Helle (1975) in Esseen (1985)
Total epiphytes, including vascular bryophytes, and lichens	Jamaica	18° N	Upper montane forests: Mor ridge forest / Mull ridge forest	1615 / 1525	2100 / 370		Tanner (1980)
Epiphytic lichens	Northwest U.S. (Cedar River Watershed, Washington)	47° N	Temperate, montane conifer, dominant *Abies amabilis*	1200	1900 / n.s.	G / Bryo	Turner and Singer (1976)
Mainly bryophytes	Colombia	5° N	Montane rain forest with *Weinmannia* spp.	3370	12,000		Veneklaas *et al.* (1990) (data from J. Wolf, personal communication)
Epiphytic lichens	Cape Breton, Nova Scotia	47° N	Boreal forest, balsam fir–black spruce	400 m ?	123		Wein and Speer (1975)
Epiphytic bryophytes in canopy only	Colombia	5° N	Wet montane forest	200 m ?	5.5	g dm^{-2} tree surface (approximate mean values taken from scatter of sub-sample points at each elevation in Fig. 9)	Wolf (1993c) Because data are given per tree surface area, not included in Fig. 2
				1000	1		
				1200	2		
				1500	1.5		
				1725	2		
				1990	2.5		
				2150	2.5		
				2460	0.5		
				2550	5.5		
				2700	7.5		
				2970	7.5		
				3200	7.5		
				3370	7.5		
				3525	8.0		
				3670	7.5		

[a] See Fig. 2 for a graphic representation of most of these data. [b] BG = cyanolichen; G = chlorolichen; Bryo = bryophyte. [c] n.s. = not sampled.

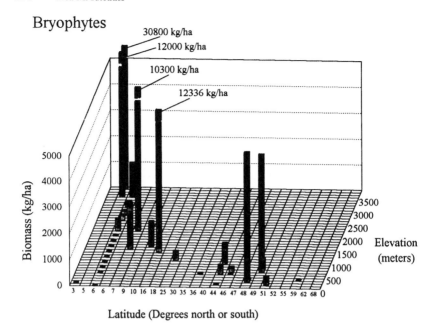

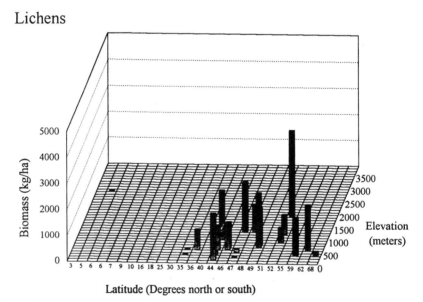

Figure 2 Distribution of worldwide epiphyte biomass by latitude (to nearest degree) and elevation (nearest 100 m). Note the uneven latitude scale. Data from sources listed in Table III.

Vascular epiphytes

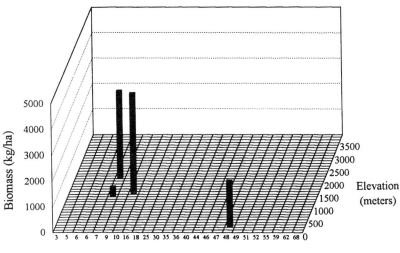

Total epiphytes

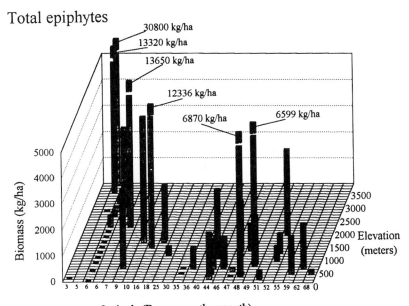

Figure 2—*Continued*

cussed biogeographical problems in lichens of Austral and Pacific tropical distributions.

Rates of evolution within the cryptogamic groups are generally considered extremely slow, because rates of regional endemism at the familial and even generic levels for both lichens and bryophytes are relatively low. Generic endemism is virtually zero for lichens in Hawaii (Smith, 1991) and in New Zealand (Galloway, 1979), and for hepatics in New Guinea (Piipo *et al.*, 1987). Endemism of species, however, is strong. In New Guinea, 48% of liverwort species and 23% of moss species are endemic (Piipo *et al.*, 1987). In lowland tropics, where endemism is high for vascular plants, there are very few endemic lichens and bryophytes (Sipman and Harris, 1989).

A problem that plagues floristic studies is the lack of worldwide monographs that define species limits within genera. When Piipo *et al.* (1987) determined endemism of mosses from habitats in New Guinea, they found lower overall endemism than previously thought (23% of species are endemic, rather than 40–50%), because of more recent synomyzing of species from different geographical areas. Until adequate study is made of all genera, considerations of evolutionary interrelationships among floras will be incomplete.

Endemic groups of families and genera occur over broad regions in the Southern Hemisphere. Regional floras of South Australia, Tasmania, New Zealand, temperate South America, and intervening islands have been discussed in light of continental evolution due to plate tectonics (Van Zanten and Pócs, 1981; Schuster, 1982; Galloway, 1988b, 1991b,c; Schofield, 1992). These areas were united in the paleocontinent Gondwanaland but remained separate from the current, north temperate areas that made up the paleocontinent Laurasia. Endemism restriction to this region occurs for liverworts (Schuster, 1982), for most mosses (although some show bipolar distributions; Catcheside, 1982; Ochi, 1982), and for lichens (Galloway, 1991c). The evolution of the moss *Macromitrium* in xerophytic canopy habitats in rain forest communities in Australia and New Zealand is complex; the present diversity of species evolved over the last 80 million years in geographic isolation from one another after Gondwanaland's breakup (Vitt and Ramsay, 1985). For the Pacific Coast of North America, Schofield (1984, 1992) suggested that prehistoric patterns of glaciation and climate change in the Holarctic region of the Laurasian supercontinent were largely responsible for present distributions.

Other factors play more immediate roles in the evolution of cryptogamic floras. The existence of long-range dispersal may explain transoceanic and other disjunct distributions of cryptogams. Van Zanten and Gradstein (1987, 1988) studied the effects of freezing on viability of Neotropical liverwort spores and gemmae and found generally better viability for propagules of species that show disjunct distributions. Long-range dispersal may function at low altitudes (spores caught in moist air currents in storms), but would not function at jet-stream altitudes that are too cold and dry. For old species with transoceanic distributions, an ancient, step-by-step mechanism of dispersal is the only explanation. The formation of the rich *Liquidambar* "cloud" forest bryophyte flora in eastern Mexico has been attributed entirely to the differential arrival of species by step-by-step mechanisms followed by elevational displacement (Delgadillo, 1979).

C. Canopy Habitats in Special Areas

As cryptogamic botanists learn more about the phenotypic and genotypic plasticity of individual species, special forest types (including natural edges of distributions) should be investigated for unique races of lichen and bryophyte species. Forests associated with thermal or volcanic activity may harbor unique genotypes that are less sensitive to sulfur dioxide than other conspecifics. The depauperate lichen floras in canopies of lodgepole pines of thermal areas in Yellowstone National Park have been attributed to particulates and dissolved materials in geyser sprays (Eversman, 1990). Effects of volcanic ash deposition from Mt. St. Helens on the elemental content of *Lobaria oregana* in surrounding forests included reductions in potassium, phosphorus, and photosynthetic capacity, and increases in silicon and aluminum, but no significant increase in sulfur (Moser *et al.*, 1983).

Concern about the dwindling supply of old-growth forests has elicited studies on their biota with a focus on old-growth indicators. Lichenologists have begun to define old-growth elements in the lichen floras (in Great Britain: Rose, 1976; Gilbert, 1984; in the northwestern United States and adjacent Canada: Lesica *et al.*, 1991; Goward, 1992; Goward and Goffinet, 1993; Thomas *et al.*, 1993). The probable ancient distribution of the rare (in Norway) lichen *Usnea longissima,* an old-growth indicator was discussed by Gauslaa *et al.* (1992). Some rare lichen species, primarily of forest canopies, are associated only with extremely old, self-perpetuating "antique" forests (Goward, 1992). It is only after 1000+ years that these forests develop optimum characteristics. Higher genetic diversity among populations of all cryptogams (even those of younger forests) may require long periods of slow introductions of asexual propagules by which many of these species reproduce (During and Van Tooren, 1987).

D. Changes in Community Structure Due to Stress

The loss of lichen and bryophyte communities in regions of polluted air and acid rain is a worldwide problem. Because of their extreme and variable sensitivities to toxic substances in the air, epiphytic lichens and bryophytes are routinely used to monitor air quality. Individuals in canopy habitats are generally the first to experience the effects of air quality changes. There is a considerable literature on these subjects (for lichens, see bibliographies by Henderson in *The Lichenologist*) (Lawrey, 1984; Glime and Keen, 1984; Farmer *et al.*, 1992). In parts of the eastern United States and Britain, lichens are reinvading canopy habitats following improvements in ambient air quality (Showman, 1981, 1990; Gilbert 1992).

E. Conservation Efforts

In conjunction with floristic study of canopy cryptogams, efforts to deal with the conservation of unique and endangered species and floras are increasing (Dalby, 1988; Gradstein, 1992; Koponen, 1992; Söderström, 1992; During, 1992a). Canopy communities of cryptogams are particularly susceptible to disruptions by human activity because of the logging of old forests that may have a unique cryptogamic flora (see the foregoing) and because the canopy environment is the most prone to the deleterious effects of fragmentation (Bierregaard *et al.*, 1992) and air pollutants (Hawksworth and Hill, 1984). A related problem is the invasion of foreign species and expansion of distributions of native species, particularly in regions of human activity (Hyvönen *et al.*, 1987; Söderström, 1992). This is largely a feature of disrupted or urbanized regions involving terrestrial species but can also happen in epiphytic communities, particularly where nonnative forests are planted. In regions where many forests have been removed, human activity in the form of tropical garden maintenance has been an important factor in preserving some tropical species (Arvidsson, 1991b). Other studies that provide insight into the conservation of cryptogams include Ramsay *et al.* (1987), Söderström (1988), Norris (1990), Smith (1991), and Rose (1992). Unfortunately, the threats to the biodiversity of cryptogamic epiphytes are seen as a critical problem only among those most directly involved in conservation (December 1992 issue of *BioScience*).

IV. Population and Community Ecology

Study of populations and communities can help clarify regional floristic distributions and the dynamics of mixed species within communities. Aside from descriptions of community components and "role-oriented approaches," these studies fall into two types: those that uncover common theoretical frameworks that underlie the relationship of species to their environment, and those that examine population dispersal and demography.

A. Common Frameworks

One approach to the study of populations is to uncover common frameworks that underlie the behavior of populations of similar species. These theoretical approaches have helped focus on important ecological relationships (e.g., competition for and capture of resources, avoidance of stress, reproductive effort) that explain cryptogam distribution and functional roles (Bowler and Rundel, 1975; During, 1979, 1992b; Rogers, 1990; Grime *et al.*, 1990). Many of the resulting generalizations are based on studies of temperate terrestrial (or low-trunk epiphytic) habitats.

Epiphytes of vertical trunks are said to fit the "short-lived shuttle" life strategy, which characterizes species with short-perennial life spans, moderate sexual and asexual reproduction at relatively early age, and large spores and asexual diaspores, each with short life spans (During, 1992b; During and Van Tooren, 1987). The "perennial stayer" strategy for epiphytes would fit the canopy species of upper surfaces of large branches, as this habitat is essentially terrestrial. This strategy is characterized by long-perennial life spans, low to nearly absent sexual and asexual reproduction, variable age of first reproduction, and small spores and diaspores. Although temperate species of epiphytes colonizing foliage and twigs fit into the "colonist" strategy (annual to short-perennial life spans, numerous, small spores), tropical species that exhibit much greater diversity of morphological and life history adaptations were not included in During's analysis. Richards (1988) discussed bryophytes of tropical forests and classified overall habitat types but not ecological strategies.

A graphical approach to generalizing ecological positions of bryophyte and lichen populations ordinates species according to various strategies on triangular plots (Grime *et al.*, 1990; Rogers, 1990). Grime's method positions species according to scores estimated for resource competition, stress tolerance, and disturbance tolerance. Bryophytes exhibit stress-tolerant and disturbance-tolerant positions, and rarely, if ever, actively compete for resources (Grime *et al.*, 1990). Most canopy species occupy stress-tolerant positions, with foliicolous species in the disturbance-tolerant positions. Rogers (1990) used a morphology index to ordinate lichen species along a stress–tolerance–competitive axis, and used relative growth rate to ordinate along the disturbance–tolerance axis. This ordination was used to analyze species patterns with regard to growth form, substratum preference, nature of asexual reproduction, and taxonomic position. No overall trends segregated by canopy habitat among the few canopy species analyzed. This type of analysis could be used to compare ecological strategies among canopy species in a given area.

B. Dispersal and Demography

The dynamics of canopy cryptogam populations is complex. Phorophyte surfaces continually add new surfaces for colonization and change the nature of the existing surfaces. Associated organisms also change. The open canopies of temperate forests can be exposed to strong winds that continually remove larger thalli, opening surfaces to recolonization. In the closed canopies of tropical rain forests, wind may be a factor less frequently. In all canopies, the weight of epiphytes may cause them to fall from the phorophyte, clearing new surface areas for colonization.

Population biologists have investigated demographic processes and structures of whole populations of individual species, including the mechanisms, rates, and directions of dispersal of sexual and asexual propagules, recruitment of new individuals, and the growth and loss of members of the populations. Lichenologists and bryologists have mainly considered population dynamics of individual cryptogams in noncanopy habitats owing to access problems (Schofield, 1985; Armstrong, 1988).

Lichen thalli and bryophyte gametophytes have evolved numerous means of asexual reproduction. For the lichens, this ensures that all partners in the thallus are dispersed together, which bypasses the vagaries of recombining the separate partners. There is likely strong adaptive pressure to reproduce rapidly and abundantly in canopy habitats. Among vascular epiphytes, the number of propagules a species produces has been related to the success that species has in colonizing new habitat (Yeaton and Gladstone, 1982). For lichens and bryophytes, asexual reproduction maximizes the colonizing potential of a single, aerially introduced spore in an environment that is likely to offer many identical habitats. Local populations of these organisms are likely to be composed of mixed collections of genetically distinct individuals and genetic clones.

The study of population biology of organisms that reproduce asexually is in its infancy (Jackson *et al.,* 1985), and for canopy cryptogams there are additional unknowns. The relative roles of sexual and asexual reproduction by most lichen and bryophyte species are unclear. Little is known of the genetic structure of lichen populations in canopies, although molecular techniques may clarify this area of study (Culberson *et al.,* 1993; DePriest, 1993). A few intriguing cases of genetic heterogeneity within lichen thalli have been shown, suggesting that the mycobionts of some lichen thalli might be better viewed as populations of nuclei than as genetic individuals (Shaw, 1992).

Bryophytes are unique among terrestrial plants in exposing haploid tissue to the rigors of perennial life (in contrast to vascular plants that have dominant sporophytes). This phenomenon, coupled with the tendency toward self-fertilization, would lead many species to exhibit genetically uniform ecotypes within local environments. In the few (terrestrial) bryophyte species that have been studied using isoenzymes and chromosome markers, however, populations show considerable genetic heterogeneity, comparable to that in flowering plant populations (During and Van Tooren, 1987; Hofman, 1991). Individual genotypes have broad tolerances, allowing species to acclimatize to environmental extremes within their geographical ranges.

Propagule dispersal and recruitment are the most difficult aspects of demographic studies to accurately determine. Propagule dispersal by lichens mainly involves establishment of new thalli by asexual propagules (Armstrong, 1988; Gradstein, 1992). Armstrong studied dispersal of soredia by the epiphytic lichen *Hypogymnia physodes* (Armstrong, 1987, 1990, 1991, 1992); asexual propagule research and deposition depend on the complex pattern of prior production, accumulation, and release from thalli. Deposition correlates with wind speed and seasonal changes in temperature and humidity. In old-growth conifers in northwestern North America, where winter storms are severe, lichen thallus dispersion is both horizontal and upwardly vertical. Pendulous, fruticose lichens seem to disperse mostly by fragmentation and wind dispersal of the fragments that reentangle in nearby branches (Rhoades, 1983). This occurs throughout the year for species of *Usnea, Alectoria,* and *Bryoria* in Sweden (Esseen, 1985).

Lawrey (1980) examined the incidence of sexual, asexual (propagule, not thallus fragmentation), and mixed reproduction in the lichen *Parmotrema,* as related to latitude. Most tropical species reproduce either sexually or asexually, but not both in the same species. Most temperate species use a mixture of asexual and sexual reproduction. The quality of the propagule rain (the relative representation of a species in the total propagules produced), and not the environment, determines where the species richness is the greatest along an altitudinal gradient in Colombia (Wolf, 1993c). Pike (1971) discussed the patterns of sexual and asexual reproduction among different species on oak and ash twigs; most species reproduce strictly asexually and begin doing so within 2 to 5 years after establishment.

Most studies of local dispersal by bryophytes have been of terrestrial species (Schofield, 1985; Gradstein, 1992; Richards, 1988). Among bryophytes of tropical canopies, spores are less important in inner canopy areas, where the air is still. Rain-splash dispersal mechanisms and asexual reproduction by gemmae and by fragmented thalli carried about by birds and other animals are more important. Schofield (1984) analyzed reproductive tendencies among bryophytes, showing disjunct and endemic distributions on the Pacific Coast of North America. Asexual reproduction is relatively insignificant and there is unusually high representation of the dioicous condition. This refutes the argument for long-distance dispersal and the more dominant role of climatic changes in the north in separating once-congeographic bryophyte floras. For terrestrial species, asexual propagation is very common among bryophytes, particularly in dioicous taxa. Long periods of time of reproduction by asexual propagules and intergrowth of different genotypes can account for the observed genetic variability in populations of bryophytes (During and Van Tooren, 1987).

Rhoades' (1978, 1983) demographic study of the large, foliose cyanolichen *Lobaria oregana* in canopies of old-growth Douglas fir in northwestern North America is the only study that details an individual lichen species in a canopy habitat. Although *Lobaria* reproduces sexually, dispersion within the habitat seems to be primarily by asexual reproduction (no thalli were observed that were smaller than the smallest asexual propagules). A demographic model of the population exhibits dynamic stability following loss of large numbers of larger thalli that mimic the effects of a catastrophic storm.

C. Community Structure and Succession

The structure of canopy cryptogam communities has been the focus of numerous studies, usually as part of the more recent floristic inventories (Table I). Such communities tend to be heterogeneous, reflecting the heterogeneity of habitats. Especially in the tropics, heterogeneity also exists within the same habitat and is probably due to the role of chance in the establishment of new individuals in newly available habitat.

Overall comparisons of community structure in different floras is beyond the scope of this article. For a few representative floras of the world, the relative species richnesses apportioned among the cryptogamic groups (minus crustose species) is compared in Table II. The general trend is for bryophytes dominating over lichens in the tropics, with greater numbers of liverworts than mosses, and lichens dominating over bryophytes in the temperate zone, with mosses being better represented than liverworts. This pattern is also reflected in biomass surveys (see Section V).

Cryptogam communities change in composition on branch surfaces as the surfaces age. This successional progression of communities has mainly been studied in temperate forests, where surfaces can be more easily aged (Smith, 1982; Lawrey, 1984). Lawrey (1991a) examined lichen succession with respect to trends that occur during the succession of higher plant communities. Studies in deciduous tree canopies in Sweden provided some of the earliest data on lichen succession (Degelius, 1964, 1978). Yarranton (1972) studied differences in lichen communities on canopy substrates of different age in black spruce in northern Ontario, Canada. Succession of cryptogams on oak and ash twigs up to 20 years of age was investigated in western Oregon (Pike, 1971; Stone, 1989). In a tropical montane rain forest, a fast-growing pioneer tree (*Brunellia*) carried distinct epiphyte vegetation in comparison to older primary rain forest trees (Wolf, 1994).

Patterns of change vary depending on the physical characteristics of the phorophyte and how these change as substrates in the canopy age. Changes include the rate at which shading increases (increasing humidity, decreasing temperature), the flakiness of the maturing bark, and if and when seasonal leaf or needle loss occurs. These changes are primarily caused by the physical conditions of the growing phorophytes and only partly by competitive interactions and environmental changes caused by the epiphytes themselves. In general, the crustose lichens are replaced by small- to medium-sized foliose and fruticose lichens that are replaced by large foliose lichens (often with blue-green photobionts) and bryophytes.

Succession occurs in whole forests as the phorophyte community changes over time. Lesica *et al.* (1991) studied differences in lichen and bryophyte (mostly liverwort) community inhabitants and structure between old-growth and second-growth conifers in Montana. Old-growth canopies are dominated by *Alectoria* and second-growth by *Bryoria;* nitrogen-fixing lichens are much more common in all strata (terrestrial and canopy) in old-growth forests. McCune (1993) documented patterns of distribution and abundance of four general groups of cryptogams (cyanolichens, alectorioid lichens, other lichens, and bryophytes) in conifer forests of different ages in western Oregon and Washington. He suggested an overall "gradient hypothesis" to relate the changes observed between forests of different age to general changes seen in succession within trees of a single age. Epiphyte species were ordered similarly among three types of spatial and temporal gradients: (1) height within a stand of given age, from high to low; (2) stands differing in moisture, from dry to wet, of a given age; and (3) stands differing in age, from young to old.

V. Biomass and Biomass Production

A. Worldwide Biomass Estimates

Estimates of total epiphyte biomass elicit strong interest and, for lichens, have been reviewed by Boucher and Stone (1992); a worldwide picture can be displayed (Table III and Fig. 2). Techniques for estimating the numbers and their statistical reliability vary, but the general patterns regarding the relative abundance of the different groups of epiphytes in forests in different worldwide regions are clear. In the tropics, epiphyte communities are dominated by vascular plants in the lowlands and bryophytes at higher altitudes, with lichens playing minor roles in the epiphyte biomass in those locations. As one moves farther north or south, lowland epiphytes become more bryophyte-dominant and in boreal regions lichen-dominant. High-elevation "cloud forests" are bryophyte-dominant in the tropics with lichens gradually taking over in importance as one moves farther north or south. In the moist tropics and subtropics, vascular epiphyte biomass is relatively greater than that for cryptogams at lower elevations; the reverse is true at higher elevations. Except for some moist coastal forest ecosystems, vascular epiphyte biomass is negligible in the temperate zone. In both temperate coastal forests and high-elevation tropical forests, epiphytic biomass may be greater than leaf biomass of the phorophytes (Nadkarni, 1983, 1984a; Hofstede *et al.,* 1993).

B. Biomass Production

Care must be taken in interpreting early reports of productivity. Some authors (Scotter, 1962) use the term productivity to refer to standing crop biomass. Nearly all information on bryophyte productivity is based on studies of mosses, and most often from terrestrial species (Smith, 1982; Schofield, 1985). Jacobsen (1978, in Smith, 1982, and in Frahm, 1990d) estimated corticolous moss production at 7.6% of the standing crop (= 4.7 kg ha^{-1} yr^{-1}), based on trunk samples to 2 m. This compares with ranges of productivity of mosses in terrestrial ecosystems of 200 to 3500 kg ha^{-1} yr^{-1} in arctic tundras, 400 to 2000 kg ha^{-1} yr^{-1} in temperate forests, and 450 to 7900 kg ha^{-1} yr^{-1} in temperate bogs (Longton, 1984). Annual production of two epiphytic mosses on montane oak and cedar in the Himalayas was estimated at 28.8 and 46.4% respectively (Pande and Singh, 1988). In a cloud forest at 1550 m in Costa Rica, the bryophyte component of fine litterfall intercepted within the canopy was measured at 240 kg ha^{-1} yr^{-1} (Nadkarni and Matelson, 1991) and the total bryophyte litterfall measured at the forest floor was 266 kg ha^{-1} yr^{-1} (Nadkarni and Matelson, 1992).

More data are available on lichen productivity and are reviewed in Pike (1978) and Boucher and Stone (1992). Values are estimated either by measuring litterfall or by calculating growth of standing crop from known rates. Litterfall estimates only approximate productivity because they do not include production resulting in net biomass increase that stays in the canopy, nor that going to prelitterfall decomposition or consumption. Production of *Alectoria sarmentosa* on downed trees in high montane spruce–fir forests in British Columbia was estimated at 6.6% of the standing crop (= 50 kg ha^{-1} yr^{-1}) (Edwards *et al.*, 1960). Other estimates for annual production (litterfall) by *A. sarmentosa* in lower forests are 10.5, 16.1, and 14% (Stevenson, 1979). Epiphyte (mostly lichens with green photobionts) productivity on oak and ash twigs up to 20 years of age in western Oregon was estimated at 480 kg ha^{-1} yr^{-1} or 27% of the standing crop (Pike, 1971). Litterfall estimates of epiphytic lichens (ranging from 27 to 94 kg ha^{-1} yr^{-1}) in montane conifer stands in New Hampshire ranged from 3.4 to 75% of standing crop with a mean of 17.1% (Lang *et al.*, 1980). Rhoades (1978, 1983) used a life-table method to estimate productivity for *Lobaria oregana* in old-growth canopies of conifers in western Oregon at 31.1% of the standing crop. This would be equivalent to 158 kg ha^{-1} yr^{-1}, when applied to standing biomass estimates of *Lobaria* in forests of western Oregon (Pike, 1981). Productivity of *Usnea longissima* in two approximately 500-m spruce forests in Sweden was 7 and 10% (116 and 162 kg ha^{-1} yr^{-1}, respectively), based on litterfall measurements (Esseen, 1985). Productivity for *Ramalina menziesii* in coastal oak forests in California was 28.8% of the standing crop (= 203 kg ha^{-1} yr^{-1}) (Boucher and Nash, 1990).

VI. Ecological Roles

A. Water Relations

Epiphytes, particularly bryophytes, play major roles in modifying canopy water regimes. They intercept rainfall, reduce the detrimental effect of torrential rainstorms, intercept fog, add a tremendous surface area to that of the phorophyte, and provide additional inputs of water. During dry periods, the water in hydrated masses of epiphytes evaporates, which maintains high humidity in the canopy long after atmospheric inputs have stopped (Redon and Lange, 1983b, for coastal fog forests in Chile; Perry, 1984, for lowland rain forests in Costa Rica; Veneklaas *et al.*, 1990, for cloud forests in the Colombian Andes; Norris, 1990, for forests on Papua, New Guinea; Pórto, 1992, for lowland rain forests in northeast Brazil; Boucher and Nash, 1990, for coastal oak forests in California).

Pócs (1980, 1982) compared water interception by phorophyte parts, epiphytes, and ground cover in two forests in Tanzania, one submontane at an elevation of 1400 m and the other a cloud-elfin forest at 2100 m. He calculated that canopy microepiphytes (mostly bryophytes) intercept 13,644 and 28,942 liters ha^{-1} per rainfall in the two forests, respectively. Assuming 200 and 250 rainfall events per year, these values represent 273 and 724 mm yr^{-1} in forests that otherwise receive approximately 2500 and 3000 mm yr^{-1}.

The role of lichens in intercepting fog in montane cloud forests in the temperate zone has not been studied, but may be considerable.

B. Nutrient Cycling

Nutrient cycling by cryptogams in terrestrial ecosystems has received considerable attention but similar studies in the canopy are just beginning. Early and continuing work focuses on roles in temperate systems (Lang *et al.*, 1976, 1980; Pike, 1978; Carroll, 1980; Knops *et al.*, 1991; Bates, 1992b; Boucher and Stone, 1992). A discussion of these roles in tropical systems is developing (Nadkarni, 1983, 1984b, 1986; Coxson, 1991; Nadkarni and Matelson, 1991, 1992; Coxson *et al.*, 1992; Matelson *et al.*, 1993).

Canopy cryptogams provide distinct sources of canopy nutrients. Although they may use (or even require) nutrients previously processed by other canopy components, they are mostly dependent on atmospheric (allochthonous) inputs for their biological needs (Table IV). Their importance in nutrient cycling in some ecosystems is dictated in part by their high biomass and surface areas. They can play important roles in nutrient cycling in systems where their biomass is low, because lichens and bryophytes have biological adaptations that allow them to garner nutrients from dilute sources. Because of their adaptations to hold nutrients, epiphytic (and ter-

Table IV   Potential Sources of Nutrient Input to Cryptogamic
Epiphyte Communities in Forest Ecosystems [a]

Autochthonous sources [b]	Allochthonous sources
I. Soil-rooted phytomass	I. Atmospheric
A. Intercepted litterfall	A. Wet deposition
B. Bark decomposition	B. Dry deposition
C. Live and dying foliage	C. Gaseous input (including nitrogen fixation)
II. Animal defecation and death	
III. Cryptogams	
A. Intercepted litterfall	
B. Live and dying thalli	

[a] Modified from Nadkarni and Matelson (1991).
[b] All sources are mediated by leachates in the canopy solution.

restrial) cryptogams may moderate overall nutrient transfer by retaining excess amounts and slowly releasing these nutrients during seasons when the nutrients are in lower supply (Nadkarni, 1984a).

Populations of cryptogams use these inputs to produce biomass and function in life. This biomass enters the forest community in several ways. Primarily, biomass is lost from the populations in the form of litterfall of whole thalli or parts. A minor proportion of the biomass is lost during reproductive loss of spores and asexual structures. These materials drop to the forest floor (or to perched canopy "soil"), where they are consumed or decomposed. *In situ* consumption and decomposition can also occur. Nadkarni and Matelson (1991) discussed the possibility of intra- and intercanopy cycling through epiphyte litterfall and decomposition. In some systems with extremely high cryptogamic epiphyte biomass, the annual input of nutrients in the form of litterfall can be comparable to that from all vascular plant litter (Nadkarni, 1983).

Canopy cryptogams lose materials in leachates that enter the complex "canopy solution" (throughfall and stemflow) through pathways that are only partly understood in a few ecosystems. The canopy solution flows and drips over all canopy surfaces during times of atmospheric water input. Because the canopy solution flows in one general direction (downward), epiphytes in the upper and lower regions of the canopy experience different nutrient regimes and receive nutrients from a different set of sources. The general model developed for nitrogen transfers in forest canopies in northwestern North America (Carroll, 1980) is useful to keep all the potential players in mind for nutrients in all canopy systems (Fig. 3). The balance of influxes to and effluxes from the canopy solution (and ultimately the forest floor) vary for each element, compound, cryptogam species, geographical location, position in the canopy, and season of the year. Nutrients tend to

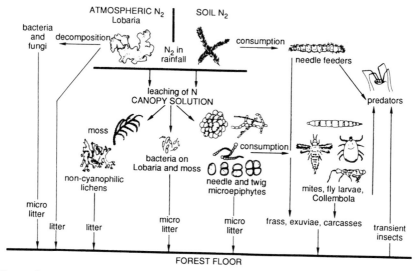

Figure 3 Scheme of nitrogen fluxes in canopies of old-growth conifers in northwestern North America. Reproduced from Carroll, 1980, with permission.

be retained in epiphyte biomass during the dry season and released during the wet season in tropical and temperate rain forest systems.

A recent finding may confound the already complex web of interrelationships of canopy cryptogams. Most lichens were thought to use their substrates only as points of commensal anchorage. Reports on the invasion of *Evernia prunastri* (Legaz *et al.,* 1988; Yague and Estevez, 1988) into the xylem of the phorophyte, and subsequent effects on the development of leaf buds, suggest that some lichens are parasitic. The extent of this relationship among other lichen species is not known. It is known that phorophytes tap into epiphyte mats with canopy roots (Nadkarni, 1981) and short-circuit the normal root input of water and nutrients. If some epiphytes tap into their phorophytes, in-canopy cycling of nutrients may be even more complex.

1. Carbon (Photosynthesis)   Compared to vascular plants, lichens and bryophytes have low photosynthetic rates and low compensation values at their optimum temperatures for photosynthesis (Kallio and Kärenlampi, 1975; Frahm, 1987b, 1990c). Some species are capable of positive net photosynthesis at extremely low light levels (e.g., Green and Lange, 1991, for *Pseudocyphellaria* and *Sticta* spp. in New Zealand rain forest habitats). Cyanobacterial lichens must be hydrated by liquid water to achieve positive assimilatory photosynthetic status (Lange *et al.,* 1993). Lichens and bryophytes are capable of becoming photosynthetically active within minutes of rehydration following a dry period. Optimum photosynthetic rates within a species often differ between individuals at different locations (Kallio and Kärenlampi, 1975) and between different species at different heights in the same tree (Hosokawa and Odani, 1947).

In bryophytes and lichens, carbon is stored in the form of sugar alcohols, partly to store fixed carbon in long-term form and apparently also to provide a mechanism to protect internal cell membranes from damage during dry periods (Coxson *et al.*, 1992). These sugar alcohols leach from thalli into the canopy solution, providing energy sources for commensal fungi, algae, and bacteria elsewhere in the canopy (Coxson *et al.*, 1992; Cooper and Carroll, 1978).

2. Nitrogen (Nitrogen Fixation)   Nitrogen is often a limiting element in forests; numerous studies of the nitrogen-fixing lichens have clarified their roles in forest ecosystems, including the canopy (Forman, 1975; Becker *et al.*, 1977; Pike, 1978; Carroll, 1980). The most complete picture of canopy lichen inputs to nitrogen cycling comes from a series of studies on old-growth conifers in the Pacific Northwest United States (Pike *et al.*, 1972; Denison, 1973; Rhoades, 1977, 1978; Carroll, 1980; Sollins *et al.*, 1980; Horstmann *et al.*, 1982; Edmonds *et al.*, 1991). Carroll's (1980) informative article documented much of the experimental background that has led to an overall model (Fig. 3). Annual fixation of nitrogen by *Lobaria oregana* is estimated at 3.5 kg N ha^{-1} yr^{-1} (Denison, 1979). Using productivity values for this lichen and a 2% thallus nitrogen content, Rhoades (1978) estimated inputs of 3.2 kg ha^{-1} yr^{-1} in the form of litterfall. Any excess fixed nitrogen flows into the canopy solution, from which other cryptogamic epiphytes and other organisms take up nitrogen. In the canopy solution, nitrogen exists as NH_4^+ and in organic form (Carroll, 1980; Edmonds *et al.*, 1991). Net fluxes of nitrogen *from* the canopy solution tend to be positive for all non-nitrogen-fixing epiphytes during dry periods (less than 0.3 cm before collection), and positive for lichens containing green algal photobionts, but negative for mosses during wet periods (Carroll, 1980). Ultimately, nitrogen in the canopy solution is used by surface bacteria, fungi, and algae and they are either consumed or drop to the forest floor as microlitter.

Nitrogen-fixing lichens are less important elements in lowland tropical forests but may be important in some cloud forest systems in the tropics (Forman, 1975). Coxson *et al.* (1992) suggested that the pulsed release of sugars and polyols by cloud forest bryophytes influences nitrogen fixation by other canopy components.

In south temperate areas, particularly the rain forests of New Zealand, Tasmania, and southern South America, the large, foliose nitrogen fixers are often dominant players (Green *et al.*, 1980). These authors suggest that as much as 1–10 kg ha^{-1} yr^{-1} is contributed by these lichens to these forests.

3. Nonnitrogen Elements A number of studies have examined the pools of elements (in addition to nitrogen, usually including P, K, Ca, Mg, Na, and often S) in canopy lichens and bryophytes, particularly in regard to the overall proportion of their ecosystem nutrient capital, their flux rates between other ecosystem components, and the retention times they remain in the epiphytes (Pike, 1971, 1978; Lang *et al.*, 1976, 1980; Sollins *et al.*, 1980; Bosserman and Hagner, 1981; Nadkarni, 1981, 1983, 1984a,b, 1986; Reiners and Olson, 1984; Brown, 1987; Boucher and Nash, 1990; Nadkarni and Matelson, 1991, 1992; Coxson, 1991; Boucher and Stone, 1992; Hofstede *et al.*, 1993). Nutrient capital in cryptogams is generally much lower than that for the remaining forest components, except in cryptogam-dominant tropical cloud forests and temperate rain forests, where they may be of equal magnitude (Nadkarni, 1983, 1984a,b; Hofstede *et al.*, 1993). In temperate forests, net fluxes of nutrient *from* the canopy solution tend to be positive for all epiphytes during dry periods and negative during wet periods. A net uptake of P by the canopy in high-elevation tropical forest has also been documented (Veneklaas, 1990).

C. Community Relationships

1. Relationships with Animals Lichens and bryophytes provide resources for invertebrate and vertebrate animals, including food, camouflage, protection, and oviposition sites (Richardson and Young, 1977; Brodo and Hawksworth, 1977; Gerson and Seaward, 1977; Lawrey, 1987; Longton, 1992; S. Sharnoff, unpublished review, 1994). Many invertebrate animals use epiphyte mats as their homes (Voegtlin, 1982). Stubbs (1989) found a high correlation between the biomass of corticolous lichens and abundance of Arthropoda, Tardigrada, and Rotifera. Liverworts, lichens, and fungi and algae live commensally on the backs of certain New Guinean weevils (Gressitt *et al.*, 1965). Other canopy-dwelling organisms often develop cryptic coloring that camouflages the organism against a naturally lichen-rich background (Seaward, 1988), for example, in England, the well-known cryptically colored, light form moth *Biston betularia* has been shown to be at a disadvantage on lichen-poor tree bark in air-polluted areas. Farkas and Pócs (1989) discussed a case of foliicolous lichen mimicry of a rain forest tree frog. Nadkarni and Matelson (1989) documented the use by birds of canopy bryophytes for water sources, nesting material, and associated invertebrates. Cao and Caihua (1991) listed species of bryophytes (all pleurocarpous mosses) in bird nests in China; S. Sharnoff (unpublished review, 1994) listed North American birds reported to use lichens as nesting material.

Consumption of lichen and bryophyte material is limited, in part because of low caloric values in comparison with vascular plant leaves (Longton, 1992) and because of the presence of toxic compounds (Lawrey, 1987; Longton, 1992). Loria and Herrnstadt (1981) reported on an unusual case of ants harvesting moss capsules of terrestrial mosses. There are other scattered reports of consumption of this potentially rich energy source by both vertebrates and invertebrates (Longton, 1992). Resource use by indigenous arthropods and other herbivores in canopies should be investigated. Sigal (1984) reported an unusual association between *Usnea strigosa* and a lichenophagous larva of the moth *Zanclognatha theralis*. The lichen acids did not deter the larvae; they must either metabolize or sequester the lichen substances (or harbor other organisms that do the same). Other such dependencies are likely among the diverse floras of the tropics where unknown invertebrates abound.

As with terrestrial lichens, arboreal lichens are consumed by vertebrates mainly during times of scarcity of other food sources. The use of arboreal lichens in the diets of caribou, black-tailed deer, white-tailed deer, mountain goat, elk, northern flying squirrel, California red-backed vole, boreal red-backed vole, and numerous other small rodents is well documented (S. Sharnoff, unpublished review, 1994). Lichens, both terrestrial and arboreal, are always a major source of nutrition for caribou, as their digestive systems include microorganisms that help digest the lichens (S. Sharnoff, unpublished review, 1994). Edwards *et al.* (1960), Scotter (1962), and Stevenson (1979) provided estimates of arboreal forage lichen production (mainly *Alectoria sarmentosa* and species of *Bryoria*) in Canadian montane conifer forests.

2. In Situ Parasitism and Decomposition   Although lichens and bryophytes are relatively resistant to attack by parasites, many species of fungi and other lichens use living lichen material as a source of nutrition (Hawksworth, 1983). Little work has assessed the roles of parasitic fungi and bacteria in the loss of individual cryptogams or their parts, or in other ecological interactions with canopy cryptogams.

Longton (1984, 1992) discussed decomposition and consumption of terrestrial bryophytes, but no information is available for canopy species. Once epiphytic lichen material is dead and as long as conditions are suitable for decomposition, rates of decomposition are relatively rapid, particularly for large nitrogen-fixing species (Longton, 1992). Rates of decomposition of species of *Pseudocyphellaria* and *Sticta* in a southern Chilean forest are between 40 and 95% weight loss per year, based on litter bag studies (Guzman *et al.*, 1990). *Lobaria oregana* and *L. pulmonaria* thalli show similar rates of decomposition in western Oregon (conifer and oak forests, respectively), decomposing completely in 1.5 to 2.0 years (Pike, 1971; Rhoades, 1978; McCune and Daly, 1994). Laboratory studies of decomposition of ground lichen material (including species of *Usnea, Parmelia, Pseudocyphellaria,* and *Sticta*) from New Zealand suggest that decomposition rates for lichens with green photobionts are similar (Greenfield, 1993). As much as 10% of the biomass of *Lobaria oregana* in old-growth conifers of western Oregon are necrotic. Rates of decomposition within the canopy may be slower because conditions are less often conducive to decomposition. Discrepancy between the growth estimates of *Lobaria oregana* obtained by population analysis and data from several litterfall studies in western Oregon suggests that *in situ* decomposition occurs (Rhoades, 1978, 1983). Vance and Nadkarni (1990) studied microbial activity in the canopy organic matter component of epiphyte (noncryptogam) mats and compared this with similar activity in the forest floor.

VII. Critically Needed Future Work

A. Personnel

It is ironic that at a time when the existence and roles of the cryptogams are suddenly becoming apparent among other biologists (particularly in applied biology fields), many academic departments are eliminating positions for cryptogamic biologists. Not only are major collections of these organisms losing curatorial support, but there are fewer places where students can be directed to the study of these organisms. This is a recursive problem. Academic departments need to continue to find positions for cryptogamic botanists on their staffs.

B. Monographs and Inventories

Work needs to continue on broad regional inventories of all tropical species and of crustose lichens worldwide. Many authors plead that crustose lichens be included in all floristic and community surveys (Schmitt and Slack, 1990). Gradstein (1991) listed proposed additions to lichen floras of the neotropics. Earlier lists also appear in the International Association of Lichenology (IAL) newsletter and a review of those gives an idea of how fast authors are actually accomplishing their goals. The IAL continues to hold symposia and small conferences on tropical lichenology.

C. Population and Community Ecology

A number of questions about the stability and dynamics of lichen and bryophyte populations and communities need answering. Of prime importance is an increased understanding of the limitations that dispersal (sexual and asexual) and the relative role of chance impose on population stability. Genetic studies of species will also help clarify many questions concerning evolution, local distribution, and mechanisms of reproduction and dispersal.

Demographic studies such as Pike (1971), Rhoades (1983), and Stone (1989) are needed to understand aspects of population stability. Continued study of community structure on common hosts within broadly similar geographical regions (e.g., Palmer, 1986; Schmitt and Slack, 1990; McCune, 1993) will clarify the interrelating environmental factors that produce cryptogam communities. Extending multivariate analyses and analysis of general ecological strategy (Grime *et al.*, 1990; Rogers, 1990; Lawrey, 1991b, 1992) to canopy species from all geographical areas may also clarify the relative importance of these factors.

Estimates of biomass and productivity of canopy cryptogams must include all groups of organisms to document their relative importance, once an adequate inventory of canopy species is made in any area. Estimates of production by lichens in all forest types, particularly in the rich cloud forests throughout the world, will clarify the importance of these organisms in those areas. There are very few studies of epiphyte (of any kind) productivity in the tropics (Nadkarni and Matelson, 1991).

D. Roles

Laboratory studies of physiological function should be coupled with *in situ* measurements of the controlling variables and, if possible, the functions themselves. Work should assess the effects of lichen substances on the establishment, growth, and health of other organisms in canopies. Epiphytic lichen and bryophyte loads in tropical montane cloud forests may have a major impact on cloud interception and water retention and one study suggests that lichens can have a similar effect in the temperate zone (Boucher and Nash, 1990).

E. Conservation

Discussions must continue on the methods to conserve existing bryophyte and lichen floras and to evaluate the consequences of their disruption (Dalby, 1988; Norris, 1990; Pittam, 1991; Söderström *et al.*, 1992; During, 1992a). Studies have shown the value of older trees in forests as habitats for certain sensitive species (Gustafsson *et al.*, 1992; Goward, 1992; Thomas *et al.*, 1993). In all geographical regions, the destruction of old forests will likely require increased attention to this area of knowledge about canopy cryptogams.

Acknowledgments

Several reviewers made helpful suggestions. I particularly thank Dr. Jan Wolf for many comments that improved sections on bryophytes and discussions of community ecology. Lastly, I acknowledge the patience and understanding of my family while I assembled this review.

References

Ahmadjian, V. (1988). The lichen alga *Trebouxia:* Does it occur free-living? *Plant Syst. Evol.* 158, 243–247.

Ahmadjian, V. (1993). The lichen photobiont—What can it tell us about lichen systematics? *Bryologist* 96, 310–313.

Ahti, T. (1977). Lichens of the boreal coniferous zone. *In* "Lichen Ecology" (M. R. D. Seaward, ed.), pp. 145–181. Academic Press, New York.

Almborn, O. (1988). Some distribution patterns in the lichen flora of South Africa. *Monogr. Syst. Bot. Mo. Bot. Gard.* 25, 429–432.

Aptroot, A. (1991). Tropical pyrenocarpous lichens, a phylogenetic approach. *In* "Tropical Lichens: Their Systematics, Conservation, and Ecology" (D. J. Galloway, ed.), pp. 253–274. Oxford Univ. Press, Oxford.

Armstrong, R. A. (1987). Dispersal in a population of the lichen *Hypogymnia physodes*. *Environ. Exp. Bot.* 27, 357–363.

Armstrong, R. A. (1988). Substrate colonization, growth and competition. *In* "Handbook of Lichenology" (M. Galun, ed.), Vol. 2, pp. 3–16. CRC Press, Boca Raton.

Armstrong, R. A. (1990). Dispersal, establishment and survival of soredia and fragments of the lichen, *Hypogymnia physodes* (L.) Nyl. *New Phytol.* 114, 239–245.

Armstrong, R. A. (1991). The influence of climate on the dispersal of lichen soredia. *Environ. Exp. Bot.* 31, 239–245.

Armstrong, R. A. (1992). Soredial dispersal from individual soralia in the lichen *Hypogymnia physodes* (L.) Nyl. *Environ. Exp. Bot.* 32, 55–63.

Arvidsson, L. (1991a). Lichenological studies in Ecuador. *In* "Tropical Lichens: Their Systematics, Conservation, and Ecology" (D. J. Galloway, ed.), pp. 123–134. Oxford Univ. Press, Oxford.

Arvidsson, L. (1991b). On the importance of botanical gardens for lichens in the Asian tropics. *In* "Tropical Lichens: Their Systematics, Conservation, and Ecology" (D. J. Galloway, ed.), pp. 193–200. Oxford Univ. Press, Oxford.

Barkman, J. J. (1958). "Phytosociology and Ecology of Cryptogamic Epiphytes." Van Gorcum, Assen.

Bates, J. W. (1992a). Influence of chemical and physical factors on *Quercus* and *Fraxinus* epiphytes at Loch Sunart, western Scotland: A multivariate analysis. *J. Ecol.* 80, 163–179.

Bates, J. W. (1992b). Mineral nutrient acquisition and retention by bryophytes. *J. Bryol.* 17, 223–240.

Bates, J. W., and Farmer, A. M., eds. (1992). "Bryophytes and Lichens in a Changing Environment." Oxford Univ. Press, New York.

Becker, V. E. (1980). Nitrogen fixing lichens in forests of the southern Appalachian Mountains of North Carolina. *Bryologist* 83, 29–39.

Becker, V. E., Reeder, J., and Stetler, R. (1977). Biomass and habitat of nitrogen-fixing lichen in an oak forest in the North Carolina Piedmont. *Bryologist* 80, 93–99.

Beever, J. E. (1984). Moss epiphytes of tree-ferns in a warm-temperate forest, New Zealand. *J. Hattori Bot. Lab.* 56, 89–95.

Biazrov, L. G. (1971). Distribution of the phytomass of epiphytic lichens in certain types of biogeocoenoses in the broad-leaved-spruce forest subzone. *Lesovedenie* 5, 85–90.

Bierregaard, R. O., Jr., Lovejoy, T. E., Kapos, V., Augusto dos Santos, A., and Hutchings, R. W. (1992). The biological dynamics of tropical rainforest fragments. *BioScience* 42, 859–866.

Bird, C. D., and Marsh, A. H. (1973). Phytogeography and ecology of the lichen family Parmeliaceae in southwestern Alberta. *Can. J. Bot.* 51, 261–288.

Bosserman, R. W., and Hagner, J. E. (1981). Elemental composition of epiphytic lichens from Okefenokee Swamp. *Bryologist* 84, 48–58.

Boucher, V. L., and Nash, T. H., III (1990). The role of the fruticose lichen *Ramalina menziesii* in the annual turnover of biomass and macronutrients in a blue oak woodland. *Bot. Gaz. (Chicago)* 151, 114–118.

Boucher, V. L., and Stone, D. F. (1992). Epiphytic lichen biomass *In* "The Fungal Community: Its Organization and Role in the Ecosystem" (G. C. Carroll and D. Wicklow, eds.), 2nd ed., pp. 583–599. Dekker, New York.

Bowler, P. A., and Rundel, P. W. (1975). Reproductive strategies in lichens. *Bot. J. Linn. Soc.* 70, 325–340.

Brodo, I. M. (1973). Substrate ecology. *In* "The Lichens" (V. Ahmadjian and M. E. Hale, eds.), pp. 401–441. Academic Press, New York.

Brodo, I. M., and Hawksworth, D. L. (1977). *Alectoria* and allied genera in North America. *Opera. Bot.* 42, 1–164.

Brown, D. H. (1987). The location of mineral elements in lichens: Implications for metabolism. *In* "Progress and Problems in Lichenology in the Eighties" (E. Peveling, ed.), pp. 361–375. J. Cramer, Berlin.

Brusse, F. A. (1991). Eight new species in the lichen genus *Parmelia* (Parmeliaceae, Ascomycotina) from southern Africa with notes on southern African lichens. *Mycotaxon* 40, 377–393.

Bryant, E. H., Crandall-Stotler, B., and Stotler, R. E. (1973). A factor analysis of the distribution of some Puerto Rican liverworts. *Can. J. Bot.* 51, 1545–1554.

Buck, W. R., and Thiers, B. M. (1989). Review of bryological studies in the tropics. *In* "Floristic Inventory of Tropical Countries" (D. G. Campbell and H. D. Hammond, eds.), pp. 484–493. New York Botanical Garden, Bronx.

Canters, K. J., Schöller, H., Ott, S., and Jahns, H. M. (1991). Microclimatic influences on lichen distribution and community development. *Lichenologist* 23, 237–252.

Cao, T., and Caihua, G. (1991). First report of bryophyte bird-nests and their bryophytes in China. *Wuyi Sci. J.* 8, 207–213.

Carroll, G. C. (1980). Forest canopies: Complex and independent subsystems. *In* "Forests: Fresh Perspectives from Ecosystem Analysis. Proceedings of the 40th Annual Biology Colloquium" (R. H. Waring, ed.), pp. 87–107. Oregon State Univ. Press, Corvallis.

Catcheside, D. G. (1982). The geographical affinities of the mosses of South Australia. *J. Hattori Bot. Lab.* 52, 57–64.

Coddington, J. A., Griswold, C. E., Dávila, D. S., Peñaranda, E., and Larcher, S. F. (1991). Designing and testing sampling protocols to estimate biodiversity in tropical ecosystems. *In* "The Unity of Evolutionary Biology: Proceedings of the Fourth International Congress of Systematic and Evolutionary Biology" (E. Dudley, ed.), Vol. 1. Dioscorides Press, Portland, OR.

Coleman, B. B., Muenscher, W. C., and Charles, D. R. (1956). A distributional study of the epiphytic plants of the Olympic Peninsula, Washington. *Am. Midl. Nat.* 56, 54–87.

Cooper, S., and Carroll, G. C. (1978). Ribitol as a major component of water-soluble leachates from *Lobaria oregana*. *Bryologist* 81, 568–572.

Cornelissen, J. H. C., and ter Steege, H. (1989). Distribution and ecology of epiphytic bryophytes and lichens in dry evergreen forest of Guyana. *J. Trop. Ecol.* 5, 131–150.

Coxson, D. S. (1991). Nutrient release from epiphytic bryophytes in tropical montane rain forest (Guadeloupe). *Can. J. Bot.* 69, 2122–2129.

Coxson, D. S., Webber, M. R., and Kershaw, K. A. (1984). The thermal operating environment of corticolous and pendulous tree lichens. *Bryologist* 87, 197–202.

Coxson, D. S., McIntyre, D. D., and Vogel, H. J. (1992). Pulse release of sugars and polyols from canopy bryophytes in tropical montane rain forest (Guadeloupe, French West Indies). *Biotropica* 24, 121–133.

Culberson, W. L., Culberson, C. F., and Johnson, A. (1993). Speciation in lichens of the *Ramalina siliquosa* complex (Ascomycotina, Ramalinaceae): Gene flow and reproductive isolation. *Am. J. Bot.* 80, 1472–1481.

Dalby, K. (1988). Lichen conservation news. *Br. Lichen Soc. Bull.* 63, 29–30.

Davis, R. B. (1964). Bryophytes and lichens of the spruce-fir forests of the coast of Maine. II. The corticolous flora. *Bryologist* 67, 194–196.

Degelius, G. (1964). Biological studies of the epiphytic vegetation on twigs of *Fraxinus excelsior*. *Acta Horti Gotob.* 27, 11–55.

Degelius, G. (1978). Further studies on the epiphytic vegetation on twigs. *Acta Univ. Gothoburgensis* 7, 1–58.

Delgadillo M., C. (1979). Mosses and phytogeography of the *Liquidambar* forest of Mexico. *Bryologist* 82, 432–449.

Delgadillo M., C., and Cárdenas S., A. (1989). Phytogeography of high-elevation mosses from Chiapas, Mexico. *Bryologist* 92, 461–466.

Denison, W. C. (1973). Life in tall trees. *Sci. Am.* 228, 74–80.

Denison, W. C. (1979). *Lobaria oregana*, a nitrogen-fixing lichen in old-growth Douglas fir forests. *In* "Symbiotic Nitrogen Fixation in the Management of Temperate Forests" (J. C. Gordon, C. T. Wheeler, and D. A. Perry, eds.), pp. 266–275. Oregon State University School of Forestry, Corvallis.

Denison, W. C., Tracy, D. M., Rhoades, F. M., and Sherwood, M. (1972). Direct, nondestructive measurement of biomass and structure in living old-growth Douglas-fir. *In* "Research on Coniferous Forest Ecosystems: First Year Progress in the Coniferous Forest Biome, US/IBP" (J. F. Franklin, L. J. Dempster, and R. H. Waring, eds.), pp. 147–158. Pacific Northwest Forest and Range Experiment Station, Forest Service, U.S.D.A., Portland, OR.

DePriest, P. T. (1993). Variation in the *Cladonia chorophaea* complex I: Morphological and chemical variation in southern Appalachian populations. *Bryologist* 96, 555–563.

Dey, J. P. (1976). Phytogeographic relationships of the fruticose and foliose lichens of the southern Appalachian Mountains. *In* "The Distributional History of the Biota of the Southern Appalachians. Part IV. Algae and Fungi. Biogeography, Systematics, and Ecology" (B. C. Parker and M. K. Roane, eds.), pp. 398–416. Univ. Press of Virginia, Charlottesville.

During, H. J. (1979). Life strategies of bryophytes: A preliminary review. *Linbergia* 5, 2–18.

During, H. J. (1992a). Endangered bryophytes in Europe. *Trends Ecol. Evol.* 7, 253–255.

During, H. J. (1992b). Ecological classifications of bryophytes and lichens. *In* "Bryophytes and Lichens in a Changing Environment" (J. W. Bates and A. M. Farmer, eds.), pp. 1–31. Oxford Univ. Press, Oxford.

During, H. J., and Van Tooren, B. F. (1987). Recent developments in bryophyte population ecology. *Trends Ecol. Evol.* 2, 89–93.

Edmonds, R. L., Thomas, T. B., and Rhodes, J. J. (1991). Canopy and soil modification of precipitation chemistry in a temperate rain forest. *Soil Sci. Soc. Am. J.* 55, 1685–1693.

Edwards, P., and Grubb, P. J. (1977). Studies of mineral cycling in a montane rain forest in New Guinea. *J. Ecol.* 65, 943–969.

Edwards, R. Y., Soos, J., and Ritcey, R. W. (1960). Quantitative observations on epidendric lichens used as food by caribou. *Ecology* 41, 425–431.

Esseen, P. (1985). Litter fall of epiphytic macrolichens in two old *Picea abies* forests in Sweden. *Can. J. Bot.* 63, 980–987.

Eversman, S. (1982). Epiphytic lichens of a ponderosa pine forest in southeastern Montana. *Bryologist* 85, 204–213.

Eversman, S. (1990). Lichens of Yellowstone National Park. *Bryologist* 93, 197–205.

Eversman, S., Johnson, C., and Gustafson, D. (1987). Vertical distribution of epiphytic lichens on three tree species in Yellowstone National Park. *Bryologist* 90, 212–216.

Farkas, E., and Pócs, T. (1989). Foliicolous lichen–Mimicry of a rainforest treefrog? *Acta Bot. Hung.* 35, 73–76.

Farmer, A. M., Bates, J. W., and Bell, J. N. B. (1992). Ecophysiological effects of acid rain on bryophytes and lichens. *In* "Bryophytes and Lichens in a Changing Environment" (J. W. Bates and A. M. Farmer, eds.), pp. 284–313. Oxford Univ. Press, New York.

Forman, R. T. T. (1975). Canopy lichens with blue-green algae: A nitrogen source in a Colombian rain forest. *Ecology* 56, 1176–1184.

Frahm, J. (1987a). Which factors control the growth of epiphytic bryophytes in tropical rainforests? *Symp. Biol. Hung.* 35, 639–648.

Frahm, J. (1987b). Ökologische Studien über die epiphytische Moosvegetation in Regenwäldern NO-Perus. *Beih. Nova Hedwigia* 88, 143–158.

Frahm, J. (1990a). The ecology of epiphytic bryophytes on Mt. Kinabalu, Sabah (Malaysia). *Nova Hedwigia* 51, 121–132.

Frahm, J. (1990b). The altitudinal zonation of bryophytes on Mt. Kinabalu. *Nova Hedwigia* 51, 133–149.

Frahm, J. (1990c). The effect of light and temperature on the growth of the bryophytes of tropical rain forests. *Nova Hedwigia* 51, 151–164.

Frahm, J. (1990d). Bryophyte phytomass in tropical ecosystems. *Bot. J. Linn. Soc.* 104, 23–33.

Frey, W., Gossow, R., and Kürschner, H. (1990). Verteilungsmuster von Lebensformen, wasserletenden und wassspeichernden Strukturen in epiphytischen Moosgesellschaften am Mt. Kinabalu (Nord-Borneo). *Nova Hedwigia* 51, 87–119.

Fulford, M. (1951). Distribution patterns of the genera of leafy Hepaticae of South America. *Evolution (Lawrence, Kans.)* 5, 243–264.

Galloway, D. J. (1979). Biogeographical elements in the New Zealand lichen flora. *In* "Plants and Islands" (D. Bramwell, ed.), pp. 201–224. Academic Press, London.

Galloway, D. J. (1987). Austral lichen genera: Some biogeographical problems. *In* "Progress and Problems in Lichenology in the Eighties" (E. Peveling, ed.), pp. 385–399. J. Cramer, Berlin.

Galloway, D. J. (1988a). Studies in *Pseudocyphellaria* (lichens). I. The New Zealand species. *Bull. Br. Mus. (Nat. Hist.), Bot. Ser.* 17, 1–267.

Galloway, D. J. (1988b). Plate tectonics and the distribution of cool temperate Southern Hemisphere macrolichens. *Bot. J. Linn. Soc.* 96, 45–55.

Galloway, D. J., ed. (1991a). "Tropical Lichens: Their Systematics, Conservation, and Ecology." Oxford Univ. Press, New York.

Galloway, D. J. (1991b). Biogeographical relationships of Pacific tropical lichen floras. *In* "Tropical Lichens: Their Systematics, Conservation, and Ecology" (D. J. Galloway, ed.), pp. 1–16. Oxford Univ. Press, Oxford.

Galloway, D. J. (1991c). Phytogeography of Southern Hemisphere lichens. *In* "Quantitative Approaches to Phytogeography" (P. L. Nimis and T. J. Crovello, eds.), Tasks Veg. Sci., No. 24, pp. 233–262. Kluwer Academic Publishers, Dordrecht, The Netherlands.

Galloway, D. J. (1992). Lichens of Laguna San Rafael, Parque Nacional 'Laguna San Rafael,' southern Chile: Indicators of environmental change. *Global Ecol. Biogeogr. Lett.* 2, 37–45.

Galloway, D. J., and Arvidsson, L. (1990). Studies in *Pseudocyphellaria* (lichens). II. Ecuadorian species. *Lichenologist* 22, 103–135.

Gauslaa, Y., Anonby, J., Gaarder, G., and Tønsberg, T. (1992). Huldrestry, *Usnea longissima*, en sjelden urskogslav på Vestlandet. *Blyttia* 50, 105–114.

Gerson, U., and Seaward, M. R. D. (1977). Lichen–invertebrate associations. *In* "Lichen Ecology" (M. R. D. Seaward, ed.), pp. 69–119. Academic Press, New York.

Gilbert, O. L. (1984). Some effects of disturbance on the lichen flora of oceanic hazel [*Corylus avellana*] woodland. *Lichenologist* 16, 21–30.

Gilbert, O. L. (1992). Lichen reinvasion with declining air pollution. *In* "Bryophytes and Lichens in a Changing Environment" (J. W. Bates and A. M. Farmer, eds.), pp. 159–177. Oxford Univ. Press, Oxford.

Glime, J. M., and Keen, R. E. (1984). The importance of bryophytes in a man-centered world. *J. Hattori Bot. Lab.* 55, 133–146.

Golley, F., McGinnis, J., and Clements, R. (1971). La biomasa y la estructura de algunos bosques de Darién, Panama. *Turrialba* 21, 189–196.

Gough, L. P. (1975). Cryptogam distributions on *Pseudotsuga menziesii* and *Abies lasiocarpa* in the Front Range, Boulder County, Colorado. *Bryologist* 78, 124–145.

Gowan, S. P., and Brodo, I. M. (1988). The lichens of Fundy National Park, New Brunswick, Canada. *Bryologist* 91, 255–325.

Goward, T. (1992). Preliminary observations on "antique" forests and epiphytic macrolichen diversity in British Columbia. *Northwest Sci.* 66, 133 (abstr.).

Goward, T., and Goffinet, B. (1993). *Nephroma silvae-veteris,* a new lichen (Ascomycotina) from the Pacific Northwest of North America. *Bryologist* 96, 242–244.

Gradstein, S. R. (1991). Flora neotropica news: Lichens. *Int. Lichenol. Newsl.* 24, 60–61.

Gradstein, S. R. (1992). The vanishing tropical rain forest as an environment for bryophytes and lichens. *In* "Bryophytes and Lichens in a Changing Environment" (J. W. Bates and A. M. Farmer, eds.), pp. 234–258. Oxford Univ. Press, Oxford.

Gradstein, S. R., and Pócs, T. (1989). Bryophytes. *In* "In Ecosystems of the World. 14B. Tropical Rain Forest Ecosystems. Biogeographical and Ecological Studies" (H. Lieth and M. J. A. Werger, eds.), pp. 311–325. Elsevier, Amsterdam.

Gradstein, S. R., Van Reenen, G. B. A., and Griffin, D., III (1989). Species richness and origin of the bryophyte flora of the Colombian Andes. *Acta Bot. Neerl.* 38, 439–448.

Gradstein, S. R., Montfoort, D., and Cornelissen, J. H. C. (1990). Species richness and phyto-geography of the bryophyte flora of the Guianas, with special reference to the lowland forest. *Trop. Bryol.* 2, 117–126.

Green, S. W., and Harrington, A. J. (1988). The conspectus of bryological taxonomic literature. Part 1. *Bryophytorum Bibl.* 35, 1–272.

Green, S. W., and Harrington, A. J. (1989). The conspectus of bryological taxonomic literature. Part 2. *Bryophytorum Bibl.* 37, 1–321.

Green, T. G. A., and Lange, O. L. (1991). Ecophysiological adaptations of the lichen genera *Pseudocyphellaria* and *Sticta* to south temperate rainforests. *Lichenologist* 23, 267–282.

Green, T. G. A., Horstmann, J., Bonnett, H., Wilkins, A., and Silvester, W. B. (1980). Nitrogen fixation by members of the Stictaceae (Lichenes) of New Zealand. *New Phytol.* 84, 339–348.

Greenfield, L. G. (1993). Decomposition studies on New Zealand and Antarctic lichens. *Lichenologist* 25, 73–82.

Gressitt, J. L., Sedlacek, J., and Szent-Ivany, J. J. H. (1965). Flora and fauna on backs of large Papuan moss-forest weevils. *Science* 150, 1833–1835.

Grime, J. P., Rincon, E. R., and Wickerson, B. E. (1990). Bryophytes and plant strategy theory. *Bot. J. Linn. Soc.* 104, 175–186.

Gustafsson, L., Fiskesjö, A., Ingelög, T., Pettersson, B., and Thor, G. (1992). Factors of importance to some lichen species of deciduous broad-leaved woods in southern Sweden. *Lichenologist* 24, 255–266.

Guzman, G., Quilhot, W., and Galloway, D. J. (1990). Decomposition of species of *Pseudocyphellaria* and *Sticta* in a southern Chilean forest. *Lichenologist* 22, 325–331.

Hale, M. E., Jr. (1952). Vertical distribution of cryptogams in a virgin forest in Wisconsin. *Ecology* 33, 398–406.

Hale, M. E., Jr. (1983). "The Biology of Lichens," 3rd ed. Edward Arnold, London.

Hallé, F. (1990). A raft atop the rain forest. *Natl. Geogr.* 178, 128–138.

Halonen, P., Hyvärinen, M., and Kauppi, M. (1991). The epiphytic lichen flora on conifers in relation to climate in the Finnish middle-boreal zone. *Lichenologist* 23, 61–72.

Hawksworth, D. L. (1983). A key to the lichen-forming, parasitic, parasymbiotic and sapro-phytic fungi occurring on lichens in the British Isles. *Lichenologist* 15, 1–44.

Hawksworth, D. L. (1988a). The fungal partner. *In* "Handbook of Lichenology" (M. Galun, ed.), Vol. 1, pp. 35–38.

Hawksworth, D. L. (1988b). The variety of fungal–algal symbioses, their evolutionary signifi-cance, and the nature of lichens. *Bot. J. Linn. Soc.* 96, 3–20.

Hawksworth, D. L., and Ahti, T. (1990). A bibliographic guide to the lichen floras of the world. *Lichenologist* 22, 1–78.

Hawksworth, D. L., and Hill, D. J. (1984). "The Lichen-Forming Fungi." Chapman & Hall, New York.

Hoffmann, M. (1987). Species associations among corticolous cryptogams. A comparative study of two sampling methods. *Symp. Biol. Hung.* 35, 527–547.

Hofman, A. (1991). Phylogeny and population genetics of the genus *Plagiothecium* (Bryopsida). Ph.D. Thesis, University of Groningen.

Hofstede, R. G. M., Wolf, J. H. D., and Benzing, D. H. (1993). Epiphytic biomass and nutrient status of a Columbian upper montane rain forest. *Selbyana* 14, 37–45.

Horstmann, J. L., Denison, W. C., and Silvester, W. B. (1982). $^{15}N_2$ fixation and molybdenum enhancement of acetylene reduction by *Lobaria* spp. *New Phytol.* 92, 235–241.

Hosokawa, T., and Kubota, H. (1957). On the osmotic pressure and resistance to desiccation of epiphytic mosses from a beech forest, south-west Japan. *J. Ecol.* 45, 579–591.

Hosokawa, T., and Odani, N. (1947). The daily compensation period and vertical ranges of epiphytes in a beech forest. *J. Ecol.* 45, 901–915.

Hosokawa, T., Odani, N., and Tagawa, T. (1964). Causality of the distribution of corticolous species in forests with special reference to the physio-ecological approach. *Bryologist* 67, 396–411.

Hyvönen, J., Koponen, T., and Norris, D. H. (1987). Human influence on the mossflora of tropical rainforest in Papua New Guinea. *Symp. Biol. Hung.* 35, 621–629.

Jackson, J. B. C., Buss, L. W., and Cook, R. E., eds. (1985). "Population Biology and Evolution of Clonal Organisms." Yale Univ. Press, New Haven, CT.

Jacobsen, N. H. G. (1978). An investigation into the ecology and productivity of epiphytic mosses. *J. S. Afr. Bot.* 44, 297–312.

John, E. (1992). Distribution patterns and interthalline interactions of epiphytic foliose lichens. *Can. J. Bot.* 70, 818–823.

Jørgensen, P. M. (1977). Foliose and fruticose lichens from Tristan da Cunha. *Skr. Nor. Vidonski.-Akad. [Kl.] I: Mat.-Naturvidensk. Kl.* [N.S.] 36, 1–40.

Jørgensen, P. M. (1979). The phytogeographical relationships of the lichen flora of Tristan da Cunha (excluding Gough Island). *Can. J. Bot.* 57, 2279–2282.

Kallio, P., and Kärenlampi, L. (1975). Photosynthesis in mosses and lichens. *In* "Photosynthesis and Productivity in Different Environments" (J. P. Cooper, ed.), pp. 393–423. Cambridge Univ. Press, Cambridge, UK.

Kantvilas, G. (1988). Tasmanian rainforest communities: A preliminary classification. *Phytocoenologia* 16, 391–428.

Kantvilas, G. (1990). The genus *Pertusaria* in Tasmanian rainforests. *Lichenologist* 22, 289–300.

Kantvilas, G., and James, P. W. (1987). The macrolichens of Tasmanian rainforest: Key and notes. *Lichenologist* 19, 1–28.

Kappen, L. (1988). Ecophysiological relationships in different climatic regions. *In* "Handbook of Lichenology" (M. Galun, ed.), Vol. 2, pp. 37–100. CRC Press, Boca Raton.

Kenkel, N. C., and Bradfield, G. E. (1981). Ordination of epiphytic bryophyte communities in a wet-temperate coniferous forest, south-coastal British Columbia. *Vegetatio* 45, 147–154.

Klinge, H., and Herrera, R. (1983). Phytomass structure of natural plant communities on spodosols in southern Venezuela. The tall Amazon caatinga forest. *Vegetatio* 53, 65–84.

Knops, J. M. H., Nash, T. H., III, Boucher, V. L., and Schlesinger, W. H. (1991). Mineral cycling and epiphytic lichens: Implications at the ecosystem level. *Lichenologist* 23, 309–321.

Koponen, T. (1992). Endangered bryophytes on a global scale. *Biol. Conserv.* 59, 255–258.

Krog, H. (1987). Altitudinal zonation of tropical lichens. *In* "Progress and Problems in Lichenology in the Eighties" (E. Peveling, ed.), pp. 379–384. J. Cramer, Berlin.

Krog, H. (1991). Lichenological observations in low montane rainforests of eastern Tanzania.

In "Tropical Lichens: Their Systematics, Conservation, and Ecology" (D. J. Galloway, ed.), pp. 85–94. Oxford Univ. Press, Oxford.

Kürschner, H. (1990a). Die epiphytischen Moosgesellschaften am Mt. Kinabalu (Nord-Borneo, Sabah, Malaysia). *Nova Hedwigia* 51, 1–75.

Kürschner, H. (1990b). Höhengliederung (Ordination) von epiphytischen Laub- und Lebermoosen in Nord-Borneo (Mt. Kinabalu). *Nova Hedwigia* 51, 77–86.

Lang, G. E., Reiners, W. A., and Heier, R. K. (1976). Potential alteration of precipitation chemistry by epiphytic lichens. *Oecologia* 25, 229–241.

Lang, G. E., Reiners, W. A., and Pike, L. H. (1980). Structure and biomass dynamics of epiphytic lichen communities of balsam fir forests in New Hampshire. *Ecology* 61, 541–550.

Lange, O. L., Büdel, B., Meyer, A., and Kilian, E. (1993). Further evidence that activation of net photosynthesis by dry cyanobacterial lichens requires liquid water. *Lichenologist* 25, 175–189.

La Roi, G. H., and Stringer, M. H. L. (1976). Ecological studies in the boreal spruce–fir forests of the North American taiga. II. Analysis of the bryophyte flora. *Can. J. Bot.* 54, 619–643.

Larson, D. W. (1987). The absorption and release of water by lichens. *In* "Progress and Problems in Lichenology in the Eighties" (E. Peveling, ed.), pp. 351–360. J. Cramer, Berlin.

Lawrey, J. D. (1980). Sexual and asexual reproductive patterns in *Parmotrema* (Parmeliaceae) that correlate with latitude. *Bryologist* 83, 344–350.

Lawrey, J. D. (1984). "Biology of the Lichenized Fungi." Praeger, New York.

Lawrey, J. D. (1986). Biological role of lichen substances. *Bryologist* 89, 111–122.

Lawrey, J. D. (1987). Nutritional ecology of lichen/moss arthropods. *In* "Nutritional Ecology of Insects, Mites, and Spiders" (F. Slansky, Jr. and J. G. Rodriguez, eds.), pp. 209–233. Wiley, New York.

Lawrey, J. D. (1991a). Biotic interactions in lichen community development: A review. *Lichenologist* 23, 205–214.

Lawrey, J. D. (1991b). The species–area curve as an index of disturbance in saxicolous lichen communities. *Bryologist* 94, 377–382.

Lawrey, J. D. (1992). Natural and randomly-assembled lichen communities compared using the species–area curve. *Bryologist* 95, 137–141.

Lawrey, J. D. (1993). Chemical ecology of *Hobsonia christiansenii*, a lichenicolous hyphomycete. *Am. J. Bot.* 80, 1109–1113.

Legaz, M. E., Perez-Urria, E., Avalos, A., and Vicente, C. (1988). Epiphytic lichens inhibit the appearance of leaves in *Quercus pyrenaica. Biochem. Syst. Ecol.* 16, 253–259.

Lesica, P., McCune, B., Cooper, S. V., and Hong, W. S. (1991). Differences in lichen and bryophyte communities between old-growth and managed second-growth forests in the Swan Valley, Montana. *Can. J. Bot.* 69, 1745–1755.

Longton, R. E. (1984). The role of bryophytes in terrestrial ecosystems. *J. Hattori Bot. Lab.* 55, 147–163.

Longton, R. E. (1992). The role of bryophytes and lichens in terrestrial ecosystems. *In* "Bryophytes and Lichens in a Changing Environment" (J. W. Bates and A. M. Farmer, eds.), pp. 32–76. Oxford Univ. Press, Oxford.

Loria, M., and Herrnstadt, I. (1981). Moss capsules as food of the harvester ant, *Messor. Bryologist* 83, 524–525.

Lowman, M. D. (1986). Light interception and its relation to structural differences in three Australian rainforest canopies. *Aust. J. Ecol.* 11, 163–170.

Lowman, M. D., and Moffett, M. (1993). The ecology of tropical rain forest canopies. *Trends Ecol. Evol.* 8, 104–107.

Makryi, T. V. (1985). The epiphytic lichens of the Baikalsky Mountain Range. *Bot. Zh. (Leningrad)* 70, 1441–1451.

Marcelli, M. P. (1991). Aspects of the foliose lichen flora of the southern-central coast of Saõ

Paulo State, Brazil. *In* "Tropical Lichens: Their Systematics, Conservation, and Ecology" (D. J. Galloway, ed.), pp. 151–170. Oxford Univ. Press, Oxford.

Massman, W. J. (1982). Foliage distribution in old-growth coniferous tree canopies. *Can. J. For. Res.* 12, 10–17.

Matelson, T. J., Nadkarni, N. M., and Longino, J. T. (1993). Longevity of fallen epiphytes in a Neotropical montane forest. *Ecology* 74, 265–269.

McCune, B. (1982). Lichens of the Swan Valley, Montana. *Bryologist* 85, 13–21.

McCune, B. (1990). Rapid estimation of abundance of epiphytes on branches. *Bryologist* 93, 39–43.

McCune, B. (1993). Gradients in epiphyte biomass in three *Pseudotsuga–Tsuga* forests of different ages in western Oregon and Washington. *Bryologist* 96, 405–411.

McCune, B., and Antos, J. A. (1982). Epiphyte communities of the Swan Valley, Montana. *Bryologist* 85, 1–12.

McCune, B., and Daly, W. J. (1994). Consumption and decomposition of lichen litter in a temperate coniferous rainforest. *Lichenologist* 26, 67–72.

McCune, B., and Lesica, P. (1992). The trade-off between species capture and quantitative accuracy in ecological inventory of lichens and bryophytes in forests in Montana. *Bryologist* 95, 296–304.

Montfoort, D., and Ek, R. (1990). Vertical distribution and ecology of epiphytic bryophytes and lichens in a lowland rain forest of French Guiana. M.Sc. Thesis, University of Utrecht, Utrecht.

Moser, T. J., Swafford, J. R., and Nash, T. H., III (1983). Impact of Mount St. Helens' emissions on two lichen species of south-central Washington. *Environ. Exp. Bot.* 23, 321–329.

Mull, I., and Liat, L. B. (1970). Vertical zonation in a tropical rainforest in Malaysia. *Science* 196, 788–789.

Nadkarni, N. M. (1981). Canopy roots: Convergent evolution in rainforest nutrient cycles. *Science* 214, 1023–1024.

Nadkarni, N. M. (1983). The effects of epiphytes on nutrient cycles within temperate and tropical rainforest tree canopies. Ph.D. Thesis, University of Washington, Seattle.

Nadkarni, N. M. (1984a). Biomass and mineral capital epiphytes in an *Acer macrophyllum* community of a temperate moist coniferous forest, Olympic Peninsula, Washington State [USA]. *Can. J. Bot.* 62, 2223–2228.

Nadkarni, N. M. (1984b). Epiphytic biomass and nutrient capital of a Neotropical elfin forest. *Biotropica* 16, 249–256.

Nadkarni, N. M. (1986). The nutritional effects of epiphytes on host trees with special reference to alteration of precipitation chemistry. *Selbyana* 9, 44–51.

Nadkarni, N. M. (1988). Use of a portable platform for observations of tropical forest canopy animals. *Biotropica* 20, 350–351.

Nadkarni, N. M., and Matelson, T. J. (1989). Bird use of epiphyte resources in neotropical trees. *Condor* 91, 891–907.

Nadkarni, N. M., and Matelson, T. J. (1991). Fine litter dynamics within the tree canopy of a tropical cloud forest. *Ecology* 72, 2071–2081.

Nadkarni, N. M., and Matelson, T. J. (1992). Biomass and nutrient dynamics of epiphytic litterfall in a Neotropical montane forest, Costa Rica. *Biotropica* 24, 24–30.

Nimis, P. L. (1985). Phytogeography and ecology of epiphytic lichens at the southern rim of the clay belt (N-Ontario, Canada). *Bryologist* 88, 315–324.

Nimis, P. L. (1991). Developments in lichen community studies. *Lichenologist* 23, 215–225.

Nishida, F. H. (1989). Review of mycological studies in the neotropics. *In* "Floristic Inventory of Tropical Countries" (D. G. Campbell and H. D. Hammond, eds.), pp. 494–522. New York Botanical Garden, Bronx.

Norris, D. H. (1990). Bryophytes in perennially moist forests of Papua New Guinea: Ecological orientation and predictions of disturbance effects. *Bot. J. Linn. Soc.* 104, 281–291.

Novichkova-Ivanova, L. N. (1983). Epiphytic synusiae of cryptogams in pistachio woodlands of the Turkmen SSR [USSR]. *Bot. Zh. (Leningrad)* 68, 1543–1550.

Ochi, H. (1982). A phytogeographical consideration of Australasian Bryoideae in relation to those in other continents. *J. Hattori Bot. Lab.* 52, 65–73.

Oksanen, J. (1988). Impact of habitat, substrate and microsite classes on the epiphyte vegetation: Interpretation using exploratory and canonical correspondence analysis. *Ann. Bot. Fenn.* 25, 59–71.

Palmer, M. W. (1986). Pattern in corticolous bryophyte communities of the North Carolina Piedmont: Do mosses see the forest or the trees? *Bryologist* 89, 59–65.

Pande, N., and Singh, J. S. (1988). Bryophyte biomass of dominant species and net production of different communities in various habitats of the Nainital Hills, NW Himalay. *Lindbergia* 14, 155–161.

Parker, G. G., Smith, A. P., and Hogan, K. P. (1992). Access to the upper forest canopy with a large tower crane. *BioScience* 42, 664–670.

Perry, D. R. (1984). The canopy of the tropical rain forest. *Sci. Am.* 251, 138–147.

Piipo, S., Koponen, T., and Norris, D. H. (1987). Endemism of the bryophyte flora in New Guinea. *Symp. Biol. Hung.* 35, 361–372.

Pike, L. H. (1971). The role of epiphytic lichens and mosses in production and nutrient cycling of an oak forest. Ph.D. Thesis, University of Oregon, Eugene.

Pike, L. H. (1978). The importance of epiphytic lichens in mineral cycling. *Bryologist* 81, 247–257.

Pike, L. H. (1981). Estimation of lichen biomass and production with special reference to the use of ratios. *In* "The Fungal Community: Its Organization and Role in the Ecosystem" (D. T. Wicklow and G. C. Carroll, eds.), pp. 533–552. Dekker, New York.

Pike, L. H., Tracy, D. M., Sherwood, M. A., and Nelson, D. (1972). Estimates of biomass and fixed nitrogen of epiphytes from old-growth Douglas-fir. *In* "Research on Coniferous Forest Ecosystems: First Year Progress in the Coniferous Forest Biome, US/IBP" (J. F. Franklin, L. J. Dempster, and R. H. Waring, eds.), pp. 177–187. Pacific Forest and Range Experiment Station, Forest Service, U.S.D.A., Portland, OR.

Pike, L. H., Denison, W. C., Tracy, D. M., Sherwood, M. A., and Rhoades, F. M. (1975). Floristic survey of epiphytic lichens and bryophytes growing on old-growth conifers in western Oregon. *Bryologist* 78, 389–402.

Pike, L. H., Rydell, R. A., and Denison, W. C. (1977). A 400-year-old Douglas fir tree and its epiphytes: Biomass, surface area, and their distribution. *Can. J. For. Res.* 7, 680–699.

Pitkin, P. H. (1975). Variability and seasonality of the growth of some corticolous pleurocarpous mosses. *J. Bryol.* 8, 337–356.

Pittam, S. K. (1991). The Rare Lichens Project. A progress report. *Evansia* 8, 45–47.

Pócs, T. (1980). The epiphytic biomass and its effect on the water balance of two rain forest types in the Uluguru Mountains (Tanzania, East Africa). *Acta Bot. Acad. Sci. Hung.* 26, 143–167.

Pócs, T. (1982). Tropical forest bryophytes. *In* "Bryophyte Ecology" (A. J. E. Smith, ed.), pp. 59–104. Chapman & Hall, London.

Pórto, K. C. (1992). Bryoflores d'une forêt de planine et d'une forêt d'altitude moyenne dans l'état de Pernambuco (Brésil). 2. Analyse écologique comparative des forêts. *Cryptogam. Bryol. Lichenol.* 13, 187–219.

Proctor, M. C. F. (1990). The physiological basis of bryophyte production. *Bot. J. Linn. Soc.* 104, 61–77.

Ramsay, H. P., Streimann, H., and Harden, G. (1987). Observations on the bryoflora of Australian rainforests. *Symp. Biol. Hung.* 35, 605–620.

Redon, J., and Lange, O. L. (1983a). Epiphytic lichens in the region of a Chilean fog oasis, Fray Jorge National Park: 1. Distributional patterns and habitat conditions. *Flora (Jena)* 174, 213–243.

Redon, J., and Lange, O. L. (1983b). Epiphytic lichens in the region of a Chilean fog oasis, Fray Jorge National Park: 2. Ecophysiological characterization of carbon dioxide exchange and water relations. *Flora (Jena)* 174, 245–284.

Reiners, W. A., and Olson, R. K. (1984). Effects of canopy components on throughfall chemistry: An experimental analysis. *Oecologia* 63, 320–330.

Rhoades, F. M. (1977). Growth rates of the lichen *Lobaria oregana* as determined from sequential photographs. *Can. J. Bot.* 55, 2226–2233.

Rhoades, F. M. (1978). Growth, production, litterfall and structure in populations of the lichen *Lobaria oregana* (Tuck.) Müll. Arg. in canopies of old-growth Douglas fir. Ph.D. Thesis, University of Oregon, Eugene.

Rhoades, F. M. (1981). Biomass of epiphytic lichens and bryophytes on *Abies lasiocarpa* on a Mt. Baker lava flow, Washington. *Bryologist* 84, 39–47.

Rhoades, F. M. (1983). Distribution of thalli in a population of the epiphytic lichen *Lobaria oregana* and a model of population dynamics and annual production. *Bryologist* 86, 309–331.

Richards, P. W. (1954). Notes on the bryophyte communities of lowland tropical rain forest, with special reference to Moraballi Creek, British Guyana. *Vegetatio* 5/6, 319–327.

Richards, P. W. (1984). The ecology of tropical forest bryophytes. *In* "New Manual of Bryology" (R. M. Schuster, ed.), Vol. 2, pp. 1233–1270. Hattori Bot. Lab., Nichinan, Miyazaki, Japan.

Richards, P. W. (1988). Tropical forest bryophytes: Synusiae and strategies. *J. Hattori Bot. Lab.* 64, 1–4.

Richardson, D. H. S., and Young, C. M. (1977). Lichens and vertebrates. *In* "Lichen Ecology" (M. R. D. Seaward, ed.), pp. 121–144. Academic Press, New York.

Rogers, R. W. (1990). Ecological strategies of lichens. *Lichenologist* 22, 149–162.

Rose, F. (1976). Lichenological indicators of age and continuity in woodlands. *In* "Lichenology: Progress and Problems" (D. H. Brown, D. L. Hawksworth, and R. H. Bailey, eds.), pp. 279–307. Academic Press, New York.

Rose, F. (1988). Phytogeographical and ecological aspects of Lobarion communities in Europe. *Bot. J. Linn. Soc.* 96, 69–79.

Rose, F. (1992). Temperate forest management: Its effects on bryophyte and lichen floras and habitats. *In* "Bryophytes and Lichens in a Changing Environment" (J. W. Bates and A. M. Farmer, eds.), pp. 211–233. Oxford Univ. Press, New York.

Rosentreter, R. (1993). Forest canopy lichens of central Idaho. *Northwest Sci.* 67, 137 (abstr.).

Rundel, P. W., Bratt, G. C., and Lange, O. L. (1979). Habitat ecology and physiological response of *Sticta filix* and *Pseudocyphellaria delisei* from Tasmania. *Bryologist* 82, 171–180.

Russell, K. W., and Miller, H. A. (1977). The ecology of an elfin forest in Puerto Rico. 17. Epiphytic mossy vegetation of Pico del Oeste. *J. Arnold Arbor., Harv. Univ.* 58, 1–24.

Schmitt, C. K., and Slack, N. G. (1990). Host specificity of epiphytic lichens and bryophytes: A comparison of the Adirondack Mountains (New York) and the southern Blue Ridge Mountains (North Carolina). *Bryologist* 93, 257–274.

Schnock, G. (1972). Évapotranspiration de la végétation épiphytique de la base des troncs de chêne et d'érable champêtre. *Bull. Soc. R. Bot. Belg.* 105, 143–150.

Schofield, W. B. (1981). Ecological significance of morphological characters in the moss gametophyte. *Bryologist* 84, 149–165.

Schofield, W. B. (1984). Bryogeography of the Pacific Coast of North America. *J. Hattori Bot. Lab.* 55, 35–43.

Schofield, W. B. (1985). "Introduction to Bryology." Macmillan, New York.

Schofield, W. B. (1992). Bryophyte distribution patterns. *In* "Bryophytes and Lichens in a Changing Environment" (J. W. Bates and A. M. Farmer, eds.), pp. 103–130. Oxford Univ. Press, Oxford.

Schuster, R. M. (1957). Boreal Hepaticae, a manual of the liverworts of Minnesota and adjacent regions. II. Ecology. *Am. Midl. Nat.* 57, 203–299.

Schuster, R. M. (1982). Generic and familial endemism in the hepatic flora of Gondwanaland: Origins and causes. *J. Hattori Bot. Lab.* 52, 3–35.

Schuster, R. M., ed. (1984). "New Manual of Bryology," Vols. 1 and 2. Hattori Bot. Lab., Nichinan, Miyazaki, Japan.

Schuster, R. M. (1988). Ecology, reproductive biology and dispersal of Hepaticae in the tropics. *J. Hattori Bot. Lab.* 64, 237–269.

Scott, G. A. M. (1982). The ecology of mosses: An overview. *J. Hattori Bot. Lab.* 52, 171–177.

Scotter, G. W. (1962). Productivity of arboreal lichens and their possible importance to barren-ground caribou (*Rangifer arcticus*). *Arch. Soc. Zool. Bot. Fenn. "Vanamo"* 16, 151–161.

Seaward, M. R. D. (1988). Contribution of lichens to ecosystems. *In* "Handbook of Lichenology" (M. Galun, ed.), Vol. 2, pp. 107–129. CRC Press, Boca Raton.

Shaw, A. J. (1985). The relevance of ecology to species concepts in bryophytes. *Bryologist* 88, 199–206.

Shaw, A. J. (1992). The evolutionary capacity of bryophytes and lichens. *In* "Bryophytes and Lichens in a Changing Environment" (J. W. Bates and A. M. Farmer, eds.), pp. 362–380. Oxford Univ. Press, New York.

Showman, R. E. (1981). Lichen recolonization following air quality improvement. *Bryologist* 84, 492–497.

Showman, R. E. (1990). Lichen recolonization in the upper Ohio River Valley. *Bryologist* 93, 427–428.

Sigal, L. L. (1984). Of lichens and lepidopterans. *Bryologist* 87, 66–68.

Sillett, S. C., and Sillett, T. S. (1992). A floristic survey of the epiphyte communities inhabiting the crowns of two old-growth conifers in Redwood National Park, California. *Northwest Sci.* 66, 132 (abstr.).

Simon, T. (1974). Estimation of phytomass dry-weight of epiphytic mosses at Sikfokut (near Eger, N. Hungary). *Acta Bot. Acad. Sci. Hung.* 20, 341–348.

Sipman, H. J. M. (1989). Lichen zonation in the Parque los Nevados transect. *Stud. Trop. Andean Ecosyst.* 3, 461–483.

Sipman, H. J. M. (1991). Notes on the lichen flora of the Guianas, a Neotropical lowland area. *In* "Tropical Lichens: Their Systematics, Conservation, and Ecology" (D. J. Galloway, ed.), pp. 135–150. Oxford Univ. Press, Oxford.

Sipman, H. J. M. (1992). The origin of the lichen flora of the Colombian paramos. *In* "Paramo: An Andean Ecosystem Under Human Influence" (H. Balslev and J. L. Luteyn, eds.), pp. 95–109. Academic Press, London.

Sipman, H. J. M., and Harris, R. C. (1989). Lichens. *In* "Ecosystems of the World. 14B. Tropical Rain Forest Ecosystems. Biogeographical and Ecological Studies." (H. Lieth and M. J. A. Werger, eds.), pp. 303–309. Elsevier, Amsterdam.

Sjögren, E. (1961). Epiphytische Moosvegetation in Laubwäldern der Insel Öland (Schweden). *Acta Phyogeogr. Suec.* 44, 1–149.

Slack, N. G. (1976). Host specificity of bryophyte epiphytes in eastern North America. *J. Hattori Bot. Lab.* 41, 107–132.

Slack, N. G. (1977). Species diversity and community structure in bryophytes. *N. Y. State Mus. Bull.* 428.

Slack, N. G. (1982). Bryophytes in relation to ecological niche theory. *J. Hattori Bot. Lab.* 52, 199–217.

Slack, N. G. (1984). A new look at bryophyte community analysis: Field and statistical methods. *J. Hattori Bot. Lab.* 55, 113–132.

Slack, N. G. (1988). The ecological importance of lichens and bryophytes. *In* "Lichens, Bryophytes and Air Quality" (T. H. Nash, III and V. Wirth, eds.), Bibl. Lichenol. No. 30, pp. 23–53. Cramer, Berlin and Stuttgart.

Smith, A. J. E. (1982). Epiphytes and epiliths. *In* "Bryophyte Ecology" (A. J. E. Smith, ed.), pp. 191–227. Chapman & Hall, London.

Smith, C. W. (1991). Lichen conservation in Hawaii. *In* "Tropical Lichens: Their Systematics, Conservation, and Ecology" (D. J. Galloway, ed.), pp. 35–46. Oxford Univ. Press, Oxford.

Söderström, L. (1988). The occurrence of epixylic bryophyte and lichen species in an old natural and managed forest stand in northeast Sweden. *Biol. Conserv.* 45, 169–178.

Söderström, L. (1992). Invasions and range expansions and contractions of bryophytes. *In* "Bryophytes and Lichens in a Changing Environment" (J. W. Bates and A. M. Farmer, eds.), pp. 131–158. Oxford Univ. Press, Oxford.

Söderström, L., Hallingbäck, T., Gustafsson, L., Cronberg, N., and Hedenäs, L. (1992). Bryophyte conservation for the future. *Biol. Conserv.* 59, 265–270.

Sollins, P., Grier, C. C., McCorison, F. M., Cromack, K., Jr., Fogel, R., and Fredricksen, R. L. (1980). The internal element cycles of an old-growth Douglas-fir ecosystem in western Oregon. *Ecol. Monogr.* 50, 261–285.

Stevens, G. N. (1987). The lichen genus *Ramalina* in Australia. *Bull. Br. Mus. (Nat. Hist.), Bot.* 16, 107–223.

Stevens, G. N. (1991). The tropical Pacific species of *Usnea* and *Ramalina* and their relationship to species in other parts of the world. *In* "Tropical Lichens: Their Systematics, Conservation, and Ecology" (D. J. Galloway, ed.), pp. 47–68. Oxford Univ. Press, Oxford.

Stevenson, S. K. (1979). "Effects of Selective Logging on Arboreal Lichens used by Selkirk Caribou," Fish Wild. Rep. No. R-2. Ministry of Forests, Fish and Wildlife Branch, Victoria, British Columbia.

Stone, D. F. (1989). Epiphyte succession on *Quercus garryana* branches in the Willamette Valley of western Oregon. *Bryologist* 92, 81–94.

Stubbs, C. S. (1989). Patterns of distribution and abundance of corticolous lichens and their invertebrate associates on *Quercus rubra* in Maine. *Bryologist* 92, 453–460.

Sulkava, S., and Helle, T. (1975). Range ecology of the domesticated reindeer in the Finnish coniferous forest area. *Biol. Pap. Univ. Alaska, Spec. Rep.* 1, 308–315.

Tanner, E. V. J. (1980). Studies on the biomass and productivity in a series of montane rain forests in Jamaica. *J. Ecol.* 68, 573–588.

ter Steege, H., and Cornelissen, J. H. C. (1988). Collecting and studying bryophytes in the canopy of standing rainforest trees. *In* "Methods in Bryology" (J. M. Glime, ed.), Proc. Bryol. Methods Workshop, Mainz, pp. 285–290. Hattori Botanical Laboratory, Nichinan, Miyazaki, Japan.

Thaithong, O. (1984). Bryophytes of the mangrove forest. *J. Hattori Bot. Lab.* 56, 85–87.

Thiers, B. M. (1988). Morphological adaptations of the Jungermanniales (Hepaticae) to the tropical rainforest habitat. *J. Hattori Bot. Lab.* 64, 5–14.

Thomas, J. W., *et al.* (1993). "Forest Ecosystem Management: An Ecological, Economic and Social Assessment," Report of the Forest Ecosystem Management Assessment Team. U.S. Dept. of Agriculture Forest Service: U.S. Dept. of Commerce, National Oceanic and Atmospheric Administration and National Marine Fisheries Service; U.S. Dept. of Interior, Bureau of Land Management, Fish and Wildlife Service and National Park Service; Environmental Protection Agency, Portland, OR.

Trynoski, S. E., and Glime, J. M. (1982). Direction and height of bryophytes on four species of northern trees. *Bryologist* 85, 281–300.

Tucker, S. C., Matthews, S. W., and Chapman, R. L. (1991). Ultrastructure of subtropical crustose lichens. *In* "Tropical Lichens: Their Systematics, Conservation, and Ecology" (D. J. Galloway, ed.), pp. 171–192. Oxford Univ. Press, Oxford.

Turner, J., and Singer, M. J. (1976). Nutrient distribution and cycling in a sub-alpine coniferous forest ecosystem. *J. Appl. Ecol.* 13, 295–301.

Vance, E. D., and Nadkarni, N. M. (1990). Microbial biomass and activity in canopy organic matter and the forest floor of a tropical cloud forest. *Soil Biol. Biochem.* 22, 677–684.

Van Leerdam, A., Zagt, R. J., and Veneklaas, E. J. (1990). The distribution of epiphyte growth-forms in the canopy of a Colombian cloud-forest. *Vegetatio* 87, 59–71.

Van Reenen, G. B. A., and Gradstein, S. R. (1984). An investigation of bryophyte distribution and ecology along an altitudinal gradient in the Andes of Colombia. *J. Hattori Bot. Lab.* 56, 79–84.

Van Zanten, B. O., and Gradstein, S. R. (1987). Feasibility of long-distance transport in Colombian hepatics, preliminary report. *Symp. Biol. Hung.* 35, 315–322.

Van Zanten, B. O., and Gradstein, S. R. (1988). Experimental dispersal geography of Neotropical liverworts. *Beih. Nova Hedwigia* 90, 41–94.

Van Zanten, B. O., and Pócs, T. (1981). Distribution and dispersal of bryophytes. *Adv. Bryol.* 1, 479–562.

Veneklaas, E. J. (1990). Nutrient fluxes in bulk precipitation and throughfall in two montane tropical rain forests. *J. Ecol.* 78, 974–992.

Veneklaas, E. J., Zagt, R. J., Van Leerdam, A., Van Ek, R., Brockhoven, A. J., and Van Genderen, M. (1990). Hydrological properties of the epiphyte mass of a montane tropical forest, Colombia. *Vegetatio* 89, 183–192.

Verhoeven, T. A., and Lee, K. L. (1976). "Techniques for Tree Canopy Research in the Tropics and Preliminary Investigations of Tree Structure and Epiphyte Distributions." Smithsonian Tropical Research Institute, Washington, DC.

Vitt, D. H. (1981). Adaptive modes of the moss sporophyte. *Bryologist* 84, 166–186.

Vitt, D. H. (1991). Distribution patterns, adaptive strategies, and morphological changes of mosses along elevational and latitudinal gradients on South Pacific Islands. *In* "Quantitative Approaches to Phytogeography" (P. L. Nimis and T. J. Crovello, eds.), Tasks Veg. Sci., No. 24, pp. 205–231. Kluwer Academic Publishers, Dordrecht, The Netherlands.

Vitt, D. H., and Ramsay, H. P. (1985). The *Macromitrium* complex in Australasia (Orthotrichaceae: Bryopsida). Part II. Distribution, ecology, and paleogeography. *J. Hattori Bot. Lab.* 59, 453–468.

Vitt, D. H., Ostafichuk, M., and Brodo, I. M. (1973). Foliicolous bryophytes and lichens of *Thuja plicata* in western British Columbia. *Can. J. Bot.* 51, 571–580.

Voegtlin, D. (1982). "Invertebrates of the H. J. Andrews Experimental Forest: A Survey of Arthropods Associated with the Canopy of Old Growth *Pseudotsuga menziessii*," For. Res. Lab. Spec. Publ. No. 4. School of Forestry, Oregon State University, Corvallis.

Wein, R. W., and Speer, J. E. (1975). Lichen biomass in Acadian and boreal forests of Cape Breton Island, Nova Scotia. *Bryologist* 78, 328–333.

Wetmore, C. M. (1967). Lichens of the Black Hills of South Dakota and Wyoming. *Publ. Mus., Mich. State Univ., Biol. Ser.* 3, 209–464.

Wilmanns, O. (1962). Rindenbewohnende Eiphytengemeinschaften in Süwestdeutschland. *Beitr. Naturk. Forsch. Sudwestdtsch.* 21, 87–164.

Wolf, J. H. D. (1993a). Epiphyte communities of tropical montane rain forests in the northern Andes. I. Lower montane communities. *Phytocoenologia* 22, 1–52.

Wolf, J. H. D. (1993b). Epiphyte communities of tropical montane rain forests in the northern Andes. I. Upper montane communities. *Phytocoenologia* 22, 53–103.

Wolf, J. H. D. (1993c). Diversity patterns and biomass of epiphytic bryophytes and lichens along an altitudinal gradient in the northern Andes. *Ann. Mo. Bot. Gard.* 84, 928–960.

Wolf, J. H. D. (1993d). Ecology of epiphytes and epiphyte communities in montane rain forests, Colombia. Ph.D. Thesis, University of Amsterdam.

Wolf, J. H. D. (1994). Factors controlling the distribution of vascular and non-vascular epiphytes in the northern Andes. *Vegetatio*. 112, 15–28.

Wolseley, P. A. (1991). Observations on the composition and distribution of the '*Lobarion*' in the forest of South-East Asia. *In* "Tropical Lichens: Their Systematics, Conservation, and Ecology" (D. J. Galloway, ed.), pp. 217–244. Oxford Univ. Press, Oxford.

Yague, E., and Estevez, M. P. (1988). The epiphytic lichen *Evernia prunastri* synthesizes a secretable cellulase system that degrades crystalline cellulose. *Physiol. Plan.* 74, 515–520.

Yarranton, G. A. (1972). Distribution and succession of epiphytic lichens on black spruce near Cochrane, Ontario. *Bryologist* 75, 462–480.

Yeaton, R. I., and Gladstone, D. E. (1982). The pattern of colonization of epiphytes on calabash trees (*Crescentia alata* HBK) in Guanacaste Province, Costa Rica. *Biotropica* 14, 137–140.

III

Processes in Tree Canopies

17

Photosynthesis in Forest Canopies

N. Michele Holbrook and Christopher P. Lund

I. Introduction

Energy and material exchanges in plant canopies occur primarily across leaf surfaces. The clearly discernible boundaries and tractable size of individual leaves has encouraged the study of photosynthesis at the leaf level and the development of techniques appropriate for measurements at this scale. Plant canopies, however, are composed of many leaves, and extending a leaf-based understanding of photosynthesis to the level of the canopy requires consideration of the collective properties of these leaves. Extrapolating from leaf to canopy photosynthesis is difficult because of the effect of canopy structure on the conditions experienced by individual leaves and nonlinearities in the response of photosynthesis to resource levels (e.g., light). Scaling from leaf to canopy is further complicated by the variety of species and growth forms found in most forests.

On an annual basis, forests are estimated to account for more than 50% of the global carbon dioxide (CO_2) flux between terrestrial ecosystems and the atmosphere (Potter *et al.*, 1993). The need to predict long-term responses of terrestrial ecosystems to climate change resulting from human activities further underscores the importance of understanding factors that govern the carbon balance of forest ecosystems. This chapter provides an introduction to some of the questions and issues that arise when the forest canopy is considered as a porous structure for the uptake of CO_2. The

Copyright © 1995 by Academic Press, Inc.
All rights of reproduction in any form reserved.

reader is encouraged to consult the many recent reviews of physiological processes at the canopy level (Russell *et al.*, 1989; Pearcy, 1990; Ehleringer and Field, 1993).

II. Leaf-Level Photosynthesis in Forest Canopies

The primary function of leaves is to convert light energy and CO_2 into carbohydrates. Within forest canopies, there can be substantial variation in the photosynthetic rates of individual leaves due to differences in light intensity at the leaf surface, intrinsic biochemical capacity to harvest light energy and fix CO_2, and ability to support the water loss inherent in CO_2 uptake. Vertical patterns in leaf photosynthetic properties within the canopy often parallel those of time-averaged light intensity. Leaves high in the canopy tend to have higher light-saturated photosynthetic rates, light compensation points, and dark respiration rates than conspecific leaves positioned lower in the canopy (e.g., Pearcy, 1987; Hollinger, 1989; Ellsworth and Reich, 1993; Table I and Figs. 1 and 2b). Within a forest canopy, light is both spatially and temporally heterogeneous, and plants in the understory may utilize intermittent periods of direct illumination (sunflecks) for a large proportion of their daily carbon gain (Pearcy, 1990). C_3 photosynthesis is the dominant assimilatory pathway among forest species. Crassulacean acid metabolism (CAM) occurs in certain epiphytes, succulent vines, and hemiepiphytes; C_4 photosynthesis has only been reported for one forest tree species (Pearcy and Troughton, 1975).

Maximum leaf-level assimilation rates (A_{max}) of canopy trees vary within

Table I   Photosynthetic Characteristics of Sugar Maple (*Acer saccharum*) Leaves in the Canopy of a Forest Stand in Southwestern Wisconsin, U.S.A. [a]

Parameter	15 m height	12 m height	9 m height	5 m height
A_{max} (μmol m^{-2} s^{-1})	10.9 ± 1.2	8.5 ± 1.2	6.4 ± 1.1	5.2 ± 0.2
Light saturation point (μE m^{-2} s^{-1})	384 ± 62	335 ± 34	280 ± 5	184 ± 22
Light compensation point (μE m^{-2} s^{-1})	8 ± 2	5 ± 0	4 ± 1	3 ± 0
Dark respiration (μmol m^{-2} s^{-1})	0.83 ± 0.25	0.39 ± 0.04	0.25 ± 0.01	0.20 ± 0.03
Chlorophyll content (mg g^{-1})	5.5 ± 0.4	7.4 ± 0.8	10.6 ± 0.6	12.3 ± 0.2

[a] Reprinted from Ellsworth and Reich (1993).

and between forests (Table II). A_{max} of broad-leaved trees ranges between 4 and 15 μmol CO_2 m^{-2} s^{-1} (Larcher, 1969; Bazzaz and Picket, 1980; Medina and Klinge, 1983; Jarvis and Leverenz, 1983). Conifers generally have lower A_{max}, with values between 5 and 10 μmol CO_2 m^{-2} s^{-1} (Jarvis *et al.*, 1976; Jarvis and Leverenz, 1983). Comparisons between broad- and needle-leaved trees may be complicated by whether the photosynthetic rates of the latter are expressed on a dry weight, two-sided, or projected area basis. A_{max} represents a leaf's capacity to utilize abundant light energy. Among canopy plants in lowland rain forest in Panama, A_{max} was linearly related to diurnal CO_2 uptake (Zotz and Winter, 1993). Factors that contribute to variation in A_{max} among canopy species include soil fertility, climate, successional status, and leaf longevity.

Seasonal variation in the light-saturated photosynthetic rate within an individual can be associated with developmental changes or with factors that influence stomatal aperture, in particular soil water availability and leaf-to-air vapor pressure deficits (VPD). The degree to which stomatal closure in the upper canopy restricts CO_2 uptake may determine whether maximum

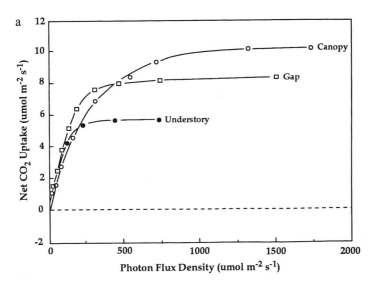

Figure 1   (a) Steady-state light response curves for canopy, gap, and understory individuals of *Argyrodendron peralatum* growing in tropical forest in northeastern Australia. (b) Diurnal data for photon flux density, photosynthesis (A), stomatal conductance (g_s), intercellular CO_2 concentration (C_i), leaf temperature (T_{leaf}; dashed line), and leaf–air vapor pressure gradient (Δw; solid line) for canopy and understory individuals of *Argyrodendron peralatum*. Reprinted from Pearcy, R. W. (1987). Photosynthetic gas exchange responses of Australian tropical forest trees in canopy, gap, and understory microenvironments. *Funct. Ecol.* 1, 169–178, by permission of Blackwell Science Ltd. (*Continues*)

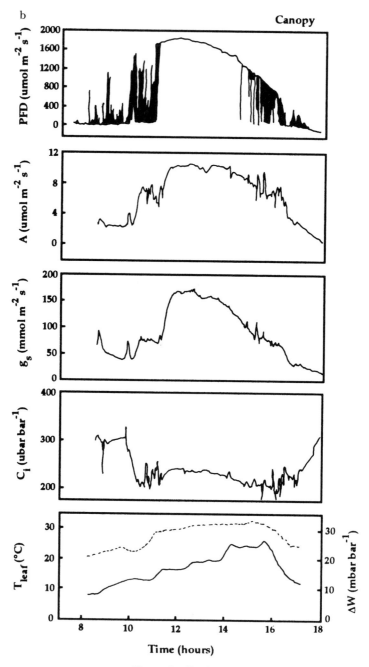

Figure 1—*Continued*

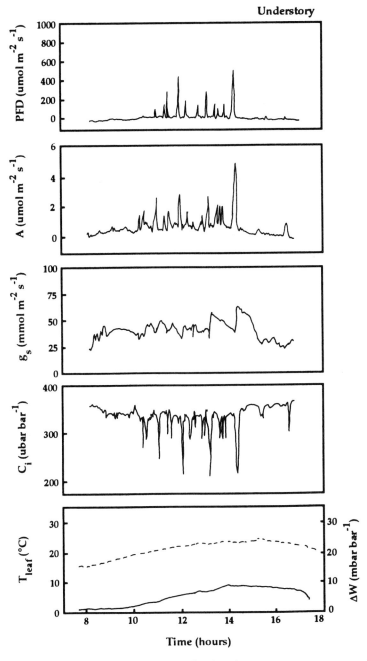

Figure 1—*Continued*

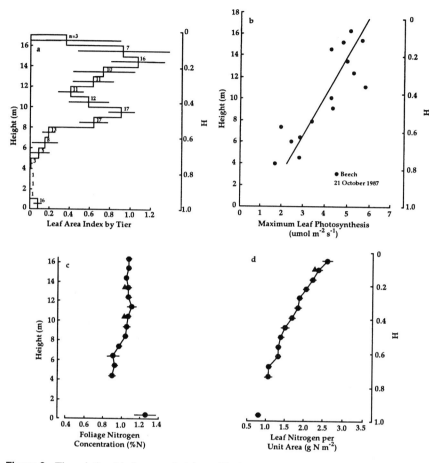

Figure 2   The relationship between height and leaf area index (a), maximum leaf photosynthesis (b), foliage nitrogen concentration (c), and leaf nitrogen per unit area (d) in a broad-leaved evergreen forest of *Nothofagus solandri* var. *cliffortiodes* (mountain beech) in New Zealand. Reprinted from Hollinger, D. Y. (1989). Canopy organization and foliage photosynthetic capacity in a broad-leaved evergreen montane forest. *Funct. Ecol.* 3, 53–62, by permission of Blackwell Science Ltd.

assimilation rates occur at the very top of the canopy or somewhat below. In coniferous forests, maximum photosynthetic rates often occur substantially below the top of the crown (Woodman, 1971; Watts *et al.*, 1976; Schulze *et al.*, 1977a; but see Helms, 1970). Midday reductions in photosynthesis due to stomatal closure in response to high VPD and leaf water stress have been reported in broad-leaved canopy trees (e.g., Doley *et al.*, 1988; Roy and Salager, 1992; Koch *et al.*, 1994).

Table II Maximum Assimilation Rate (A_{max}) for Selected Canopy Species

Species	Family	A_{max}	Site	Forest type	Reference
Acer saccharum	Aceraceae	$11 \ \mu\text{mol m}^{-2} \text{s}^{-1}$	Northern U.S.A. (Wisconsin)	Closed canopy Deciduous forest	Ellsworth and Reich (1993)
Ceiba pentandra	Bombacaceae	$15 \ \mu\text{mol m}^{-2} \text{s}^{-1}$	Panama	Lowland rain forest	Zotz and Winter (1993)
Dialium pachyphyllum	Caesalpiniaceae	10 to $12 \ \mu\text{mol m}^{-2} \text{s}^{-1}$	Cameroon	Lowland rain forest	Koch *et al.* (1994)
Nothofagus solandri var. *cliffortioides*	Fagaceae	$6 \ \mu\text{mol m}^{-2} \text{s}^{-1}$	New Zealand	Broad-leaved mon-tane forest	Hollinger (1989)
Quercus agrifolia	Fagaceae	$8.3 \ \mu\text{mol m}^{-2} \text{s}^{-1}$	Western U.S.A. (California)	Coastal woodland	Hollinger (1992)
Quercus coccifera	Fagaceae	$15 \ \mu\text{mol m}^{-2} \text{s}^{-1}$	Portugal	Macchia	Caldwell *et al.* (1986)
Quercus lobata	Fagaceae	$16.8 \ \mu\text{mol m}^{-2} \text{s}^{-1}$	Western U.S.A. (California)	Coastal woodland	Hollinger (1992)
Pentaclethra macroloba	Mimosaceae	$6.5 \ \mu\text{mol m}^{-2} \text{s}^{-1}$	Costa Rica	Lowland rain forest	Oberbauer and Strain (1986)
Eucalyptus maculata	Myrtaceae	$9.5 \ \mu\text{mol m}^{-2} \text{s}^{-1}$	Southeastern Australia	Coastal region	Wong and Dunin (1987)
Eucalyptus pauciflora	Myrtaceae	10 to $12 \ \mu\text{mol m}^{-2} \text{s}^{-1}$	Southeastern Australia	Montane forest	Slatyer and Morrow (1977)
Picea abies	Pineaceae	$4 \ \text{mg CO}_2 \text{g}^{-1} \text{h}^{-1}$	Northern Germany	Evergreen, mon-tane forest	Schulze *et al.* (1977a)
Picea sitchensis	Pineaceae	$10 \ \mu\text{mol m}^{-2} \text{s}^{-1}$	Scotland	25-year-old stand	Watts *et al.* (1976)
Pinus radiata	Pineaceae	$6.4 \ \mu\text{mol m}^{-2} \text{s}^{-1}$	New Zealand	9-year-old stand	Benecke (1980)
Pseudotsuga menziesii	Pineaceae	$3.5 \ \text{mg CO}_2 \text{g}^{-1} \text{h}^{-1}$	Northwest U.S.A. (Washington)	38-year-old stand	Woodman (1971)
Argyrodendron peralatum	Sterculiaceae	$11 \ \mu\text{mol m}^{-2} \text{s}^{-1}$	Queensland	Rain forest	Pearcy (1987); Doley *et al.* (1988)
Qualea rosea	Vochysiaceae	$1.4 \ \mu\text{mol m}^{-2} \text{s}^{-1}$	French Guiana	Lowland rain forest	Roy and Salager (1992)

III. Canopy Organization in Relation to Photosynthetic Capacity

Understanding the photosynthetic properties of the canopy as a whole requires examination of the environmental variables and physiological responses that determine the photosynthetic properties of individual leaves throughout the canopy. Light is the primary environmental factor that influences the distribution of photosynthetic surfaces in forests, and the factor that varies most dramatically through the canopy (e.g., Jarvis and Leverenz, 1983; Pearcy, 1987; Ellsworth and Reich, 1993). Gradients in air temperature, water vapor, and CO_2 concentration within the canopy are small in comparison to gradients in light (Chapters 3 and 4), and modeling studies indicate that they have only minimal effects on canopy CO_2 uptake (Sinclair *et al.*, 1976; Baldocchi, 1993).

Average light levels decline exponentially with depth in the canopy, with light penetration dependent on (in order of decreasing importance): (1) total amount of leaf area displayed; (2) vertical distribution of leaves; (3) leaf angle; (4) leaf spectral properties (reflectance and transmittance of visible wavelengths); (5) spatial aggregation (clumping); and (6) azimuthal orientation (Jarvis and Leverenz, 1983). The vertical distribution of average light levels within a forest appears to be well approximated by

$$Q(z) = Q(0) \exp^{-\kappa \cdot LAI(z)}$$

where $Q(z)$ is the average light level at depth z in the canopy, $Q(0)$ is the light incident on the top of the canopy, κ is an empirically determined extinction coefficient, and $LAI(z)$ is the cumulative leaf area (leaf area index: m^2 leaf area/m^2 ground surface, LAI) to depth z (Norman, 1975; Fig. 3; Chapter 4). The magnitude of κ depends on leaf angle as well as the spectral properties of individual leaves. κ values of 0.4–0.6 and 0.5–0.8 have been reported for needle- and broad-leaved forests, respectively (Jarvis and Leverenz, 1983). According to this model, LAIs sufficient to intercept 95% of the light incident on the canopy surface should fall between 3.8 and 6 for broad-leaved forests and between 4.6 and 7.5 for needle-leaved trees.

Forests differ substantially in the total amount of leaf area they support per unit of ground area. Leaf area indices range from 1 to 3 in dry sclerophyll eucalypt forests (Carbon *et al.*, 1979) to values that exceed 15 in coniferous forests of northwestern North America (Jarvis and Leverenz, 1983). In contrast, LAIs of forests consisting primarily of broad-leaved trees rarely exceed 7 (Jarvis and Leverenz, 1983; Medina and Klinge, 1983). In both coniferous and broad-leaved forests, LAI is positively correlated with precipitation and soil fertility (Grier and Running, 1977; Waring *et al.*, 1978; Jarvis and Leverenz, 1983; Medina and Klinge, 1983). Upper limits on leaf area in broad-leaved forests (including tropical evergreen forests in which leaf longevity is typically 11 to 13 months; Medina and Klinge, 1983) may

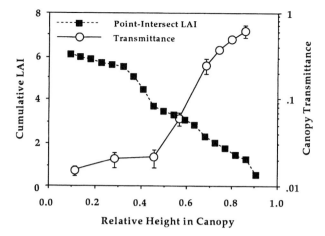

Figure 3 Vertical pattern of cumulative leaf area index (LAI) and canopy radiation transmittance [$= \exp(^{-\kappa \cdot LAI(z)})$] as a function of relative height (z) in a deciduous forest canopy in Wisconsin. A relative height of 1.0 denotes the top of the canopy. Reprinted from Ellsworth and Reich (1993), by permission of Springer-Verlag.

be due to the annual costs of replacing the entire collection of leaves (Schulze *et al.*, 1977b; Jarvis and Leverenz, 1983). The highest LAIs occur in forests with overlapping generations of leaves (i.e., conifers). In these forests, the photosynthetic advantage of being evergreen may be due primarily to the longevity of individual needles rather than the capacity for carbon uptake during the winter months (Schulze *et al.*, 1977b).

Leaf angle also plays an important role in light capture by leaves, and hence in patterns of light penetration through the canopy. Forests composed of species with typically steep leaf angles (e.g., *Eucalyptus*) tend to have substantial light penetration to the forest floor despite high LAIs (Oren *et al.*, 1986). In many forests, leaf angles are greatest (i.e., steepest) at the top of the canopy (e.g., Caldwell *et al.*, 1986; Hollinger, 1989). This pattern may be due in part to the greater water demands and risk of photoinhibition associated with leaves receiving high levels of illumination; steep leaf angles reduce the flux density of light that reaches the leaf surface. Lower in the canopy, leaves oriented closer to the horizontal are better able to intercept intermittent patches of direct beam radiation (sunflecks).

Why do some forests support substantially more leaf area than appears necessary for light interception? Because assimilation rates of individual leaves tend to saturate at high light levels, the photosynthetic capacity of a collection of leaves will be greatest if all leaves receive an intermediate, nonsaturating level of illumination such that the rate of change of assimilation with light intensity (i.e., $\delta A/\delta Q$) is maximized for each leaf. Within an individual tree, this can be achieved by reducing the light incident on surfaces near the top of the crown (e.g., with steeper leaf angles) and thus increasing light arriving at leaves farther down (Hollinger, 1989). Although the spatial arrangement of leaves may have only a small effect on total light interception, it may allow trees to support more leaf area by changing the effective distribution of light within the canopy. Architectural constraints associated with leaf support limit a tree's ability to distribute light evenly throughout its crown and may explain why light saturation does not appear to occur at the canopy level (Larcher, 1969; Fig. 4).

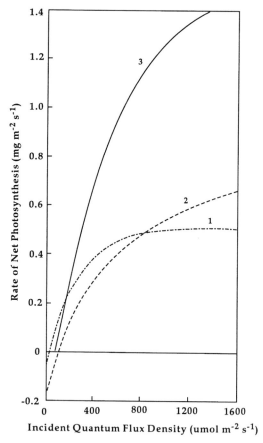

Figure 4 The relationship between rate of net photosynthesis (A_n) and incident quantum flux density (Q_0) for *Picea sitchensis* for needles (curve 1), shoot (curve 2), and forest canopy (curve 3). Needles were illuminated bilaterally (without any mutual shading), the shoot was illuminated bilaterally with some mutual shading, and the forest canopy was illuminated naturally from above. In each case, A_n and Q_0 are expressed on the same area basis (needle area in curve 1, shoot silhouette area in curve 2, and ground area in curve 3). Reprinted from Jarvis and Leverenz (1983), by permission of Springer-Verlag.

Leaf nitrogen (N) content generally scales to the time-averaged light level, with the highest N concentrations occurring at the top of the canopy (Field, 1983; DeJong and Doyle, 1985; Hirose and Werger, 1987; Hirose *et al.*, 1989). A_{max} is closely related to leaf N content in a wide variety of plant species (Field and Mooney, 1986). Models of canopy photosynthesis can be used to determine the distribution of N among leaves that maximizes canopy photosynthesis and to compare estimates of photosynthetic capacity of an existing stand with a hypothetical canopy in which the same amount of total N is distributed equally among leaves (Field, 1983; Hirose and Werger, 1987; Leuning *et al.*, 1991; Ellsworth and Reich, 1993). At high LAIs, these models predict more than a twofold difference in canopy photosynthesis between optimal and uniform N distribution (Field, 1991). This difference is substantially reduced with greater light penetration through the canopy (Field, 1983; Leuning *et al.*, 1991). In general, observed patterns of foliar N among leaves are similar or only slightly more uniform than the optimal distribution (Field, 1983; Ellsworth and Reich, 1993).

Trends in photosynthetic capacity with height in forest canopies are closely correlated with variation in leaf structure (Figs. 2 and 5). In particular, leaf mass per area (LMA) declines with depth through the canopy (Hollinger, 1989; Ellsworth and Reich, 1993). Variation in LMA within forest canopies is thought to be due to the direct effect of light levels on leaf development (Bowes *et al.*, 1972; Bjorkman, 1981). Variation in LMA within the canopy appears to be one way in which a balance is achieved between surface area for light capture and biochemical capacity for CO_2 uptake (Oren *et al.*, 1986; Gutschick and Wiegel, 1988; Ellsworth and Reich, 1993). Photosynthetic capacity is highly correlated with LMA among annual crops (Bowes *et al.*, 1972; Bjorkman, 1981) and forest species (Hollinger, 1989; Ellsworth and Reich, 1993). Because LMA is easier to measure than photosynthetic capacity, it may be useful as an integrated measure of photosynthetic capacity. Oren *et al.* (1986) suggested a photosynthetic index of leaf biomass · LMA (normalized to the maximum value within the crown) as a simple means of estimating the relative contribution of different portions of the canopy to total annual carbon uptake.

IV. Canopy-Level Flux Measurements

Scaling leaf-level gas exchange measurements to the canopy level is greatly complicated by spatial and temporal heterogeneity (Baldocchi, 1993). Direct measurement of gas exchange at the canopy level integrates over such heterogeneity and has become increasingly common over the past decade. Limitations inherent in this approach include fairly stringent site requirements, a lack of information regarding the contribution of spe-

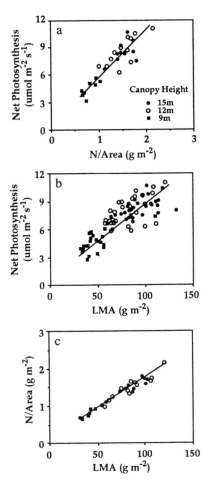

Figure 5 Relationships between net photosynthesis and nitrogen per leaf area (a), net photosynthesis and leaf mass per area (b), and nitrogen per leaf area and leaf mass per area (c) for sugar maple leaves (*Acer saccharum*) at three heights in the canopy. Reprinted from Ellsworth and Reich (1993), by permission of Springer-Verlag.

cific elements (e.g., soil, species, understory versus overstory), and the difficulty of isolating particular environmental variables for the purpose of identifying and characterizing controls over gas flux.

Canopy-level gas exchange can be measured with a variety of micrometeorological techniques, including the commonly employed eddy correlation method (Baldocchi *et al.*, 1988). This technique involves the simultaneous determination of vertical wind speed and concentration of the gas of interest (typically water vapor and/or CO_2) at a fixed height above the canopy. Because the movement of gas above the canopy is governed by the swirling motion of eddies, canopy fluxes are given by the covariance of vertical wind speed and concentration (Baldocchi *et al.*, 1988; Chapter 3). For CO_2, this canopy flux includes photosynthesis, plant respiration, and soil respiration and is typically referred to as instantaneous net ecosystem production (NEP). Gross canopy photosynthesis can be estimated by separating the flux into a daytime rate (photosynthesis and respiration) and a nighttime rate (respiration only). By estimating the contribution of plant respiration to the nighttime CO_2 flux, a similar approach can be used to estimate net canopy photosynthesis (Verma *et al.*, 1992). Estimations of canopy photosynthesis based on ecosystem flux data assume that plant and soil respiration rates do not vary diurnally once temperature differences are accounted for. This is a critical assumption as ecosystem respiration and gross photosynthesis are typically of similar magnitude.

Using two years of continuous eddy correlation data, Wofsy *et al.* (1993) reported an annual net uptake rate of 3.7 $\pm$ 0.7 Mg of carbon ha^{-1} y^{-1} for a 50- to 70-year-old deciduous forest in the Northeast United States. This is substantially higher than the previous estimates of 1.7 to 2.7 Mg for aggrading forest ecosystems (Houghton *et al.*, 1991). Maximum ecosystem CO_2 uptake rates determined by direct measurement of canopy flux are surprisingly consistent across a variety of forest types. Values of 8.6, 9.8, 9.5, 9.3, and 8.8 kg C ha^{-1} h^{-1} have been reported for southern beech, oak–maple, macchia, tropical forest, and oak–hickory canopies, respectively (Hollinger *et al.*, 1994; Wofsy *et al.*, 1993; Valentini *et al.*, 1991; Fan *et al.*, 1990; Verma *et al.*, 1986).

When water is nonlimiting, ecosystem CO_2 uptake, a measure of canopy photosynthesis, appears to be driven by the flux density of photosynthetically active radiation (PPFD) (Fan *et al.*, 1990; Baldocchi *et al.*, 1987; Fig. 4). Maximum quantum efficiencies of 0.055 and 0.051 mol CO_2 per mol absorbed photons have been reported for temperate deciduous forest canopy and tropical forest canopies, respectively (Wofsy *et al.*, 1993; Fan *et al.*, 1990). These values are similar to quantum yields reported for C_3 plants in laboratory studies (Ehleringer and Bjorkman, 1977). Ecosystem CO_2 uptake typically tracks PPFD over the course of the day, becoming negative (i.e., CO_2 efflux) when PPFD approaches zero (Valentini *et al.*, 1991; Verma *et al.*, 1986; Fig. 6). Ecosystem CO_2 uptake is also sensitive to soil moisture, vapor pressure deficit, and temperature (Baldocchi *et al.*, 1987). For a given PPFD, higher CO_2 uptake rates occur during the morning than the afternoon. This may be caused by diurnal water stress, feedback inhibition due to carbohydrate accumulation, increased respiration in response to higher afternoon temperatures, or lower canopy CO_2 concentrations during the afternoon (Wofsy *et al.*, 1993).

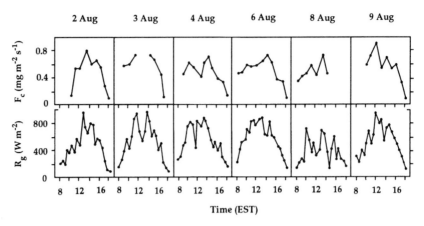

Figure 6   Diurnal courses of ecosystem CO_2 flux (F_c) and solar radiation (R_g) over an oak–hickory forest canopy. Reprinted from Verma *et al.* (1986) by permission of Kluwer Academic Publishers.

In addition to providing basic information on ecosystem gas exchange, whole-canopy flux measurements have also been used to test aspects of canopy photosynthesis models. Baldocchi *et al.* (1987) compared eddy correlation data from a temperate deciduous forest with results from a coupled leaf photosynthesis/canopy radiative transfer model. Although agreement was good, the model tended to underestimate canopy photosynthesis at low PPFD and overestimate canopy photosynthesis at high PPFD. The latter may result from an overestimation of PPFD deep within the canopy (Baldocchi *et al.*, 1987). Amthor *et al.* (1994) reported a strong correlation between predicted and measured hourly CO_2 fluxes for an oak–maple forest. The big-leaf model they used, however, overestimated mean daytime CO_2 uptake by 13% as a result of overestimated afternoon CO_2 uptake rates. As with the morning/afternoon difference reported by Wofsy *et al.* (1993), this overestimation may be due to the effect of accumulated leaf carbohydrates on photosynthesis, a feedback that is not included in the model.

V. Modeling Canopy Photosynthesis

Models are an important tool for understanding canopy photosynthesis. They estimate canopy level fluxes, address the contribution of specific components of the canopy to the total flux (e.g., Caldwell *et al.*, 1986), provide a tool to examine which processes, variables, and feedbacks are most important (e.g., Sinclair *et al.*, 1976), and predict canopy response to alteration of climatic conditions (e.g., Running and Nemani, 1991). Two basic approaches are used: (1) mechanistic, process-driven models, in which information on small spatial and short temporal scales is used to derive predictions on larger and longer scales; and (2) "lumped-parameter" models based on empirically derived relationships (Jarvis, 1993). These are referred to as "bottom-up" and "top-down" models, respectively. Bottom-up models can be sophisticated in their representation of mechanistic interactions (e.g., Sellers *et al.*, 1992; Norman, 1993). However, they generally require detailed information about canopy structure and physiological properties, are potentially incomplete because many of the interactions at the canopy level may not be known, and are open-ended in that model predictions are not constrained and thus sensitive to errors in input parameters (Jarvis, 1993). Top-down models rely on the strength of an empirically determined relationship, but predictions for conditions that lie outside of the range used in determining this underlying relationship may be unreliable (Jarvis, 1993).

A major challenge facing bottom-up, process-oriented models lies in addressing vertical gradients in light availability and photosynthetic capacity through the canopy (Norman, 1993). One approach involves dividing the canopy into a collection of horizontal layers. The distribution of shortwave radiation, longwave radiation, leaf energy balance, water vapor conductance, and photosynthesis is iteratively computed for each layer. This allows interactions between layers and calculation of total canopy flux by summation of fluxes from individual layers (e.g., Caldwell *et al.*, 1986; Norman, 1993; Fig. 7). Because of the nonlinear relationship between leaf-level photosynthesis and PPFD, attempts to use average levels of illumination have proven unsatisfactory (Sinclair *et al.*, 1976; Norman, 1993). Most multilayer models keep track of the vertical distribution of both beam (direct) and diffuse radiation, so that the photosynthesis of sunlit and shaded leaf area in each layer can be determined separately.

An alternative approach is to consider the canopy as a single integrated leaf and make use of the coordination between photosynthetic capacity and light levels that occurs through the canopy (Running and Coughlan, 1988; Field, 1991; Sellers *et al.*, 1992; Amthor, 1994). One such "big-leaf" model involves scaling the photosynthetic rate of the uppermost leaves to that of the whole canopy (Sellers *et al.*, 1992). In this case, the equation for canopy CO_2 uptake consists of the product of the photosynthetic rate of the uppermost leaves (primarily determined by leaf N levels) and a canopy scaling factor. The latter scales the performance of the "top" leaf to that of the canopy and is given by the time-averaged measure of the amount of radiation intercepted by the canopy divided by the average extinction coefficient for direct (solar) beam flux within the canopy.

Top-down models are based on empirical relationships and thus generally require fewer input parameters than a more explicitly mechanistic

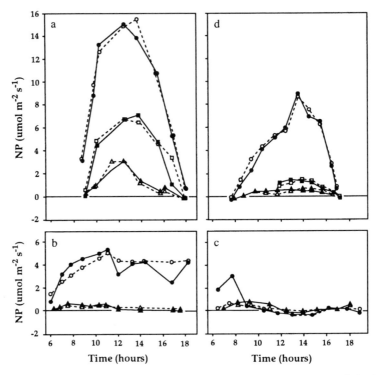

Figure 7 Measured (solid symbols, continuous lines) and simulated (open symbols, dashed lines) leaf net photosynthesis (NP) under prevailing ambient conditions for different canopy layers of *Quercus coccifera* during four seasons of the year. Reprinted from Caldwell *et al.* (1986), by permission of Springer-Verlag.

model. One example of a top-down model that has been used for estimating canopy photosynthesis and net primary production (NPP) draws on the relationship between absorbed radiation (APAR) and biomass accumulation (Monteith, 1977; Field, 1991; Jarvis, 1993). Field (1991) suggested that APAR may function as an activity-weighted measure of stand leaf area, and Potter *et al.* (1993) used this approach to estimate global NPP for terrestrial ecosystems. They assumed that the relationship between potential biomass accumulation and APAR is constant across ecosystems, with realized biomass accumulation being down-regulated by water or temperature limitations to account for vegetation types that may restrict carbon uptake without altering leaf displays (e.g., many evergreen species). The effects of other changes in resource availability or environmental conditions on NPP are assumed primarily to influence APAR (Field, 1991; Potter *et al.*, 1993).

A substantial difficulty in estimating photosynthesis of forest ecosystems is how to deal with spatial heterogeneity. Most forests are made up of many species, and significant variation in forest structure and composition generally occurs at the regional scale. This means that models that require detailed information about canopy structure or photosynthetic properties may not be tractable when addressing large spatial scales. Models that are driven primarily by remote-sensing data (e.g., Running and Coughlan, 1988; Sellers *et al.*, 1992; Potter *et al.*, 1993) are required at the global scale. Spatially explicit models (e.g., Wang and Jarvis, 1990) require detailed information about stand geometry but include heterogeneity of leaf displays within and between individuals.

VI. Summary

Obtaining a comprehensive understanding of canopy photosynthesis is a key goal of both forest ecology and global change studies. Accurately predicting the response of terrestrial ecosystems to rising atmospheric CO_2 concentrations requires consideration of the factors governing exchange of CO_2, water, and energy between forest canopies and the atmosphere. Although leaf-level studies are valuable in that they provide information regarding the contribution of specific components of the forest canopy to ecosystem gas fluxes, they become more powerful when used in conjunction with canopy-level flux measurements. The scaling rules derived from studies addressing multiple levels of organization should prove valuable in improving models of the global carbon budget.

Acknowledgments

Valuable comments were provided by C. B. Field, A. L. Fredeen, G. W. Koch, M. T. Lerdau, and C. Malmström on an earlier version of this manuscript. C.P.L. was supported by the DOE-GFGC program administered by ORISE.

References

Amthor, J. S. (1994). Scaling CO_2–photosynthesis relationships from the leaf to the canopy. *Photosynth. Res.* 39, 321–350.

Amthor, J. S., Goulden, M. L., Munger, J. W., and Wofsy, S. C. (1994). Testing a mechanistic model of forest-canopy mass and energy exchange using eddy correlation: Carbon dioxide and ozone uptake by a mixed oak–maple stand. *Aust. J. Plant Physiol.* 21 (in press).

Baldocchi, D. D. (1993). Scaling water vapor and carbon dioxide exchange from leaves to a canopy: Rules and tools. *In* "Scaling Physiological Processes: Leaf to Globe" (J. R. Ehleringer and C. B. Field, eds.), pp. 77–114. Academic Press, San Diego.

Baldocchi, D. D., Verma, S. B., and Anderson, D. E. (1987). Canopy photosynthesis and water-use efficiency in a deciduous forest. *J. Appl. Ecol.* 24, 251–260.

Baldocchi, D. D., Hicks, B. B., and Meyers, T. P. (1988). Measuring biosphere–atmosphere exchange of biologically related gases with micrometeorological methods. *Ecology* 69, 1331–1340.

Bazzaz, F. A., and Picket, S. T. A. (1980). Physiological ecology of tropical succession: A comparative review. *Annu. Rev. Ecol. Syst.* 11, 287–310.

Benecke, U. (1980). Photosynthesis and transpiration of *Pinus radiata* D. Don under natural conditions in a forest stand. *Oecologia* 44, 192–198.

Bjorkman, O. (1981). Responses to different quantum flux densities. *Encycl. Plant Physiol., New Ser.* 12A, 57–107.

Bowes, G., Ogren, W. L., and Hageman, R. H. (1972). Light saturation, photosynthesis rate, RuBP carboxylase activity, and specific leaf weight in soybeans grown under different light intensities. *Crop Sci.* 12, 77–79.

Caldwell, M. M., Meister, H.-P., Tenhunen, J. D., and Lange, O. L. (1986). Canopy structure, light microclimate and leaf gas exchange of *Quercus coccifera* L. in a Portuguese macchia: Measurements in different canopy layers and simulations with a canopy model. *Trees* 1, 25–41.

Carbon, B. A., Bartle, G. A., and Murray, A. M. (1979). Leaf area index of some eucalypt forests in south-west Australia. *Aust. For. Res.* 9, 323–326.

DeJong, T. M., and Doyle, J. F. (1985). Seasonal relationships between leaf nitrogen content (photosynthetic capacity) and leaf canopy light exposure in peach (*Prunus perspicata*). *Plant, Cell Environ.* 8, 701–706.

Doley, D., Unwin, G. L., and Yates, D. J. (1988). Spatial and temporal distribution of photosynthesis and transpiration by single leaves in a rain forest tree, *Argyrodendron peralatum*. *Aust. J. Plant Physiol.* 15, 317–326.

Ehleringer, J. R., and Bjorkman, O. (1977). Quantum yields for CO_2 uptake in C_3 and C_4 plants. *Plant Physiol.* 59, 86–90.

Ehleringer, J. R., and Field, C. B., eds. (1993). "Scaling Physiological Processes: Leaf to Globe." Academic Press, San Diego.

Ellsworth, D. S., and Reich, P. B. (1993). Canopy structure and vertical patterns of photosynthesis and related leaf traits in a deciduous forest. *Oecologia* 96, 169–178.

Fan, S.-M., Wofsy, S. C., Bakwin, P. S., Jacob, D. J., and Fitzjarrald, D. R. (1990). Atmosphere–biosphere exchange of CO_2 and O_3 in the central Amazon forest. *J. Geophys. Res.* 95, 16851–16864.

Field, C. B. (1983). Allocating leaf nitrogen for the maximization of carbon gain: Leaf age as a control on the allocation program. *Oecologia* 56, 341–347.

Field, C. B. (1991). Ecological scaling of carbon gain to stress and resource availability. *In* "Response of Plants to Multiple Stresses" (H. A. Mooney, W. E. Winner, and E. J. Pell, eds.), pp. 35–65. Academic Press, San Diego.

Field, C. B., and Mooney, H. A. (1986). The photosynthesis–nitrogen relationship in wild plants. *In* "On the Economy of Plant Form and Function" (T. J. Givnish, ed.), pp. 25–55. Cambridge Univ. Press, Cambridge, UK.

Grier, C. C., and Running, S. W. (1977). Leaf area of mature north-western coniferous forests: Relation to site water balance. *Ecology* 58, 893–899.

Gutschick, V. P., and Wiegel, F. W. (1988). Optimizing the canopy photosynthetic rate by patterns of investment in specific leaf area. *Am. Nat.* 132, 67–86.

Helms, J. A. (1970). Summer net photosynthesis of ponderosa pine in its natural environment. *Photosynthetica* 4, 243–253.

Hirose, T., and Werger, M. J. A. (1987). Maximizing daily canopy photosynthesis with respect to the leaf nitrogen allocation pattern in the canopy. *Oecologia* 72, 520–526.

Hirose, T., Werger, M. J. A., and van Rheenen, J. W. A. (1989). Canopy development and leaf nitrogen distribution in a stand of *Carex acutiformis*. *Ecology* 70, 1610–1618.

Hollinger, D. Y. (1989). Canopy organization and foliage photosynthetic capacity in a broad-leaved evergreen montane forest. *Funct. Ecol.* 3, 53–62.

Hollinger, D. Y. (1992). Leaf and simulated whole-canopy photosynthesis in two co-occurring tree species. *Ecology* 73, 1–14.

Hollinger, D. Y., Kelliher, F. M., Byers, J. N., Hunt, J. E., McSeveny, T. M., and Weir, P. L. (1994). Carbon dioxide exchange between an undisturbed old-growth temperate forest and the atmosphere. *Ecology* 75, 134–150.

Houghton, J. T., Jenkins, G. J., and Ephraums, J. J., eds. (1991). "Climate Change: The IPCC Scientific Assessment." Cambridge Univ. Press, Cambridge, UK.

Jarvis, P. G. (1993). Prospects for bottom-up models. *In* "Scaling Physiological Processes: Leaf to Globe" (J. R. Ehleringer and C. B. Field, eds.), pp. 115–126. Academic Press, San Diego.

Jarvis, P. G., and Leverenz, J. W. (1983). Productivity of temperate, deciduous and evergreen forest. *Encycl. Plant Physiol., New Ser.* 12D, 233–280.

Jarvis, P. G., James, G. B., and Landsberg, J. J. (1976). Coniferous forest. *In* "Vegetation and the Atmosphere" (J. L. Monteith, ed.), Vol. 2, pp. 171–240. Academic Press, London.

Koch, G. W., Amthor, J. S., and Goulden, M. L. (1994). Diurnal patterns of leaf photosynthesis, conductance, and water potential at the top of a lowland rain forest canopy in Cameroon: Measurements from the Radeau des Cimes. *Tree Physiol.* 14, 347–360.

Larcher, W. (1969). The effect of environmental and physiological variables on the carbon dioxide gas exchange of trees. *Photosynthetica* 3, 167–198.

Leuning, R., Wang, Y. P., and Cromer, R. N. (1991). Model simulations of spatial distributions and daily totals of photosynthesis in *Eucalyptus grandis* canopies. *Oecologia* 88, 494–503.

Medina, E., and Klinge, H. (1983). Productivity of tropical forests and tropical woodlands. *Encycl. Plant Physiol., New Ser.* 12D, 281–303.

Monteith, J. L. (1977). Climate and the efficiency of crop production in Britain. *Philos. Trans. R. Soc. London, Ser. B* 281, 277–294.

Norman, J. M. (1975). Radiative transfer in vegetation. *In* "Heat and Mass Transfer in the Biosphere" (D. A. De Vries and N. H. Afgan, eds.), pp. 187–205. Scripta Book, Washington, D.C.

Norman, J. M. (1993). Scaling processes between leaf and canopy levels. *In* "Scaling Physiological Processes: Leaf to Globe" (J. R. Ehleringer and C. B. Field, eds.), pp. 41–76. Academic Press, San Diego.

Oberbauer, S. F., and Strain, B. R. (1986). Effects of canopy position and irradiance on the leaf physiology and morphology of *Pentaclethra macroloba* (Mimosaceae). *Am. J. Bot.* 73, 409–416.

Oren, R., Schulze, E.-D., Matyssek, R., and Zimmermann, R. (1986). Estimating photosynthetic rate and annual carbon gain in conifers from specific leaf weight and leaf biomass. *Oecologia* 70, 187–193.

Pearcy, R. W. (1987). Photosynthetic gas exchange responses of Australian tropical forest trees in canopy, gap and understory microenvironments. *Funct. Ecol.* 1, 169–178.

Pearcy, R. W. (1990). Sunflecks and photosynthesis in plant canopies. *Annu. Rev. Plant Physiol. Plant Mol. Biol.* 41, 421–453.

Pearcy, R. W., and Troughton, J. H. (1975). C_4 photosynthesis in tree form *Euphorbia* species from Hawaiian rain forest sites. *Plant Physiol.* 55, 1054–1056.

Potter, C. S., Randerson, J. T., Field, C. B., Matson, P. A., Vitousek, P. M., Mooney, H. A., and Klooster, S. A. (1993). Terrestrial ecosystem production: A process model based on global satellite and surface data. *Global Biogeochem. Cycles* 7, 811–841.

Roy, J., and Salager, J.-L. (1992). Midday depression of net CO_2 exchange of an emergent rain forest tree in French Guiana. *J. Trop. Ecol.* 8, 499–504.

Running, S. W., and Coughlan, J. C. (1988). A general model of forest ecosystem processes for regional applications. I. Hydrologic balance, canopy gas exchange and primary production processes. *Ecol. Model.* 42, 125–154.

Running, S. W., and Nemani, R. R. (1991). Regional hydrologic and carbon balance responses of forests resulting from potential climate change. *Clim. Change* 19, 349–368.

Russell, G., Marshall, B., and Jarvis, P. G., eds. (1989). "Plant Canopies: Their Growth, Form and Function." Cambridge Univ. Press, Cambridge.

Schulze, E.-D., Fuchs, M. I., and Fuchs, M. (1977a). Spatial distribution of photosynthetic capacity and performance in a mountain spruce forest of northern Germany. I. Biomass distribution and daily CO_2 uptake in different crown layers. *Oecologia* 29, 43–61.

Schulze, E.-D., Fuchs, M., and Fuchs, M. I. (1977b). Spatial distribution of photosynthetic capacity and performance in a mountain spruce forest of northern Germany. III. The significance of the evergreen habit. *Oecologia* 30, 239–248.

Sellers, P. J., Berry, J. A., Collatz, C. J., Field, C. B., and Hall, F. G. (1992). Canopy reflectance, photosynthesis, and transpiration. III. A re-analysis using improved leaf models and a new canopy integration scheme. *Remote Sens. Environ.* 42, 187–216.

Sinclair, T. R., Murphy, C. E., Jr., and Knoerr, K. R. (1976). Development and evaluation of simplified models for simulating canopy photosynthesis and transpiration. *J. Appl. Ecol.* 13, 813–829.

Slatyer, R. O., and Morrow, P. A. (1977). Altitudinal variation in the photosynthetic characteristics of snow gum, *Eucalyptus pauciflora* Sieb. ex Spreng. I. Seasonal changes under field conditions in the snowy mountains area of south-eastern Australia. *Aust. J. Bot.* 25, 1–20.

Valentini, R., Scarascia Mugnozza, G. E., DeAngelis, P., and Bimbi, R. (1991). An experimental test of the eddy correlation technique over a Mediterranean macchia canopy. *Plant, Cell Environ.* 14, 987–994.

Verma, S. B., Baldocchi, D. D., Anderson, D. E., Matt, D. R., and Clement, R. J. (1986). Eddy fluxes of CO_2, water vapor, and sensible heat over a deciduous forest. *Boundary-Layer Meteorol.* 36, 71–91.

Verma, S. B., Kim, J., and Clement, R. J. (1992). Momentum, water vapor, and carbon dioxide exchange at a centrally located prairie site during FIFE. *J. Geophys. Res.* 97, 18629–18639.

Wang, Y. P., and Jarvis, P. G. (1990). Description and validation of an array model—MAESTRO. *Agric. For. Meteorol.* 51, 257–280.

Waring, R. H., Emmingham, W. H., Gholz, H. L., and Grier, C. C. (1978). Variation in maximum leaf area in coniferous forests in Oregon and its ecological significance. *For. Sci.* 24, 131–140.

Watts, W. R., Neilson, R. E., and Jarvis, P. G. (1976). Photosynthesis in sitka spruce (*Picea sitchensis* (Bong.) Carr.). VII. Measurements of stomatal conductance and $^{14}CO_2$ uptake in a forest canopy. *J. Appl. Ecol.* 13, 623–638.

Wofsy, S. C., Goulden, M. L., Munger, J. W., Fan, S.-M., Bakwin, P. S., Daube, B. C., Bassaw, S. L., and Bazzaz, F. A. (1993). Net exchange of CO_2 in a mid-latitude forest. *Science* 260, 1314–1317.

Wong, S. C., and Dunin, F. X. (1987). Photosynthesis and transpiration of trees in a *Eucalyptus* forest stand: CO_2, light and humidity responses. *Aust. J. Plant Physiol.* 14, 619–632.

Woodman, J. N. (1971). Variation of net photosynthesis within the crown of a large forest-grown conifer. *Photosynthetica* 5, 50–54.

Zotz, G., and Winter, K. (1993). Short-term photosynthesis measurements predict leaf carbon balance in tropical rain-forest canopy plants. *Planta* 191, 409–412.

18

Herbivory as a Canopy Process in Rain Forest Trees

Margaret D. Lowman

There awaits a rich harvest for the naturalist who overcomes the obstacles—gravitation, ants, thorns, rotten trunks—and mounts to the summits of jungle trees.
—William Beebe, "Tropical Wild Life" (1917)

I. Introduction

It has long been assumed that forests represented vast expanses of homogeneous green tissue, but this assumption is oversimplified. When walking through a forest, we usually focus our observations on a narrow band of green foliage, from ground level to about 2 m in height. This represents at most 10% of the foliage in mature forests, with the rest often high above our heads and consequently beyond our observations. Because the majority of plant–herbivore relationships occur where the foliage is located, it is obvious that herbivory as a forest process remains relatively unknown. Fifteen years of observations on herbivory in canopies have given me a different perspective: I no longer see forests as expanses of green, but instead I see them as great mosaics of holes in leaves!

As human beings making observations over relatively short time spans, we usually fail to appreciate the dynamic processes that separate life from death in complex canopy ecosystems. The life of a leaf, which comprises the building block of the forest canopy, is no exception. In a leaf's life span, it

Copyright © 1995 by Academic Press, Inc.
All rights of reproduction in any form reserved.

is critical to survive the vulnerable weeks of foliar expansion without being eaten. From a plant's point of view, there exists an evolutionary roulette of rendering one's green foliage less susceptible to successive generations of defoliators. Viewed from the perspective of a herbivore, a complex world of different bites must be recognized: soft versus tough, nutritious versus nonnutritious, old versus young, apparent versus nonapparent, rare versus common, and probably other choices that have not yet been detected by biologists.

Unlike Jack Putz, who can successfully carry out vine studies from his "rocking chair perspective" (Chapter 14), I have been forced to climb into the canopy to study herbivory. Alas, no combination of monkeys, telephoto lenses, helicopters, or low-flying planes offer the quality of detail to substitute for being up there. I have become resigned to a lifetime of aerial exploits, with high hopes (no pun intended) that during my waning years I would be capable of hobbling into the forest to address all my remaining hypotheses on seedlings.

The importance of understanding insect pests in forests has led to increased research on canopy defoliation (e.g., Barbosa and Schultz, 1987; Wong *et al.*, 1991; Lowman and Heatwole, 1992). The current controversies surrounding the biodiversity of tropical forests have stimulated ecological interest in the numbers of invertebrates in tropical tree crowns (Erwin, 1991; Gaston, 1991; Wilson, 1992). These two topics—insect pests and the biodiversity debate—have fostered a lively literature on herbivory and herbivores in forests, but little research has included the upper crowns.

This review will address the spatial and temporal heterogeneity of herbivory in rain forest tree crowns, with brief coverage of the characterization of their insect herbivore communities. I also emphasize the methodological challenges associated with herbivory as a canopy process, because the reliability of methods has an enormous impact on the accuracy of the results. Studies on understory shrubs and herbs in forests are not included, except where findings are relevant to canopy foliage.

I have also outlined possible protocols for future ecological studies of herbivory as a canopy process. Whereas biologists have successfully counted and measured the abundance of herbivorous mollusks on a two-dimensional system such as intertidal rocky shores (e.g., Underwood and Denley, 1984), the height and structural complexity make it more difficult to count and measure the grazing impacts of invertebrates in a forest canopy. Because of these logistic constraints, biologists have tended to study herbivory in systems other than forest canopies (e.g., Harper, 1977). However, the small number of studies reported here have used canopy access techniques, resulting in a plethora of ecological questions concerning the complexity of arthropod/canopy foliage interactions.

II. Herbivory as a Component of Forest Ecosystems

A. History of Herbivory Studies in Forest Canopies

Historically, most herbivory studies have involved the measurement of levels of defoliation in forests at one point in time. Foliage was typically sampled near ground level in temperate deciduous forests, where annual losses of 3–10% leaf surface area were reported (reviews in Bray and Gorham, 1964; Landsberg and Ohmart, 1989). Most studies, however, could not be extrapolated to evergreen rain forests for three reasons: (1) temperate deciduous forests have a comparatively simply phenology with an annual turnover of leaves; (2) measurements were sometimes made from senescent leaves retrieved from the forest floor; and (3) replicated stratified sampling was rarely attempted. In short, defoliation was treated as a discrete, snapshot event (Diamond, 1986), accounting for neither temporal nor spatial variability.

More recent studies have expanded in scope to include temporal and spatial factors to explain heterogeneity of herbivory throughout the canopy. Five noteworthy discoveries in the history of herbivory research are:

1. An important attribute affecting levels of foliage consumption is age of leaf tissue, with soft, young leaves being preferred over old, tough leaves (e.g., Coley, 1983; Lowman, 1985).
2. The most abundant herbivores in forests are insects in terms of both number and estimated impacts (reviewed in Schowalter *et al.*, 1986; Lowman and Moffett, 1993). In some ecosystems, however, mammals are also important, for example, monkeys, koalas, and tree kangaroos (review in Montgomery, 1978).
3. Canopy grazing levels are not homogeneous *between* forests, but range from negligible losses to total foliage losses, and are heterogeneous *within* forests, varying with plant and herbivore species, height, light regime, phenology, age of leaves, and individual crown (e.g., Lowman, 1992).
4. The assumptions common in the 1960s (i.e., that herbivory averaged 5–10% annual leaf area loss and was homogeneous throughout forests) were oversimplified and underestimated, particularly for evergreen forests (e.g., Fox and Morrow, 1983).
5. Foliage feeders are featured in the ecological literature as the most common type of herbivore. However, sap-suckers may also be important although they have not been as well studied. Foliage consumption is reputedly easier to measure than sap consumption, yet even for measurement of folivory, standard protocols are not well established (see Lowman, 1984b).

During the 1980s, canopy access was developed, which expanded the scope of foliage sampling, and studies of herbivory in evergreen tropical forests subsequently increased.

B. Role of Herbivory in Canopy Processes

The consumption of plant material by herbivores is a subject of great economic and ecological importance (e.g., Barbosa and Schultz, 1987; Price *et al.*, 1991). Because forest canopies contain the bulk of terrestrial photosynthetic material that is involved in the production of oxygen and maintenance of global biogeochemical cycles (review in Wilson, 1992), the processes affecting canopy foliage have direct consequences on the health of our planet. Leaf predation is an example of such a process. The loss of foliage by predators can occur by direct consumption or by less obvious impacts such as mining, sap-sucking, and leaf-tying. Herbivory affects foliage during almost all stages in the life of a leaf, and leaves have subsequently become adapted with defenses against predation (Fig. 1). Levels of herbivory range from negligible grazing to the mortality of leaves, branches, entire crowns, and sometimes whole forest stands.

The study of herbivory as an integrated process throughout a forest stand requires information on many aspects of forest biology, including plant phenology, demography of insect populations, leaf growth dynamics, tree architecture, foliage quality, physical environment, nutrient cycling, and plant succession. Recent interest in plant–insect relationships has emphasized single-species relationships with fewer investigations at the ecosystem level. It has also centered around studies of shrubs and herbs. Few studies of trees exist, with even fewer that involve entire forest stands and almost none that includes the canopy. Ironically, forest canopies comprise an ecological arena where plant–insect relationships may be most complex in terms of spatial, taxonomic, temporal, structural, or any other measurable factor.

Herbivory is an ecological process that affects all canopy components (Fig. 2), either directly or indirectly. Foliage is removed directly by herbivores (called "primary consumption") or else it escapes, resulting in "secondary consumption," whereby foliage senesces and then decomposes via arthropods on the forest floor. Both fates comprise pathways that link herbivory to nutrient cycling in the forest ecosystem, either via frass or via leaf litter. Foliage that is partially grazed by herbivores is called "herbivory," whereas foliage that is grazed in entirety (or grazed extensively and then senesces) is classified as "defoliated." It is important to recognize that herbivory is the direct effect of grazing, whereas defoliation results in mortality that may only be partially a consequence of the grazing mechanism.

Herbivore populations fluctuate in the canopy and in turn affect the populations of other invertebrates and of birds and mammals that feed

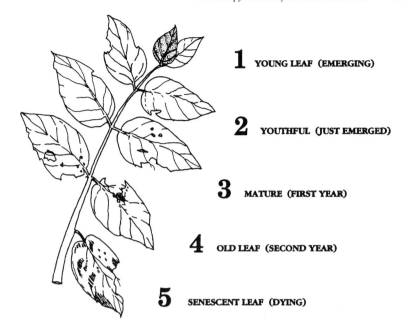

1 YOUNG LEAF (EMERGING)

2 YOUTHFUL (JUST EMERGED)

3 MATURE (FIRST YEAR)

4 OLD LEAF (SECOND YEAR)

5 SENESCENT LEAF (DYING)

age class	POTENTIAL FACTORS OF MORTALITY	MECHANISMS OF PROTECTION
1.	Desiccation	Morphology, phenology
	Herbivory	Morphology, phenology, chemistry
	Physical Damage	Morphology, phenology
2.	Desiccation	Morphology, phenology
	Herbivory	Morphology, phenology, chemistry
	Physical Damage	Morphology, phenology
3.	Herbivory	Morphology, phenology, chemistry
	Physical Damage	Morphology, phenology
4.	Herbivory	Morphology, phenology, chemistry
	Epiphylly	Morphology, chemistry
5.	Decay	Morphology, phenology, chemistry
	Age	

Figure 1　Leaf age classes as defined for use in rain forest canopy growth and herbivory studies, with factors affecting survival at each stage.

on these herbivorous organisms (e.g., Woinarski and Cullen, 1984). Even stand growth and dynamics may be ultimately affected by herbivory, and by the susceptibility of a species to grazing (review in Schowalter *et al.*, 1986). The impact of leaf consumption on herbs, seedlings, and shrubs has been quantified in terms of mortality, succession, and compensatory growth (e.g., Lowman, 1982; Coley, 1983; Marquis, 1991). Such factors are more

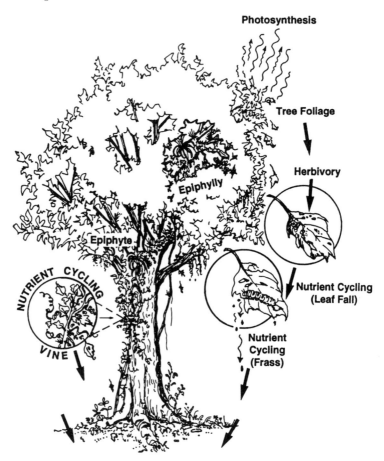

Figure 2　Canopy components and processes that are affected by herbivory in a forest stand.

difficult to measure for tall trees and across entire forest stands. Examples of herbivory that have been integrated with other aspects of forest dynamics include studies on the spatial distribution of canopy insect populations in the Australian rain forest tree *Argyrodendron actinophyllum* (Basset, 1991), nutrient cycling via frass or litterfall pathways (Lowman, 1988), pest outbreaks and stand mortality (e.g., Lowman and Heatwole, 1992), and herbivory in relation to phenology of stands (Schultz and Baldwin, 1982).

III. Canopy Foliage and Estimates of Its Defoliation

In many temperate forests and some seasonal tropical forests, annual and cumulative herbivory are the same: every year, a full cohort of leaves emerges and senesces (Kikuzawa, 1983). Herbivory in evergreen forest canopies may be more complicated, because the cycles of leaf turnover are not always seasonally distinct (e.g., Lowman, 1992). Consequently, ecologists in these habitats are confronted with many cohorts or leaf populations within one crown, requiring a more complex sampling design to ascertain both annual defoliation and cumulative herbivory over a leaf's life span.

In evergreen forests, leaf longevity is extremely variable, and the canopy is a complex mosaic of different-aged leaves, with different susceptibilities to herbivores. Evergreen rain forest leaves have life spans ranging from as short as 4–6 months (e.g., *Dendrocnide excelsa,* Urticaceae; Lowman, 1992) up to 25 years (e.g., *Araucaria* sp., Aracaceae; Molisch, 1928). The average age of an Australian subtropical rain forest canopy leaf ranged from 2–4 years (sun) to 4–12 years (shade) (Lowman, 1992).

In recent years, the complex temporal and spatial patterns of leaves in forest canopies have caused ecologists to expand their sampling designs. For example, the traditional methods of measuring herbivory by destructive sampling of small quantities of leaves have been expanded (reviews in Lowman, 1984b; Landsberg and Ohmart, 1989). Whereas earlier measurements of forest herbivory were conducted over short time spans, were restricted to understory foliage, and involved very little replication within and between crowns, more recent studies have incorporated larger sampling regimes. And when herbivory was monitored over longer periods (>1 year) and included wider ranges of leaf cohorts (including different age classes, species, and heights), higher levels of grazing were reported (Coley, 1983; Lowman, 1985; Lowman and Heatwole, 1992). Long-term measurements have also illustrated the high variability of herbivory, both temporally and spatially, within a stand (e.g., Coley, 1983; Lowman, 1985; Brown and Ewel, 1987, 1988).

Herbivory in canopies ranges from negligible (e.g., Schowalter *et al.,* 1981; Ohmart *et al.,* 1983) to over 300% of annual foliage production in cases where eucalypts refoliated three successive times after defoliation (Lowman and Heatwole, 1992). The comparison of herbivory measurements from several forest canopies presented in the following illustrates the variability in levels of defoliation *between* forest types, although the different methods employed may be a source of some of the apparent variation (Table I).

Herbivory levels vary significantly both between species and between forest types, as illustrated by my long-term studies in Australian rain forests (Fig. 3). I originally hypothesized that evergreen forests with lower diversity would have higher herbivory than neighboring evergreen forests with higher diversity (Lowman, 1982). The cool temperate rain forest, where *Nothofagus moorei* dominated over 75% of the canopy, averaged an annual 26% leaf area loss to grazing insects (Selman and Lowman, 1983). The majority of that was due to a host-specific chrysomelid beetle that fed

Table I Studies of Herbivory in Forests (Including Understory)

Location	Forest type	Level of grazing	Technique[a]	Source
Costa Rica	Tropical forest	7.5% (new leaves)	1	Stanton (1975)
	Tropical evergreen	30% (old)	1	
Panama	Tropical evergreen	13%	1	Wint (1983)
Panama (BCI)	Tropical evergreen	8% (6% insect;		
		1–2% vertebrates)	2,5	Leigh and Smythe (1978)
		15%	2,5	Leigh and Windsor (1982)
	Understory only	21% (but up to 190%)	3	Coley (1983)
Puerto Rico	Tropical evergreen	7.8%	2	Odum and Ruiz Reyes (1970)
		5.5–16.1%	4	Benedict (1976)
		2–5%	4	Schowalter (1994)
New Guinea	Tropical evergreen	9–12%	1	Wint (1983)
Australia	Montane or cloud	26%	4	Lowman (1984b)
	Warm temperate	22%	4	Lowman (1984b)
	Subtropical	14.6%	4	Lowman (1984b)
Africa–Cameroon	Tropical evergreen	8–12%	4	Lowman (1993)
North America	Temperate deciduous	7–10%	1	Bray (1961)
	Temperate deciduous	1–5%	4	Schowalter et al. (1981)
	Coniferous	<1%	7	Schowalter (1989)
Australia	Evergreen	15–300%	4	Lowman and Heatwole (1992)
	Dry	11–60%	1	Fox and Morrow (1983)
		3–6%	6	Ohmart et al. (1983)
Europe	Deciduous	7–10%	1	Nielson (1978)

[a] 1, Visual ranking; 2, litter trap; 3, graph paper or template squares; 4, LA meter; 5, other calculations; 6, insect frass; 7, estimation of missing or truncated needles.

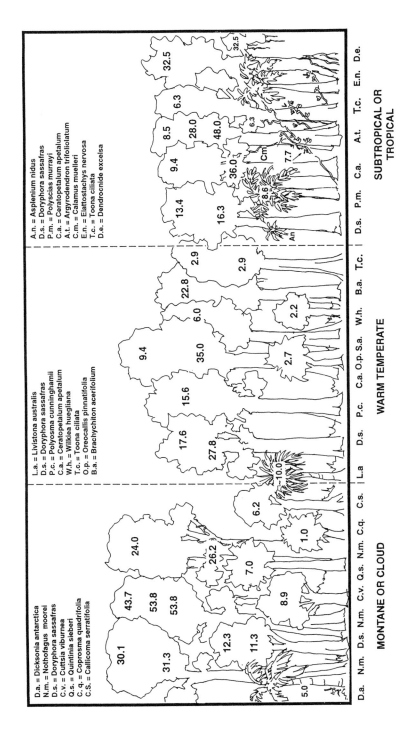

Figure 3 Variation in herbivory between species, heights, and sites along an elevational gradient in Australia.

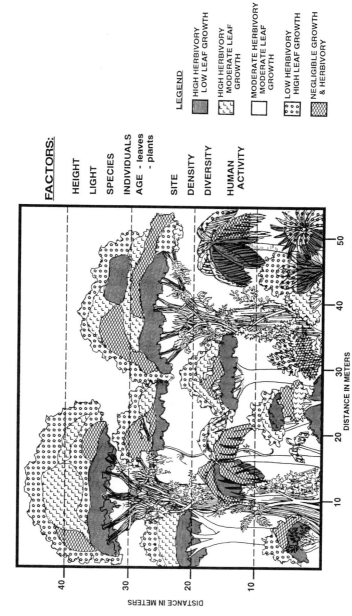

FACTORS:

HEIGHT

LIGHT

SPECIES

INDIVIDUALS

AGE - leaves
      - plants

SITE

DENSITY

DIVERSITY

HUMAN
ACTIVITY

LEGEND

HIGH HERBIVORY
LOW LEAF GROWTH

HIGH HERBIVORY
MODERATE LEAF
GROWTH

MODERATE HERBIVORY
MODERATE LEAF
GROWTH

LOW HERBIVORY
HIGH LEAF GROWTH

NEGLIGIBLE GROWTH
& HERBIVORY

DISTANCE IN METERS

Figure 4 Schematic representation of "hot spots" in the canopy, where herbivores are attracted to foliage that is more susceptible. It should be possible to sample for insect herbivores in specific regions of the canopy rather than throughout this complex array of foliage. "Hot spots" will vary over time, due to the differences in phenology and foliage qualities among species.

exclusively on the young leaves of *N. moorei* during the spring flush. In contrast, the subtropical rain forests, where no species occupied more than 5% of the canopy within a stand (Lowman, 1985), averaged only 15% annual leaf area losses to insect grazers. Herbivory also varied significantly among species. *Toona australis* (Meliaceae), which is relatively rare and is annually deciduous, averaged <5% losses, whereas neighboring canopies of *Dendrocnide excelsa* (Urticaceae) that colonize gaps averaged over 40% annual leaf area losses (Lowman, 1992).

In addition to variation in grazing levels between species and stands, herbivory is variable within individual crowns. The heterogeneity of defoliation is a consequence of a leaf's environment and phenology, with different leaf cohorts exhibiting different susceptibilities to grazing (*sensu* Whittam, 1981). From long-term studies, I was able to pinpoint "hot spots" in the canopy where grazing will be predictably higher (Fig. 4). These "hot spots" included areas of predictable susceptibility to herbivores, such as new leaf flushes, colonizing species that are characterized by soft tissue, lower shade regions of the canopy where insects aggregate to feed in the absence of predators, and canopy regions that attracted more insects due to the presence of flowers, epiphytes, or vines (e.g., Lowman, 1992; Lowman *et al.*, 1993b).

For example, long-term measurements of *Nothofagus moorei* showed that crowns exhibited different grazing levels with leaf age, different stands, and time (Selman and Lowman, 1983). *Nothofagus moorei* had approximately eight cohorts of leaves present in the canopy at one point in time (Fig. 5), each with varying levels of susceptibility to insect attack. Young leaves that emerged during spring (Oct.–Nov.) were the most preferred by the beetle larvae who emerged synchronously with flushing, whereas old leaves (>1 year) from summer flushes and from the previous year were highly resistant to grazing. In addition, herbivory varied significantly between branches and individual crowns, but not with light regime or height (Selman and Lowman, 1983).

More comparisons between forests are needed to better understand the impact of herbivory on forests. For example, the annual levels of defoliation in Australian trees ranged from as low as 8–15% in subtropical rain forests to as high as 300% in nearby dry sclerophyll (*Eucalyptus*) stands (Lowman, 1992; Lowman and Heatwole, 1992). In this case, different mechanisms are clearly regulating insect defoliators and subsequent foliage responses in two forests. Gentry cited the need for comparisons between Central and South American forests as the stimulus for a recent Association for Tropical Biology symposium and the subsequent production of a volume on four neotropical forests (see Gentry, 1990). Similarly, a recent American Institute of Biological Sciences session was held to stimulate comparisons between

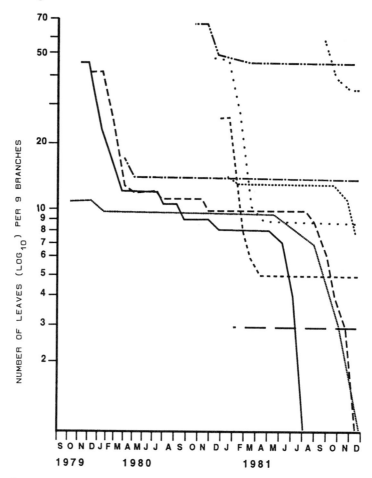

Figure 5   Leaf survivorship curves for leaves of *Nothofagus moorei* (Fagaceae) in cool temperate rain forests of Australia that flushed during a 2-year period (September 1979–September 1981) expressed on a $\log_{10}$ scale. Leaves were tallied from nine branches and illustrate the complexity of cohorts throughout the canopy of a tree that exhibits a bimodal seasonal flush.

Neotropical and Paleotropical forests (Lowman, 1993). The prospect of increased ecological comparisons between forests is an incentive to develop better protocols for field sampling of events such as grazing. For example, what species are appropriate to sample? Is there greater variation within or between forests? And how do we tackle these questions with statistical and biological accuracy?

IV. Dynamics of Herbivores in Forest Canopies

Although mammals are obviously important herbivores in grasslands and other ecosystems, there is relatively little research on the impacts of mammals grazing in forest canopies. Koalas, tree kangaroos, sloths, birds, monkeys, and porcupines are some of the well-known herbivorous mammals that frequent the canopy (review in Montgomery, 1978). In this chapter, insects remain the focus, because their role as herbivores is purportedly more widespread throughout forests. More information on canopy mammals is available in Chapters 10 and 11.

A. Patterns of Distribution and Abundance of Insect Herbivores in the Canopy

In a classic study, Southwood (1961) examined the species of insects in tree canopies in Great Britain. For several species of temperate deciduous trees, he found that more diversity was associated with trees that had been established over a long time period as compared to species introduced more recently. His insecticidal knockdown procedure has since been altered to include "misting" (Kitching *et al.*, 1993), "fogging" (Erwin, 1982), and "restricted canopy fogging" (Basset, 1992), as more biologists become curious about the variety and numbers of insects in tree canopies. Similar to the artifacts of sampling for foliage consumption measurements, the variability in methodology to assess the diversity of insects in trees also leads to discrepancies in our estimates of herbivorous insects.

Just how many insects are found in tree canopies, and what proportions are herbivorous? (This topic is reviewed more comprehensively in Chapter 5.) Over the past 15 years, the hypothesized numbers of species of arthropods in trees has expanded from 1–2 million (Southwood, 1978) to far higher estimates of 8–10 million (Adis, 1990), 10–30 million (Erwin, 1982, 1991), and as high as 100 million (Wilson, 1992), of which just under 50% of insect species classified are reputed to be herbivorous (Wilson, 1992). From these predictions, one can speculate that as many as 49 million herbivore species may exist!

The major contributions to our knowledge of the canopy herbivore community have been extrapolations using ground-based studies of saplings and shrubs. It is obvious that woody plants sustain more insect species than herbaceous species, although few studies have quantified these differences (but see Basset and Burckhardt, 1992). In herbaceous plants, hosts with greater chemical complexity appear to support herbivores with higher degrees of specificity (Berenbaum, 1981). Will canopy trees show similar patterns to herbs? Seedlings of a Neotropical tree with high tannin concentrations sustained lower herbivory levels than seedlings with low tannin levels

(Coley, 1986), but can we extrapolate from seedlings to adult trees? It is daunting to design a rigorous sampling regime that would replicate within and among tall trees; but as canopy access becomes easier, the opportunities for this work may arise.

Herbivores, as with other invertebrate groups, are assumed to attain higher diversity in tropical forest canopies, with progressive decreases in temperate forests (e.g., Kitching *et al.*, 1993). Four hypotheses are posed to explain this population decline away from the equator. These are: (1) historical—higher latitudes are out of equilibrium due to recent glaciation (Wallace, 1878); (2) structural—the tropical environment contains more niches (MacArthur and MacArthur, 1961); (3) dynamic—predation and competition in the tropics permit more species to coexist (Janzen, 1970; Connell, 1978); and (4) energetic—the higher stability of productivity in the tropics permits greater specialization within ecological niches (Connell and Orians, 1964). In canopies, more data are required to test hypotheses to explain the underlying reasons for the distribution of insect herbivores.

The density of herbivores in the canopy might be predicted to be higher in the upper crown where new foliage is relatively more abundant, as insects prefer to feed on young foliage (Coley, 1983; Lowman, 1985). However, herbivores also aggregate in the lower shade regions of crowns during leaf-flushing periods (Lowman, 1985). The methodological problems associated with quantifying such densities of small, numerous organisms are enormous. Basset *et al.* (1992), using branch-clipping techniques, obtained three times higher arthropod densities in the canopy than in the shrub layer of a lowland rain forest in Cameroon. In a geographical comparison, they obtained 20 arthropods per m^2 in tropical foliage as compared to a range of 19–78 herbivores per m^2 in temperate trees (Basset and Burckhardt, 1992) and 11 herbivores per m^2 in a subtropical tree (Basset and Arthrington, 1992). As predicted by Elton (1973), densities of arthropods may be higher in temperate than in tropical canopies, but only during peaks such as in midsummer.

In temperate deciduous forest canopies, the abundance of invertebrates was found to be higher near ground (0–2 m) than in the canopy (>20 m) (Lowman *et al.*, 1993a), probably due to the increase in the number of niches and more favorable microclimate nearer to the ground. This pattern is reversed in the tropics (Erwin, 1982; and Chapter 5).

Homoptera are a relatively well-studied group of herbivores, but even their abundance is difficult to interpret. Using nets, Elton (1973) measured a density of 0.044–0.078 Homoptera per m^2 ground surface, whereas Wolda (1979), by fogging at 0–16 m, measured 3.5–11.8 Homoptera per m^2 of ground surface. Wolda also found more Homoptera in a canopy with vines as compared to canopy trees without vines, and he suggested that the more complex canopy architecture may support more insects.

In initial results of a long-term fogging analysis in Australian rain forests, proportions of herbivorous groups (of the total invertebrate catch) were 25% in tropical forest, 13% in subtropical forest, and 23% in cool temperate forest (Kitching *et al.*, 1993). In dry sclerophyll forest canopies, pest outbreaks occur with greater frequency, leading to greater fluctuations in insect numbers. For example, during an outbreak of psyllids, herbivores comprised as high as 71.3% of the total catch (Majer and Recher, 1988). The next decade of research, with improvements in the field techniques and analyses of canopy insects, should bring a better understanding of the distribution of herbivores in tree crowns.

B. Crown Phenology and Stand Dynamics in Relation to Herbivores

The seasonality of growth in the canopy is critical to the process of herbivory. Leaves in tree crowns are more susceptible to being eaten when they are young, soft, situated lower in the crown, grown in shade conditions of the lower canopy, and of a mesophyllous, pioneering species as compared to a sclerophyllous, climax species (Coley, 1983; Lowman, 1985). For herbivores, the availability of evergreen, continuously flushing foliage differs from that of deciduous or intermittently flushing crowns.

In the canopies of five Australian rain forest tree species, I found that most defoliators fed as part of a guild, that is, several insect species were repeatedly observed eating leaves of several neighboring tree species, but they concentrated on leaves of similar phenology comprising the same age and texture of leaf tissue (Lowman, 1985). Similar to Basset's (1992) results on booyong trees, I found few herbivores exhibiting host specificity. However, those that did specialize consumed large amounts of foliage on their host tree. For example, a chrysomelid beetle (*Novocastria nothofagi* Selman) annually consumed >50% of the new leaves of *Nothofagus moorei* (Fagaceae). The beetle larvae emerged synchronously in spring with the annual flush of this beech. *Nothofagus moorei* leaves live for 2 years; consequently, annual spring leaf flushes of beech replace half of the canopy each year in this montane or cool temperate rain forest canopy tree (Fig. 5). Because *N. moorei* grows in single-species stands, the beetle's host-specific behavior has a significant impact on the canopy in this ecosystem. No canopy trees were observed to die from defoliation during 15 years of study, although this is too short a duration in the lifetime of a tree that reputedly persists for many hundreds of years.

In summary, a growing body of literature on forest insects makes it possible to predict regions of higher diversity or "hot spots" of arthropods in the canopy, namely, forests situated in the tropics, young leaves, single-species stands, shade leaves in the lower canopy with their softer texture and lower toxicity, and pioneering species with soft leaves.

C. Feeding Behavior of Herbivores in Tree Crowns

Much has changed since Hairston *et al.* (1960) claimed that green foliage was equally available to herbivores, and that the impact of insects on forests was negligible. Research on the relationships between insects and foliage has expanded into chemistry, phenology, seasonality, species, succession, age of leaf tissues, toughness, and a multitude of factors—in short, the green world is complex!

Despite the enormous numbers of forest Lepidoptera (over 120,000 species described worldwide), fewer than 15 species have been documented as causing outbreaks (Mason, 1987). Insect outbreaks are most frequently associated with temperate forests, probably because their economic impacts are more closely monitored (although intermittent accounts in the literature attest to the existence of pest epidemics in tropical forests; e.g., Wolda and Foster, 1978; Selman and Lowman, 1983; Wong *et al.*, 1991). Foliage loss during outbreaks can reach 100% defoliation (reviewed in Schowalter *et al.*, 1986) or even 300% annually in evergreen species such as *Eucalyptus* that repeatedly undergo reflushing of their canopies after defoliation (Lowman and Heatwole, 1992).

In his classic studies, Janzen (1970) suggested that high levels of host specificity exist in tropical plant–insect relationships, but his assumptions were primarily based on observations of seed predators. In later reports, he found that the majority of caterpillars in a Costa Rican deciduous forest were monophagous (Janzen, 1988). Of the few studies of the host specificity of other herbivores in tropical tree canopies, lower levels of host specificity were observed. Erwin (1982) found 20% host specificity in Coleoptera in *Luehea seemanii* in Panama. In his comprehensive studies of the invertebrates of *Argyrodendron actinophyllum* in Australian subtropical rain forest, Basset found only 11% host-specific herbivores (1992), results similar to earlier collections on five Australian canopy trees (Lowman, 1985). Of the relatively few lepidopteran defoliators on Asian dipterocarps, most were polyphagous (Holloway, 1989). There is still too little information to fully explain the feeding behavior of herbivores in trees.

Few data exist on other types of herbivores such as miners and sap-suckers. Approximately 31 mines per m^2 leaf area were calculated for a subtropical canopy tree (Basset, 1991), whereas less than 2% of leaves in nearby dry sclerophyll canopies contained any galls (Lowman and Heatwole, 1992). Although the impact of foliage feeders may be directly measured in terms of leaf area lost, the damage to a tree from mining or sap-sucking is less apparent without detailed physiological studies.

V. Impact of Ecological Methods on Our Assessment of Herbivory

A. Methods to Measure Herbivory: Are They Accurate?

The turnover of photosynthetic tissue not only has direct impact on the growth and maintenance of the trees, vines, epiphytes, and herb layers of the forest, but the production of green tissue is indirectly responsible for the maintenance of all animal life in the canopy. The ability of biologists to measure foliage and predict photosynthetic activity in forests has become an important topic in advocating the conservation of forests. It is obviously important to make accurate measurements of both the production of foliage in a forest and of the removal of leaf material by herbivores in order to assess forest productivity.

The methods used to measure herbivory have direct consequence on our understanding of herbivore dynamics. As biologists historically became interested in defoliation, a plethora of literature on herbivory was produced, much of which utilized different (and not entirely comparable) methods. For example, the techniques used to measure foliage losses include visual estimates (e.g., Wint, 1983), graph paper tracings (Lowman, 1984b), templates in the field (Coley, 1983), and leaf area meters (Lowman, 1984b). Similarly, the sampling designs varied, including leaves collected in litter traps (Odum and Ruiz-Reyes, 1970), leaf selection undefined (Bray, 1961), leaves marked in the understory (Coley, 1983) and upperstory (Lowman, 1985), leaves marked along a vertical transect (Lowman, 1992), and frass collections (Ohmart *et al.*, 1983). Obviously, such sampling designs may be adequate for the particular hypothesis they address, but they are not conducive to intersite comparisons (e.g., Lowman, 1987). In some cases, literature reviews have misquoted herbivory levels, perhaps because it is so difficult to interpret the various methods employed by different studies (e.g., Landsberg and Ohmart, 1989).

Comparison of discrete versus long-term measurement techniques in rain forest canopies revealed discrepancies of up to fivefold, with long-term studies producing significantly higher measurements. For example, estimates of herbivory levels in Neotropical saplings were three times higher than in previous studies that used discrete, harvested leaves (21% in Coley, 1983, versus 7% in Odum and Ruiz-Reyes, 1970). Similarly, long-term measurements of coral cay shrubs produced estimates of 21% area missing (Heatwole *et al.*, 1981) compared to levels of 2–3% measured by discrete sampling (Lowman, 1984a). Grazing in some plant communities may be higher than recorded previously from discrete measurements of missing-leaf area, which results in an underestimation of the impact of herbivory. Temporal variability in levels of herbivory further complicates our ability to monitor this canopy process.

In addition to the potential errors from methods that do not account for heterogeneity of foliage throughout the canopy, other methods must be used with caution. Daily rates of defoliation are useful but can be misleading if measured only over short durations, as they will not account for seasonal differences. The use of grazing categories (e.g., ranking 1–5) are useful for quick, rapid assessment, but such information may not be transformable into numbers that can be statistically tested. And the assumption that the presence of leaf petioles indicates 100% defoliation may be misleading, because physical factors are also responsible for loss of a leaf blade. The extent to which methods may alter results remains a critical issue in the literature on herbivory.

B. Possible Protocols for Future Sampling

I pursued my studies of rain forest herbivory at Sydney University, where I shared office space with the graduate students of Tony Underwood, whose concepts of experimental design on rocky intertidal organisms have greatly improved scientific methods in that ecosystem. How, I wondered, could one quantify and sample with similar statistical rigor in the canopy? Obviously, the forest canopy has several obvious differences from the rocky intertidal, namely:

1. it is extremely three dimensional with heights of up to 50–60 m (versus two dimensional on the rocky shore);
2. it has organisms ranging a 100-fold in size, for example, seedlings and adult trees, thrips and sloths (versus a more homogeneous range); and
3. it has an air substrate (versus water) that is difficult for human mobility.

The logistics of counting and manipulating herbivores in the forest canopy may be more complicated than on an intertidal rock platform, but the advantages of implementing a sound sampling protocol are enormous.

Different components of a forest canopy (Fig. 1) must be quantified to measure a specific canopy process. In the case of herbivory, all foliage components plus active herbivores require measurement. Initial observations, using ropes or a platform, are ideal for determining the organisms involved in foliage grazing. It should be emphasized that nocturnal surveys are also important for evaluating herbivore activity. Sampling protocols are illustrated at different spatial scales, ranging from ecosystem (Fig. 6) to site (Fig. 7) to individual tree (Fig. 8). At least seven spatial scales (Fig. 6) are important for a thorough ecological understanding of herbivory as a canopy process, although different studies may prefer to approach research at the level of species or of ecosystem.

During 13 years of canopy research in Australia, I developed protocols for measuring herbivory in rain forest tree canopies, using replicate sites, trees, canopy heights, branches, and leaves. In the subtropical hammocks

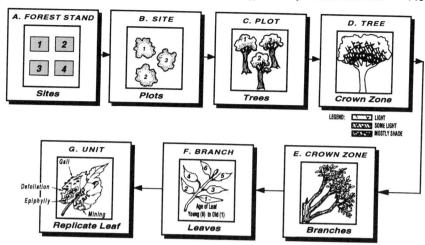

Figure 6 Experimental design for canopy foliage studies, illustrating the replication recommended at the spatial scales of forest stand, site, plot, tree, crown zone, branch, and unit (leaf).

of Florida, a group from The Marie Selby Botanical Gardens is currently working out protocols to measure herbivory and growth dynamics of vines and epiphytes. The ideal replication for ecological sampling in the canopy involves spatial replication (Fig. 6) with a temporal component (Fig. 1). Monthly sampling is ideal for herbivory studies, with weekly or daily measurements during periods of active leaf flushing and flowering. Because this

Figure 7 Generalized subtropical oak–pine hammock site regime for canopy foliage studies, illustrating the structural regions of understory, midcanopy, and upper canopy that vary in light levels and other microclimatic factors. Shade leaves are indicated by gray shading, and sun leaves are shown in white.

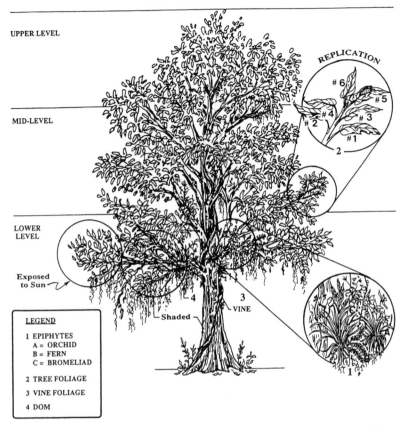

Figure 8 Schematic diagram of the components of a forest canopy to sample, including epiphytes (orchid, fern, and bromeliad), tree foliage, vines, and DOM (dead organic matter).

frequency of sampling is not possible for most studies that are situated in remote sites, techniques that are more rapid and time-efficient must be developed, as with most ecological studies involving dynamic interactions.

VI. Conservation of Forest Canopies and Future Implications of Herbivory as an Ecological Process

As habitat destruction continues to reduce the world's forests, canopies will become reduced both in area and in diversity of species. It is believed that many canopy organisms have already disappeared before they were ever scientifically described. Most of them are presumed to be insects (Wilson, 1992), including many herbivores. The complex interactions between canopy foliage and defoliators is an arena for ecological change as a consequence of human activities. The concept of a forest pest usually denotes a foliage-feeding insect and such outbreaks are often the result of human perturbation (e.g., gypsy moth, reviewed by Elkington and Liebold, 1990). Another example is the death of millions of eucalypt trees in Australia, the result of a complex impact of human activities in the rural regions resulting in outbreaks of a scarab beetle (Lowman and Heatwole, 1992). Although pest outbreaks are still regarded as relatively rare events in forests, it is obvious that the natural processes regulating canopy foliage and their defoliators require further study to fully understand the implications of imbalances that result from human impacts.

Long-term studies and quantified maps of herbivore distributions and of the patchiness of herbivory in tree canopies are needed. Some of the many unanswered questions in canopy foliage–insect interactions include:

1. How do spatial and temporal factors affect the distribution of insect herbivores throughout the canopy—both *within* and *between* different tree crowns?
2. What is the variation in levels of herbivory over longer time scales in canopies?
3. How does stand diversity in forests affect the feeding patterns of herbivores?
4. What are the small-scale and large-scale consequences of patchy herbivore attack within one region of an individual canopy?
5. What is the trophic structure of canopy insects?
6. Ultimately, what effect does herbivory have on other processes such as photosynthesis and nutrient cycling in the canopy? Does it influence succession?

Herbivores and the process of herbivory are a complex and integral component of the forest ecosystem. In the tropics, the opportunities to study these ecological interactions under natural conditions are becoming more reduced, as human beings increasingly destroy and alter tropical forests. However, studies of insect pests in disturbed or regenerating forests will become more commonplace. In *The Lorax* (1971, Random House), Dr. Seuss presented the potential problems concerning loss of forests and their canopy inhabitants, and suggested a solution,

Plant a new Truffula. Treat it with care.
Give it clean water. And feed it fresh air.
Grow a forest. Protect it from axes that hack.
Then the Lorax and all of his friends may come back. . . .

References

Adis, J. (1990). Thirty million arthropods—Too many or too few? *J. Trop. Ecol.* 6, 115–118.

Barbosa, P., and Schultz, J. C. (1987). "Insect Outbreaks." Academic Press, Orlando, FL.

Basset, Y. (1991). The spatial distribution of herbivory, mines and galls within an Australian rain forest tree. *Biotropica* 23, 271–281.

Basset, Y. (1992). Host specificity of arboreal and free-living insect herbivores in rain forests. *Biol. J. Linn. Soc.* 47, 115–133.

Basset, Y., and Arthrington, A. H. (1992). The arthropod community associated with an Australian rainforest tree: Abundance of component taxa, species richness and guild structure. *Aust. J. Ecol.* 17, 89–98.

Basset, Y., and Burckhardt, D. (1992). Abundance, species richness, host utilization and host specificity of insect folivores from a woodland site, with particular reference to host architecture. *Rev. Suisse Zool.* 99, 771–791.

Basset, Y., Aberlenc, H.-P., and Delvare, G. (1992). Abundance and stratification of foliage arthropods in a lowland rain forest of Cameroon. *Ecol. Entomol.* 17, 310–318.

Benedict, F. (1976). "Herbivory Rayes and Leaf Properties in Four Forests in Puerto Rico and Florida." PhD Thesis, University of Florida, Gainesville.

Berenbaum, M. (1981). Patterns of furanocoumarin distribution and insect herbivory in the Umbelliferae: Plant chemistry and community structure. *Ecology* 62, 1254–1266.

Bray, J. R. (1961). Primary consumption in three forest canopies. *Ecology* 45, 165–167.

Bray, J. R., and Gorham, E. (1964). Litter production in forests of the world. *Adv. Ecol. Res.* 2, 101–157.

Brown, B. J., and Ewel, J. J. (1987). Herbivory in complex and simple tropical successional ecosystems. *Ecology* 68, 108–116.

Brown, B. J., and Ewel, J. J. (1988). Responses to defoliation of species-rich and monospecific tropical plant communities. *Oecologia* 75, 12–19.

Coley, P. D. (1983). Herbivory and defensive characteristics of tree species in a lowland tropical forest. *Ecol. Monogr.* 53, 209–233.

Coley, P. D. (1986). Costs and benefits of defense by tannins in a Neotropical tree. *Oecologia* 70, 238–241.

Connell, J. H. (1978). Diversity in tropical rain forests and coral reefs. *Science* 199, 1302–1310.

Connell, J. H., and Orians, G. (1964). The ecological regulation of species diversity. *Am. Nat.* 98, 399–414.

Diamond, J. (1986). Overview: Laboratory experiments, field experiments and natural experiments. *In* "Community Ecology" (J. Diamond and T. J. Case, eds.), pp. 3–22. Harper & Row, New York.

Elkington, J. S., and Liebold, A. M. (1990). Population dynamics of gypsy moth in North America. *Annu. Rev. Entomol.* 35, 571–596.

Elton, C. S. (1973). The structure of invertebrate populations inside Neotropical rain forests. *J. Anim. Ecol.* 42, 55–104.

Erwin, T. L. (1982). Tropical forests: Their richness in Coleoptera and other arthropod species. *Coleopterists Bull.* 36, 74–75.

Erwin, T. L. (1991). How many species are there? Revisited. *Conserv. Biol.* 5, 330–333.

Fox, L. R., and Morrow, P. A. (1983). Estimates of damage by insect grazing on *Eucalyptus* trees. *Aust. J. Ecol.* 8, 139–147.

Gaston, K. J. (1991). The magnitude of global insect species richness. *Conserv. Biol.* 5, 283–296.

Gentry, A. H., ed. (1990). "Four Neotropical Rainforests." Yale Univ. Press, New Haven, CT.

Hairston, N. G., Smith, F. E., and Slobodkin, L. B. (1960). Community structure, population control, and competition. *Am. Nat.* 94, 421–425.

Harper, J. L. (1977). "Population Biology of Plants." Academic Press, New York.

Heatwole, H., Done, T., and Cameron, E. (1981). "Community Ecology of a Coral Cay, a Study of One Tree Island, Great Barrier Reef." Dr. W. Junk Publ., The Hague, The Netherlands.

Holloway, J. D. (1989). Moths. *In* "Tropical Rain Forest Ecosystems" (H. Lieth and M. J. A. Werger, eds.), pp. 437–453. Elsevier, Amsterdam.

Janzen, D. H. (1970). Herbivores and the number of tree species in tropical forests. *Am. Nat.* 104, 501–528.

Janzen, D. H. (1988). Ecological characterization of a Costa Rican dry forest caterpillar fauna. *Biotropica* 20, 120–135.

Kikuzawa, K. (1983). Leaf survival of woody plants in deciduous broad-leaved forests. I. Tall trees. *Can. J. Bot.* 61, 2133–2139.

Kitching, R. L., Bergelsohn, J. M., Lowman, M. D., MacIntyre, S., and Carruthers, G. (1993). The biodiversity of arthropods from Australian rainforest canopies: General introduction, methods, sites and ordinal results. *Aust. J. Ecol.* 18, 181–191.

Landsberg, J., and Ohmart, C. P. (1989). Levels of defoliation in forests: Patterns and concepts. *Trends Ecol. Evol.* 4, 96–100.

Leigh, E. G., and Smythe, N. (1978). Leaf production, leaf consumption, and the regulation of folivory on Barro Colorado Island. *In* "The Ecology of Arboreal Folivores" (G. G. Montgomery, ed.), pp. 33–50. Smithsonian Institution Press, Washington, DC.

Leigh, E. G., and Windsor, D. M. (1982). Forest production and regulation of primary consumers on Barro Colorado Island. *In* "The Ecology of a Tropical Forest" (E. G. Leigh, A. S. Rand, and D. M. Windsor, eds.), pp. 109–123. Smithsonian Institution Press, Washington, DC.

Lowman, M. D. (1982). Leaf growth dynamics and herbivory in Australian rain forest canopies. Ph.D. Thesis, University of Sydney, Sydney, Australia.

Lowman, M. D. (1984a). Grazing of *Utethesia pulchelloides* larvae on its host plant, *Argusia argentea*, on coral cays of the Great Barrier Reef. *Biotropica* 16, 14–18.

Lowman, M. D. (1984b). An assessment of techniques for measuring herbivory: Is rainforest defoliation more intense than we thought? *Biotropica* 16, 264–268.

Lowman, M. D. (1985). Temporal and spatial variability in insect grazing of the canopies of five Australian rainforest tree species. *Aust. J. Ecol.* 10, 7–24.

Lowman, M. D. (1987). Insect herbivory in Australian rain forests—Is it higher than in the neotropics? *In* "ESA Symposium Proceedings: Are Australian Ecosystems Different?" Vol. 14, pp. 109–121. Blackwell, Oxford.

Lowman, M. D. (1988). Litterfall and leaf decay in three Australian rainforest formations. *J. Ecol.* 76, 451–465.

Lowman, M. D. (1992). Leaf growth dynamics and herbivory in five species of Australian rainforest canopy trees. *J. Ecol.* 80, 433–447.

Lowman, M. D. (1993). Forest canopy research: Old World, New World comparisons. *Selbyana* 14, 1–2.

Lowman, M. D., and Heatwole, H. (1992). Spatial and temporal variability in defoliation of Australian eucalypts. *Ecology* 73, 129–142.

Lowman, M. D., and Moffett, M. (1993). The ecology of tropical rain forest canopies. *Trends Ecol. Evol.* 8, 103–108.

Lowman, M. D., Taylor, P., and Block, N. (1993a). Vertical stratification of small mammals and insects in the canopy of a temperate deciduous forest: A reversal of tropical forest distribution? *Selbyana* 14, 25–26.

Lowman, M. D., Moffett, M., and Rinker, H. B. (1993b). A new technique for taxonomic and ecological sampling in rain forest canopies. *Selbyana* 14, 75–79.

MacArthur, R. H., and MacArthur, J. W. (1961). On bird species diversity. *Ecology* 42, 594–598.

Majer, J. D., and Recher, H. F. (1988). Invertebrate communities on Western Australian eucalypts: A comparison of branch clipping and chemical knockdown procedures. *Aust. J. Ecol.* 13, 269–278.

Marquis, R. (1991). Herbivore fauna of *Piper* (Piperaceae) in a Costa Rican wet forest: Diversity, specificity and impact. *In* "Plant–Animal Interactions, Evolutionary Ecology in Tropical and Temperate Regions" (P. Price, T. M. Lewinsohn, G. W. Fernandes, and W. W. Benson, eds.), pp. 179–208. Wiley, New York.

Mason, R. R. (1987). Non outbreak species of forest Lepidoptera. *In* "Insect Outbreaks" (P. Barbosa and J. Schultz, eds.), Chapter 2. Academic Press, Orlando, FL.

Molisch, H. (1928). "The Longevity of Plants." E. H. Fulling, New York.

Montgomery, G. G. (1978). "The Ecology of Arboreal Folivores." Smithsonian Institution Press, Washington, DC.

Nielson, B. O. (1978). Above ground food resources and herbivory in a beech forest. *Oikos* 31, 273–279.

Odum, H. T., and Ruiz-Reyes, J. (1970). Holes in leaves and the grazing control mechanism. *In* "A Tropical Rain Forest" (H. T. Odum and R. F. Pigeon, eds.), pp. I-69–I-80. U.S. Atomic Energy Commission, Oak Ridge, TN.

Ohmart, C. P., Stewart, L. G., and Thomas, J. L. (1983). Phytophagous insect communities in the canopies of three *Eucalyptus* forest types in south-eastern Australia. *Aust. J. Ecol.* 8, 395–403.

Price, P. W., Lewinsohn, T. M., Fernandes, G. W., and Benson, W. W. (1991). "Plant–Animal Interactions, Evolutionary Ecology in Tropical and Temperate Regions." Wiley, New York.

Schowalter, T. (1989). Canopy arthropod community structure and herbivory in old-growth and regenerating forests in western Oregon. *J. For. Res.* 19, 318–322.

Schowalter, T. D. (1994). Invertebrate community structure and herbivory in a tropical rain-forest canopy in Puerto Rico, following Hurricane Hugo. *Biotropica* 26, 312–320.

Schowalter, T. D., Webb, J. W., and Crossley, D. A. (1981). Community structure and nutrient content of canopy arthropods in clearcut and uncut forest ecosystems. *Ecology* 62, 1010–1019.

Schowalter, T. D., Hargrove, W. W., and Crossley, D. A., Jr. (1986). Herbivory in forested ecosystems. *Annu. Rev. Entomol.* 31, 177–196.

Schultz, J., and Baldwin, I. (1982). Oak leaf quality declines in response to defoliation by gypsy moth larvae. *Science* 217, 149–151.

Selman, B., and Lowman, M. D. (1983). The biology and herbivory rates of *Novacastria nothofagi* Selman (Coleoptera: Chrysomelidae), a new genus and species on *Nothofagus moorei* in Australian temperate rain forests. *Aust. J. Zool.* 31, 179–191.

Southwood, T. R. E. (1961). The number of species of insect associated with various trees. *J. Anim. Ecol.* 30, 1–8.

Southwood, T. R. E. (1978). The components of diversity. *Symp. R. Entomol. Soc. London* 9, 19–40.

Stanton, N. (1975). Herbivore pressure on 2 types of forests. *Biotropica* 7, 8–11.

Underwood, A. J., and Denley, E. J. (1984). Paradigms, explanations, and generalizations in models for the structure of intertidal communities on rocky shores. *In* "Ecological Communities: Conceptual Issues and the Evidence" (D. R. Strong, D. Simberloff, L. G. Able, and A. B. Thistle, eds.), pp. 151–180. Princeton Univ. Press, Princeton, NJ.

Wallace, A. R. (1878). "Tropical Nature, and Other Essays." Macmillan, London.

Whittam, T. G. (1981). Individual trees at heterogeneous environments: Adaptation to herbivory or epigenetic noise? *In* "Species and Life History Patterns: Geographic and Habitat Variations" (R. F. Denno and H. Dingle, eds.), pp. 9–27. Springer, New York.

Wilson, E. O. (1992). "The Diversity of Life." Harvard Univ. Press, Cambridge, MA.

Wint, G. R. W. (1983). Leaf damage in tropical rain forest canopies. *In* "Tropical Rain Forest: Ecology and Management" (S. L. Sutton, T. C. Whitmore, and A. C. Chadwick, eds.), pp. 229–240. Blackwell, Oxford.

Woinarski, J. C. Z., and Cullen, J. M. (1984). Distribution of invertebrates on foliage in forests of southeastern Australia. *Aust. J. Ecol.* 9, 207–233.

Wolda, H. (1979). Abundance and diversity of Homoptera in the canopy of a tropical forest. *Ecol. Entomol.* 4, 181–190.

Wolda, H., and Foster, R. B. (1978). *Zunacetha annulata* (Lepidoptera: Dioptidae), an outbreak insect in a Neotropical forest. *Geo-Eco-Trop* 2, 443–454.

Wong, M., Wright, S. J., Hubbell, S. P., and Foster, R. B. (1991). The spatial pattern and reproductive consequences of outbreak defoliation in *Quararibea asterolepis*, a tropical tree. *J. Ecol.* 78, 579–588.

19

Reproductive Biology and Genetics of Tropical Trees from a Canopy Perspective

Darlyne A. Murawski

I. Introduction

With few exceptions, trees reproduce in their crowns. Their reproduction involves two mobile stages: pollination and seed dispersal, which are mediated biotically (by animals) or abiotically (by wind or water). In tropical forests, the complex, multilayered and open canopy poses conditions for the dispersal of a tree's pollen and seeds that are qualitatively distinct from closed understory or midstory levels. Trees that occupy different canopy strata should likewise exhibit reproductive adaptations that ensure adequate and reliable gene dispersal and seed set in relation to the conditions in which they reproduce.

Initial curiosity about genetic processes and reproductive patterns in tropical trees stemmed mainly from observations of high species diversity and low average population density, and how such rarity relates to speciation processes. Most of our knowledge of tree reproduction is based on three types of studies: surveys or descriptions of floral morphology and phenology; observations of pollinators and their foraging behavior; and breeding system experiments that test for self-incompatibility.

More recently, genetic markers have been used to quantify levels of inbreeding versus outcrossing and pollen dispersal distances. New advances in phylogenetic analysis provide the opportunity to interpret reproductive patterns in light of evolutionary history. Surveys of genetic diversity in tropical trees are yielding some generalities on how genetic variation is struc-

Copyright © 1995 by Academic Press, Inc.
All rights of reproduction in any form reserved.

tured within and between populations of trees. Very few studies, however, have addressed the actual mechanisms that produce the observed diversity. With current awareness of the alarming rates of tropical deforestation, recent studies focus more on the effects of human disturbance on reproductive patterns, with a new emphasis on implications for conservation.

Methods for sampling reproductive data on trees have depended on whether there is a need to return to specific inflorescences rather than to merely collect material for examination or genetic analysis. Since pole-pruners, shotguns, slingshots, tree-felling, and local tree climbers have been used to obtain samples rapidly and efficiently, certain areas of reproductive biology are better developed than those that require repeated trips to particular inflorescences or detailed within-canopy observations.

Some of the topics in this chapter have been covered in other reviews (e.g., Hamrick *et al.,* 1992; Loveless, 1992; and volumes edited by Bawa and Hadley, 1990, and Fleming and Estrada, 1993; McDade *et al.,* 1993). I expand on that literature and emphasize reproduction from the point of view of the canopy environment (in contrast to the understory). I integrate different areas of reproductive biology (e.g., floral and fruiting phenology, pollination, mating systems) with studies of genetics and recommend challenging directions for future research.

II. Floral Phenology of Tropical Trees

Studies of floral phenology in tropical rain forests have centered mainly on overall community patterns; few have examined individual patterns of periodicity. One notable exception is a 12-year study of the flowering patterns of 254 individual trees in 173 species (excluding palms) at La Selva, Costa Rica (Newstrom *et al.,* 1991). Canopy and subcanopy strata were compared with respect to four major flowering frequencies: continual, subannual, annual, and supra-annual. Considering all trees, 7% flowered continually, 55% subannually, 29% annually, and 9% supra-annually. Of all patterns, annual flowering was the most predictable in terms of the timing of individual anthesis. A larger percentage of canopy trees flowered annually than subcanopy trees. The more prevalent continual or subannual flowering pattern was attributed to relatively more stable microclimate in the understory.

Although many long-term studies of communitywide patterns have employed classifications similar to that of Newstrom *et al.* (e.g., Koelmeyer, 1959; Medway, 1972; Dieterlen, 1978), terminology varies widely. Different forms of subannual flowering are labeled "multiple-bang" (e.g., species of Bignoniaceae with very short flowering duration; Gentry, 1974), "episodic" (e.g., *Guarea rhopalocarpa*; Bullock *et al.,* 1983), and "periodic" (e.g., species pollinated by hawkmoths; Haber and Frankie, 1989). The serious lack of standardized terminology and methods in the literature on tropical trees prevents a synthesis of data from different sites. Newstrom *et al.* (1993, 1994) addressed the confusion of concepts and terminology by providing specific definitions for variables describing flowering patterns.

At the population level, most fig species (*Ficus*) flower continuously. Individual trees, however, have been described as having a subannual flowering pattern with a high degree of synchrony. Thus, their short-lived symbiotic wasp pollinators are encouraged to move between trees in search of receptive flowers, a strategy that promotes interplant pollen dispersal (Galil and Eisikowitch, 1968; Janzen, 1979; Wharton *et al.*, 1980; Bronstein, 1992).

Seasonal trends in flowering phenology have been reported by numerous authors (e.g., Medway, 1972; Frankie *et al.*, 1974; Gentry, 1974; Putz, 1979; Opler *et al.*, 1980; Augspurger, 1982, 1983a; Wright and van Schaik, 1994; van Schaik *et al.*, 1993). Many trees produce both flowers and leaves during the dry season, whether or not there is strong seasonality in rainfall, whereas others predictably produce them in the wet season. Janzen (1972) noted that understory plants flowered and flushed new leaves when the canopy was deciduous in seasonal deciduous forest of Guanacaste, Costa Rica. Wright and van Schaik (1994) reviewed the growing evidence for seasonal light limitation in tropical rain forest canopies and evaluated the role of irradiance on seasonal phenologies in eight tropical forests. Their results were consistent with the hypothesis that both leaf and flower production coincide with seasonal peaks of irradiance. They examined the relationship of drought sensitivity (from minimum midday leaf water potential) to phenology and irradiance. The drought-tolerant species (associated with deep root systems) tended to flower and flush leaves during the dry season irradiance peak. In contrast, drought-sensitive species (associated with shallow root systems) consistently produced flowers and leaves in the wet season. Specific cues for flower initiation for specific phenological patterns are still poorly understood.

An unusual flowering pattern, reminiscent of some members of the Agavaceae, has been documented for the semelparous canopy-emergent tree *Tachigali versicolor* (Caesalpiniaceae), in which individuals flower once, set fruit, and die. Four-year intervals separate major synchronous flowering episodes in a population (Foster, 1977). Other members of this genus may also be semelparous, but such an extreme pattern has not been documented in other tropical trees.

Synchronous supra-annual flowering is especially striking in tropical Asian forests. Participants include the majority of Dipterocarpaceae, certain canopy species in the Bombacaceae, Papilionaceae, and Polygalaceae, and understory species in the Myristicaceae and Euphorbiaceae (Appanah, 1990). Mast years vary in flowering intensity (Yap and Chan, 1990) and alternate with periods of 2–8 years with no or relatively little reproductive activity (Medway, 1972; Appanah, 1979, 1985; Curran, 1994). In West Kalimantan, Borneo, from 1968 to 1992, mast years for illipe nut occurred on average every 3.8 years (Curran, 1994).

In the Pasoh forest of peninsular Malaysia, species of *Shorea* section *Mutica* (Dipterocarpaceae) flower in a predictable overlapping sequence (Chan and Appanah, 1980). Ashton *et al.* (1988) demonstrated that the overlapping pattern of flowering is significantly nonrandom. This pattern may serve to reduce competition among closely related species for pollinators and reduce clogging of stigmas with interspecific pollen.

The supra-annual flowering of many Malaysian and West Bornean dipterocarps correlates with the El Niño climatic event in the Pacific (Ashton *et al.,* 1988; Curran, 1994). New data on the flowering phenology of Sri Lankan dipterocarps indicate that certain species of *Shorea* section *Doona* likewise flower sequentially in multiannual years (Ashton and Gunatilleke, personal communication).

Despite the extensive literature emerging on flowering phenologies of tropical trees at the population and community levels, future research must narrow in on the mechanisms underlying phenological events and their relation to the pollinator community.

III. Pollination

The tropical rain forest canopy is a dynamic environment reflected not only in the temporal and spatial patterns of anthesis, but also in the types of flowers, pollinators, and foraging behaviors found there. The vast literature on tropical pollination biology has been summarized in several excellent reviews (especially Bawa and Hadley, 1990; Bawa, 1990).

A. Pollination Systems

In tropical rain forests, animals play the major role in the pollination of flowering trees, with wind dispersal confined to very few species (Bawa *et al.,* 1985a; Kress and Beach, 1993). Pollinators are often grouped taxonomically or by their tendency to visit flowers of particular morphologies having characteristic rewards (e.g., pollen, nectar, resins) and timing of anthesis, the so-called "guilds" of pollinators and floral "syndromes" (Proctor and Yeo, 1973; Faegri and van der Pijl, 1979; Baker *et al.,* 1982; Prance, 1985; Bawa, 1990). The largest category of pollinators by far are bees (Bawa *et al.,* 1985b; Roubik, 1989), but of course many other groups play a part.

The degree of specialization of a plant for particular pollinators and of a pollinator for particular plants is highly variable and poorly understood.

Among tree species, only figs are known to have obligate one-on-one relationships (for figs with individual species of agaonid wasps, see Berg and Wiebes, 1992; Bronstein, 1992). At the opposite end of the spectrum are plants with a generalist pollination strategy. However, these species, with small flowers that are physically accessible to a wide range of small insect pollinators (e.g., Sri Lankan *Shorea* section *Doona*) (Dayanandan *et al.*, 1990), may be effectively pollinated by only one or a few species (Bawa, 1990; Bawa *et al.*, 1985a).

In a few studies, the canopy environment has been partitioned. Kress and Beach (1993) summarized pollination systems of flowering plants in the lowland wet forest of La Selva, Costa Rica. They divided the forest into three strata: understory, subcanopy, and canopy. The canopy was dominated by two major categories of pollination systems: those of large- and medium-sized bees, mostly Anthophoridae (37% of species), and "small diverse insects," including wasps, flies, beetles, butterflies, moths, and thrips (27%). Moth, wasp, and bat pollination systems, although present in the canopy, were more typical of subcanopy trees; small bee and beetle systems were almost totally confined to subcanopy and understory strata. Hummingbird plants comprised 24% of understory species, but were atypical of upper strata.

Assertions that large tropical bees "prefer" to forage in the forest canopy were challenged by Roubik (1993). By using chemical and light traps simultaneously in the canopy and understory of a Panamanian moist forest, he tested whether there were consistent stratum associations in bees. His evidence did not support the hypothesis that big bees prefer higher flowers. The only canopy specialists found were nocturnal bees (genus *Megalopta*) and one stingless bee (*Partamona*). Other bee species that were caught more than once were taken at both levels and exhibited a large variance in stratum association that was shown in some cases to be unrelated to nest height. The majority of euglossine bees were caught in the understory, where their concentrations increased dramatically during the dry season. Only the largest euglossine species appeared to tolerate the apparently greater heat and exposure of the canopy (also see Armbruster and Berg, 1994).

Not all sites exhibit similar patterns of pollination systems. In Madagascar, where flower-visiting bats are rare, lemurs appear to be major pollinators of a diversity of canopy trees, including species of *Symphonia* (Clusiaceae) and certain Bombacaceae and Lecythidaceae (Sussman and Raven, 1978).

In tropical Australia, beetle pollination plays a prominent role (Irvine and Armstrong, 1990), and together with flies replaces bees as the most numerous pollinator groups. In North Queensland, 24% of trees, many of which are canopy species, are described as beetle pollinated. Numerous examples of beetle pollinated trees were also reported in the Lambir Hills of Sarawak, Borneo (Momose and Inoue, 1994).

Several species-rich taxa of tropical Asian trees are pollinated primarily by thrips (Thysanoptera). Their behavior is especially well documented for species of *Shorea* section *Mutica* (Appanah and Chan, 1981; Appanah, 1990), the lowland Malaysian dipterocarps with multiannual flowering patterns discussed earlier. The thrips breed in the flowers they pollinate, a situation that has been documented in several insect groups: fig-wasps on *Ficus* (Janzen, 1979), weevils on oil palms (*Elaeis*) and in cycad cones (Norstog *et al.*, 1986; Tang, 1987), *Tegiticula* moths on *Yucca* (Pellmyr and Thompson, 1992), and *Chiastocheta* flies on *Trollius europaeus* (Pellmyr, 1992).

In a dry forest in Guanacaste, Costa Rica, bees and moths were found to be the most important pollinators of trees (Frankie *et al.*, 1990). Hawkmoths (Sphingidae) in particular were primary pollinators of about 10% of tree species, their abundance corresponding to the seasonal abundance of hawkmoth flowers in the early part of the wet season (Haber and Frankie, 1989).

B. Pollinator Foraging

Pollinator foraging ranges and the degree of fidelity of individuals to certain plant species have a major impact on conspecific pollen dispersal. The upper limit of a typical foraging range for large bees has been documented to be between 10 and 20 km from the nest. Smaller bees (stingless and honeybees) also can have large foraging ranges of 3–10 km from the nest (Roubik, 1992). Other strong fliers certainly capable of ranging over many kilometers include bats (Heithaus *et al.*, 1975) and hawkmoths (Haber and Frankie, 1989).

Janzen (1971a) distinguished two extreme patterns of flower visitation in Neotropical bees: "traplining" and "opportunism." The former was originally observed in female euglossine bees and subsequently described for certain anthophorid bees (Eickwort and Ginsberg, 1980), hummingbirds in the Phaethornidae (Feinsinger 1976, 1978), bats (Fleming, 1982; Lemke, 1984), butterflies (Gilbert, 1980), and thought to occur in hawkmoths (Baker, 1973). Traplining involves regular and repeated visits to different plants. Species that encourage traplining typically produce one to a few flowers per day over long periods (e.g., the "steady-state" pattern of Gentry, 1974), whereas species that attract opportunistic pollinators (such as small social bees) are characterized by lavish floral displays (Baker, 1973). But many trees with lavish displays attract specialized nontraplining pollinators such as thrips (Appanah and Chan, 1981) and beetles (Rogstad, 1986).

Traplining and foraging over long distances should not only promote outcrossing, but should increase the number of individual plants in an interbreeding population (effective population size). At the other extreme, opportunistic foraging and territorial defense of nectar sources (typical of some male hummingbirds) should result in inbreeding or reduced seed set through self-pollination and mating with kin (Baker, 1973). In reality, most long-lived pollinators exhibit some ability to survey their environments and adjust their foraging behavior according to changing resource availability (Feinsinger, 1976, Wolf *et al.,* 1976; Heinrich, 1979; Fleming, 1982; Waddington, 1983; Murawski, 1987; Roubik, 1992), whereas short-lived species respond to such changes by increasing or decreasing their numbers (e.g., Appanah and Chan, 1981).

IV. Sexual Systems and Mating Patterns

The level of inbreeding (versus outcrossing) in plants is partly determined by the sexuality of their flowers. Dioecy, identified by distinct male and female plants, enforces outcrossing although inbreeding may occur through kin mating. Dioecious trees have been associated with (but not limited to having) small bee-pollinated flowers and fleshy fruits whose seeds are animal-dispersed (Bawa, 1980, 1994; Bawa *et al.,* 1985b; Renner and Feil, 1993). Although dioecy has been described as more common in canopy trees than in subcanopy species (Bawa, 1980), this difference may be due at least in part to terminology. The Bawa *et al.* study (1985b) defined all trees over 25 m tall as canopy species. At another site, Ashton (1969) compared only trees greater than 10 cm dbh and found that dioecy had the highest relative abundance among the smallest trees. In the latter study, it appears that the combination of dioecy and fleshy fruits may be a feature mainly of larger subcanopy trees, not upper-canopy or emergent species that exhibit a propensity for hermaphroditism and wind-dispersed fruits. The significance of this finding may also relate to the canopy structure and species composition of a given site.

The proportion of dioecious trees and shrubs relative to the total flora ranged between 16 and 24% in lowland and montane, wet and dry Neotropical forests (Tanner, 1982; Bawa and Opler, 1975; Ruiz and Arroyo, 1978; Croat, 1979; Flores and Schemke, 1984; Bullock, 1985; Bawa *et al.,* 1985b; Sobrevilla and Arroyo, 1982; Ramirez and Brito, 1990) and from 26 to 28% in Southeast Asian forests (Ashton, 1969; Manokaran *et al.,* 1990; Kochummen *et al.,* 1991). Dioecious species appear to be very well represented in early-successional stages in Amazonian varzea (whitewater floodplain forest; W. D. Hamilton, personal communication; but see Rohwer, 1986, for comments on the Lauraceae). In contrast, flowering species of lianas and epiphytes, which account for a large part of the flora of most lowland tropical rain forests, are almost entirely hermaphroditic (Schatz, 1990). A variety of other systems (gyno- and androdioecy; gyno- and andromonoecy) are common in tropical canopy trees, especially in the Anacardiaceae, the Sapindaceae and allied families, and the Fagaceae.

Monoecy, in which individuals produce unisexual flowers of both sexes, is mainly confined to palms, figs, and euphorbs (Kress and Beach, 1993), although discrete male and female flowering phases in many species precludes self-fertilization.

Hermaphroditism remains the most common sexual system found in tropical tree species, regardless of canopy stratum (Kress and Beach, 1993). For hermaphroditic trees, the degree of outcrossing is influenced by many factors, such as the presence or intensity of self-incompatibility, the tendency of pollinators to forage within and between crowns, the selective abortion of selfed fruit or seed, and the timing of pollen presentation relative to stigma receptivity. For example, some hermaphroditic tropical trees develop their flowers in cohorts that synchronize sexual phase. In extreme examples, entire sets of trees are functionally male while others are functionally female, and over time the functional status of individuals reverses (e.g., Rogstad, 1994; Kubitzki and Kurz, 1984; McDade, 1986, for species in the Annonaceae, Lauraceae, and Rubiaceae, respectively).

A. Self-Incompatibility

Self-incompatibility (SI), the rejection of self (and sometimes related) pollen, is under genetic control. It can occur on the stigma surface (sporophytic SI), in the style (gametophytic SI), or inside the ovary (called "late-acting" LSI) (de Nettancourt, 1977). In the latter case, incompatibility may be difficult to distinguish from inbreeding depression (the mortality of inbred progeny), because rejection of selfed flowers can take place before or after fertilization (but see Waser and Price, 1991, for argument against inbreeding depression). LSI may be more common in tropical trees than previously thought, occurring in diverse taxa such as the Bombacaceae, Bignoniaceae, and Mimosaceae (Kenrick and Knox, 1985; Seavey and Bawa, 1986; Gibbs and Bianchi, 1993). The genetic control mechanisms of LSI are still unknown.

Experimental pollination of hermaphroditic trees using self-pollen, outcross pollen, and open pollination treatments can test for the presence or absence of self-incompatibility. Kress and Beach (1993) summarized breeding system results in relation to forest stratum for species of flowering plants at La Selva, Costa Rica. The highest proportion of self-incompatible species occurred in the subcanopy (90.9%), followed by the canopy (75.0%), and last by understory flowering plants (34.2%). The majority of hermaphroditic tropical trees at low altitudes in the neotropics and paleotropics possess strong barriers to self-fertilization (e.g., Bawa and Opler, 1975; Zapata and Arroyo, 1978; Bawa *et al.*, 1985b; Dayanandan *et al.*, 1990). High-altitude forests, however, have a much higher proportion of self-compatible hermaphroditic trees (Sobrevilla and Arroyo, 1982; Tanner, 1982; Hernandez and Abud, 1987), perhaps paralleling trends toward more apomixis in alpine and arctic habitats generally (Bierzychudek, 1985). The proportion of dioecious species, however, remains similar to that found in the lowlands (Sobrevilla and Arroyo, 1982; Tanner, 1982).

Classic breeding system experiments, however, may greatly over- or underestimate the naturally occurring levels of inbreeding in certain species. For example, some hermaphroditic plants have a flexible mating strategy that allows them to produce mostly outcrossed progeny when presented with mixtures of self-pollen and outcross pollen, while retaining the ability to produce selfed seed in the absence of outcross pollen (Bateman, 1956; Bowman, 1987; Cruzan and Barrett, 1993). The greater pollen tube growth rate of outcross pollen over self-pollen may explain this differential siring ability (Cruzan and Barrett, 1993). Other forms of incomplete self-incompatibility might be inferred from observations of relatively low levels of fruit set following self-pollination, and from tree-to-tree differences in levels of self-compatibility (Bawa *et al.,* 1985b). Bertin and Sullivan (1988) and Bertin *et al.* (1989), after having established that the "trumpet creeper" (*Campsis radicans,* Bignoniaceae) was self-incompatible following self-pollination, found by using an allozyme marker that in fact a sizable proportion of selfed progeny (2–30% of total) could be produced by applying mixed pollen loads of both self-pollen and outcross pollen. What causes rejection of pure self-pollen loads in this species is not understood, nor whether other species with multiseeded fruits behave similarly. Although such experiments may reveal the presence or absence of mechanisms that favor outcrossing, other methods are needed to measure levels of inbreeding and outcrossing under natural conditions. Furthermore, virtually all of these breeding system studies fail to distinguish apomixis as a process separate from self-fertilization and outcrossing.

B. Allozyme Analysis: A Definition

By examining the electrophoretic mobilities of the enzyme products of a sample of genetic loci, researchers have been able to characterize individual genotypes as well as genetic properties of populations and species. The term "allozyme" refers to the variant forms of an enzyme resulting from the presence of different alleles; the similar term "isozyme," often used interchangeably with "allozyme," refers to the enzyme product of a particular locus when more than one locus codes for enzymes of similar structure and function. Allozymes are convenient and reliable genetic markers that exhibit various useful properties, such as codominant expression of alleles, Mendelian inheritance patterns, and absence of confounding gene interactions such as epistasis and pleiotropy (Weeden and Wendel, 1989). Although allozymes do not reveal all underlying genetic variation, they have obvious temporal and economic advantages for population-level studies (allozyme techniques are reviewed in Hedrick, 1985; Liengsiri *et al.*, 1990).

C. Mating Systems

Allozyme analysis has made it possible to quantify outcrossing rates and to test for apomixis using samples of progeny (embryos/seedlings) of different individuals. Multilocus mixed-mating models (e.g., Ritland and Jain, 1981; Ritland, 1990) are used to estimate several parameters: population and individual outcrossing rates, pollen and ovule allele frequencies, and even maternal genotype (a convenient feature in the event of poor maternal allozyme expression or difficulty in collection of maternal material, e.g., when trees are deciduous). Certain assumptions of the models may be unrealistic in natural populations, such as random outcrossing by all maternal genotypes to a homogeneous pollen pool. However, individual assumptions can be tested, as can the fit of each locus to expected genotypic frequencies based on Hardy–Weinberg equilibrium. Fortunately, significant departures from the model have little effect on the multilocus outcrossing estimate (Ritland, 1983).

Multilocus outcrossing estimates for tropical trees are summarized in Table 1. [Outcrossing values exceeding 1.0 (100%) are statistically possible.] Of 22 species examined, the majority had outcrossing rates above 0.9. The notable exceptions displaying mixed mating systems were in the families Bombacaceae, Dipterocarpaceae, and Sterculiaceae. Outcrossing in the Bombacaceae ranged between 21 and 100%. Large canopy trees comprise the bulk of tropical tree species examined, some of which were chosen for their economic and conservation implications. Although breeding system experiments show self-compatability to be more common in tropical understory trees than in canopy trees, population outcrossing rates will need to be quantified for a proper comparison of levels of inbreeding in different forest strata.

Most tropical tree species occur at low densities, often at less than one individual of reproductive size per hectare (Hubbell and Foster, 1986; Ashton, 1988). If a species is partially or completely self-compatible, then inbreeding might be expected to increase at low flowering tree densities. This hypothesis was tested by sampling the same population (and when possible the same individuals) over different years (Murawski *et al.*, 1990), and by sampling populations differing in tree density (Murawski and Hamrick, 1991).

As predicted, elevated levels of inbreeding were correlated with a decrease in flowering tree density in hawkmoth-pollinated *Cavanillesia platanifolia* on Barro Colorado Island, Panama (Murawski *et al.*, 1990; Murawski and Hamrick, 1992a). Outcrossing rates for individual trees were similarly lower in years when nearby conspecifics failed to flower (Murawski *et al.*, 1990). In a broader comparison of nine species from Barro Colorado Island, outcrossing rates declined significantly with decreasing density of flowering trees (Murawski and Hamrick, 1991). In the La Selva forest (Costa Rica), hermaphroditic trees were found at lower average population densities than either monoecious trees or dioecious trees (Lieberman and Lieberman, 1993). This independently suggests that strict outcrossing might be a disadvantage at low densities.

Human disturbance through selective logging can artificially reduce tree densities, potentially altering pollinator densities and behaviors as well as population genetic structure. In a study of an endemic tree from Sri Lanka (*Shorea megistophylla*), outcrossing rates were compared in selectively logged and unlogged forest (Murawski *et al.*, 1994b). Although the species is primarily outcrossed, higher levels of inbreeding were found in trees from the logged forest. Moreover, a single isolated tree outside the study area produced seed through a combination of self-fertilization and apomixis. Establishing sound concepts of sustainable yield management of forest products will require more study of the effects of management practices such as selective logging on inbreeding and on reproductive output.

Mating systems play an important role in determining the genotypic proportions of subsequent generations (Hamrick and Godt, 1989). But forces such as heterotic selection (heterozygote advantage) or inbreeding depression can alter the genotypic proportions of developing seeds and of a seedling cohort as it matures. Although relative fitnesses of selfed progeny can be studied in annual plants (Dole and Ritland, 1993), time constraints prevent such direct studies in trees. If we were to examine the fixation indices of adults and their progeny, we could at least infer whether the adult genotypic proportions were determined by mating system alone or whether inbreeding depression and perhaps heterotic selection operate between the time of seed dispersal and reproductive maturity.

The progeny fixation rate (or inbreeding coefficient, F) can be calculated directly from the multilocus outcrossing rate or by pooling progeny of all trees examined, whereas the adult fixation rate is measured indirectly using the observed and Hardy–Weinberg heterozygosities. In either case, if F is zero, the genotypic frequencies of the adults or progeny are in Hardy–Weinberg proportions. If F is positive or negative, then there is a deficit or excess of heterozygotes, respectively. If mating is random, there is complete outcrossing, and the expected progeny F would be 0; with several generations of complete selfing, the inbreeding coefficient should approach 1.0.

For several tree species examined, the increased heterozygosity of adults over their progeny suggests that some homozygote disadvantage occurs during recruitment into the breeding population (Table II and references therein). Further support for this tendency comes from the observation

Table I Multilocus Outcrossing Rates and Standard Errors for Populations of Tropical Trees

Family	Species	Site	Outcrossing rate t_m (±SE)	Source
Arecaceae	Astrocaryum mexicanum	Mexico, Los Tuxtlas	0.933–1.050	1
Bombacaceae	Cavanillesia platanifolia	Panama, BCI (3 years)	0.213 (0.052)	2,3,4
			0.347 (0.025)	
			0.569 (0.024)	
	Ceiba pentandra	Panama, Darien	0.661 (0.074)	4
	Pseudobombax munguba	Panama, BCI	0.689 (0.03)	3,5
	Quararibea asterolepis	Brazil, Tefe	0.97	6
		Panama, BCI	1.008 (0.010)	2,3
Boraginaceae	Cordia alliodora	Costa Rica, La Selva	0.96	7
Dipterocarpaceae	Shorea congestiflora	Sri Lanka, Sinharaja	0.852 (0.023)	8
	Shorea megistophylla	Sri Lanka, Sinharaja	0.737 (0.019)	9
	Shorea trapezifolia	Sri Lanka, Sinharaja (2 years)	0.519 (0.035)	8
			0.602 (0.032)	
	Stemonoporus oblongifolius	Sri Lanka, Peak Wilderness	0.844 (0.021)	10
Fabaceae	Acacia auriculiformis	Queensland and Papua, New Guinea	0.93 (0.040)	11
			0.92 (0.023)	
	Acacia crassicarpa	Queensland and Papua, New Guinea	0.99 (0.032)	11
			0.93 (0.03)	
	Pithecellobium pedicellare	Costa Rica, La Selva (3 sites)	0.951 (0.021)	12
			0.970 (0.028)	17
			0.986 (0.014)	17

468

Family	Species	Location		Source
	Platypodium elegans	Panama, BCI (2 years)	0.924 (0.043) 0.898 (0.043)	3
Lauraceae	Tachigali versicolor	Panama, BCI	0.937 (0.044)	3
Lecythidaceae	Beilschmedia pendula	Panama, BCI	0.918 (0.058)	3
	Bertholletia excelsa	Brazil, Acre	0.849 (0.033)	13
Meliaceae	Carapa guianensis	Costa Rica (2 sites)	0.967 (0.022) 0.986 (0.028)	14
	Trichilia tuberculata	Panama, BCI	1.077 (0.028)	3
Moraceae	Brosimum alicastrum	Panama, BCI	0.875 (0.035)	3
	Sorocea affinis	Panama, BCI (2 sites)	1.089 (0.045) 0.969 (0.020)	3
Pinaceae	Pinus caribaea	Bahamas, Belize	0.85 to 0.93	18
	P. kesiya	Thailand, northern	0.68 to 0.97	15
	P. oocarpa	Nicaragua, Belize	0.81 to 0.96	18
	P. maximinoi	Honduras	0.65	18
Rubiaceae	Psychotria faxucens	Mexico, Los Tuxtlas	0.995 (0.086) 1.015 (0.061)	16

Sources: 1: Eguiarte *et al.* (1992); 2: Murawski *et al.* (1990); 3: Murawski and Hamrick (1991); 4: Murawski and Hamrick (1992a); 5: Murawski and Hamrick (1992b); 6: D. A. Murawski, W. N. do Amaral, and W. D. Hamilton, unpublished data; 7: Boshier *et al.* (1995); 8: Murawski *et al.* (1994a); 9: Murawski *et al.* (1994b); 10: Murawski and Bawa (1994); 11: Moran *et al.* (1989); 12: O'Malley and Bawa (1987); 13: O'Malley *et al.* (1988); 14: P. Hall, L. C. Orell, and K. S. Bawa, unpublished data; 15: Boyle *et al.* (1991) (range of estimates from four populations); 16: Perez-Nasser *et al.* (1993) (estimates for pin and thrum plants, respectively); 17: P. Hall, K. S. Bawa, and S. Walker, unpublished data; 18: Matheson *et al.* (1989).

Table II   Comparison of Adult and Progeny Fixation Indices
for Several Species of Tropical Trees [a]

Family/species	F_{adult}	F_{progeny}
Arecaceae		
Astrocaryum mexicanum[1]	− 0.41	− 0.19
Bombacaceae		
Cavanillesia platanifolia[2]		
1987	− 0.198	0.275
1988	− 0.198	0.475
1989	− 0.198	0.649
Ceiba pentandra[3]	− 0.297	0.184
Quararibea asterolepis[2]	− 0.046	0.00
Dipterocarpaceae		
Shorea congestiflora[4]	0.088	0.067
Shorea megistophylla[5]	− 0.247	0.151
Shorea trapezifolia[4]		
1990	− 0.06	0.295
1991	− 0.30	0.237
Stemonoporus oblongifolius[6]	− 0.101	0.085
Lauraceae		
Beilschmedia pendula[3]	− 0.300	0.043
Meliaceae		
Trichilia tuberculata[3]	0.074	− 0.034
Moraceae		
Brosimum alicastrum[3]	− 0.271	0.067
Rubiaceae		
Psychotria faxucens[7]	0.075	0.128

Sources: 1: Eguiarte *et al.* (1992); 2: Murawski *et al.* (1990); 3: Murawski and Hamrick, unpublished data; 4: Murawski *et al.* (1994a); 5: Murawski *et al.* (1994b); 6: Murawski and Bawa (1994); 7: Perez-Nasser *et al.* (1993).

[a] If genotypic frequencies are determined solely by the population's mating system, then progeny and adults should have the same fixation index. If values for adults are lower than those of their progeny, then there is an indication for higher survivorship of heterozygous individuals between the seed and reproductive adult stage.

that in three Panamanian tree species, larger-diameter classes were biased toward heterozygotes (Hamrick *et al.*, 1993). However, the negative fixation indices for adults of most species in Table II indicate further selection favoring heterozygosity. A similar excess of heterozygosity above Hardy–Weinberg expectations has been reported for north-temperate conifer species (review in Mitton and Jeffers, 1989). Whether allozyme loci themselves are under heterotic selection is debatable, as there are other possible explanations (e.g., linkage to other loci under direct selection or advantages from new homozygous epistatic interactions; Hamilton, 1993).

In summary, evidence is accumulating that outcrossing rates in natural populations of rain forest trees tend to be high, although inbreeding (at least in hermaphroditic species) may increase with declining population densities. Population thinning may encourage inbreeding and possibly apomixis, although inbreeding depression may reduce seed set even if estimated outcrossing rates remain nearly the same. Generalities about the effects of selective logging and other alterations of the habitat such as forest fragmentation will require detailed studies of many more species. If inbreeding depression is as common in rain forest trees as implied by the few species studied, it would appear to be an important mechanism for maintaining genetic diversity in tropical forests. On the other hand, habitat alterations that enhance self-fertilization could have devastating effects on the survival of future generations.

V. Fruit and Seed Dispersal to Seedling Establishment

Following the maturation of seeds, several processes (dispersal, predation/herbivory, and dormancy) are crucial to seedling establishment. On a larger scale, these processes influence population genetic structure and determine the composition of the forest community.

A. Fruiting Phenology

Time of fruit maturation is ultimately dependent on time of anthesis, although the time between fertilization and maturation may take anywhere from several weeks to a year or more. Strongly seasonal fruiting phenologies have been observed in most tropical forest trees, independent of the degree of seasonality of rainfall (Terborgh, 1986, 1990). Different hypotheses have been postulated to explain these fruiting rhythms (review in Gautier-Hion, 1990; Terborgh, 1990).

The "competition avoidance hypothesis" proposes that fruiting seasons of species sharing common dispersal agents should be staggered to avoid competition for dispersers (Wheelwright, 1985). There appears to be no solid support for this hypothesis, and it also suffers from the impossibility of defining sets of species that share dispersers.

According to the "predator satiation hypothesis," species that fruit in unison are more likely to satiate hungry seed predators, leaving a portion of their seeds safe to germinate. This hypothesis is supported in Southeast Asian forests with mast fruiting years. Many dipterocarps and other unrelated taxa participate in simultaneous mast fruiting (Chan and Appanah, 1980). A large proportion of flowering trees successfully develop mature fruits in mast years (Yap and Chan, 1990), whereas those fruiting outside the community peak are selected against by insect and vertebrate seed predators (Curran, 1994; M. Leighton, unpublished data).

The "optimal time of seed dispersal/germination" hypothesis predicts that climatic conditions determine the ideal time of fruit maturation and germination success. In seasonally inundated floodplain forests of Central Amazonia, the community peak in fruiting coincides with the high water phase, when the primary seed dispersers (seed-eating fish) invade them (Kubitzki and Ziburski, 1994). In another study, though, Bornean dipterocarps appeared to bypass years considered most suitable for germination (Curran, 1994). Long-term studies and experimental manipulation are required to tease apart the evidence for the different hypotheses.

B. Seed Dispersal

Unlike pollen dispersal systems of tropical trees, seed dispersal systems are much more diverse and include animals, wind, water, and combinations of primary and secondary dispersal agents. Janzen and Vasquez-Yanes (1991) estimate that more than 75% of woody plants in species-rich tropical forests have animal-dispersed seeds. Many characters such as fruit color, palatability, nutritional value, accessibility, and digestive capacity have been interpreted as coadapted with particular groups of animals (Janson, 1983). However, some researchers have evidence that relationships between disperser groups and fruit types are loose at best (Herrera, 1986; Gautier-Hion *et al.,* 1985; Gautier-Hion, 1990; Howe, 1990). Animals that disperse seeds in one situation may destroy them in another, blurring the distinction between seed disperser and predator (Gautier-Hion *et al.,* 1993). Fleming *et al.* (1993) discussed conditions under which evolution would favor specialized "syndromes" versus coteries of mixed dispersers.

Different tropical forests display different patterns of fruit dispersal systems. Dispersal by fish, although prevalent in the inundated forests of Amazonia (Gottsburger, 1978; Goulding *et al.,* 1988; Kubitzki and Ziburski, 1994; Andrea Pires, personal communication) has never been documented in Paleotropical forests. Malaysian forests have an overstory of gyration or wind-dispersed seeds (due to the dominance of dipterocarps) and an understory of large bird- and mammal-dispersed seeds (Curran, 1994; Terborgh, 1990, based on M. Leighton, unpublished data). In contrast, Costa Rican dry forest canopies (Guanacaste) exhibit a greater mix of animal and wind dispersal, and small birds play a much larger role (Janzen and Vazquez-Yanes, 1991).

Comparison of dispersal distances for different systems is difficult because of the diverse methods employed by various researchers and because the tail end of seed shadows cannot be accurately quantified (Willson, 1993). However, secondary dispersal agents may considerably enhance dispersal distances in tropical forests (Forget, 1991; Willson, 1993; Janzen and Vazquez-Yanes, 1991).

C. Seedling Establishment

Family structure, or the spatial distribution of relatives, depends not just on pollen and seed dispersal distance, but also on factors determin-

ing seedling establishment and juvenile survivorship, such as seed predation, pathogens, and herbivory. Following proposals by Janzen (1970) and
Connell (1971) that seedling establishment near parent trees in tropical
forests might be inhibited by seed predation or seedling herbivory, various researchers have found some evidence for either distance- or density-
dependent seed predation (Burkey, 1994; Clark and Clark, 1984; Connell *et
al.*, 1984; Howe, 1989; Janzen, 1971b, 1972; Terborgh *et al.*, 1993) and seedling herbivory (Clark and Clark, 1987; Howe, 1990).

Also conforming to the Janzen–Connell hypothesis is the damping-off by
fungal pathogens, which may similarly alter the spatial distribution of dispersed seeds relative to their parents (Augspurger, 1983b,c; 1990). Disease
levels have been shown to be greatest at high seed densities (Augspurger,
1983b) and at distances closest to the parent tree (Augspurger and Kelly,
1984). The effect of density-dependent seed mortality, whether from pathogens or seed predators, is often that seeds do not survive near the parent
tree. This affects not only the spatial distribution of juvenile recruitment,
and thus adult conspecific trees, but also the fine-scale genetic structure in
a local area.

Not all species exhibit such patterns of seed(ling) mortality. For example,
four out of five Peruvian tree species studied by Terborgh *et al.* (1993)
showed no distance effect of seed predation. Observations indicated that
the seeds of those four species were eaten primarily by mammals, which
may search the forest floor more thoroughly than invertebrate seed predators. In another study, Condit *et al.* (1992) found little evidence supporting
the Janzen–Connell hypothesis from a recensus of tagged trees on Barro
Colorado Island. However, their data base excludes individuals less than
1 cm dbh—the group with the highest presumed risk of mortality.

The distribution of conspecific trees is sometimes highly aggregated, as
in mast-fruiting dipterocarp communities. From long-term studies on regeneration and recruitment of Bornean dipterocarps, Curran (1994) developed
a general ectomycorrhizal theory based on evidence that ectomycorrhizal
associations are relatively host-specific and spore dispersal is low. Short-
distance seed dispersal coupled with predator satiation, or inverse-density
dependence, may explain clustering of conspecific adult dipterocarps.

Seed dormancy also plays a role in establishment. Most pioneer tree species have small seeds that are deeply dehydrated and dormant upon dispersal. Dormancy can be broken when particular requirements such as light,
temperature, or heat are met (Vazquez-Yanes and Orozco-Segovia, 1990;
Garwood, 1989). Certain mature-phase forest trees similarly have small
seeds with low moisture content. These are often dispersed by wind or
ingested by vertebrate fruit dispersers, and may remain dormant until germination requirements are met. However, the majority of mature-phase
forest trees have large seeds with a high moisture content that allow them
to germinate readily. Their seeds typically do not undergo dormancy, and

because they are not found in the soil seed bank, they do not regenerate as readily following disturbances as do seeds of pioneer species.

The literature on seed dispersal and establishment in the tropics, like that on pollination, is too large to cover in this review. I recommend that the reader consult volumes edited by Fleming and Estrada (1993) and Bawa and Hadley (1990).

VI. Genetic Diversity

The genetic composition of individuals, populations, and species is determined by countless ecological and historical processes. Genetic variation and its structuring in time and space therefore are composite attributes that can be compared among populations or species with differing ecologies, distributions, and histories. Some of the processes affecting genetic diversity, such as gene flow, inbreeding, and genetic drift, can be inferred through hierarchical analyses of genetic structure and knowledge of the reproductive biology of the organism.

Studies of the genetic diversity of tropical tree populations have traditionally used allozyme (isozyme) genetic markers. Three relatively recent techniques take advantage of portions of the genome with high levels of length polymorphism. Randomly amplified polymorphic DNA (RAPDs) has been used to assay population-level genetic variability (e.g., Chalmers *et al.*, 1992) and to identify clones. Multilocus and single-locus fingerprinting (which employ micro- or minisatellite DNA) has additional applications for genetic analysis (Queller *et al.*, 1993). Single-locus fingerprinting has the highest level of resolution of the three techniques and the data generated by it are most amenable to mathematical inference because allelism and homology among loci are known. This latter technique may therefore be used in combination with allozyme data in studies requiring high levels of genetic polymorphism, such as paternity analyses and genealogy reconstruction.

Although few genetic studies of tropical trees have focused on canopy-related issues, they are nonetheless crucial to our understanding of reproductive processes and biodiversity in general.

A. The Individual

What is an individual? To look at a tree, the answer seems obvious. But apomixis, clonal growth, somatic mutations, and fusion of different genetic entities can force the revision of one's concept of individuality.

In higher plants, apomixis (or agamospermy) is the formation of seeds without fertilization of the ovule, although pseudogamous apomicts require fertilization of the endosperm for seeds to develop. Apomixis results in progeny that are genetically identical to the maternal plant. The occurrence of apomixis spans diverse families and genera, many of which are perennials found in disturbed habitats (Asker and Jerling, 1992). Plants from such environments are periodically reduced to small population sizes that are subject to stochastic events such as founder effects, genetic drift, and inbreeding depression. Facultative apomixis may counteract the effect of inbreeding depression and genetic drift by prolonging the life span of a genotype and thereby conserving the heterozygosity gained from sexual reproduction (Gustafsson, 1946).

The observation that many tropical trees have low population densities and apparently few interbreeding individuals led biologists to speculate about their mode of speciation and manner of reproduction (Corner, 1954; Baker, 1959; Federov, 1966; Ashton, 1969; van Steenis, 1969). Apomixis was deemed one potential solution in tropical trees, stemming from the notions of perpetuating successful gene combinations and colonizing new areas (Rollins, 1967).

Although most breeding system surveys of tropical trees have failed to test for apomixis (see earlier discussion), there is embryological evidence that it occurs facultatively in several Malaysian dipterocarps in the genera *Shorea* and *Hopea* (Kaur *et al.*, 1978, 1986), in species of *Clusia* (Maguire, 1976), *Syzygium* (formerly *Eugenia*; Tiwary, 1926), and *Citrus* (Swingle and Reece, 1967), and in various members of the dioecious genus *Garcinia* including mangosteen (Nygren, 1967; Richards, 1990). A consequence of apomixis in dioecious plants is an increase in females in the population; an extreme example of this has been discovered in *Garcinia parvifolia*, which forms entirely female populations (Ha *et al.*, 1988; S. C. Thomas, personal communication). Of four species of Sri Lankan dipterocarps tested by allozyme analysis, one species of *Shorea* (Murawski *et al.*, 1994a) and one of *Stemonoporus* (Murawski and Bawa, 1994) were highly outcrossed, but both showed evidence for apomixis in particular maternal individuals. The few Neotropical species examined, however, showed no evidence for apomixis from allozyme data. Too little is known to generalize about the role of apomixis in tropical trees, its occurrence in relation to forest strata, or its evolutionary significance.

Asexual reproduction through vegetative propagation and aerial roots is not commonly reported in tropical trees. One exception is the "great banyan tree" (*Ficus bengalensis*) in Calcutta, whose 1000 trunks covered an area of four acres. Sagers (1993) reported another form of clonal spread, through breakage and rerooting of leaves or stems, in shrub and tree species in the tropical understory. She postulated that the unusually high nitrogen and carbon content of abscised leaves in leaf-rooting species may be an adaptation to enhance survival of plant fragments in moist tropical habitats. However, the high cost of investing in additional nitrogen and carbon may generally limit the evolution of this form of vegetative reproduction to plants with small crowns, especially understory species. Such clonal spread shares certain advantages of apomictic reproduction, except that fragments would lack the dispersal agents of apomictically formed seeds, and fragments are more certain to perpetuate any viruses with which the parent is infected.

Sprouting from stem and subterranean tissues is a common response of rain forest trees to various types of disturbance. In eastern Amazonia, human-caused fires induced a large proportion of tree species (46%) to resprout from subterranean tissues (Kauffman, 1991). Bulldozing of forest on a bauxite mine site stimulated a similar regeneration from root suckers in a diverse array of tree taxa, including large canopy species (W. D. Hamilton and O. H. Knowles, personal communication). Following large-scale hurricane damage, hardwood trees resprouted rapidly from stem suckers on standing and toppled trees (Yih *et al.,* 1991; Frangi and Lugo, 1991). Resprouting was the main form of vegetation recovery in the early stages of secondary succession in nutrient-poor podzol soils (kerangas) in East Kalimantan, Indonesia, but not in nearby mixed dipterocarp forest (Riswan, 1982). It is not generally known if sprouting leads to successful recovery following various types of disturbance. Root suckering leading to the clonal spread of trees from multiple trunks does not necessarily require disturbance. The various implications of the different forms of asexual reproduction merit further study.

Perhaps one of the most unusual concepts of "individual" in tropical trees stems from the recent discovery that strangler figs can form composite individuals. An allozyme analysis of leaves from different portions of crowns of six stranglers from Barro Colorado Island in Panama revealed intracrown genetic differences in all species and in all but one individual sampled (Thomson *et al.,* 1991). Strangler figs begin life as hemiepiphytes, germinating in the crowns of host trees. As aerial roots reach ground and proliferate, they fuse, eventually forming the meshwork of stems that strangle the host tree (Berg and Wiebes, 1992). Apparently, this fusion is not necessarily limited to an individual genotype, but can comprise various individuals (genets). The interpretation of genetic mosaicism is also corroborated by phenological observation; crowns of the nonstrangler *Ficus* species typically flower synchronously, whereas different branches within the crowns of adult stranglers often flower asynchronously (C. Handley, personal communication). The occurrence of genetic mosaicism in other sites and other species of hemiepiphytes is unknown. Future research should elaborate on the extent and consequences of genetic mosaicism, as well as the developmental mechanisms that produce it.

Somatic mutation is another potential source of genetic variation within an "individual." A somatic mutation hypothesis has been postulated based on evidence that trees present highly variable environments to their herbivores (Edmunds and Alstad, 1978, 1981; Whitham, 1981; Whitham *et al.,* 1984). The assumption that somatic mutations create the observed variation remains untested. However, even with conservative estimates of somatic mutation per cell division, one can deduce that large organisms, especially long-lived trees, harbor many mutations, some of which may affect loci controlling herbivory (Antolin and Strobeck, 1985). Screening for genetic variation within tree crowns (e.g., using RAPDs or DNA fingerprinting) should yield basic information on the occurrence of somatic mutations.

B. The Population

In natural systems, populations can be continuous over vast distances. In such cases, effective isolation of an individual from conspecifics increases with distance. Alternatively, populations can be discrete (e.g., true islands). Most often, a population is hard to define without prior knowledge of gene dispersal distances. This forces the researcher to rely somewhat on intuition to choose a sampling unit and a consistent method of sampling specimens within each unit. In this case, populations are best described as sampling sites.

Several measures of genetic diversity are typically applied to variation at either the population or species level (review in Hamrick, 1989, and Liengsiri *et al.,* 1990): (1) the proportion of polymorphic loci (P), (2) the average number of alleles per polymorphic locus (A_p), (3) the effective number of alleles per locus (designated as A_e or n_e)(Crow and Kimura, 1970), and (4) the observed and expected heterozygosities (H_o and H_e) (Nei, 1975), the latter of which is also referred to as the polymorphic index (PI) (Allard *et al.,* 1978). Unlike the last three measures, the first two (P and A) give no indication of the relative frequency of alleles at a locus, but are typically reported for comparison with other published values. The proportion of all genotypes observed to be heterozygous (H_o) can be compared to the proportion expected to be heterozygous (H_e) given Hardy–Weinberg equilibrium. The fixation index $F = (H_e - H_o)/H_e$ (Wright, 1951) serves a similar purpose.

The first survey of allozyme diversity in tropical trees was made on 16 commonly occurring trees from Barro Colorado Island (Hamrick and Loveless, 1987, 1989; Loveless and Hamrick, 1987). Although there was considerable population-level variation in diversity estimates among species (range of P: 29.6–89.3%; range of H: 0.073–0.340), the means of P and H (60.9% and 0.211) were high compared with estimates from the literature for other plants (Hamrick and Loveless, 1987). No difference in diversity estimates between understory and canopy species was found ($P = 62.5$ and 58.9%, $H = 0.214$ and 0.212 for understory and canopy taxa, respectively).

Subsequent to the study of common tropical tree species, 16 uncommon species were surveyed from the same site (Hamrick and Murawski, 1991). To facilitate comparisons of the genetic parameters between common and uncommon species, a random sample of individuals of the common species matched the average sample size of the (much rarer) uncommon species. On average, significantly less genetic variation (mean P = 33.6% and H = 0.124) was found in the populations of uncommon trees, perhaps due to inbreeding, genetic drift, founder effect, or less selection for polymorphism.

In the most recent review of genetic diversity within populations of woody taxa, which includes an additional 43 species (Hamrick, 1994), the mean estimate of heterozygosity (H = 0.125) is revised downward and is lower than that of temperate woody species (H = 0.146). This could be due to a new focus on low-density species. However, the lack of standardization of methodology among laboratories might also affect diversity estimates. Difficulties can arise when lack of family data and presence of multiple bands on gels preclude a proper understanding of the number of genetic loci present. For suggestions on methodology and interpretation of gels, new students should see Soltis and Soltis (1989) and consult with major population genetics labs for advice on specific problems.

Species of the genus *Stemonoporus,* endemic to Sri Lanka, are exceptionally high in population-level heterozygosity compared to other tropical trees. The genus exhibits a range of sizes (from canopy to understory) and a range of distributions (from lowlands to cloud forest). The observed heterozygosities for the 12 species studied range from 0.159 to 0.509, with an average heterozygosity of 0.301 (D. A. Murawski, I. A. U. N. Gunatilleke, P. S. Ashton, and K. S. Bawa, unpublished data; Murawski and Bawa, 1994). The highest heterozygosities were detected in midcanopy to upper-canopy species that also tended to occur at higher elevations. These high levels of diversity are especially surprising owing to the highly localized distributions of many of the species. Historical factors, retention of ancient polymorphisms, hybridization, substantial pollen dispersal distances, and apomixis may all play some role in maintaining the high observed levels of genetic polymorphism in this species group.

The amount of genetic differentiation among population subunits is referred to as genetic structuring. The degree of structuring depends on a variety of factors such as gene flow, population size, and microhabitat selection. An extension of the fixation index, Wright's F-statistics (1965), allows us to partition the relative distribution of genetic variation among, as compared to within, subpopulations. F_{IT} is the total deviation from Hardy–Weinberg expectations in the total population, F_{ST} is that part of the deviation due to heterogeneity in allele frequencies among subpopulations, and F_{IS} is the deviation from Hardy–Weinberg equilibrium within subpopulations. Nei's G_{ST} (1973) is equivalent to a multiallelic F_{ST}, which merely extends the model to include more than two alleles per locus. An F_{ST} (or G_{ST}) of 0.1, for example, would mean that 10% of the total genetic variation is due to differences among subpopulations, while the remaining variation is found within subpopulations.

On the local scale of Barro Colorado Island, there is little heterogeneity among sample sites separated by distances of up to 2 km (average G_{ST} = 0.055) (Hamrick and Loveless, 1989). Suspecting that the range of G_{ST} values were due to differences in the extent of gene flow, the authors ranked species for gene flow potential based on knowledge of pollen and seed dispersal agents. They found a positive correlation between the degree of differentiation and the rank of predicted gene dispersal potential for the different species. This supported the hypothesis that the level of gene dispersal is a major contributor to the presence or lack of site differentiation. Understory species did not differ from canopy species in the degree of population (or site) differentiation (G_{ST} = 0.056 and 0.054 for understory and canopy species, respectively).

Several studies of tropical trees surveyed genetic diversity at a larger scale. Three species occurring in the La Selva Biological Reserve were examined both from the reserve and also from other locales in Costa Rica. Eight to twelve populations were sampled over a maximum distance of 70 km apart. These included the canopy-emergent, swamp specialist *Carapa guianensis* (Hall *et al.*, 1994), a midcanopy tree of high local density, *Pentaclethra macroloba* (Hall *et al.*, 1994), and the emergent tree *Pithecellobium pedicellare* (= *P. elegans*) (P. Hall, personal communication). F_{ST} values for the three species were 0.046, 0.219, and 0.101, respectively. The greater population differentiation in *Pentaclethra* was probably due to restricted pollen dispersal. On a scale that included all of Central America and Mexico, Chalmers *et al.* (1992) found that 60% of variation in RAPD markers occurs between populations of the leguminous tree *Gliricidia sepium*.

In Sri Lanka, ten populations of the canopy-emergent tree *Shorea trapezifolia*, sampled over a maximum distance of 43 km, had an F_{ST} of 0.111 (Dayanandan, 1994). Another Sri Lankan dipterocarp of montane elevations, *Stemonoporus oblongifolius*, sampled over its present known range with a maximum distance between sampling sites of 10 km, had a similar G_{ST} of 0.163. The proportion of the total genetic diversity among populations of these canopy species is similar to the average reported in the literature for tropical woody species (0.135), but is somewhat higher than the average for temperate woody species (0.099) (Hamrick, 1994).

At a more local scale, genetic heterogeneity among patches of new seedlings can be studied to see how seed dispersal patterns shape genetic structure. Hamrick *et al.* (1993) examined the fine-scale genetic structure of three tree species on Barro Colorado Island that differed in adult densities and seed dispersal mechanisms. The low-density, wind-dispersed species (*Platypodium elegans*) had the strongest and coarsest genetic structuring. In contrast, the high-density, bird-dispersed species (*Swartzia simplex*) had both the finest and weakest structuring. Presumably the fruits dispersed by birds from individual trees were more evenly distributed than those of the wind-dispersed species whose seedling shadows consisted primarily of the maternal trees' own offspring. The high-density, wind-dispersed species (*Alseis blackiana*) was intermediate between the other two species. In both wind-dispersed species, the genetic structure seen in the smaller size classes disappears in more mature trees. This deterioration of patch structure with age should prevent biparental inbreeding (kin mating) in adult trees.

Pollen dispersal also plays a major role in shaping the genetic composition of populations and determining effective population size (the number of interbreeding individuals). Various methods have been used to estimate the extent of pollen flow in tropical canopy and understory trees. Pollinator movement patterns (Janzen, 1971a; Frankie *et al.*, 1976; Augspurger, 1980, 1981; Frankie and Haber, 1983) and fluorescent dyes (Webb and Bawa, 1983; Murawski, 1987; Eguiarte *et al.*, 1993) give rough approximations; results suggest a tremendous range of dispersal distances. Genetic markers have the added advantage of estimating effective dispersal (resulting in fertilization). By studying the dispersal distance of a rare pollen allele into the progeny of an isolated cluster of *Tachigali versicolor* trees that were monomorphic for the common allele, Hamrick and Loveless (1989) determined that long-distance pollen dispersal (>500 m) resulted in a significant proportion (25%) of effective pollinations in this low-density population.

An extension of the mating system analysis is paternity exclusion analysis (Devlin *et al.*, 1988; Devlin and Ellstrand, 1990), which has been used to determine distances between mating pairs and the proportion of seeds sired by individuals outside a study population. The method requires a great deal of genetic polymorphism in order to exclude all but the unique father of a given seed. Not all plant species have the required level of allozyme polymorphism, which is why some scientists are turning to highly variable microsatellite DNA loci for additional polymorphism. Only one paternity study has been published on a tropical rain forest tree (Hamrick and Murawski, 1990). *Platypodium elegans*, a canopy emergent, like *Tachigali versicolor*, showed considerable long-distance pollen dispersal movement on Barro Colorado Island, with a significant proportion of seeds (17–40% over 3 years) sired by trees outside an 84-ha plot. New data on several *Ficus* species indicate that the wasp pollinators fly long distances among trees, such that all conspecific individuals on Barro Colorado Island (ca. 12 km^2) may form a single breeding unit (J. Nason, A. Herre, and J. Hamrick, unpublished data). These results help explain the observations of low levels of genetic structuring and high levels of intrapopulation genetic diversity in previous surveys of tropical rain forest trees.

Gene flow among populations (pollen plus seed dispersal) can be estimated indirectly from allele frequency data (Wright, 1951; Slatkin, 1981, 1985; Barton and Slatkin, 1986; Slatkin and Barton, 1989). Estimates of gene flow among populations, expressed as Nm (the number of migrants per generation), varied widely in 14 canopy and understory tree species from Panama (Hamrick and Loveless, 1989) and in one palm species from Mexico (Eguiarte *et al.*, 1993). The mean Nm of 5.38 for Barro Colorado Island is considered high (by Slatkin's standards, $Nm > 1.0$ is high), indicating that gene flow must be extensive over an area of at least 6 km². One notable exception to this generalization is the observation of apparent near-neighbor matings in the hawkmoth-pollinated, canopy-emergent *Cavanillesia platanifolia* (Murawski *et al.*, 1990). Perhaps at very low densities, certain species do not experience reliable long-distance pollen dispersal. However, preliminary results on species with low densities on Barro Colorado Island indicate that not all pollinations are between near-neighbors; some long-distance pollen flow occurs (E. Stacy, personal communication). It remains to be demonstrated for hyperdispersed species that gene flow is sufficiently great and effective population sizes are large enough to prevent genetic drift from shaping population structure.

Although genetic diversification is presumably a continuous process, methods most commonly used to examine it have been in the realms of the distinct disciplines of population genetics and systematics. It is now becoming possible to take a phylogenetic approach to population genetics, creating gene trees to estimate gene flow and effective population size (Hudson, 1990; Slatkin and Maddison, 1990; Felsenstein, 1992; Maddison, 1995; Donoghue, 1994). By implementing such challenging new approaches, progress will be made in understanding the process of biological diversification and its implications for canopy biology.

C. The Species

Taxa with high species richness constitute ideal test subjects for hypotheses of species divergence and the course of evolutionary change in reproductive characteristics. Examples of species-rich taxa abound in many tropical forests, where notably large congeneric woody taxa occur in such families as the Clusiaceae, Dipterocarpaceae, Ebenaceae, Euphorbiaceae, Fagaceae, Lauraceae, Lecythidaceae, Meliaceae, Moraceae, and Myrtaceae (Heywood, 1993), many of which are canopy trees. With the recent advances in phylogenetic analysis, morphological and molecular data can be used separately and in combination to infer the history of extant species. Population genetic data and studies of reproductive isolating mechanisms can lend additional clues regarding the mechanisms by which a group diverged.

The evolutionary history of particular traits can be inferred by mapping them onto a best estimate of the true phylogeny. The resulting inference is itself a hypothesis that can be tested using additional biologically relevant data (Coddington, 1988; Donoghue, 1989; Armbruster, 1992). For example, Armbruster (1992) inferred from phylogenetic data that pollination by fragrance-collecting male euglossine bees evolved independently three or four times in the genus *Dalechampia* (Euphorbiaceae). He predicted that the fragrance-reward system evolved by different pathways in the different lineages. His subsequent study of floral rewards substantiated that prediction, lending further support to the hypothesis of multiple origins of male euglossine pollination in the genus.

Thus, phylogeny can be used to understand the evolution of features such as sexual systems, plant–pollinator interactions, and fruit characteristics (e.g., Donoghue, 1989; Eriksson and Bremer, 1991; Armbruster, 1992; Chase and Hills, 1992; McDade, 1992; Stein, 1992). For instance, Weller *et al.* (1995) determined that self-compatibility is most likely ancestral in angiosperms, with different forms of self-incompatibility evolving independently on many occasions. Plant–pollinator interactions themselves can be used to reconstruct phylogenies, given that certain specialized interactions are conserved within higher taxonomic levels and thereby act as evolutionary constraints (Anderson, 1979; Howe, 1984; Stein, 1992). Future comparative studies of reproductive ecology in relation to canopy position should take phylogeny into consideration in order to differentiate between canopy- and lineage-related explanations.

VII. Summary

Reproduction in tropical trees is a dynamic process influencing population genetic structuring and the course of biological diversification, and also affecting the myriad of animal and plant species that interact in different ways with the complex canopy environment. Because the trees are the structural framework for all canopy biology, their functional requirements and behaviors should be a principal concern for all biologists involved in ecology, evolution, and conservation.

Growing evidence points to strong barriers to inbreeding in most tropical trees based on floral sexual dimorphism in some and on self-incompatibility and high outcrossing rates in others. Evidence from studies of genetic structuring and pollen-mediated gene flow (e.g., paternity analysis), together with observations of various long-ranging pollen and seed dispersers, indicates substantial levels of gene flow.

High species richness and the preponderance for low population densities distinguish many tropical rain forests from most other forest types. The handful of studies that have focused on species with characteristically low-density populations present a slightly different picture from the one described above. That is, self-fertilization increases, sexual dimorphism decreases, individuals appear to mate with fewer partners, and their populations contain less average genetic diversity than populations of common trees. Their diversity, however, is not low for all species. While some species may exhibit patterns of near-neighbor matings, others may thrive on reliable long-distance pollinator service. It appears that diversity in the low-density populations can be maintained through selection favoring outbred progeny (as in *Cavanillesia*) and perhaps heterosis and facultative apomixis.

Alteration of rain forest canopies affects the physiological conditions of all forest strata, and very likely alters the level of inbreeding as well as the composition, abundance, and behaviors of pollinators, frugivores, seed dispersers, pathogens, and herbivores. Given that the majority of tree species in a community are represented by very low population densities, more emphasis should be focused on the consequences of rarity on the maintenance of species and genetic diversity.

Acknowledgments

For constructive advice on the manuscript, I thank D. Ackerly, P. S. Ashton, L. Curran, M. J. Donoghue, W. D. Hamilton, J. L. Hamrick, M. Lowman, N. Nadkarni, R. Primack, and L. Vawter, and a Plant Evolutionary Ecology discussion group at Harvard.

References

Allard, R. W., Miller, R. D., and Kahler, A. L. (1978). The relationship between degree of environmental heterogeneity and genetic polymorphism. *In* "Structure and Functioning of Plant Populations" (A. J. Freysen and J. Woldendorp, eds.), pp. 49–73. North-Holland Publishing House, New York.

Anderson, W. R. (1979). Floral conservativism in neotropical Malpighiaceae. *Biotropica* 11, 219–223.

Antolin, M. F., and Strobeck, C. (1985). The population genetics of somatic mutation in plants. *Am. Nat.* 126, 52–62.

Appanah, S. (1979). The ecology of insect pollination of some tropical rain forest trees. Ph.D. Thesis, University of Malaya, Kuala Lumpur.

Appanah, S. (1985). General flowering in the climax rain forests of South-east Asia. *J. Trop. Ecol.* 1, 225–240.

Appanah, S. (1990). Plant–pollinator interactions in Malaysian rain forests. *In* "Reproductive Ecology of Tropical Forest Plants" (K. S. Bawa and M. Hadley, eds.), pp. 85–101. UNESCO, Paris, and Parthenon Publishing Groups, Carnsforth, UK.

Appanah, S., and Chan, H. T. (1981). Thrips: The pollinators of some dipterocarps. *Malaysian Forester* 44, 234–252.

Armbruster, W. S. (1992). Phylogeny and the evolution of plant–animal interactions. *BioScience* 42, 12–20.

Armbruster, W. S., and Berg, E. E. (1994). Thermal ecology of male euglossine bees in a tropical wet forest: Fragrance foraging in relation to operative temperature. *Biotropica* 26, 50–60.

Ashton, P. S. (1969). Speciation among tropical forest trees: Some deductions in the light of recent evidence. *Biol. J. Linn. Soc.* 1, 155–196.

Ashton, P. S. (1988). Dipterocarp biology as a window to the understanding of tropical forest structure. *Annu. Rev. Ecol. Syst.* 19, 347–370.

Ashton, P. S., Givnish, P. S., and Appanah, S. (1988). Staggered flowering in the Dipterocarpaceae: New insights into floral induction and the evolution of mast fruiting in the aseasonal tropics. *Am. Nat.* 132, 44–66.

Asker, S. E., and Jerling, L. (1992). "Apomixis in Plants." CRC Press, Ann Arbor.

Augspurger, C. K. (1980). Mass-flowering of a tropical shrub (*Hybanthus prunifolius*): Influence on pollination attraction and movement. *Evolution* 34, 475–488.

Augspurger, C. K. (1981). Reproductive synchrony of a tropical shrub: Experimental studies on effects of pollinators and seed predators on *Hybanthus prunifolius* (Violaceae). *Ecology* 62, 775–788.

Augspurger, C. K. (1982). A cue for synchronous flowering. *In* "The Ecology of a Tropical Forest" (E. G. Leigh, Jr., A. S. Rand, and D. M. Windsor, eds.), pp. 133–149. Smithsonian Institution Press, Washington, DC.

Augspurger, C. K. (1983a). Phenology, flowering synchrony, and fruit set of six neotropical shrubs. *Biotropica* 15, 257–267.

Augspurger, C. K. (1983b). Seed dispersal by the tropical tree, *Platypodium elegans,* and the escape of its seedlings from fungal pathogens. *J. Ecol.* 71, 759–771.

Augspurger, C. K. (1983c). Offspring recruitment around tropical trees: Changes in cohort distance with time. *Oikos* 40, 189–196.

Augspurger, C. K. (1990). The potential impact of fungal pathogens on tropical plant reproductive biology. *In* "Reproductive Ecology of Tropical Forest Plants" (K. S. Bawa and M. Hadley, eds.), pp. 237–245. UNESCO, Paris, and Parthenon Publishing Group, Carnsforth, UK.

Augspurger, C. K., and Kelly, C. A. (1984). Pathogen mortality of tropical tree seedlings: Experimental studies of the effects of dispersal distance, seedling density, and light conditions. *Oecologia* 61, 211–217.

Baker, H. G. (1959). Reproductive methods as factors in speciation in flowering plants. *Cold Spring Harbor Symp. Quant. Biol.* 24, 177–190.

Baker, H. G. (1973). Evolutionary relationships between flowering plants and animals in American and African tropical forests. *In* "Tropical Forest Ecosystems in Africa and South America: A Comparative Approach" (B. J. Meggers, E. S. Ayensu, and W. D. Duckworth, eds.), pp. 145–159. Smithsonian Institution Press, Washington, DC.

Baker, H. G., Bawa, K. S., Frankie, G. W., and Opler, P. A. (1982). Reproductive biology of plants in tropical forests. *In* "Tropical Rain Forest Ecosystems: Structure and Function" (F. B. Golley, ed.), "Ecosystems of the World," Vol. 14A, pp. 183–215. Elsevier, Amsterdam.

Barton, N. H., and Slatkin, M. (1986). A quasi-equilibrium theory of the distribution of rare alleles in a subdivided population. *Heredity* 56, 409–415.

Bateman, A. J. (1956). Cryptic self-incompatibility in the wildflower: *Cheiranthus cheiri* L. *Heredity* 10, 257–261.

Bawa, K. S. (1980). Evolution of dioecy in flowering plants. *Annu. Rev. Ecol. Syst.* 11, 15–39.

Bawa, K. S. (1990). Plant–pollinator interactions in tropical rain forests. *Annu. Rev. Ecol. Syst.* 21, 399–422.

Bawa, K. S. (1992). Mating systems, genetic differentiation and speciation in tropical rain forest plants. *Biotropica* 24, 250–255.

Bawa, K. S. (1994). Pollinators of tropical dioecious angiosperms: A reassessment? No, not yet. *Am. J. Bot.* 81, 456–460.

Bawa, K. S., and Hadley, M. (1990). "Reproductive Ecology of Tropical Forest Plants." UNESCO, Paris, and Parthenon Publishing Group, Carnsforth, UK.

Bawa, K. S., and Opler, P. A. (1975). Dioecism in tropical forest trees. *Evolution* 29, 167–179.

Bawa, K. S., Bullock, S. H., Perry, D. R., Coville, R. E., and Grayum, M. H. (1985a). Reproductive biology of tropical lowland rain forest trees. II. Pollination systems. *Am. J. Bot.* 72, 346–356.

Bawa, K. S., Perry, D. R., and Beach, J. H. (1985b). Reproductive biology of tropical lowland rain forest trees. I. Sexual systems and incompatibility mechanisms. *Am. J. Bot.* 72, 331–345.

Berg, C. C., and Wiebes, J. T. (1992). "African Fig Trees and Fig Wasps." North-Holland, Amsterdam.

Bertin, R. I., and Sullivan, M. (1988). Pollen interference and cryptic self-fertility in *Campsis radicans*. *Am. J. Bot.* 75, 1140–1147.

Bertin, R. I., Barnes, C., and Guttman, S. I. (1989). Self-sterility in *Campsis radicans* (Bignoniaceae). *Bot. Gaz.* 150, 397–403.

Bierzychudek, P. (1985). Patterns in plant parthenogenesis. *Experientia* 41, 1255–1264.

Boshier, D., Chase, M. R., and Bawa, K. S. (1995). The mating system of *Cordia alliodora*: Significant levels of inbreeding observed in an outcrossing species of tropical forest trees. *Am. J. Bot.* (in press).

Bowman, R. N. (1987). Cryptic self-incompatibility and the breeding system of *Clarkia unguiculata* (Onagraceae). *Am. J. Bot.* 74, 471–476.

Boyle, T. J. B., Liengsiri, C., and Piewluang, C. (1991). Genetic studies in a tropical pine—*Pinus kesiya*. III. The mating system in four populations from northern Thailand. *J. Trop. For. Sci.* 4, 37–44.

Bronstein, J. L. (1992). Seed predators as mutualists: Ecology and evolution of the fig/pollinator interaction. *In* "Insect-Plant Interactions" (E. Bernays, ed.), pp. 1–44. CRC Press, Ann Arbor.

Bullock, S. H. (1985). Breeding systems in the flora of a tropical deciduous forest in Mexico. *Biotropica* 17, 287–301.

Bullock, S. H., Beach, J. H., and Bawa, K. S. (1983). Episodic flowering and sexual dimorphism in *Guarea rhopalocarpa* Radlk, (Meliaceae) in a Costa Rican rain forest. *Ecology* 64, 851–862.

Burkey, T. (1994). Tropical tree species diversity: A test of Janzen's spacing mechanism. *Oecologia* 97, 533–540.

Chalmers, K. J., Waugh, R., Sprent, J. I., Simons, A. J., and Powell, W. (1992). Detection of genetic variation between and within populations of *Gliricidia sepium* and *G. maculata* using RAPD markers. *Heredity* 69, 465–472.

Chan, H. T., and Appanah, S. (1980). Reproductive biology of some Malaysian dipterocarps. I. Flowering biology. *Malaysian For.* 43, 132–143.

Chase, M. W., and Hills, H. G. (1992). Orchid phylogeny, flower sexuality, and fragrance-seeking. *BioScience* 42, 43–49.

Clark, D. A., and Clark, D. B. (1984). Spacing dynamics of a tropical rain forest tree: Evaluation of the Janzen–Connell model. *Am. Nat.* 124, 769–788.

Clark, D. A., and Clark, D. B. (1987). Population ecology and microhabitat distribution of *Dipteryx panamensis*, a Neotropical rain forest emergent tree. *Biotropica* 19, 236–244.

Coddington, J. (1988). Phylogenetic tests of adaptational hypotheses. *Cladistics* 4, 3–22.

Condit, R., Hubbell, S. P., and Foster, R. B. (1992). Recruitment near conspecific adults and the maintenance of tree and shrub diversity in a Neotropical forest. *Am. Nat.* 140, 261–286.

Connell, J. H. (1971). On the role of natural enemies in preventing competitive exclusion in some marine animals and in rain forest trees. *In* "Dynamics of Numbers in Populations" (P. J. den Boer and G. R. Gradwell, eds.), pp. 2989–312. Center for Agricultural Publication and Documentation, Wageningen, The Netherlands.

Connell, J. H., Tracey, J. G., and Webb, L. J. (1984). Compensatory recruitment, growth, and mortality as factors maintaining rain forest tree diversity. *Ecol. Monogr.* 54, 141–164.

Corner, E. J. H. (1954). The evolution of tropical forest. *In* "Evolution as a Process" (J. Huxley, A. C. H. Hardy, and E. B. Ford, eds.), pp. 34–46. Allen & Unwin, London.

Croat, T. B. (1979). The sexuality of the Barro Colorado Island flora (Panama). *Phytologia* 42, 319–348.

Crow, J. F., and Kimura, M. (1970). "An Introduction to Population Genetics Theory." Harper & Row, New York.

Cruzan, M. B., and Barrett, S. C. H. (1993). Contribution of cryptic incompatibility to the mating system of *Eichhornia paniculata* (Pontederiaceae). *Evolution* 47, 925–934.

Curran, L. M. (1994). The ecology and evolution of mast-fruiting in Bornean Dipterocarpaceae: A general ectomycorrhizal theory. Ph.D Thesis, Princeton University, Princeton.

Dayanandan, B. (1994). Genetic diversity and mating system of *Shorea trapezifolia* Thw. Ashton. Masters Thesis, University of Massachusetts, Boston.

Dayanandan, S., Attygalla, D. N. C., Abeygunasekera, A. W. W. L., Gunatilleke, I. A. U. N., and Gunatilleke, C. V. S. (1990). Phenology and floral morphology in relation to pollination of some Sri Lankan dipterocarps. *In* "Reproductive Ecology of Tropical Forest Plants" (K. Bawa and M. Hadley, eds.), pp. 103–134. UNESCO, Paris, and Parthenon Publishing Group, Carnsforth, UK.

de Nettancourt, D. (1977). "Incompatibility in Angiosperms." Springer-Verlag, New York.

Devlin, B., and Ellstrand, N. C. (1990). The development and application of a refined method for estimating gene flow from angiosperm paternity analysis. *Evolution* 44, 248–259.

Devlin, B., Roeder, K., and Ellstrand, N. C. (1988). Fractional paternity assignment: Theoretical development and comparison to other methods. *Theor. Appl. Genet.* 76, 369–380.

Dieterlen, F. (1978). Zur Phanologie des Aquatorialen Regenwaldes im Ost-Zaire (Kivu). *Diss. Bot.* 47, 5–120.

Dole, J., and Ritland, K. (1993). Inbreeding depression in two *Mimulus* taxa measured by multigenerational changes in the inbreeding coefficient. *Evolution* 47, 361–373.

Donoghue, M. J. (1989). Phylogenies and the analysis of evolutionary sequences, with examples from seed plants. *Evolution* 43, 1137–1156.

Donoghue, M. J. (1994). Progress and prospects in reconstructing plant phylogeny. *Ann. Missouri Bot. Gard.* 81, 405–418.

Edmunds, G. F., and Alstad, D. N. (1978). Coevolution of insect herbivores and conifers. *Science* 199, 941–945.

Edmunds, G. F., and Alstad, D. N. (1981). Responses of black pine-leaf scale to host plant variability. *In* "Insect Life History Patterns: Habitat and Geographic Variation" (R. F. Denno and J. Dingle, eds.), pp. 29–38. Springer-Verlag, New York.

Eguiarte, L. E., Perez-Nasser, N., and Pinero, D. (1992). Genetic structure, outcrossing rate and heterosis in *Astrocaryum mexicanum* (tropical palm): Implications for evolution and conservation. *Heredity* 69, 217–228.

Eguiarte, L. E., Burquez, A., Rodriguez, J., Martinez-Ramos, M., Sarukhan, J., and Pinero, D. (1993). Direct and indirect estimates of neighborhood and effective population size in a tropical palm, *Astrocaryum mexicanum*. *Evolution* 47, 75–87.

Eickwort, G. C., and Ginsberg, H. S. (1980). Foraging and mating behavior in apoidea. *Annu. Rev. Entomol.* 25, 421–446.

Eriksson, O., and Bremer, B. (1991). Fruit characteristics, life forms and species richness in the plant family Rubiaceae. *Am. Nat.* 138, 751–761.

Faegri, K., and van der Pijl, L. (1979). "The Principles of Pollination Ecology," 3rd ed. Pergamon Press, Oxford.

Federov, A. A. (1966). The structure of the tropical rain forest and speciation in the humid tropics. *J. Ecol.* 54, 1–11.

Feinsinger, P. (1976). Organization of a tropical guild of nectarivorous birds. *Ecol. Monogr.* 46, 257–291.

Feinsinger, P. (1978). Ecological interactions between plants and hummingbirds in a successional tropical community. *Ecol. Monogr.* 48, 269–287.

Felsenstein, J. (1992). Estimating effective population size from samples of sequences: Inefficiency of pairwise and segregating sites as compared to phylogenetic estimates. *Genet. Res.* 59, 139–147.

Fleming, T. H. (1982). Foraging strategies of plant-visiting bats. *In* "Ecology of Bats" (T. H. Kunz, ed.), pp. 287–325. Plenum, New York.

Fleming, T. H., and Estrada, A., eds. (1993). "Frugivory and Seed Dispersal: Ecological and Evolutionary Aspects." Kluwer Academic Publishers, Dordrecht, The Netherlands.

Fleming, T. H., Venable, D. L., and Herrera M., L. G. (1993). Opportunism vs. specialization: The evolution of dispersal strategies in fleshy-fruited plants. *Vegetatio* 107/108, 107–120.

Flores, S., and Schemske, D. W. (1984). Dioecy and monoecy in the flora of Puerto Rico and the Virgin Islands: Ecological correlates. *Biotropica* 16, 132–139.

Forget, P.-M. (1991). Evidence for secondary seed dispersal by rodents in Panama. *Oecologia* 87, 596–599.

Foster, R. B. (1977). *Tachigalia versicolor* is a suicidal Neotropical tree. *Nature* 268, 624–626.

Frangi, J. L., and Lugo, A. E. (1991). Hurricane damage to a flood plain forest in the Luquillo Mountains of Puerto Rico. *Biotropica* 23, 324–335.

Frankie, G. W., and Haber, W. A. (1983). Why bees move among mass-flowering Neotropical tree species. *In* "Handbook of Experimental Pollination Biology" (C. J. Jones and R. J. Little, eds.), pp. 360–372. Van Nostrand–Reinhold, New York.

Frankie, G. W., Baker, H. G., and Opler, P. A. (1974). Comparative phenological studies of trees in tropical wet and dry forests in the lowlands of Costa Rica. *J. Ecol.* 62, 881–919.

Frankie, G. W., Opler, P. A., and Bawa, K. S. (1976). Foraging behavior of solitary bees: Implications for outcrossing for a Neotropical forest tree species. *J. Ecol.* 64, 1049–1057.

Frankie, G. W., Vinson, S. B., Newstrom, L. E., Barthell, J. F., Haber, W. A., and Frankie, J. K. (1990). Plant phenology, pollination ecology, pollinator behaviour and conservation of pollinators in Neotropical dry forest. *In* "Reproductive Ecology of Tropical Forest Plants" (K. S. Bawa and M. Hadley, eds.). UNESCO, Paris, and Parthenon Publishing Group, Carnsforth, UK.

Galil, J., and Eisikowitch, D. (1968). On the pollination ecology of *Ficus sycomorus* in East Africa. *Ecology* 49, 259–269.

Garwood, N. (1989). Tropical soil seed banks: A review. *In* "Ecology of Seed Banks" (L. A. Leck, R. L. Simpson, and V. T. Parker, eds.), pp. 149–190. Academic Press, London.

Gautier-Hion, A. (1990). Interactions among fruit and vertebrate fruit-eaters in an African tropical forest. *In* "Reproductive Ecology of Tropical Forest Plants" (K. S. Bawa and M. Hadley, eds.), pp. 219–232. UNESCO, Paris, and Parthenon Publishing Group, Carnsforth, UK.

Gautier-Hion, A., Duplantier, J.-M., Quris, R., Feer, F., Sourd, C., Decoux, J.-P., Dubost, G., Emmons, L., Erard, C., Hecketsweiler, P., Moungazi, A., Roussilhon, C., and Thiollay, J.-M. (1985). Fruit characters as a basis of fruit choice and seed dispersal in a tropical forest vertebrate community. *Oecologia* 65, 324–337.

Gautier-Hion, A., Gautier, J.-P., and Maisels, F. (1993). Seed dispersal versus seed predation: An inter-site comparison of two related African monkeys. *In* "Frugivory and Seed Dispersal: Ecological and Evolutionary Aspects" (T. H. Fleming and A. Estrada, eds.), pp. 237–244. Kluwer Academic Publishers, Dordrecht, The Netherlands.

Gentry, A. H. (1974). Flowering phenology and diversity in tropical Bignoniaceae. *Biotropica* 6, 64–68.

Gibbs, P. E., and Bianchi, M. (1993). Post-pollination events of *Chorisia* (Bombacaceae) and *Tabebuia* (Bignoniaceae) with late-acting self-incompatibility. *Bot. Acta* 106, 64–71.

Gilbert, L. E. (1980). Coevolution of animals and plants: A 1979 postscript. *In* "Coevolution of

Animals and Plants" (L. E. Gilbert and P. H. Raven, eds.), pp. 247–263, rev. ed. Univ. of Texas Press, Austin.

Gottsberger, G. (1978). Seed dispersal by fish in the inundated regions of Humaita. *Biotropica* 10, 170–183.

Goulding, M., Carvalho, M. L., and Ferreira, E. G. (1988). "Rio Negro, Rich in Life in Poor Water." SPB Academic Publishing BV, The Hague.

Gustafsson, A. (1946, 1947). Apomixis in higher plants. I–III. *Lunds Universitets Arsskrift* 42, 1–67; 43, 69–179; 183–370.

Ha, C. O., Sands, V. E., Soepadmo, E., and Jong, K. (1988). Reproductive patterns of selected understory trees in the Malaysian rain forest: The apomictic species. *Bot. J. Linn. Soc.* 97, 317–331.

Haber, W. A., and Frankie, G. W. (1989). A tropical hawkmoth community: Costa Rican dry forest Sphingidae. *Biotropica* 21, 155–172.

Hall, P., Chase, M. R., and Bawa, K. S. (1994). Low genetic variation but high population differentiation in a common tropical forest tree species. *Conserv. Biol.* 8, 471–482.

Hamilton, W. D. (1993). Inbreeding in Egypt and this book: A childish viewpoint. *In* "The Natural History of Inbreeding and Outbreeding: Theoretical and Empirical Perspectives" (N. W. Thornhill, ed.), pp. 429–450. Univ. of Chicago Press, Chicago.

Hamrick, J. L. (1989). Isozymes and the analysis of genetic structure in plant populations. *In* "Isozymes in Plant Biology" (D. E. Soltis and P. S. Soltis, eds.), pp. 87–105. Dioscorides Press, Portland.

Hamrick, J. L. (1994). Genetic diversity and conservation in tropical forests. *In* "Proceedings, International Symposium on Genetic Conservation and Production of Tropical Tree Seed" (R. M. Drysdale, S. E. T. John, and A. C. Yapa, eds.), pp. 1–9. ASEAN–Canada Forest Tree Seed Centre, Mauk Lek, Saraburi, Thailand.

Hamrick, J. L., and Godt, M. J. (1989). Allozyme diversity in plant species. *In* "Population Genetics and Germplasm Resources in Crop Improvement" (A. H. D. Brown, M. T. Clegg, A. L. Kahler, and B. S. Weir, eds.), pp. 43–63. Sinauer Associates, Sunderland.

Hamrick, J. L., and Loveless, M. D. (1987). Distribucion de la variacion en especies de arboles tropicales. *Rev. Biol. Trop.* 35, 165–175.

Hamrick, J. L., and Loveless, M. D. (1989). The genetic structure of tropical tree populations: Associations with reproductive biology. *In* "The Evolutionary Ecology of Plants" (J. H. Bock and Y. B. Linhart, eds.), pp. 129–149. Westview Press, Boulder.

Hamrick, J. L., and Murawski, D. A. (1990). The breeding structure of tropical trees. *Plant Species Biol.* 5, 157–165.

Hamrick, J. L., and Murawski, D. A. (1991). Levels of allozyme diversity in populations of uncommon Neotropical tree species. *J. Trop. Ecol.* 7, 395–399.

Hamrick, J. L., Godt, M. J., and Sherman-Broyles, S. L. (1992). Factors influencing levels of genetic diversity in woody plant species. *New Forests* 6, 95–124.

Hamrick, J. L., Murawski, D. A., and Nason, J. D. (1993). The influence of seed dispersal mechanisms on the genetic structure of tropical tree populations. *Vegetatio* 107/108, 281–297.

Hedrick, P. W. (1985). "Genetics of Populations." Jones and Bartlett Publishers, Boston.

Heinrich, B. (1979). Resource heterogeneity and patterns of movement in foraging bumblebees. *Oecologia* 40, 235–246.

Heithaus, E. R., Fleming, T. H., and Opler, P. A. (1975). Foraging patterns and resource utilization in seven species of bats in a seasonal tropical forest. *Ecology* 56, 841–854.

Hernandez, H. M., and Abud, Y. C. (1987). Notas sobre la ecologia reproductiva de arboles en un bosque mesofilo de montana en Michoacan, Mexico. *Bol. Soc. Bot. Mex.* 47, 5–35.

Herrera, C. M. (1986). Vertebrate-dispersed plants: Why they don't behave the way they should. *In* "Frugivores and Seed Dispersal" (A. Estrada and T. H. Fleming, eds.), pp. 5–18. Dr. W. Junk Publishers, Dordrecht.

Heywood, V. H. (1993). "Flowering Plants of the World." Oxford Univ. Press, New York.

Howe, H. F. (1984). Constraints on the evolution of mutualisms. *Am. Nat.* 123, 756–777.

Howe, H. F. (1989). Scatter- and clump-dispersal and seedling demography: Hypothesis and implications. *Oecologia* 79, 417–426.

Howe, H. F. (1990). Survival and growth of juvenile *Virola surinamensis* in Panama: Effects of herbivory and canopy closure. *J. Trop. Ecol.* 6, 259–280.

Hubbell, S. P., and Foster, R. B. (1986). Commonness and rarity in a neotropical forest: Implications for tropical tree conservation. *In* "Conservation Biology: The Science of Scarcity and Diversity" (M. E. Soulé, ed.), pp. 205–231. Sinauer Associates, Sunderland.

Hudson, R. R. (1990). Gene genealogies and the coalescent process. *Oxford Surv. Evol. Biol.* 7, 1–44.

Irvine, A. K., and Armstrong, J. E. (1990). Beetle pollination in tropical forests of Australia. *In* "Reproductive Ecology of Tropical Forest Plants" (K. S. Bawa and M. Hadley, eds.), pp. 135–147. UNESCO, Paris, and Parthenon Publishing Group, Carnsforth, UK.

Janson, C. H. (1983). Adaptation of fruit morphology to dispersal agents in a Neotropical forest. *Science* 219, 187–189.

Janzen, D. H. (1970). Herbivores and the number of tree species in tropical forests. *Am. Nat.* 104, 501–528.

Janzen, D. H. (1971a). Euglossine bees as long distance pollinators of tropical plants. *Science* 171, 203–205.

Janzen, D. H. (1971b). The fate of *Scheelea rostrata* fruits beneath the parent tree: Predispersal attack by bruchids. *Principes* 15, 89–101.

Janzen, D. H. (1972). Escape in space by *Sterculia apetala* seeds from the bug *Dysdercus fasciatus* in a Costa Rican deciduous forest. *Ecology* 53, 350–361.

Janzen, D. H. (1979). How to be a fig. *Annu. Rev. Ecol. Syst.* 10, 13–51.

Janzen, D. H., and Vazquez-Yanes, C. (1991). Aspects of tropical seed ecology of relevance to management of tropical forested wildlands. *In* "Rain Forest Regeneration and Management" (A. Gomez-Pompa, T. C. Whitmore, and M. Hadley, eds.), pp. 137–157. UNESCO, Paris, and Parthenon Publishing Group, Carnsforth, UK.

Kauffman, J. B. (1991). Survival by sprouting following fire in tropical forests of the eastern Amazon. *Biotropica* 23, 219–224.

Kaur, A., Ha, C. D., Jong, K., Sands, V. E., Chan, H. T., Soepadmo, E., and Ashton, P. S. (1978). Apomixis may be widespread among trees of the climax rain forest. *Nature* 271, 440–441.

Kaur, A., Jong, I., Sands, V. E., and Soepadmo, E. (1986). Cytoembryology of some Malaysian dipterocarps, with some evidence of apomixis. *Bot. J. Linn. Soc.* 92, 75–88.

Kenrick, J., and Knox, R. B. (1985). Self-incompatibility in the nitrogen-fixing tree *Acacia retinodes*: Quantitative cytology of pollen tube growth. *Theor. Appl. Genet.* 69, 481–488.

Kochummen, K. M., LaFrankie, J. V., and Manokaran, N. (1991). Floristic composition of Pasoh Forest Reserve, a lowland rain forest in peninsular Malaysia. *J. Trop. For. Sci.* 3, 1–13.

Koelmeyer, K. O. (1959). The periodicity of leaf change and flowering in the principal forest communities of Ceylon. *Ceylon For.* 4, 157–189; 308–364.

Kress, W. J., and Beach, J. H. (1993). Flowering plant reproductive systems at La Selva Biological Station. *In* "La Selva: Ecology and Natural History of a Neotropical Rain forest" (L. McDade, K. S. Bawa, G. Hartshorn, and H. A. Hespenheide, eds.), pp. 161–182. Univ. of Chicago Press, Chicago.

Kubitzki, K., and Kurz, H. (1984). Synchronized dichogamy and dioecy in Neotropical Lauraceae. *Plant Syst. Evol.* 147, 253–266.

Kubitzki, K., and Ziburski, A. (1994). Seed dispersal in flood plain forests of Amazonia. *Biotropica* 26, 30–43.

Lemke, T. O. (1984). Foraging ecology of the long-nosed bat, *Glossophaga soricina*, with respect to resource availability. *Ecology* 65, 538–548.

Lieberman, M., and Lieberman, D. (1993). Patterns of density and dispersion of forest trees. *In* "La Selva: Ecology and Natural History of a Neotropical Rain Forest" (L. McDade, K. S. Bawa, G. S. Hartshorn, and H. A. Hespenheide, eds.), pp. 106–119. Univ. of Chicago Press, Chicago.

Liengsiri, C., Piewluang, C., and Boyle, T. J. B. (1990). "Starch Gel Electrophoresis of Tropical Trees: A Manual." ASEAN–Canada Forest Tree Seed Centre, Mauk Lek, Saraburi, Thailand.

Loveless, M. D. (1992). Isozyme variation in tropical trees: Patterns of genetic organization. *New Forests* 6, 67–94.

Loveless, M. D., and Hamrick, J. L. (1987). Distribucion de la variacion genetica en especies de arboles tropicales. *Rev. Biol. Trop.* 35, 165–176.

Maddison, W. P. (1995). Phylogenetic histories within and between species. *In* "Experimental and Molecular Approaches to Plant Biosystematics" (P. C. Hoch and A. G. Stephenson, eds.), Monographs in Systematic Botany. Missouri Botanical Gardens, St. Louis (in press).

Maguire, B. (1976). Apomixis in the genus *Clusia* (Clusiaceae)—A preliminary report. *Taxon* 25, 241–244.

Manokaran, N., LaFrankie, J. V., Kochummen, K. M., Quah, E. S., Klahn, J., Ashton, P. S., and Hubbell, S. P. (1990). "Methodology for the 50-ha Research Plot at Pasoh Forest Reserve," Research Pamphlet Number 104. Forest Research Institute of Malaysia, Kepong, Malaysia.

Matheson, A. C., Bell, J. C., and Barnes, R. D. (1989). Breeding systems and genetic structure in some Central American pine populations. *Silvae Genetica* 38, 107–113.

McDade, L. A. (1986). Protandry, synchronized flowering and sequential phenotypic unisexuality in Neotropical *Pentagonia macrophylla* (Rubiaceae). *Oecologia* 68, 218–223.

McDade, L. A. (1992). Pollinator relationships, biogeography, and phylogenetics. *BioScience* 42, 21–26.

McDade, L. A., Bawa, K. S., Hartshorn, G., and Hespenheide, H. A. (1993). "La Selva: Ecology and Natural History of a Neotropical Rain Forest." Univ. of Chicago Press, Chicago.

Medway, L. (1972). Phenology of a tropical rain forest in Malaya. *Biol. J. Linn. Soc.* 4, 117–146.

Mitton, J. B., and Jeffers, R. M. (1989). The genetic consequences of mass selection for growth rate in Engelmann spruce. *Silvae Genetica* 38, 6–12.

Momose, K., and Inoue, T. (1994). Pollination syndromes in the plant–pollinator community in the lowland mixed dipterocarp forests of Sarawak. *In* "Plant Reproductive Systems and Animal Seasonal Dynamics: Long-Term Study of Dipterocarp Forests in Sarawak" (T. Inoue and A. A. Hamid, eds.), pp. 119–141.

Moran, G. F., Muona, O., and Bell, J. C. (1989). Breeding systems and genetic diversity in *Acacia auriculiformis* and *A. crassicarpa*. *Biotropica* 21, 250–256.

Murawski, D. A. (1987). Floral resource variation, pollinator response, and potential pollen flow in *Psiguria warscewiczii*. *Ecology* 68, 1273–1282.

Murawski, D. A., and Bawa, K. S. (1994). Genetic structure and mating system of *Stemonoporus oblongifolius* (Dipterocarpaceae) in Sri Lanka. *Am. J. Bot.* 81, 155–160.

Murawski, D. A., and Hamrick, J. L. (1991). The effects of the density of flowering individuals on the mating systems of nine tropical tree species. *Heredity* 67, 167–174.

Murawski, D. A., and Hamrick, J. L. (1992a). The mating system of *Cavanillesia platanifolia* under extremes of flowering-tree density. *Biotropica* 24, 99–101.

Murawski, D. A., and Hamrick, J. L. (1992b). Mating system and phenology of *Ceiba pentandra* (Bombacaceae) in Central Panama. *J. Heredity* 83, 401–404.

Murawski, D. A., Hamrick, J. L., Hubbell, S. P., and Foster R. B. (1990). Mating systems of two bombacaceous trees of a Neotropical moist forest. *Oecologia* 82, 501–506.

Murawski, D. A., Dayanandan, B., and Bawa, K. S. (1994a). Outcrossing rates of two endemic *Shorea* species from Sri Lankan tropical rain forests. *Biotropica* 26, 23–29.

Murawski, D. A., Gunatilleke, I. A. U. N., and Bawa, K. S. (1994b). The effects of selective logging on inbreeding in *Shorea megistophylla* (Dipterocarpaceae) from Sri Lanka. *Conserv. Biol.* 8, 997–1002.

Nei, M. (1973). Analysis of gene diversity in subdivided populations. *Proc. Nat. Acad. Sci. USA* 70, 3321–3323.

Nei, M. (1975). Molecular population genetics and evolution. *In* "Frontiers of Biology" (A. Neuberger and E. L. Tatum, eds.), Vol. 40. Elsevier, New York.

Newstrom, L. E., Frankie, G. W., and Baker, H. G. (1991). Survey of long-term flowering patterns in lowland tropical rain forest trees at La Selva, Costa Rica. *In* "L'arbre. Biologie et Developpement" (C. Edelin, ed.), pp. 345–366. Naturalia Monspeliensia, France.

Newstrom, L. E., Frankie, G. W., Baker, H. G., and Colwell, R. K. (1993). Diversity of long-term flowering patterns. *In* "La Selva: Ecology and Natural History of a Neotropical Rain Forest" (L. A. McDade, K. S. Bawa, H. A. Hespenheide, and G. S. Hartshorn, eds.), pp. 142–160. Univ. of Chicago Press, Chicago.

Newstrom, L. E., Frankie, G. W., and Baker, H. G. (1994). A new classification for plant phenology based on flowering patterns in lowland tropical rain forest trees at La Selva, Costa Rica. *Biotropica* 26, 141–159.

Norstog, K. J., Stevenson, D. W., and Niklas, K. J. (1986). The role of beetles in the pollination of *Zamia furfuracea* L. fil. (Zamiaceae). *Biotropica* 18, 300–306.

Nygren, A. (1967). Apomixis in the angiosperms. *Handbuch der Pflanzenphysiol.* 18, 551–596.

O'Malley, D. M., and Bawa, K. S. (1987). Mating system of a tropical rain forest tree species. *Am. J. Bot.* 74, 1143–1149.

O'Malley, D. M., Buckley, D. P., Prance, G. T., and Bawa, K. S. (1988). Genetics of Brazil nut (*Bertholletia excelsa* Humb. Bonpl.: Lecythidaceae): 2. Mating system. *Theor. Appl. Genet.* 76, 929–932.

Opler, P. A., Frankie, G. W., and Baker, H. G. (1980). Comparative phenological studies of treelet and shrub species in tropical wet and dry forests in the lowlands of Costa Rica. *J. Ecol.* 68, 167–188.

Pellmyr, O. (1992). The phylogeny of a mutualism: Evolution and coadaptation between *Trollius* and its seed-parasitic pollinators. *Biol. J. Linn. Soc.* 47, 337–365.

Pellmyr, O., and Thompson, J. N. (1992). Multiple occurrences of mutualism in the yucca moth lineage. *Proc. Nat. Acad. Sci. USA* 89, 2927–2929.

Perez-Nasser, N., Eguiarte, L. E., and Pinero, D. (1993). Mating system and genetic structure of the distylous tropical tree *Psychotria faxlucens* (Rubiaceae). *Am. J. Bot.* 80, 45–52.

Prance, G. T. (1985). The pollination of Amazonian plants. *In* "Key Environments: Amazonia" (G. T. Prance and T. E. Lovejoy, eds.), pp. 166–191. Pergamon Press, Terrytown.

Proctor, M. C. F., and Yeo, P. (1973). "The Pollination of Flowers." Collins, London.

Putz, F. E. (1979). Aseasonality in Malaysian tree phenology. *Malaysian Forester* 42, 1–24.

Queller, D. C., Strassmann, J. E., and Hughes, C. R. (1993). Microsatellites and kinship. *Trends Ecol. Evol.* 8, 285–288.

Ramirez, N., and Brito, T. (1990). Reproductive biology of a tropical palm swamp community in the Venezuelan llanos. *Am. J. Bot.* 77, 1260–1271.

Renner, S. S., and Feil, J. P. (1993). Pollinators of tropical dioecious angiosperms. *Am. J. Bot.* 80, 1100–1107.

Richards, A. J. (1990). Studies in *Garcinia*, dioecious tropical forest trees: Agamospermy. *Bot. J. Linn. Soc.* 103, 233–250.

Riswan, S. (1982). Ecological studies on primary, secondary and experimentally cleared mixed dipterocarp forest and kerangas forest in East Kalimantan, Indonesia. Ph.D Thesis, University of Aberdeen, Scotland.

Ritland, K. (1983). Estimation of mating systems. *In* "Isozymes in Plant Genetics and Breeding" (S. D. Tanksley and T. J. Orton, eds.), pp. 289–302. Elsevier, Amsterdam.

Ritland, K. (1990). A series of FORTRAN computer programs for estimating plant mating systems. *J. Heredity* 81, 235–237.

Ritland, K., and Jain, S. K. (1981). A model for the estimation of out-crossing rate and gene frequencies based on *n* independent loci. *Heredity* 47, 37–54.

Rogstad, S. H. (1986). A biosystematic investigation of the *Polyalthia hypoleuca* complex (Annonaceae) of Malesia. Ph.D Thesis, Harvard University, Cambridge.

Rogstad, S. H. (1994). The biosystematics and evolution of the *Polyalthia hypoleuca* species complex (Annonaceae) of Malesia. III. Floral ontogeny and breeding systems. *Am. J. Bot.* 81, 145–154.

Rohwer, J. (1986). Some aspects of dioecy in *Ocotea* (Lauraceae). *Plant Syst. Evol.* 152, 47–48.

Rollins, R. C. (1967). The evolutionary fate of inbreeders and nonsexuals. *Am. Nat.* 101, 343–351.

Roubik, D. W. (1989). "Ecology and Natural History of Tropical Bees." Cambridge Univ. Press, New York.

Roubik, D. W. (1992). Loose niches in tropical communities: Why are there so few bees and so many trees? *In* "Effects of Resource Distribution on Animal–Plant Interactions" (M. D. Hunter, T. Ohgushi, and P. W. Price, eds.), pp. 327–354. Academic Press, San Diego.

Roubik, D. W. (1993). Tropical pollinators in the canopy and understory: Field data and theory for stratum "preferences." *J. Insect Behav.* 6, 659–673.

Ruiz, Z. T., and Arroyo, M. T. K. (1978). Plant reproductive ecology of a secondary tropical deciduous forest in Venezuela. *Biotropica* 10, 221–230.

Sagers, C. (1993). Reproduction in Neotropical shrubs: The occurrence and some mechanisms of asexuality. *Ecology* 74, 615–618.

Schatz, G. E. (1990). Some aspects of pollination biology in Central American forests. *In* "Reproductive Ecology of Tropical Forest Plants" (K. S. Bawa and M. Hadley, eds.), pp. 69–84. UNESCO, Paris, and Parthenon Publishing Group, Carnsforth, UK.

Seavey, S. R., and Bawa, K. S. (1986). Late-acting self-incompatibility. *Bot. Rev.* 52, 196–217.

Slatkin, M. (1981). Estimating levels of gene flow in natural populations. *Genetics* 99, 332–335.

Slatkin, M. (1985). Rare alleles as indicators of gene flow. *Evolution* 39, 53–65.

Slatkin, M., and Barton, N. H. (1989). A comparison of three indirect methods of estimating average levels of gene flow. *Evolution* 43, 1349–1368.

Slatkin, M., and Maddison, W. P. (1990). Detecting isolation by distance using phylogenies of alleles. *Genetics* 126, 249–260.

Sobrevilla, C., and Arroyo, M. T. K. (1982). Breeding systems in a montane tropical cold forest in Venezuela. *Plant Syst. Evol.* 140, 19–38.

Soltis, D. E., and Soltis, P. S. (1989). "Isozymes in Plant Biology." Dioscorides Press, Portland.

Stein, B. A. (1992). Sicklebill hummingbirds, ants, and flowers. *BioScience* 42, 27–33.

Sussman, R. W., and Raven, P. H. (1978). Pollination by lemurs and marsupials: An archaic coevolutionary system. *Science* 200, 731–736.

Swingle, W. T., and Reece, P. C. (1967). The botany of *Citrus* and its wild relatives. *In* "The Citrus Industry" (W. Reuther, H. J. Webber, and L. D. Batchelor, eds.), Vol. 1, pp. 190–430. Univ. of California, Division of Agricultural Sciences, Berkeley.

Tang, W. (1987). Insect pollination in the cycad *Zamia pumila* (Zamiaceae). *Am. J. Bot.* 74, 90–99.

Tanner, E. V. J. (1982). Species diversity and reproductive mechanisms in Jamaican trees. *Biol. J. Linn. Soc.* 18, 263–278.

Terborgh, J. (1986). Keystone plant resources in the tropical forest. *In* "Conservation Biology: The Science of Scarcity and Diversity" (M. E. Soulé, ed.), pp. 330–344. Sinauer Associates, Sunderland.

Terborgh, J. (1990). Seed and fruit dispersal—Commentary. *In* "Reproductive Ecology of Tropical Forest Plants" (K. S. Bawa and M. Hadley, eds.), pp. 181–190. UNESCO, Paris, and Parthenon Publishing Group, Carnsforth, UK.

Terborgh, J., Losos, E., Riley, M. P., and Bolanos Riley, M. (1993). Predation by vertebrates and invertebrates on the seeds of five canopy tree species of an Amazonian forest. *Vegetatio* 107/108, 375–386.

Thomson, J. D., Herre, E. A., Hamrick, J. L., and Stone, J. L. (1991). Genetic mosaics in strangler fig trees: Implications for tropical conservation. *Science* 254, 1214–1216.

Tiwary, N. K. (1926). On the occurrence of polyembryony in the genus *Eugenia. J. Indian Bot. Soc.* 5, 124–136.

van Schaik, C. P., Terborgh, J. W., and Wright, S. J. (1993). The phenology of tropical forests: Adaptive significance and consequences for primary consumers. *Annu. Rev. Ecol. Syst.* 24, 353–377.

van Steenis, C. G. G. J. (1969). Plant speciation in Malesia with special reference to the theory of non-adaptive saltatory evolution. *Biol. J. Linn. Soc.* 1, 97–133.

Vazquez-Yanes, C., and Orozco-Segovia, A. (1990). Seed dormancy in the tropical rain forest. *In* "Reproductive Ecology of Tropical Forest Plants" (K. S. Bawa and M. Hadley, eds.), pp. 247–259. UNESCO, Paris, and Parthenon Publishing Groups, Carnsforth, UK.

Waddington, K. D. (1983). Foraging behavior of pollinators. *In* "Pollination Biology" (L. A. Real, ed.), pp. 213–239. Academic Press, Orlando.

Walker, L. R. (1991). Tree damage and recovery from Hurricane Hugo in Luquillo Experimental Forest, Puerto Rico. *Biotropica* 23, 379–385.

Waser, N. M, and Price, M. V. (1991). Reproductive costs of self-pollination in *Ipomopsis aggregata* (Polemoniaceae): Are ovules usurped? *Am. J. Bot.* 78, 1036–1043.

Webb, C. J., and Bawa, K. S. (1983). Pollen dispersal by hummingbirds and butterflies: A comparative study of two lowland tropical plants. *Evolution* 37, 1258–1270.

Weeden, N. F., and Wendel, J. F. (1989). Genetics of plant isozymes. *In* "Isozymes in Plant Biology" (D. E. Soltis and P. S. Soltis, eds.), pp. 46–72. Dioscorides Press, Portland.

Weller, S. G., Donoghue, M. J., and Charlesworth, D. (1995). The evolution of self-incompatibility in flowering plants: A phylogenetic approach. *In* "Experimental and Molecular Approaches to Plant Biosystematics" (P. C. Hoch and A. G. Stephenson, eds.), Monographs in Systematic Botany, Missouri Botanical Gardens, St. Louis (in press).

Wharton, R. A., Tilson, J. W., and Tilson, R. L. (1980). Asynchrony in a wild population of *Ficus sycomorus. S. Afr. J. Sci.* 76, 478–480.

Wheelwright, N. T. (1985). Competition for disperser, and the timing of flowering and fruiting in a guild of tropical trees. *Oikos* 44, 465–477.

Whitham, T. A. (1981). Individual trees as heterogeneous environments: Adaptation to herbivory or epigenetic noise? *In* "Insect Life History Patterns: Habitat and Geographic Variation" (R. F. Denno and H. Dingle, eds.), pp. 9–28. Springer-Verlag, New York.

Whitham, T. A., Williams, A. G., and Robinson, A. M. (1984). The variation principle: Individual plants as temporal and spatial mosaics of resistance to rapidly evolving pests. *In* "A New Ecology" (P. W. Price, C. N. Slobodchikoff, and W. S. Gaud, eds.), pp. 15–52. Wiley, New York.

Willson, M. F. (1993). Dispersal mode, seed shadows, and colonization patterns. *Vegetatio* 107/108, 261–280.

Wolf, L. L., Stiles, F. G., and Hainsworth, F. R. (1976). Ecological organization of a tropical, highland hummingbird community. *J. Anim. Ecol.* 45, 349–379.

Wright, S. (1951). The genetical structure of populations. *Ann. Eugenics* 15, 323–354.

Wright, S. (1965). The interpretation of population structure by F-statistics with special regard to systems of mating. *Evolution* 19, 395–420.

Wright, S. J., and van Schaik, C. P. (1994). Light and the phenology of tropical trees. *Am. Nat.* 43, 192–199.

Yap, S., and Chan, H. T. (1990). Phenological behaviour of some *Shorea* species in peninsular Malaysia. *In* "Reproductive Ecology of Tropical Forest Plants" (K. Bawa and M. Hadley, eds.), pp. 21–35. UNESCO, Paris, and Parthenon Publishing Group, Carnsforth, UK.

Yih, K., Boucher, D. H., Vandermeer, J. H., and Zamora, N. (1991). Recovery of the rain forest of southeastern Nicaragua after destruction by Hurricane Joan. *Biotropica* 23, 106–113.

Zapata, J. R., and Arroyo, M. T. K. (1978). Plant reproductive ecology of a secondary deciduous tropical forest in Venezuela. *Biotropica* 10, 221–230.

20

Ecological Roles of Epiphytes in Nutrient Cycles of Forest Ecosystems

D. S. Coxson and N. M. Nadkarni

I. Introduction

A. Background

The burgeoning scientific interest in forest canopies focuses on both the enormous biodiversity of canopy organisms and their contributions to ecosystem-level processes. In the past two decades, exciting changes in canopy research perspectives and approaches have transformed the orientation of many canopy studies, moving from autecology studies on individual organisms to ecosystem studies on the complex ways by which the canopy influences the processing of nutrients and water. By illuminating the "black box" of the canopy, these studies enhance understanding of nutrient cycles in primary forests and identify potential effects of canopy disturbance or removal due to natural and human activities.

In this chapter, we focus on the roles that epiphytic organic matter play in nutrient acquisition, storage, and release within temperate and tropical forest ecosystems. We focus on epiphytic material (hereafter designated as EM), which consists of live and dead epiphytic vascular and nonvascular plants, associated detritus, microbes, invertebrates, fungi, and "crown humus" (Jenik, 1973). Reviews have focused on components of EM (Nieboer *et al.,* 1978; Pike, 1978; Brown and Brown, 1991; Knops *et al.,* 1991). Pools and fluxes of terrestrially rooted canopy components (e.g., tree foliage) are reviewed elsewhere (e.g., Parker, 1983; Waring and Schlesinger, 1985; Vitousek and Sanford, 1986). The water regimes of canopy components are described to the extent that they affect nutrient cycling, but general effects of forest canopies on hydrological cycles are treated elsewhere (e.g.,

Copyright © 1995 by Academic Press, Inc.
All rights of reproduction in any form reserved.

Bruijnzeel, 1983; Veneklaas and Ek, 1991; Bruijnzeel *et al.*, 1993). We first review historical aspects of canopy nutrient cycling research and summarize patterns that indicate the effects of EM on nutrient cycling. Second, we present a summary of the amounts and characteristics of nutrients stored in canopies. Third, we describe fluxes of nutrients into, out from, and within forest canopies. Finally, we discuss the ecological roles that are performed by EM and discuss priorities for future research in this area.

B. Historical Perspectives and Research Approaches

By definition, epiphytes are nutritionally independent of their host trees, deriving only support from their hosts. However, this definition does not preclude epiphytes having an effect on nutrient cycles. In an early review on EM and nutrient cycling, Pike (1978) wrote: "epiphytes contain a pool of stored minerals. Minerals are added as they grow and are removed as they die and decompose. Through this uptake, storage, and release of minerals, epiphytes can affect the overall pattern of mineral cycling for any ecosystem in which they occur."

Early work focused on specific relationships between epiphytes and their supporting host tree. Certain epiphytes were called "semi-parasites" by Ruinen (1953), who ascribed some degree of parasitism to plants that were previously considered to be nutritionally independent. Other epiphytes were called indirect parasites by Benzing and Seeman (1978), who referred to the epiphytic vegetation Spanish moss (*Tillandsia usneoides*) on host trees (oaks) at their dry sites in the southeastern United States as "nutritional pirates." They presented evidence that epiphytes took up nutrients dissolved in precipitation before they became available to host trees, thus robbing their hosts of essential mineral elements and causing host tree decline. Experimental studies showed that epiphytes can significantly alter precipitation chemistry. In laboratory leaching experiments, Lang *et al.* (1976) demonstrated that epiphytic lichens lost Ca, Mg, H^+, and K to the leaching solution, while N was absorbed. They suggested that because of their mode of mineral nutrition (direct uptake from surface films of water), epiphytes are capable of significant alteration of throughfall and that minerals leached from lichens would be more readily available than nutrients added through litterfall and decomposition.

Other work focused on the canopy subsystem of the whole forest (Carroll, 1980). Among the most important studies that put EM into an ecosystem-level context were those of L. Pike, G. Carroll, W. Denison, and other researchers at the University of Oregon and Oregon State University in the 1970s. They pioneered the use of mountain-climbing techniques in old-growth Douglas fir forests and integrated field measurements of epiphyte biomass, decomposition, and distribution with laboratory and modeling studies (Denison 1973, 1979; Carroll, 1980). They quantified the nitrogen input from the N-fixing lichen *Lobaria oregana* as $1-7$ kg N ha^{-1} yr^{-1} and suggested that this canopy component, which is small in terms of biomass, may have a disproportionately large "fertilizer" effect on the forest.

C. Modeling Approaches

The roles of epiphytes in altering precipitation chemistry of simple-structured forests were investigated using a modeling approach by eco-system ecologists in the northeastern United States (Lovett, 1981, 1984; Reiners and Olson, 1984; Lovett *et al.*, 1985). In subsequent studies, many of these approaches have been integrated (e.g., Boucher and Nash, 1990; Veneklaas, 1991a; Coxson *et al.*, 1992; Nadkarni and Matelson, 1992). The primary goal has been to quantify pools and fluxes of EM and place them in an ecosystem- and landscape-level context. This demands measurement not only of EM, but also other components (Fig. 1). These pioneering stud-

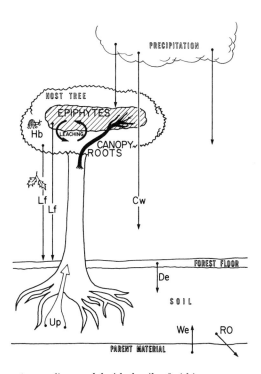

Figure 1 Forest nutrient cycling model with details of within-canopy components depicted. Nutrients enter the canopy ecosystem via weathering (We) of parent material and precipitation that can be intercepted and retained by live and dead parts of epiphyte mats. Nutrients are subsequently transferred to other ecosystem parts within the canopy and to the forest floor via litterfall (Lf), herbivory (Hb), crownwash (Cw), and, in some forest tree species, canopy roots. They can then circulate within the ecosystem via soil-rooted vegetation uptake (Up) and decomposition (De), be stored in the soil, or move from the system via runoff (RO). (From Nadkarni, 1983.)

ies have suggested five reasons for the importance of canopy components to ecosystem function:

1. Pools of nutrients and organic matter generated by epiphytes contained in the canopy can be high, even exceeding host tree foliage (e.g., Hofstede *et al.*, 1993).
2. The contribution of epiphytes to complex canopy structure may enhance wet and dry deposition by increasing the physical area of canopy surfaces for impaction and sedimentation (Lovett, 1981; Reiners and Olson, 1984), and by increasing the biotic uptake by nutrient-efficient epiphytic plants (Benzing and Seeman, 1978; Benzing 1983, 1990).
3. Free-living and symbiotic biota in the canopy can fix atmospheric nitrogen (Pike, 1978; Bentley and Carpenter, 1980, 1984; Roskowski, 1980; Sengupta *et al.*, 1981; Yatazawa *et al.*, 1983).
4. Certain microbial activities (e.g., nitrification) in the canopy appear to be suppressed, relative to the forest floor, which tends to foster nutrients in a form that is less mobile and leachable (Vance and Nadkarni, 1990).
5. Pools of canopy-held nutrients are by no means static; nutrients move from the canopy to the other ecosystem members by four routes: (a) as litterfall via abscission and by "riding down" fallen branches and whole trees, which is subsequently mineralized at variable rates, (b) as crownwash in the form of throughfall and stemflow (Lang *et al.*, 1976; Carroll, 1980), (c) via herbivory and predation of epiphytes by vertebrates and invertebrates (Nadkarni and Matelson, 1989), and (d) directly to host trees via uptake by canopy roots (Nadkarni and Primack, 1989).

II. Storage of Nutrients in Epiphytic Material

A. Biomass of Nutrients

1. General Considerations   Studies that have partitioned the mass of materials in forest ecosystems have consistently documented that soil organic matter and standing biomass account for nearly half of total nutrient pools (Louisier and Parkinson, 1976; Auclair and Rencz, 1982). However, a large portion of this is held in woody components that are not readily available to other organisms. The pools of labile nutrients are canopy foliage elements, from both terrestrially rooted and epiphytic sources.

2. Methods   Standard methods to quantify EM mass do not yet exist, and researchers have calculated the magnitude of canopy components with a variety of techniques. The most accurate method involves climbing trees in a wide range of size classes and destructively sampling subsets of the EM community. Because of the difficulties of access, certain portions of the canopy (e.g., outer branch tips) are usually poorly estimated. Few studies have been able to ascertain the accuracy of estimates, and a high priority for canopy research is the establishment of standardized measurement techniques of ecosystem-level EM biomass. Recent research in a Costa Rican montane forest, Monteverde, Costa Rica, involved cutting and lowering whole branches from the canopy to the forest floor by a professional arborist, where a ground crew took samples from all portions of the branches. These measurements were coupled with counts of branch area, number of branches per tree, and tree density to extrapolate whole-ecosystem estimates of epiphyte biomass (Nadkarni *et al.,* 1993).

A wide range of values of epiphyte biomass have been reported in temperate and tropical ecosystems (Tables I and II), but only a few of these

Table I Biomass of Epiphytes in Various Ecosystems[a]

Ecosystem/location	Biomass (Mg/ha)	Source
Temperate		
Rain forest	6.8	Grier and Nadkarni (1987)
Olympic National Park		
Hoh River Valley, U.S.A.		
Bukk Mountains, Hungary	0.0043	Smith (1982)
Tropical		
Dry forest, Guanica, Puerto Rico	0.14	Murphy and Lugo (1986)
Moist forest, Manaus, Brazil	0.05	Klinge *et al.* (1975)
Lower montane wet forest	0.05	Weaver and Murphy (1990)
LEF, Puerto Rico[b]	1.4	Golley *et al.* (1971)
Tropical moist		
Panama	1.6	Golley *et al.* (1971)
Lower montane wet	2.0–4.0	Frangi and Lugo (1986, 1992)
LEF, Puerto Rico		
Submontane forest	2.13	Pócs (1980, 1982)
Uluguru Mountains, Tanzania		
Lower montane rain forest	3.4	Edwards and Grubb (1977)
New Guinea		
Elfin woodland	4.7	Nadkarni (1984)
Monteverde, Costa Rica		
Cloud forest, Pico Oeste	5.0	Lyford (1969)
LEF, Puerto Rico		
Mossy elfin forest	13.6	Pócs (1980, 1982)
Uluguru Mountains, Tanzania		
Mangrove	21.2	Golley *et al.* (1971)
Panama		
Upper montane cloud forests	33.0	Hofstede *et al.* (1993)
Columbia		
Lower montane cloud forests	44.0	Nadkarni *et al.* (1993)
Monteverde, Costa Rica		
Riverine	0.0785	Golley *et al.* (1971)
Panama		
Premontane wet	1440.0	Golley *et al.* (1971)
Panama		

[a]From Nadkarni *et al.* (1993).
[b]Luquillo Experimental Forest.

Table II Biomass of Epiphytic Lichens in Several Coniferous Forest Communities[a]

Community	Lichen biomass (kg/ha)	Source	Notes
Picea engelmannii/Abies lasiocarpa	3300	Edwards *et al.* (1960)[b]	Study area 4, 1800 m elevation
Pinus banksiana	2000	Scotter (1962)[c]	
Picea englemannii	1200	Scotter (1962)	
Pseudotsuga menziesii/Picea engelmannii/ Abies lasiocarpa	840	Edwards *et al.* (1960)	Study area 1, 800 m elevation
Picea engelmannii/Abies lasiocarpa	760	Edwards *et al.* (1960)	Study area 3, 1100 m elevation
Picea excelsa	700	Rudnova *et al.* (cited by Rodin and Barilevich, 1967)	Archangel Province, USSR
Pinus contorta/Pseudotsuga menziesii/ Populus tremuloides, etc.	280	Edwards *et al.* (1960)	Study area 2, 800 m elevation

[a]From Pike (1971).
[b]Wells Gray Park, British Columbia, Canada (52°N, 120°W).
[c]Black Lake District, Saskatchewan, Canada (57°N, 105°W).

Table III   Apportionment and Biomass (kg Dry Weight) of the Different Components of a *Weinmannia mariauitae* Tree in the Upper Canopy Layer and Its Epiphytic Load [a]

	Trunk foot	Trunk	Inner branches	Middle branches	Outer branches	Total
EPIPHYTIC						
Living						
Green						
Bryophytes	3.88	0.14	4.27	4.89	1.59	14.77
Lichens	0.03	0.01	0.10	0.17	1.59	1.90
Ferns	0.15	0.01	1.19	0.83	0.00	2.18
Flow, plants	0.59	0.00	0.63	0.29	0.00	1.15
Total green	4.65	0.16	6.19	6.18	3.18	20.36
Nongreen						
Roots and wood	15.39	0.02	2.92	0.85	0.00	19.18
Bryophytes and bases	1.04	0.01	1.22	0.86	0.00	3.13
Total nongreen	16.43	0.03	4.14	1.71	0.00	21.31
Total living	21.08	0.19	10.33	7.89	3.18	42.67
Dead						
Coarse dead	27.60	0.11	13.40	7.88	0.08	49.07
Fine earth	21.14	0.01	1.40	0.60	0.08	23.23
Total dead	48.74	0.12	14.80	8.48	0.16	72.30
Total epiphytic	49.82	0.31	25.13	16.37	3.26	114.97
PHOROPHYTE						
(g dm^{-3} bark surface area)	37.8	4.0	130.4	32.4	1.5	32.7
Wood						755.13
Leaves						15.38

[a]From Hofstede *et al.* (1993).

have been placed into an ecosystem-level context (e.g., Lange *et al.*, 1978). Nadkarni (1983, 1984) pointed out that although canopy epiphytes comprised less than 2% of aboveground biomass in north temperate rain forest, their contribution to labile nutrient pools was four times greater than that of combined host tree foliar biomass. A similar pattern for EM exists in Colombian upper-montane rain forest (Table III) (Hofstede *et al.*, 1993), where living green epiphytic material within the canopy of *Weinmannia mariquitae* was 20.36 kg, compared to only 15.38 kg dry weight of leaf biomass on the phorophyte host. Total nutrient pools within canopy epiphytes at this site represented 51% of total canopy nutrient pools for N, 58% for P, and 80% for K (Fig. 2). Nutrient accumulation in epiphytic bryophyte communities has also been examined in forested tropical floodplain environments. Frangi and Lugo (1992) found that biomass and nutrient accumulation in epiphytic bryophyte communities was on the same order of magnitude as that in fine litter soil surface pools. These data point to the

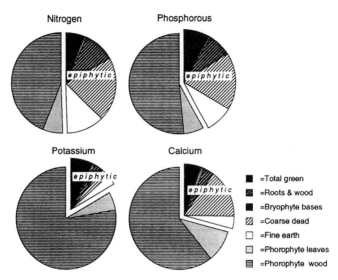

Figure 2   The apportionment of some macronutrients comprising a *Weinmannia mariquitae* tree in the upper-canopy layer and its epiphyte load in an upper-montane Colombian cloud forest (3700 m). (From Hofstede *et al.*, 1993.)

importance of considering nontree components when calculating nutrient budgets in forest ecosystems.

B. Composition of EM Pools

1. General Considerations   Composition of pools has been quantified in some studies of EM biomass. The general technique is to remove samples of whole EM communities, separate them into a small number (usually 3–8) of constituent classes, and express on a percentage of total dry weight basis (Nadkarni, 1984; Veneklaas, 1991a; Hofstede *et al.*, 1993). The distribution of constituents has been shown to vary with substrate type and location within the crown, with more dead organic matter on interior branches and more bryophytes, lichens, and xerophytic epiphytes on outer branches (Figs. 3 and 4).

2. Nutrient Content   Nutrient content of EM varies widely with component type and ecosystem type. In general, it is similar in nutrient content to host tree foliage and forest floor dead organic matter (Table IV).

3. Dead Organic Matter   Dead organic matter (DOM) in the canopy is of special interest because of its capacity to capture incoming nutrients by storing them on the negatively charged sites, its large pool of recalcitrant organically bound nutrients, its support of invertebrates (Nadkarni and Longino, 1990), and its ability to retain water for long periods of time

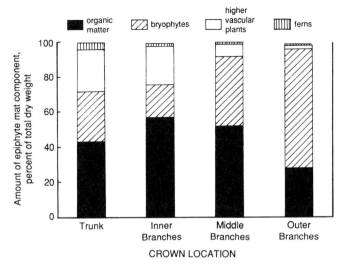

Figure 3 Distribution and composition of epiphyte mat components on a representative host tree in the Monteverde Cloud Forest, Costa Rica (1850 m), expressed as percentages of total epiphyte mat dry weight. Proportions were calculated by multiplying ratios of epiphyte dry weight-to-substrate surface area by the tree surface area of that crown location. Data given are for a representative *Clusia alata* trees (diameter = 121 cm, height = 13 m). (From Nadkarni, 1984.)

(Veneklaas, 1992; Veneklaas *et al.*, 1991). Concentrations of macronutrients are similar in forest floor organic matter (Table V).

Although comparisons between canopy organic matter and forest floor organic matter have been explored in several forests (Putz and Holbrook, 1989; Vance and Nadkarni, 1990; Lesica and Antibus, 1991), long-term data on the abiotic characteristics of DOM are few. Bohlman *et al.* (1994) re-

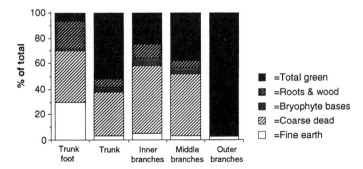

Figure 4 Distribution of different epiphytic components within a *Weinmannia mariquitae* tree in an upper-montane Colombian cloud forest (3700 m). (From Hofstede *et al.*, 1993.)

Table IV   Nutrient Concentrations (mg g⁻¹) of Tree Leaves and Epiphytes
of Some Tropical Montane Rain Forests[a]

Reference	Location	Total N	P	K	Ca
TREE LEAVES					
Lower-montane rain forests					
Grubb (1977)	Malaya	13.8	0.7	7.7	2.1
Tanner (1977)	Jamaica (gap)	17.6	1	14.9	12.6
Upper-montane rain forests					
Grubb (1977)	Australia	11.1	0.48	6.2	6
	Malaya (mean)	9.8	0.33	6.4	5.3
	Puerto Rico	9.7	0.53	3.9	7.6
Edwards and Grubb (1977)	New Guinea	14.2	0.84	10.1	10.9
Tanner (1977)	Jamaica (mor)	10.5	0.5	5.5	6.2
	Jamaica (mull)	16.1	0.7	12.3	9.3
	Jamaica (wet slope)	12.7	0.8	14.9	11.4
Veneklaas (1991a)	Colombia	14.7	0.90	6.1	—
Hofstede *et al.* (1993)	Colombia	10.2	0.90	5.03	6.45
Subalpine rain forests					
Grubb (1977)	New Guinea	9.7	0.54	6.4	9.7
EPIPHYTES					
Nadkarni (1984)					
Bryophytes		16.9	0.62	3.66	1.37
Dead organic matter		19.9	0.66	1.73	2.23
Ferns		15.8	0.79	18	9.5
Hofstede *et al.* (1993)					
Bryophytes		7.06	0.86	5.34	2.66
Dead organic matter		11.21	0.80	0.54	2.92
Ferns (mean)		8.73	1.00	7.22	4.87
Flowering plants		7.05	0.88	6.08	3.62

[a]From Hofstede *et al.* (1993).

corded moisture and temperature of organic matter in the canopy (C-O) and forest floor (FF-O) of trees in the montane forest of Monteverde, Costa Rica. During the three years of their measurements, the temperatures of the C-O differed little from those of FF-O (Fig. 5). In contrast to the temperature patterns, moisture fluctuated seasonally and showed striking differences between the canopy and the forest floor. The FF-O moisture pattern was constant throughout the year (at or above 55% moisture content), but C-O showed great extremes in wetting and drying (Fig. 6). Although C-O maintained a consistently and significantly higher moisture content during the wet and misty seasons (above 70%), C-O had periods of very low soil moisture (20–40% moisture content) during the dry season. C-O could dehydrate substantially, losing 30% of its water content within a week. This moisture pattern may play a role in determining the composition and ac-

Table V Concentrations (mg kg^{-1}) of H$_2$O-Extractable Nutrients of Suspended Soil and Terrestrial Soil[a]

	pH	NO$_2$	S.D.	NO$_3$	S.D.	NH$_4$	S.D.	PO$_4$	S.D.	K	S.D.	Ca	S.D.
Suspended soil													
Trunk foot	3.7–4.1	1.0	*0.4*	1.5	*0.3*	78.6	*16.9*	100.7	*48.3*	115.7	*38.1*	33.8	*12.7*
Inner branches	3.9–4.7	0.7	*0.3*	0.7	*0.2*	41.3	*19.0*	128.3	*9.2*	197.5	*51.5*	72.9	*25.7*
Middle branches	4.2–4.3	2.4	*1.5*	1.2	*0.8*	34.3	*27.0*	144.3	*2.6*	323.7	*2.8*	81.9	*2.1*
Terrestrial soil	3.9–4.1	0.9	*0.2*	1.3	*0.5*	92.8	*6.2*	5.0	*1.4*	179.3	*9.4*	53.0	*10.9*

[a]Samples are bulked from 10 subsamples and analyzed in triplicate. Standard deviations are given in italics. From Hofstede *et al.* (1993).

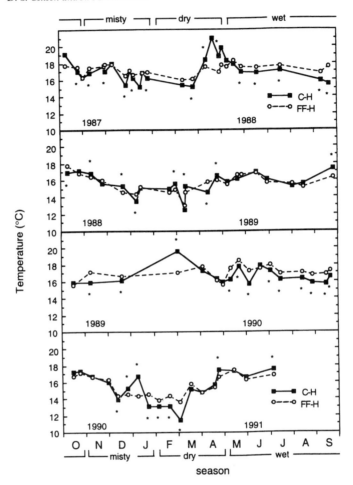

Figure 5 Soil temperature of canopy organic matter (C-O) at 5 cm depth and the forest floor O horizon (FF-O) at 0–10 cm depth in the Monteverde Cloud Forest, Costa Rica (1850 m). Seasons are indicated at the top of the graph. Asterisks represent a significant difference ($P <$ 0.05) between C-O and FF-O. (From Bohlman *et al.*, 1995.)

tivity of the canopy-dwelling organisms, although no data exist on these relationships. The lower densities of invertebrates in the canopy relative to forest floor [particularly the absence of isopods and amphipods in C-O reported by Nadkarni and Longino (1990)], may be a response to low moisture (Levings and Windsor, 1984). Drought adaptations in epiphytes that

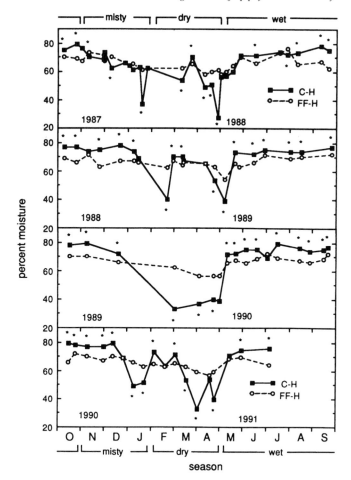

Figure 6   Moisture content of canopy organic matter (C-O) at 5 cm depth and the forest floor O horizon (FF-O) at 0–10 cm depth in the Monteverde Cloud Forest, Costa Rica (1850 m). Moisture content, determined from 10-g samples taken from the top 10 cm of soil, is expressed as water content/total wet weight. Asterisks represent a significant difference ($P < 0.05$) between C-O and FF-O. (From Bohlman *et al.*, 1995.)

are rooted in C-O may be related to periodic low soil moisture. Rates of decomposition (mediated by the density, composition, and life cycles of organisms responsible for them) may also be affected by these dry-downs. Further studies examining the relationships between moisture and temperature patterns and biological activity in C-O and FF-O are needed.

III. Intercanopy Nutrient Fluxes

A. Nutrient Input into Canopy Environments

1. General Considerations   Because epiphytic plants have no direct vascular connection to the bank of nutrients in the forest floor, they must rely on morphological and physiological attributes such as litter-impounding pools, foliar trichomes, insectivory, myrmecochory, and poikilohydric foliage to acquire and conserve nutrients in an environment that may deliver nutrients only sporadically and in dilute concentrations (Chapter 11). Some specialized epiphytes gain nutrients exclusively from precipitation and dry deposition, but many epiphytes have no apparent specialized adaptations for directly obtaining nutrients from atmospheric sources.

Potential sources of nutrients for epiphytes include an array of both autochthonous and allochthonous sources (Table VI). The relative contribution of each of these sources to epiphyte communities is unknown. From an ecosystem standpoint, it is important to distinguish between these two source types in forests where nutrient availability may limit productivity of the overall system. If epiphytes were to obtain all of their nutrients from autochthonous sources, then they would simply be diverting nutrients from the tree-to-ground flux pathway for some length of time; they would not increase the total pool, but merely change the form or compartment in which nutrients are stored. Alternatively, if they were to sequester nutrients from outside the system, this would potentially increase the total nutrient input to the ecosystem in addition to altering the form and location of these nutrients.

2. Atmospheric Deposition   One of the major avenues for input of "new" nutrients and water in forest canopy environments is uptake from atmospheric sources in the form of wet and dry deposition. Lindberg *et al.* (1986) estimated that atmospheric depositon can supply 40% of Ca and N requirements and 100% of S requirements for annual woody increment growth in deciduous forest environments in the eastern United States

Table VI   Potential Sources of Nutrient Input to Epiphyte Communities in Forest Ecosystems [a]

Autochthonous sources	Allochthonous sources
I. Soil-rooted phytomass	I. Atmospheric
A. Intercepted litterfall	1. Wet deposition
B. Bark decomposition	2. Dry deposition
C. Leachate of live foliage	3. Gaseous input (including nitrogen fixation)

[a]From Nadkarni and Matelson (1991).

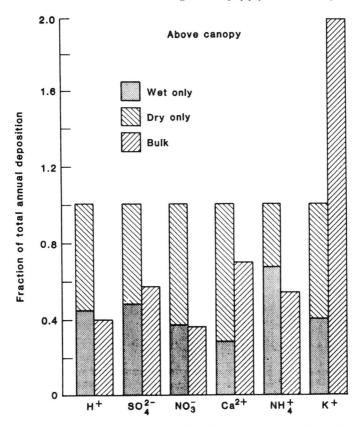

Figure 7   Comparison of annual atmospheric deposition to a mixed deciduous forest, as determined by bulk deposition measurements and by separate collection of wet and dry deposition. All values have been normalized to the total annual wet plus dry fluxes. (From Lindberg *et al.,* 1986.)

(Fig. 7). For elements such as K and Ca, dry deposition sources may include soil or biological material of local origin, so the determination of the source depends on the spatial scale at which nutrient cycling is being considered.

A major source of nutrients in precipitation is prior suspension of ions from marine aerosols. In nearshore environments, the deposition of salts can have a major impact on nutrient loading (Clayton, 1972). The differential loss of elements due to particulate sorting provides a clear signature of the importance of marine sources in regional nutrient input. Ionic ratios (equivalents) between Na and Cl (0.85) in seawater can be used to assess seasonal differences in precipitation sources. Veneklaas (1991b) found that precipitation in Colombian montane rain forest had Na/Cl ratios ranging from 0.85 in the dry season to 3.5 in the wet season. In continental areas, the resuspension of soil particles plays a larger role in determining ionic composition. Point source natural emissions such as volcanic eruptions, however, can play a large role in determining rainwater composition (Kellman *et al.*, 1982; Veneklaas, 1991b). The eruption of "Nevado del Ruiz" in 1985 resulted in a large pulse of Ca, Mg, and SO_4 in rain forest sites monitored by Veneklaas (1991b), and Kellman *et al.* (1982) noted a large influx of K, Ca, Mg, and Na, thought to derive from the Volcan Fuego, some 300 km west of their study sites in Honduras. The absence of measurable N or P provides a check that measured nutrient spikes reflected volcanic inputs rather than a resuspension of local biotic material. Veneklaas (1991b) found that excess sulfate deposition accounted for over 97% of sulfates in bulk throughflow and correlated closely with the acidity of local rainfall.

Although cloud water or occult precipitation generally accounts for only 10–20% of precipitation in forest canopies, inputs of major ions from occult precipitation may equal or exceed that from direct precipitation alone, especially for ions such as NH_4^+ and NO_3^- (Lovett *et al.*, 1985; Sigmon *et al.*, 1989). Cloud water can be two to four times more concentrated for most ions, although the order of dominance for ions parallels that of bulk precipitation. One of the most notable differences between cloud water and bulk precipitation is the greater acidification of cloud water. Sigmon *et al.* (1989) found that cloud water was four times more acidic than bulk deposition (251 versus 63 μequiv H^+/L) in eastern deciduous forest in North America. This trend has been the focus of numerous studies on anthropogenic influence on atmospheric chemistry (Weathers *et al.*, 1988; see also Section III, A, 5).

3. Abscised Plant Litter   A potential autochthonous source to epiphyte communities is abscised plant litter that is intercepted within the tree canopy by inner branches and their epiphytes. The decomposition and mineralization of nutrients from abscised foliage are now standard measurements in studies of forest nutrient cycling (Vitousek and Sanford, 1986; Vitousek and Matson, 1988). Mineralization of nutrients in forest ecosystems is dependent on substrate quality, including C/N ratio, secondary metabolites, and local microclimate patterns.

The forest canopy has traditionally been viewed as a substrate source for decomposition processes that occur after transfer of foliage elements to the forest floor. However, most decomposition processes that have previously been described for forest floor environments also occur within the canopy. Fine litter dynamics within the canopy, however, may differ from litter dynamics on the forest floor for three reasons. First, canopy litter may be ephemeral, as it can be removed from branches by within-canopy disturbances such as wind, rain, and arboreal animals. In contrast to fallen leaves on the forest floor, which can shift their position on the ground with only minor consequences for plants rooted in soil (Orndorff and Lang, 1981; Welbourn *et al.*, 1981), leaf movement in the canopy may remove a potentially substantial contribution to epiphyte nutrition. Second, leaf litter in the canopy may be deposited in smaller amounts than is leaf litter in the forest floor due to lack of input from subcanopy and understory vegetation. Third, decomposition rates of litter deposited and retained within the canopy may differ from that of litter on the forest floor because of microclimate and substrate differences between the canopy and forest floor, as well as differences in community structure and density of macroinvertebrate detritivores and microbial decomposers.

In the montane forest of Monteverde, Costa Rica, Nadkarni and Matelson (1991) quantified the amounts of litter stored in the canopy and subsequently released to be potentially used by epiphyte communities. The investigators differentiated between "gross fine litter" (the amount of litter that passes through the canopy) and "net fine litter" (the amount of litter that accumulates and is mineralized within the canopy after disturbances such as wind have occurred). Their measurements of leaf litter attrition (whole leaf loss from branches due to wind and other disturbances) with marked leaves documented that 70% of all leaves deposited on branches were lost within two weeks and nearly all were gone in 16 weeks. Certain branch characteristics (branch angle, number of epiphyte stems and clumps) affected the amount of litter retained at particular microsites. Decomposition of tethered, dead leaves within the canopy was half that of leaves on the forest floot (turnover time in the canopy = 2.7 yr versus 1.7 yr in the forest floor), which may in part be due to the dry environmental conditions and low densities of canopy macroinvertebrates discussed earlier. Based on their assumptions that litter accumulation within the canopy is at a steady state, the biomass of fine litter retained and decomposed within the canopy was calculated as only 2.0 g m^{-2} yr^{-1} for major macronutrients. They concluded that nutrient replenishment of epiphyte communities appears to be decoupled from the litterfall pathway, because input from litterfall retained within the canopy was small relative to epiphyte productivity and nutrient requirements reported in other studies.

Because the replenishment of nutrients to many cloud forest epiphytes at this site appears to be only partially derived from litterfall decomposition, the balance must be derived from the other sources (Table VI). Two sources, foliar leachate (autochthonous) and atmospheric deposition (allochthonous), seem the most likely candidates for most of the balance. Nutrients derived from foliar leachates (especially mobile nutrients K^+ and NO_3^-) may be important, as throughfall concentrations collected at the forest floor are occasionally lower than that in bulk precipitation (K. Clark, personal communication), indicating that at least some of the canopy components are "scavenging" nutrients from bulk precipitation. The majority of host tree foliage at this site, however, is sclerophyllous, with waxy cuticles, and many cloud forest trees retranslocate high proportions of N and P, which presumably minimizes water and nutrient transfer through foliage (Grubb, 1977; Tanner, 1980; Tanner *et al.*, 1992).

Atmospheric deposition is an allochthonous source that is likely to contribute nutrients to all epiphytes. On outer branches, poikilohydric epiphytes such as bryophytes and filmy ferns capture atmospheric nutrients and incorporate them into their biomass. When they die, their detritus contributes to the development of humus buildup, which no doubt increases nutrient retention. The vascular epiphytes that occupy inner branch areas and that do not have morphological adaptations for direct atmospheric uptake must acquire nutrients by root uptake from nutrients sequestered in the accumulated mats. These plants may obtain at least some of their nutrients by physically intercepting precipitation (especially wind-blown mist) with their shoots and redirecting it to the humus mats, which are permeated with their root systems. Understanding the ultimate nutrient sources of canopy solutions and sinks of nutrients by quantifying the uptake and release of specific canopy components is needed to differentiate these two sources.

4. Canopy Nitrogen Fixation   The use of the nitrogenase enzyme system to convert atmospheric N to mineral N can occur at several points in the canopy environment. The most widely distributed association for biological N fixation is a symbiosis between several families of higher plants and *Rhizobium* root-nodulating bacteria. The possession of symbiotic root nodules frequently confers a significant ecological advantage in low-nutrient environments and can result in major changes to processes of autogenic succession (Vitousek *et al.*, 1987). Although we normally associate nodulating root systems with terrestrial environments, the description of nodulating root systems in aboveground adventitious roots (AAR) by Nadkarni (1981, 1983) provides a significant new pathway for canopy N input that must now be considered. Nodulated AAR of red alder (*Alnus rubra*) were shown to fix atmospheric N at comparable rates to belowground roots in temperate wet forests of the Pacific Northwest (Nadkarni, 1983). Further work should be done to quantify the amounts, phenology, and fate of N fixed by nodulated canopy roots.

Another source of canopy N fixation derives from asymbiotic and lichenized N fixation, usually associated with free-living or lichenized cyanobacteria. Both typically share common constraints of poikilohydrous existence and limitations of carbon supply (from photosynthetic activity) at the site of nitrogenase activity. The association of free-living cyanobacteria with forest floor moss mats and lichenized cyanobacterial associations provides significant N input at the soil surface in many forests (Weber and Van Cleve, 1981; Crittenden, 1983). Further evidence from $^{15}N_2$-labeling studies shows that asymbiotically fixed N from moss mat hummocks is rapidly utilized by a range of other plant and animal species, from both digestion of free-living cyanobacteria and assimilation from N enriched substrates following mineralization of organically bound N (Jones and Wilson, 1978; see also Chapter 15 for further discussion of commutation by forest microfauna). Lichenized cyanobacterial associations and associated N fixation have been recognized in forest canopy environments (Forman, 1975; Kelly and Becker, 1975; Pike, 1978; Kershaw and MacFarlane, 1982).

Additionally, epicaulous and epiphyllous cyanobacterial crusts have been reported in forest canopies, including north temperate and tropical rain forest, mangroves, and tropical deciduous forests (Fritz-Sheridan and Portecop, 1987; Mann and Steinke, 1989; Carpenter, 1992). Rates of $^{15}N_2$ fixation by leaf epiphyllae of the understory palm *Welfia gorgii* reached 37.5 μg N cm^{-2} h^{-1} in full sunlight and averaged 1.8 μg cm^{-2} h^{-1} in low-light environments (Carpenter, 1992). This compares with average $^{15}N_2$ fixation rates of 50 μg N cm^{-2} h^{-1} for *Welfia* fronds measured by Bentley and Carpenter (1984) at La Selva field station (Costa Rica), with point values exceeding 400 μg N cm^{-2} h^{-1}. Epicaulous crusts have also been observed in mangrove communities (Steinke and Naidoo, 1990), where they are thought to provide a key N source in a nutrient-limited ecosystem (Mann and Stieke, 1989; Sheridan, 1991).

The fate and environmental control of processes of canopy N fixation are less clear. Throughfall nutrient budgets on canopy epiphyte mats (Nadkarni, 1983, 1988) show recurring episodes of N release in studies that compared stripped and intact branches in tropical forest canopies (Fig. 8). Bentley and Carpenter (1984) provided the first direct field evidence that newly fixed asymbiotic nitrogen can be transferred directly from leaf surface epiphylls to the host leaf, up to 25% of total N in rain forest canopy leaves may be derived from this source. The transfer of carbon supply from host leaves, both directly and in throughfall (Bentley, 1987; Coxson *et al.,* 1992), may exert a controlling influence on the amount of total N gain from canopy epiphytes. Rates of N fixation in tropical forest canopies are higly correlated with age, reflecting time available for colonization by leaf-surface epiphylls (Bentley, 1987; Fritz-Sheridan and Portecop, 1987). A major limiting factor controlling rates of N fixation in canopy environments is duration of hydration episodes. Numerous investigators have shown that nitrogenase activity is quite sensitive to moisture level (Bentley and Carpenter, 1980; Fritz-Sheridan and Coxson, 1988), reflecting both the lag period for resynthesis of the nitrogenase enzyme itself as well as the time required for restoration of phytosynthetically derived reductant pools (Coxson and Kershaw, 1983). Epicaulous diazotrophic crusts in dry tropical forest and mangrove communities may be particularly vulnerable to repeated desiccation episodes (Sheridan, 1991). Even in wet tropical forests, leaf epiphyll surfaces can face repeated desiccation episodes (Coxson, 1991).

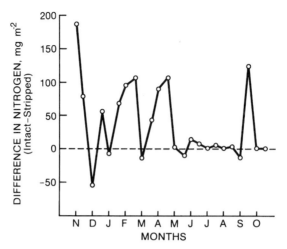

Figure 8   Branch frame budgets for analyzing effects of epiphyte mats on throughfall nitrogen flux in the Olympic Peninsula rain forest, Washington, U.S.A. Points are the amount of nitrogen (rainfall volume multiplied by sample concentration) collected from beneath an intact branch minus the amount of nitrogen collected beneath a paired stripped branch. A positive value indicates nitrogen release by epiphytes relative to stripped branches. (From Nadkarni, 1983.)

5. Anthropogenic Inputs   Anthropogenic inputs, especially H^+ ion deposition, have major implications for nutrient dynamics in canopy environments. Lovett *et al.* (1985) demonstrated increased leaching of cations from canopies as a result of H^+ deposition. High-elevation forests are especially vulnerable to pollutant loading from cloud water sources, which can be four times more acidic than bulk precipitation (Sigmon *et al.,* 1989). Reviews summarize the impacts of pollutant loading on forest canopies (Hosker and Lindberg, 1982; Lechowicz, 1982; Nieboer and MacFarlane, 1984; Chamberlain, 1986; Runeckles, 1986; Mooney *et al.,* 1987; Linzon, 1988; Murach and Ulrich, 1988; Weathers *et al.,* 1988; Aber *et al.,* 1989; Sehmel, 1980).

The sensitivity of nonvascular canopy epiphytes such as lichens to pollutant loading has long been recognized (Chapter 16). Their ability to accumulate and concentrate ions at surface exchange sites is valuable in oligotrophic canopy environments, but makes them especially vulnerable to even dilute atmospheric concentrations of air pollutants. Numerous reviews deal with the sensitivity of lichens to air pollutants and related biomonitoring applications (Lechowicz, 1982; Hutchinson *et al.,* 1986; Muir and McCune, 1988; Richardson, 1988; Richardson and Nieboer, 1983a). The series "Literature on Air Pollution and Lichens" provides a comprehensive ongoing listing of work in this area (Henderson, 1993).

B. Nutrient Release in Canopy Environments

1. General Considerations   Nutrients from live and dead EM are released into the nutrient cycles of terrestrially rooted vegetation by two pathways: (a) epiphytic material falls to the forest floor, dies, and decomposes, and (b) epiphyte mats are leached by precipitation and the nutrients are transferred to the forest floor via stemflow and throughfall. Various processes cause epiphytic material to fall to the forest floor, including senescence, wind, disruption by birds and mammals, and the falling of supporting branches and whole trees.

2. Litterfall   The death, deposition, decomposition, and mineralization of fallen litter is a major pathway for transferring nutrients and energy from vegetation to soils, and is the most frequently measured nutrient flux in forest ecosystems (Bray and Gorham, 1964; Proctor *et al.*, 1983; Vitousek, 1982, 1984; Vitousek and Sanford, 1986). Nearly all litterfall studies have focused on the biomass and nutrient composition of "fine litter" deposited by terrestrially rooted trees and understory plants, as their abscised leaves, twigs, and reproductive parts constitute the major component of labile nutrients in most forest types.

In forests where epiphytes constitute a large portion of the canopy biomass, EM litterfall can be an important constituent in the litterfall pathway. The contribution of the epiphyte community to nutrient transfer in forests is poorly understood. Measurements of nutrient fluxes from epiphytes to the forest floor in boreal and north temperate environments have largely focused on nitrogen fixation and nitrogen input from corticolous cyanolichens such as *Lobaria* and *Pseudocyphellaria* (Rhoades, 1977; Pike, 1978; Carroll, 1980; Reiners and Olson, 1984; Esseen, 1985). The few reports in which epiphyte litterfall has been reported in tropical forests have been anecdotal or based on small collectors designed to trap tree fine litter (e.g., Tanner, 1980; Songwe *et al.*, 1988). Only a few tropical studies have estimated fallen EM input by separating EM from material collected in fine litter collectors (Table VII). As a consequence, estimates of the total input of nutrients to the forest floor are probably inaccurate in forests where epiphytes are a substantial canopy component.

In the montane cloud forest of Monteverde, Costa Rica, Nadkarni and Matelson (1992) focused on the EM component of total litterfall. They characterized EM litterfall as being extremely sporadic both spatially and temporally. Individual collections of extremely large samples were infrequent; only 2% of all collections exceeded 10 g m^{-2} per collection interval (equivalent to 2.6 tons ha^{-1} yr^{-1}). There was no apparent seasonal trend for these sporadic pulses of large EM deposition (Fig. 9), which comprised 53% of the total EM input. A steady input of small amounts of material fell throughout the year. During the entire study period, only 8% of the indi-

Table VII   Litterfall (tons ha⁻¹) of Forests Where Epiphyte Components Were Separated

Location	Elevation (m)	Litterfall Total	Litterfall Epiphyte	Source
Sabah	280	6.5	0.01	Proctor *et al.* (1983)
	330	7.4	0.00	
	480	5.2	0.004	
	610	5.6	0.11	
	790	5.5	0.03	
	870	4.8	0.04	
Cameroon	150	1.3	0.105	Songwe *et al.* (1988)
Costa Rica	1480	7.5	0.5	Nadkarni and Matelson (1992)
Jamaica	1550	6.6	0.18	Tanner (1980)
		5.5	0.004	
		5.6	0.034	
		6.5	0.06	
Venezuela	2300	7.0	0.23	Steinhardt (1979)
Colombia	3370	4.3	0.23	Veneklaas (1992)

vidual collections had no EM litter. The temporally variable nature of the deposition of EM was manifested by the observation that over half of the EM fell in less than 2% of the collections. In a similar study of Colombia, Veneklaas (1992) found little discernible pattern in the proportions of constituents in EM litterfall collectors (Fig. 10).

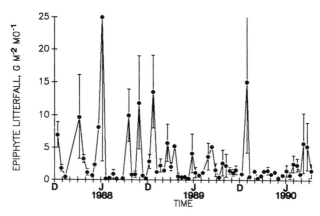

Figure 9   Epiphyte litterfall (g m⁻² month⁻¹), collected from 5-×-5-m plots and fine litter collectors between December 1987 and September 1990 in the Monteverde Cloud Forest, Costa Rica (1850 m). Error bars represent one standard error of the mean. (From Nadkarni and Matelson, 1992.)

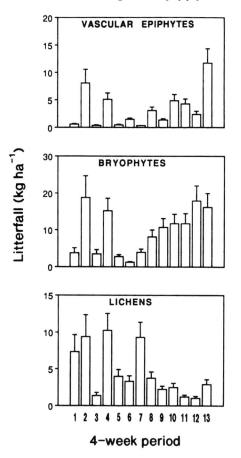

Figure 10 Epiphyte litterfall (kg ha^{-1}) in a Colombian cloud forest (3370 m) for three different groups (vascular epiphytes, bryophytes, and lichens) between August 20, 1986 and August 19, 1987. Bars indicate means and standard errors for 4-week periods ($n = 20$ except for periods 1–3, where $n = 10$). (Reprinted from Veneklaas, 1991a, by permission of Cambridge University Press.)

These patterns have at least two implications for plants rooted in the forest floor. First, because of the "clumpy" nature of fallen EM, nutrient deposition from EM concentrates input in particular but unpredictable locations. This contrasts with terrestrially rooted material, which is distributed fairly evenly across the forest floor (Nadkarni and Matelson, 1992). Second, nutrients deposited in EM that ride down on treefalls and large branchfalls co-occur with higher levels of light associated with resulting gaps. We would speculate that this pulse of nutrients released from EM may alter nutrient availability in the immediate vicinity of regenerating gap species.

Results from related research on litterfall from terrestrially rooted material indicate that EM should be considered in future nutrient cycling studies in forests that support appreciable amounts of canopy epiphytes for three reasons. First, epiphyte material falls to the ground in large clumps, which include whole plants that are intertwined with a variety of materials such as nutrient-rich dead organic matter and bryophytes. Bryophytes have higher nutrient contents than terrestrially rooted material and show faster cycling of retained nutrients (Nadkarni and Matelson, 1992). Second, in contrast to many tree species whose abscised leaves undergo considerable retranslocation before abscission, the foliage of EM may not go through retranslocation, as it is the unpredictable and sudden dropping of limbs and whole trees (rather than senescence of epiphytes themselves) that causes much of the EM to descend to the forest floor. Third, nutrient sources of epiphytes may partially or wholly differ from those of terrestrial plants. Forest plants that are rooted in the soil and the forest floor derive the majority of their nutrients from soil parent material, whereas epiphytic plants derive a major portion of their nutrients from atmospheric sources (Nadkarni and Matelson, 1991).

3. EM Death   The physical movement of live epiphytes from the canopy to the forest floor is a frequent event in epiphyte communities. Live epiphytes fall to the forest floor because they are dislodged by wind or animals or because branches and trees break and fall (Strong, 1977). Some epiphytes with poorly developed root systems (e.g., tank bromeliads) tend to fall as individuals. However, live epiphytes in the form of contiguous mats, connected by interwoven root systems and a layer of crown humus (Jenik, 1973), often fall intact as "clumps."

The continued association of individual epiphytes with their original canopy organic material may affect the survival of these individuals. The fate of an epiphyte falling as part of an intact mat, in contrast to falling as an individual, may differ because roots imbedded in mats may be less disturbed than the unprotected roots of individual plants. Also, the spongelike mats retain considerable amounts of water under drying conditions, which affects the water status of the plants and conditions for associated pathogens and mutualists. Anecdotal observations of fallen epiphytes include a range of responses; some epiphytes vanish within weeks and others persist and even thrive for months to years. Apparently, diverse factors allow or limit the survival of epiphytes after they land on the forest floor. The nutrients in live epiphytes that fall to the ground will ultimately be mineralized and absorbed by terrestrial vegetation. However, their prolonged survival on the ground would delay mineralization with consequent effects on storage, cycling, and potential loss of nutrients from the ecosystem.

In the montane cloud forest of Monteverde, Costa Rica, Matelson *et al.* (1993) studied the fate of fallen epiphytes to ascertain how long different

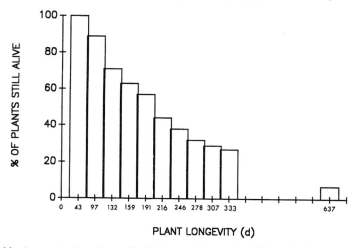

Figure 11 Longevity of a cohort of individual epiphytes after falling to the forest in the Monteverde Cloud Forest, Costa Rica (1850 m). Longevity is defined as the time between Day 1 and the last sampling day that a plant was recorded alive after placement on the forest floor. $N =$ 223 plants (including nonvascular plants). (Reproduced from Matelson *et al.*, 1993, with permission.)

taxa of epiphytes survive. By the end of the first year, 27% of the plants remained alive, and by the end of the study (21 months), 7% had survived (Fig. 11). The branches to which clumps had been originally attached were intact but had deteriorated. Of the cases in which the clumps were unattached to branches, some clumps had disappeared, whereas for others, lumps of dead organic matter knit together with remains of the root mat were detectable on the forest floor. All plant taxa exhibited similar rates of mortality.

These results have implications in nutrient cycling. Before nutrients in epiphytes can be released through decomposition, the live plants must die. Thus, fallen epiphytic material probably affects nutrient cycles differently than litterfall from terrestrially rooted plants, whose nutrients can be mineralized faster because that material is already dead (Vitousek and Sanford, 1986). For fallen live epiphytes, then, there is a potential lag time in nutrient release via mineralization. In forests with well-developed canopy communities, epiphytes can profoundly affect both the amounts of nutrient storage and the timing of nutrient release. Further investigations should pursue the spatial and temporal distribution of fallen epiphytes in relation to microhabitat characteristics to determine the role of epiphytes at an ecosystem level and to reveal insights into the mechanisms that foster epiphytism.

4. Decomposition of EM Decomposition of EM has been measured only very rarely (e.g., Pike, 1971), as the litterbag techniques commonly used for tree litter do not lend themselves to epiphytic material. There is a tendency for material placed in litterbags to resprout and actually grow and gain weight instead of decomposing! Therefore, most EM-specific decomposition rates have been calculated indirectly.

In the EM litterfall study discussed earlier, Nadkarni and Matelson (1992) assumed that the primary forest was in a steady state, so that over an annual cycle, litter decomposition equaled litter deposition (Olson, 1963). The annual decay rate (K_a) of this material was the annual litter input divided by the forest floor pool and was calculated as 1.3 for EM biomass. The fractional turnover time $(1/K_a)$ for EM biomass would be 0.8, or approximately 10 months. They found a more rapid decay rate and shorter turnover time for EM than were measured for terrestrially rooted material (TM) at the same time and place $(K_a = 0.7, 1/K_a = 1.4)$ (Nadkarni and Matelson, 1991). However, they found a much slower release rate for nutrients in EM litter (except K) than for nutrients in TM fine litter. Turnover time for all nutrients except K was four to six times longer for EM than TM. The turnover time for K was 10-fold faster in EM than TM.

These results indicate that at least a portion of EM is recalcitrant and resistant to decomposition and mineralization. However, certain components of EM decompose very rapidly, whereas other components are more resistant to mineralization. Bryophytes appeared to decompose very rapidly; input was 76% of fallen EM, but only 22% of the EM standing crop, which suggests that this component decomposes quickly. The calculated decay rate for bryophytes was 4.3, with a turnover time of only 0.23 yr, or less than 3 months. Conversely, dead organic matter appears to have a much slower decay rate, comprising only 13% of input, but 58% of the EM standing crop; K_a for this material is 0.28, and the turnover time is 3.6 yr. Vascular plants are intermediate; K_a is 0.68 and turnover time is 1.5 yr. Further studies of particular components are needed to determine the timing of eventual nutrient mineralization and nutrient release from this material.

The inhibition of soil nitrifying bacteria by phenols leaching from canopy foliage may also limit net loss of N from organic matter horizons (Chandler and Goosem, 1982). This is especially true in environments such as tropical montane rain forest, where the accumulation of phenols in canopy foliage has been associated with leaf longevity (Grubb, 1977). Avalos *et al.* (1986) reported surface uptake of lichen phenolics from canopy epiphytes into supporting host trees, with subsequent translocation via xylem element to surrounding foliage. They noted that a reduction in vigor has commonly been associated with heavy epiphyte cover in oak forests. Photosynthetic activity was not reduced in comparisons of epiphyte-laden branches with canopy branches that had been stripped of epiphyte cover. These comparisons, however, are not conclusive, as the time course for phenol retention from previous epiphyte cover on experimental branches is unknown. A reduction of vigor with epiphyte cover through agents introduced by the epiphyte cover itself would provide a mechanism for action in the paludification (forest senescence due to the growth of forest floor and epiphytic byrophyte communities) of forest environments. The trend toward more efficient nutrient cycling in mature or late-successional forests may be attributed to increases in phenol concentrations in litter at the forest surface and accompanying decrease in microbial nitrification potential (Grubb, 1977). Circumstantial evidence for this shift is the increasing predominance of ammonia-N over nitrate-N in mature ecosystems, with ammonia being less easily leached from soils (Rice and Pancholy, 1974). The role of phenolics in controlling decomposition processes in EM horizons is an area for fruitful investigation.

The lack of direct studies on soil N transformation in EM soil profiles also necessitates comparison with nutrient studies in forest floor organic soils. Vitousek and Denslow (1986) reported that rates of net N mineralization are quite high (averaging $50-80$ μg g^{-1} month^{-1}) in lowland tropical rain forest soils. These rates decrease markedly with increased elevation (Marrs *et al.*, 1988; Vitousek, and Matson, 1988). The balance between mineralization and immobilization is highly sensitive to substrate C/N ratios. Perturbation experiments in north temperate environments have shown that addition of carbon-rich substrates to N-limited forests can result in net immobilization of soil mineral N (White *et al.*, 1988). The release of soluble carbohydrates into throughfall solutions in the cloud forest canopy may thus trigger the immobilization of available mineral nitrogen as C/N ratios rise. An alternative scenario is predicted by the work of Bosatta and Berendse (1984), which provides evidence for repeated oscillations between net mineralization and immobilization when forest systems are perturbed by amendments of C or N. Their models predict that addition of C to N-limited environments should result in the short-term stimulation of microbial respiration and net mineralization, followed by a series of oscillations until steady-state conditions are reached. This trend may be reinforced even further in EM by the stimulation of nitrogen fixation in leaf-surface cyanobacteria on exposure to carbohydrate-rich throughfall solutions. With release of newly fixed N, declining C/N ratios should favor even greater net mineralization within the epiphyte community.

One of the difficulties of assessing decomposition processes in canopy environments is the extreme variability of the canopy environment. Not only do we have spatial variability of the canopy environment due to diversity of taxa and physiognomy (Veneklaas, 1991a; Wolf, 1993; see also Chapter 2), we must also contend with numerous physical gradients, including temperature, moisture, light, and variation in the intensity, duration, and distribution of incident precipitation during storm events (Clements and Colons, 1975; Chapter 4). Lesica and Antibus (1991) noted that nonequilibrium conditions in EM horizons may be an important prerequisite for maintaining high species diversity in Neotropical forests. Nonequilibrium processes that characterize many of the physical attributes of the canopy environment may play an important role in suppressing competitive exclusion between members of the canopy flora.

Reinforcing these trends toward nonequilibrium nutrient transfer in canopy environments are major disturbance events such as hurricanes and cyclones. Lodge *et al.* (1991) found that about 1000 g m^{-2} of litter was deposited to the ground from canopy environments by Hurricane Hugo in lower-montane rain forest in Puerto Rico. Nutrient input from storm events was qualitatively more important than "regular" litterfall, with N and P concentrations of storm-deposited litter 1.5 to 3.3 times higher. This may reflect the absence of normal sequestration processes that remove nutrients from foliage prior to leaf-fall. The opening of canopy environments following disturbance pulses (e.g., hurricanes) would also promote increased mineralization due to higher surface temperatures (D. Coxson, unpublished data). After Hurricane Hugo hit the study site of Lodge *et al.* in Puerto Rico, NO$_3$ concentrations in surface stream water initially declined, reflecting short-term nutrient immobilization. This was followed by a subsequent large increase in stream nitrate concentration, following resumption of mineralization activity. A subsequent shift from mineralization to dentrification might also be expected, with this shift in soil ammonium/ nitrate concentrations accompanying a hurricane-induced litterfall pulse (Lodge *et al.*, 1991).

5. Leaching and Throughfall Loss   The development of canopy EM usually occurs in environments with abundant rainfall and few periods of prolonged water deficit (Jacobs, 1988). Many canopy epiphytes, however, show physiological adaptations more often associated with drought-tolerant plants in xeric environments (Ong *et al.*, 1986; Nasrulhaq-Boyce and Hajimohamed, 1987; Veres, 1990; Ball *et al.*, 1991). The adoption of these traits reflects their exposure to full-insolation conditions. With no access to soil moisture pools, canopy epiphytes are highly susceptible to even short periods without rainfall. These conditions of physiological drought exposure are even more severe for nonvascular epiphytes. Lacking roots and stomata to compensate for water loss, they are highly sensitive to drying conditions and must depend on periods of direct precipitation or mist interception for metabolic activity (Bentley and Carpenter, 1980; Coxson *et al.*, 1984; Larson, 1987). Under these conditions of alternating desiccation and hydration, the preservation of membrane integrity assumes paramount importance. The accumulation of sugars and polyols in intracellular pools is thought to play a key role in the stabilization of membrane function in these nonvascular canopy epiphytes (Bewley, 1979; Davison and Reed, 1985), with polyhydroxyl compounds replacing membrane hydration water during periods of desiccation. Internal sugar pools thus function as a "physiological buffer," promoting structural integrity and supporting metabolic demands during periods of physiological stress (Lechowicz, 1981; Eickmeier, 1982; Scott and Larson, 1985).

Although sugar reserves in canopy epiphytes contribute to the restoration of membrane function, they are susceptible to leaching during the initial rewetting phase (before membrane function is restored). Intercellular amino acids and ionic components can also be lost during the rewetting phase. The efflux of solutes from cellular pools has previously been demonstrated in many species of lichens and mosses (Lewis and Smith, 1967; Gupta, 1976; Bewley, 1979) and may account for the loss of up to 10% of standing sugar reserves in some species (Dudley and Lechowicz, 1987). This loss has important implications for both community structure and ecosystem function (Alpert and Oechel, 1987; Dudley and Lechowicz, 1987). In laboratory studies the release of solutes to standing solution on rewetting is typically followed by a period of slow reassimilation, if experimental material is held in the same solution during the recovery period (Buck and Brown, 1977). Under field conditions, however, released sugars may potentially be removed by throughfall solution before reassimilation can occur, especially for nonvascular epiphytes in the canopy. This "pulse release" of organic solutes is not confined to nonvascular plants. Canopy epiphytes such as the resurrection plant (*Craterostigma plantagineum*) also show an accumulation of organic solutes that function in the protection of membrane structure during periods of desiccation exposure and are labile to leaching during the rewetting phase (Gupta, 1976; Bianchi *et al.*, 1991).

Solute loss on rewetting has now been confirmed for canopy bryophyte mats in tropical montane rain forest in Guadeloupe, where experimental trials demonstrated rewetting-induced solute efflux of 80.1, 1.4 and 11.8 kg ha^{-1} yr^{-1} during "pulse release episodes," largely in the form of organically bound N (Coxson, 1991). These results are consistent with previous laboratory studies demonstrating efflux of N-amino acids on rewetting of bryophytes (Gupta, 1977). The efflux of sugars and polyols induced by canopy wetting–drying cycles (Coxson, 1991) reached 122 kg ha^{-1} yr^{-1} (Fig. 12). Net release patterns of Ca and Mg were somewhat different; initial efflux was followed by a period of apparent reassimilation from throughfall solutions. Initial release of Ca and Mg may reflect ion displacement (by wetting solutions) at surface exchange sites. Schwarzmaier and Brehm (1975) showed that cation exchange capacity in *Sphagnum* is linked to the chemistry of hydrophilic cell-wall carbohydrates. The release of Ca and Mg may thus reflect conformation changes (and loss of function) during

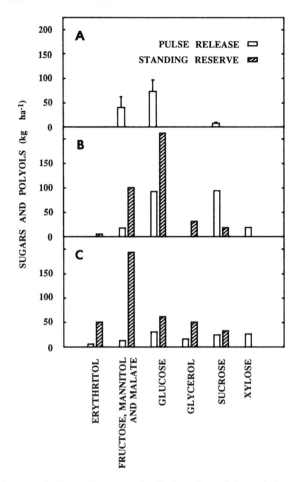

Figure 12 Net annual release of sugars and polyols to throughflow solution on rewetting of canopy bryophytes and standing reserves for canopy bryophytes in a tropical montane rain forest (Guadeloupe, French West Indies). (A) Field release from upper-canopy bryophyte mats. Bars denote standard error ($N = 15$). (B) Experimental release and standing reserves for upper-canopy bryophyte mats. (C) Experimental release and standing reserves for lower-canopy bryophyte mats. Pulse release estimates were scaled from the mean of individual experiments to net annual release based on 135 drought episodes annually in upper-canopy exposures and four episodes annually in lower-canopy exposure (from 1988/1989 microclimate measurements). HPLC resolution required presentation of fructose, mannitol, and malate as combined values. (From Coxson *et al.*, 1992.)

periods of desiccation. The examination of surface exchange complexes during transitional periods between desiccated and fully hydrated states remains an area for further study.

One of the major questions posed by these findings of "pulse nutrient release" in canopy environments is the origin of solute efflux during rewetting episodes. Do these solutes originate from intracellular epiphyte pools or do they simply reflect the resuspension of surface-deposited ions during the rewetting phase? This issue was addressed by Coxson (1991) in experimental trials where leaf segments of the canopy epiphyte *Phyllogonium fulgens* were exposed to pretreatment rinse to remove surface particulates and freely soluble ions. This pretreatment had little effect on net release of most ions and increased P efflux on rewetting. These results confirm that leaching from intracellular pools and/or exchange from surface cation-exchange complexes is a plausible mechanism for pulse nutrient release on rehydration of nonvascular canopy epiphytes.

This pulse release of previously sequestered nutrients from nonvascular canopy epiphytes differs fundamentally from pulse release episodes that have previously been documented in air pollution studies in north temperate forests (Alcock and Morton, 1985; Evans *et al.,* 1977; Gaber and Hutchinson, 1988). The pulse of nutrients released to throughfall solutions after drought exposure in north temperate forests derives largely from the resuspension of surface-deposited nutrients (dry deposition) and from exchange processes between surface water films and mesophyll cells (with released nutrients ultimately diffusing through pores or cracks in the cuticle into throughfall streams). The mesophyll cells under these conditions remain fully functional. The exchange of nutrients with surface water films does not usually result from the loss of membrane function, typically seen in bryophytes and lichens during the rewetting period. This does not exclude dry deposition as a posible mechanism for nutrient accumulation in lower plant communities. In environments where lichens and mosses are exposed to longer drought episodes, resuspension of surface deposits can be a major component of nutrient exchange with throughfall solutions (Scott and Hutchinson, 1987). Nonetheless, the comparison between "pulse nutrient release" from nonvascular epiphytes and "pulse release" estimates from foliage elements of higher plants must recognize that cellular function in higher plants is not normally impaired during dry deposition episodes and thus normal diffusion/exchange processes continue during the rewetting periods. In contrast, the rewetting response of lower plants must include a transitional period where the disruption of membrane function allows nutrient efflux to external throughfall solutions. Net nutrient balance will subsequently depend on the degree of reassimilation before throughfall streams drain from surface exchange sites.

6. Gaseous Loss   Although the potential for N gain through asymbiotic and symbiotic N fixation is high in many canopy environments, these gains may be balanced by processes of denitrification or ammonia volatilization. The loss of N in the form of nitrous oxides is highly correlated with soil NO_3^- concentrations. Thus, although denitrification is often a severe problem in agroecosystems (where soil N levels have been maintained at artificially high levels), organic forest soils typically have much lower rates of denitrification. Vermes and Myrold (1992) suggested that little N is lost through denitrification from organic soils in north temperate rain forest environments, although they did find that significant denitrification potential existed on addition of soil NO_3^- Nadelhoffer *et al.* (1984) found that NO_3^- supplied most of the mineral N taken up by forest systems in nine north temperate ecosystems. Although potential for soil denitrification exists, competition for labile nitrate-N may limit net N loss through gaseous flux. In wet canopy environments, N input may initially take the form of ammonia, the immediate end product of asymbiotic N fixation. This presents the possibility of direct N loss through ammonia volatilization, a common phenomenon in alkaline soils of grassland and desert environments. In EM soil horizons, however, solution pH is more likely to be neutral or acidic (Lesica and Antibus, 1991), reflecting the equilibrium of soil solutions with carbonic acid in throughfall rainwater and the production of organic acids from humus complexes. Rapid immobilization of soil N pools by microbial flora or conversion to soil mineral nitrite/nitrate are more likely outcomes for ammonia-N pools in EM horizons.

IV. Intracanopy Nutrient Fluxes

A. Surface Nutrient Exchanges

The two major determinants of nutrient exchange in canopy environments are the surface area available for rainwater interception and the physiological and structural adaptations that allow nutrient uptake and retention.

Typical leaf area indices (LAI) in north temperate deciduous forest range between 3 and 6, whereas values of up to 20 have been reported for coniferous forest canopies (Waring and Schlesinger, 1985). In contrast, LAIs of 160 have been measured for canopy epiphyte communities in tropical montane rain forest (Coxson, 1991) and may be as high in other wet forest environments. Though these numbers seem large compared to those of higher plant canopies, when one considers the complicated bilobed leaf structure of many pendulous epiphytic liverworts and their adaptations for holding water, for example, succubous leaf structure (Schofield, 1981), nonvascular epiphytes can be seen to provide a major surface area for precipitation interception. As a group, the ability of nonvascular plants to remove solutes from throughfall solutions at extremely low concentrations through processes of active uptake provides a major pathway for nutrient sequestration in canopy environments (Brown and Beckett, 1984). Jordan *et al.* (1980) noted that rainwater passage through epiphyte-rich canopies resulted in net uptake by the canopy of P, S, and Ca. They called this uptake "nutrient scavenging" and hypothesized that it represented a nutrient conservation mechanism that enabled ecosystem survival on nutrient-poor soils and subsoils. Nadkarni (1983) extended this work, focusing specifically on epiphyte loading by using paired comparisons between branches with intact epiphyte cover and canopy branches that had been artificially stripped of all epiphyte cover. Again, the finding was one of net nutrient retention, at least during the dry season, with net release of nutrients occurring during the wet season. These findings suggest that resuspension of nutrients following periods of dry deposition may be a major pathway for nutrient input in nonvascular canopy epiphyte communities. Direct nutrient uptake (incorporation into cellular pools) by nonvascular canopy epiphytes during these periods of dry deposition would be severely restricted by reduced metabolic activity associated with low cell water content (Coxson, 1991), although canopy epiphytes would still be capable of particulate trapping and accumulation of ions at surface ion-exchange sites (Richardson and Nieboer, 1983b). Nutrient incorporation in nonvascular components of EM communities in Nadkarni's study may thus be limited to periods in the dry season when rainfall events allow the resuspension of surface-deposited ions in water films. These limitations would not apply to vascular canopy epiphytes such as tank bromeliads, where active uptake and adsorption of surface-deposited nutrients could continue during periods of dry deposition.

The retention of nutrients by canopy epiphytes during periods of dry deposition parallels findings of Boonpragob and Nash (1989, 1990) in transplant studies with *Ramalina menziesii*. Leachable NO_3^- showed the highest accumulation in dry intervals between storm events, with NH_4^+, H^+, and Cl^- occurring at intermediate concentrations. One complicating factor in this study was the presumed release of elements from intracellular pools due to pollutant damage. Rewetting of nonvascular epiphytes can also induce nutrient release from cellular pools, although ion profiles for release from cellular pools differ significantly from those of release patterns from surface exchange sites (Coxson, 1991).

The technique of Bates (1987) for selective ion displacement addresses this differential storage of ionic components. Using these techniques, Coxson (1991) found that Ca and Mg in the canopy epiphyte *Phyllogonium fulgens* was divided between surface exchange sites and residual organic complexes. In contrast, the greatest proportion of K in *P. fulgens* was localized in intracellular pools. These results suggest that K loss may be the best indicator for desiccation-induced damage when membrane integrity is disrupted. The evidence for translocation of nutrients, once retained in nonvascular epiphytes, is mixed. Bates (1989) did not find evidence of internal translocation of nutrients (P and K) in forest floor swards previously exposed to periods of luxury consumption (nutrient supply in excess of immediate growth requirements). The use of fertilizer applications in these experiments, however, may be problematic in determining the *in situ* response of organisms whose life-history strategy is typically one of stress tolerance, where periods of resource availability do not lead to immediate growth response.

In vascular epiphytes, structural adaptations against water loss (e.g., thick epidermal cells with evaporation-retarding cuticles) may act as a limitation on surface nutrient exchange processes (Benzing, 1990). Vascular epiphytes, however, especially those rooted in EM, can gain access to soil nutrient pools through the action of root-surface enzymatic activity. Antibus and Lesica (1990) provided an interesting correlation between epiphyte habitat and demonstratable activity of root-surface phosphatase. Biotic elements in the canopy environment possess many counterpart adaptations to those traditionally associated with forest floor soil processes.

The release of nutrients in throughfall pulses (sugars and ions) during rewetting episodes (discussed earlier) has important implications for intra-canopy nutrient cycling. An obvious candidate for the reassimilation of nutrients from throughfall solutions are the EM communities associated with bryophyte mats at greater depth within the canopy. Heterotrophic nutrition has previously been observed in many different bryophytes (Proctor, 1982) and enzymes of sucrose metabolism play an important part in the cellular metabolism of mosses and liverworts (Galloway and Black, 1989). Another major candidate for reassimilation of nutrients is leaf-surface epiphylls Cyanobacterial mats are ubiquitous on leaf surfaces in both upper- and lower-canopy rain forest environments (Fritz-Sheridan and Portecop, 1987) and typically show rapid response to the addition of sugars. Bentley (1987) showed a doubling in rates of asymbiotic N fixation on application of exogenous glucose solutions to leaf epiphylls in Panamanian lowland rain forest. The release of sugars and polyols to throughfall solutions may also have direct effects on fungal hyphae and adventitious root mats in soil pockets and humus layers on branches in the canopy.

Although the influence of sugar release on fungal communities has not been examined in tropical montane rain forest environments, pulsed release of sugars to throughfall solution in oak woodland results in enhanced growth and increased mineralization by soil fungal communities (deBoois and Jansen, 1975). The influence of sugar release on adventitious root mats may be more indirect, through influence on carbon metabolism of associated mycorrhizal and rhizosphere organisms (Niederer *et al.*, 1989). In each of these cases, the concurrent release of inorganic ions and nutrients during wetting/drying cycles should result in even greater metabolic response by other community members (Coxson, 1991).

B. Canopy Roots

Direct uptake of nutrients by higher plants in the canopy environment can also occur from adventitious roots, especially from DOM horizons in the canopy. Vance and Nadkarni (1992) found that fine root biomass in EM horizons at canopy branch junctions was about 20% higher than comparable organic soil horizons on the forest floor (per unit soil volume). Overall fine root biomass (<2 mm) comprised about 45% of total root biomass in the canopy of tropical montane forest at their study site in Monteverde. The size of nutrient pools associated with fine root production provides an additional important labile nutrient pool in forest ecosystems (Nadelhoffer and Raich, 1992). These canopy roots are highly significant in that they have access to potentially large fluxes of nutrients from throughfall precipitation and from direct mineralization from within EM.

Herwitz (1991) found that aboveground adventitious roots (AAR) were particularly well developed in montane rain forest tree species that had high rates of leaching from canopy foliage. Direct cycling of previously released nutrients may thus be possible even at the level of individual forest canopy trees. Nutrient-rich stemflow often derives from sap exudates in canopy species with well-developed AAR. It is not clear whether species that possess abundant AAR are more efficient at scavenging "new" nutrients from throughfall or are simply compensating for poorly developed mechanisms of nutrient conservation. Efficiency of nutrient use is not simply the inverse of plant nutrient concentration. Rather, it is more closely related to the inverse of nutrient concentration in aboveground litterfall and organic matter increment (Vitousek, 1982). Translocation studies with AAR using the radionuclides ^{75}Se, ^{138}Cs, ^{54}Mn, and ^{65}Zn showed that most nutrients acquired by AAR in *Salix* tend to be immobilized in leaves and branches closest to the point of nutrient uptake (Nadkarni and Primack, 1989). This contrasts to the more general distribution of nutrients from belowground sources. Although we associate the development of AAR with wet tropical environments, Nadkarni (1981) points out that convergent evolution has led to the adoption of AAR morphology in many tree species in both tropical and temperate environments. Among the diverse taxa with AAR are members of the Salicaceae, Aceraceae, Lauraceae, Flacourtiaceae, Aralizaceae, and Cunoniaceae (Nadkarni, 1981).

C. Mycorrhizal Interactions

The importance of soil mycorrhizal fungi to seedling establishment in low-nutrient or drought-stressed environments has been repeatedly demonstrated in a variety of forest environments (Fitter, 1985; Harvey *et al.*, 1987; Vogt *et al.*, 1991). The action of vesicular-arbuscular mycorrhizae (VAM) in improving water relations and increasing phosphate uptake seems especially beneficial for plants that are rooted in canopy environments (both

AAR systems and canopy epiphyte root masses). Unfortunately, there is very little information on the ecological significance of canopy mycorrhizal associations. Many the canopy epiphytes (e.g., Orchidaceae) require mycorrhizal associations for at least germination. This is thought to be a key factor in habitat selection by canopy phorophytes (Benzing, 1990). Reports of epiphyte mycorrhizae are few, however, and attempts at identification are cursory. See (1990) noted that although ectomycorrhizae have been reported from 13 of 16 dipterocarp tree genera, we have few studies on the role of mycorrhizae in their survival and establishment.

Further areas that warrant investigation include the issue of nutrient storage in canopy mycorrhizal communities (Vogt *et al.*, 1991). Dispersal agents for mycorrhizal communities are also an area of concern, particularly with anthropogenic influences on small and large mammal fauna of many forests. Numerous studies note that canopy mammals are a particularly important dispersal agent for mycorrhizal inoculum through the vehicle of fecal pellets, especially for coniferous forests of the Pacific Northwest (United States), where flying squirrels promote the distribution of mycorrhizal fungal spores (Maser *et al.*, 1978).

A further role of fungal–plant interactions is the association of Basidiomycete fungi and canopy bryophytes. Kost (1988) described an apparent interaction of regulated cohabitation that can be interpreted as symbiotic. Fungal contact frequently produces canopy associations where liverwort rhizoids are branched and swollen at their apex (Pocock *et al.*, 1985). This presumably provides a nutritional relationship equivalent to that of mycorrhizal relationships, and has been called mycorrhizoid (Kost, 1988).

V. Implications of EM in Nutrient Cycling

Multiple nutrient sources are required to sustain ecosystem function, including rainfall, weathering, litterfall, herbivory, and, in some cases, fire release (Westman, 1978). The corollary of this statement is that standing vegetation in forest canopies can in one sense be regarded as "fossil" in nature, with accumulated nutrient pools (in standing biomass) much greater than could be sustained from immediate regrowth from soil nutrient pools. These ideas are now finding greater support in the design of forest management strategies (Hilton, 1987; Benson, 1990; Gillis, 1990; Maser, 1990). The recent emphasis on canopy epiphytes provides the basis for a significant reappraisal of ecosystem function in temperate and tropical forest environments.

The patterns identified for ecosystems where epiphytes constitute a significant part of the canopy biomass and where atmospheric nutrient input may be important for maintaining the nutrient and water cycles suggest that

EM may play a "keystone" role. The idea of "keystone plant resources" was developed in the animal ecology literature. Terborgh (1986) suggested that the carrying capacity of mammalian and avian frugivores in tropical wet forests may depend on certain keystone plant resources that provide crucial food and energy resources that maintain animal populations during "lean" (often dry season) times of the year. In his research, the keystone plant resources (e.g., palm nuts, nectar) constituted only a very small percentage of total plant diversity (less than 1% of tree species in a Peruvian lowland forest) but supported up to 80% of the mammalian frugivore biomass and much of the avian biomass. He characterized such resources in three ways: (1) these resources make up only a small or unapparent portion of the total species diversity or biomass of the forest, (2) the resources are reliably and consistently produced on a year-to-year basis, and (3) their dependent organisms utilize and presumably depend on these resources. Although the experiment of removing such resources has never been done, the presumption is that were they reduced or removed, the frugivore populations would be negatively affected.

Only a few examples of particular species or components of ecosystems have been documented as acting in this "keystone" role to control major functional characteristics of terrestrial ecosystems. For instance, Lugo *et al.* (1973) suggested that the energy and nutrient concentration by leaf-cutter ant mounds has an ecosystem-level effect on nutrient distribution in tropical lowland forests. Vitousek *et al.* (1987) described long-term successional effects of an introduced N-fixing shrub, *Myrica faya,* on Hawaiian substrates. Whitford *et al.* (1982) suggested that termites may control processes of nutrient and carbon transfers in arid ecosystems.

Because epiphytes have no vascular connection to the forest floor and are known to depend solely on atmospheric sources, it is tempting to allocate a keystone role to some epiphytes. However, few studies have rigorously examined this hypothesis. One such study was by Nash and his colleagues in oak woodland communities of coastal California. They did a canopy manipulation (removing all epiphytes from a set of short-statured experimental trees) to investigate the role of the epiphytic lichen *Ramalina menziesii* in ecosystem nutrient cycling (Boucher and Nash, 1990; Knops *et al.,* 1991). They found greater volume of rainfall and throughfall underneath trees that had been stripped of their epiphytes compared to intact trees (Tables VIII and IX). These patterns demonstrate that removal of epiphytes had a discernible effect on nutrient cycling.

Evidence reviewed in this chapter has demonstrated that the presence of epiphytes can result in the recruitment and retention of atmospheric nutrients, which are subsequently cycled to other ecosystem members. This enhancement of allochthonous nutrients may more speedily swell and maintain nutrient reserves of the ecosystem as a whole than if the epiphytes were

Table VIII Mean Throughfall Volume under Trees with and without Canopy Lichens, and Precipitation in the Open[a]

Month	Mean	S.E.	n	Mean	S.E.	n	Mean	S.E.	n	Mean	n
March 1990	40.1	1.5	100	14.8	1.1	100	n.d.[b]	49.8	2		2
May 1990	20.2	0.7	100	26.4	0.8	100	n.d.	34.8	2		2
October 1990	2.5	0.1	100	3.1	0.1	100	3.4	0.2	12	6.2	2
December 1990	18.4	0.7	99	23.3	0.6	99	27.1	0.8	15	30.2	2
January 1991	9.3	0.3	100	10.9	0.3	100	14.1	0.5	16	27.8	2
February 1991	19.8	0.6	100	21.5	0.6	100	24.8	1.2	16	25.5	2

[a]Throughfall and precipitation were collected with 216 15-cm-diameter funnels, each placed directly on 2-l polypropylene bottles. Wet deposition was collected in two wetfall dryfall separating collectors (Aerochem Metrics). Values are quoted as mean, standard error, and number of determinations. nd = not determined. From Knops et al. (1991).

Table IX   Chemistry of Throughfall under Trees with and without Canopy Lichens[a]

Element	Trees with *Ramalina*		Trees without *Ramalina*	
	Mean	S.E.	Mean	S.E.
Cl	371.7	24.0	325.5	14.2
SO$_2$	29.8	1.3	37.2	1.4
NA$^-$	140.4	8.2	128.0	5.3
K$^-$	143.2	15.4	109.9	10.1
Ca$^-$	83.1	9.9	64.0	6.0
Mg$^-$	28.4	1.8	22.9	1.2
N in NO	2.1	0.1	2.3	0.1
N in NH$^-$	1.0	0.1	1.1	0.3
Total N	14.7	0.9	11.8	1.5
Organic N	11.6	0.8	8.5	1.3
Total P	7.2	1.2	2.3	0.6

[a] Data collected from February 15 to March 15, 1990, are in Mg m^{-2} = 100. From Knops *et al.* (1991).

not there, or were reduced in abundance. In certain forests, then, EM may act in the specifically keystone role described here because of three attributes: (1) EM makes up a minor portion of most forest ecosystems, less than 5% of total ecosystem biomass in most cases; (2) EM functions to reliably recruit nutrients from allochthonous atmospheric sources because of their morphology and physiology (e.g., poikilohydric foliage and water-impounding tanks; in montane forests, where EM is most abundant, this occurs especially during the dry season, when most atmospheric nutrients are in the form of wind-driven mist and fog that may otherwise "blow by" the ecosystem); and (3) other ecosystem members gain access to nutrients via leaching, decomposition, and mineralization.

Studies of forest nutrient cycling and hydrology should therefore not regard the canopy merely as a static "black box," acting as a filter of precipitation, for several reasons. First, between storm events, the canopy interacts with and accumulates dry and wet aerosols that dissolve in canopy fluids and are assimilated into living tissues. Second, canopy effects on precipitation chemistry vary seasonally, change during forest succession (epiphyte biomass tends to increase as succession proceeds; Hale, 1967), and fluctuate according to forest turnover rates (Lang *et al.*, 1978). Third, the specific biological composition of forest communities has marked effects on the way canopies interact with atmospheric inputs, for instance, through active metabolic uptake from, and pulsed release of nutrients to, throughfall precipitation by nonvascular epiphytes. As a component of many forests, then, EM may play an important regulatory role in nutrient conservation. Nutrient input, mineralization, and release all occur in EM. Therefore epiphytic material may be viewed as analogous to the rhizosphere, in terms of community function, and deserves attention from ecologists and soil scientists interested in understanding the patterns and processes of nutrient and water cycling and their relationship to forest productivity.

VI. Summary and Priorities for Future Research

The ecological role of any ecosystem component is difficult to define. It is even more tenuous to do so for such variable and poorly understood entities as epiphytes. The data that have been gathered so far reveal that in most ecosystems, epiphyte biomass is small relative to total ecosystem pools, but that in some ecosystems (e.g., tropical montane forests and temperate wet forests), biomass of epiphytes can exceed host tree foliage. Because of their physiology and their location at the atmosphere–vegetation interface, epiphytes can affect the ecosystem in many ways. By considering these effects, we can gain a better understanding of forest ecosystems.

Research directions of highest priority include: (1) carry out more process-oriented studies to understand the patterns identified to date; (2) develop mutually agreeable standardized measurements for canopy nutrient cycling (e.g., EM biomass, EM litterfall, EM decomposition) so that measurements among habitats will be comparable; (3) learn how to scale up from within-tree to ecosystem- and landscape-level spatial scales by developing and using statistically sound techniques; (4) carry out experimental work (e.g., fertilization studies, removal experiments) to rigorously test hypotheses concerning canopy nutrient cycling; (5) establish long-term studies to quantify unknown properties such as epiphyte growth rates, death and decomposition rates, and colonization and successional regimes; and (6) initiate simultaneous studies in a wide range of forest ecosystems to generate a broadly based and comparable pool of data on epiphytic material and its effects on nutrient cycles.

Acknowledgments

We thank two reviewers for helpful comments on the manuscript and Dr. Ken Kershaw for stimulating discussion on lichenized canopy epiphytes. This work was supported by research grants from the National Science Foundation (Washington, D.C.), the Natural Sciences and Engineering Research Council (Ottawa), the Whitehall Foundation, and the National Geographic Society Committee on Research and Exploration. Teri Matelson, Doug Schaefer, and Ken Clark contributed to the ideas presented here.

References

Aber, J. D., Nadelhoffer, K. J., Stuedler, P., and Melillo, J. M. (1989). Nitrogen saturation in northern forest ecosystems. *BioScience.* 39, 378–386.

Alcock, M. R., and Morton, A. J. (1985). Nutrient content of throughfall and stemflow in woodland recently established on heathland. *J. Ecol.* 73, 625–632.

Alpert, P., and Oechel, W. C. (1987). Comparative patterns of net photosynthesis in an assemblage of mosses with contrasting microdistributions. *Am. J. Bot.* 74, 1787–1796.

Antibus, R. K., and Lesica, P. (1990). Root surface phosphatase activities of vascular epiphytes of a Costa Rican rain forest. *Plant Soil* 128, 233–240.

Auclair, A. N. D., and Rencz, A. N. (1982). Concentration, mass, and distribution of nutrients in a subarctic *Picea mariana–Cladonia alpestris* ecosystem. *Can. J. For. Res.* 12, 947–968.

Avalos, A., Legaz, M. E., and Vincente, C. (1986). The occurrence of lichen phenolics in the xylem sap of *Quercus pyrenaica,* their translocation to leaves and biological significance. *Biochem. Syst. Ecol.* 14, 381–384.

Ball, E., Hann, J., Kluge, M., Lee, H. S. J., Lüttge, U., Orthen, B., Poop, M., Schmitt, A., and Ting, I. P. (1991). Ecophysiological comportment of the tropical CAM-tree *Clusia* in the field. *New Phytol.* 117, 473–481.

Bates, J. W. (1987). Nutrient retention by *Pseudoscleropodium purum* and its relation to growth. *J. Bryol.* 14, 565–580.

Bates, J. W. (1989). Interception of nutrients in wet deposition by *Pseudoscleropodium purum:* An experimental study of uptake and retention of potassium and phosphorus. *Lindbergia* 15, 93–98.

Benson, C. A. (1990). The potential for integrated resource management with intensive or extensive forest management: Reconciling vision with reality. *For. Chron.* 66, 457–460.

Bentley, B. L. (1987). Nitrogen fixation by epiphylls in a tropical rainforest. *Ann. Mo. Bot. Gard.* 74, 234–241.

Bentley, B. L., and Carpenter, E. J. (1980). Effects of desiccation and rehydration on nitrogen fixation by epiphylls in a tropical rainforest. *Microb. Ecol.* 6, 109–113.

Bentley, B. L., and Carpenter, E. J. (1984). Direct transfer of newly-fixed nitrogen from free-living epiphyllous microorganisms to their host plant. *Oecologia* 63, 52–56.

Benzing, D. H. (1983). Vascular epiphytes: A survey with special reference to their interactions with other organisms. *In* "Tropical Rain Forest: Ecology and Management" (S. L. Sutton, T. C. Whitmore, and A. C. Chadwick, eds.), pp. 11–24. Blackwell, Oxford.

Benzing, D. H. (1990). "Vascular Epiphytes." Cambridge Univ. Press, Cambridge, UK.

Benzing, D. H., and Seeman, J. (1978). Nutritional piracy and host decline: A new perspective on the epiphyte–host relationship. *Selbyana* 2, 133–148.

Bewley, J. D. (1979). Physiological aspects of desiccation tolerance. *Annu. Rev. Plant Physiol.* 30, 195–238.

Bianchi, G., Gamba, A., Murelli, C., Salamini, F., and Bartells, D. (1991). Novel carbohydrate metabolism in the resurrection plant *Craterostigma plantagineum. Plant J.* 1, 355–359.

Bohlman, S., Matelson, T., and Nadkarni, N. (1995). Moisture and temperature patterns of dead organic matter in the canopy and soils on the forest floor of a montane cloud forest, Costa Rica. *Biotropica* 27, 13–19.

Boonpragob, K., and Nash, T. H., III (1989). Seasonal deposition of acidic ions and ammonium to the lichen *Ramalina menziesii* Tayl. in southern California. *Environ. Exp. Bot.* 29, 187–197.

Boonpragob, K., and Nash, T. H., III (1990). Seasonal variation of elemental status in the lichen *Ramalina menziesii* Tayl. from two sites in southern California: Evidence for dry deposition accumulation. *Environ. Exp. Bot.* 30, 415–428.

Bosatta, E., and Berendse, F. (1984). Energy or nutrient regulation of decomposition: Implications for the mineralization–immobilization response to perturbations. *Soil Biol. Biochem.* 16, 63–68.

Boucher, V. L., and Nash, T. H., III (1990). The role of the fruticose lichen *Ramalina menziesii* in the annual turnover of biomass and macronutrients in a blue oak woodland. *Bot. Gaz. (Chicago)* 151, 114–118.

Bray, J., and Gorham, E. (1964). Litter production of forests of the world. *Adv. Ecol. Res.* 2, 101–157.

Brown, D. H., and Beckett, R. P. (1984). Uptake and effect of cations on lichen metabolism. *Lichenologist* 16, 173–188.

Brown, D. H., and Brown, R. M. (1991). Mineral cycling and lichens: The physiological basis. *Lichenologist* 23, 293–307.

Bruijnzeel, L. A. (1983). Hydrological and biogeochemical aspects of man-made forests in south-central Java, Indonesia. Ph.D. Thesis, Free University, Amsterdam.

Bruijnzeel, L. A., Waterloo, M. J., Proctor, I., Kuiters, A. T., and Kotterink, B. (1993). Hydrological observations in montane rain forests on Gunung Silam, Sabah, Malaysia, with special reference to the 'Massenerhebung' effect. *J. Ecol.* 81, 145–167.

Buck, G. W., and Brown, D. H. (1977). Cation analysis of bryophytes; The significance of water content and ion location. *Bryophytorum Bib.* 13, 735–750.

Carpenter, E. J. (1992). Nitrogen fixation in the epiphyllae and root nodules of trees in the lowland tropical rainforest of Costa Rica. *Acta Oecol.* 13, 153–160.

Carroll, G. (1980). Forest canopies: Complex and independent subsystems. *In* "Forests: Fresh Perspectives from Ecosystem Analysis" (R. H. Waring, ed), pp. 87–107. Oregon State Univ. Press, Corvallis.

Chamberlain, A. C. (1986). Deposition of gases and particles on vegetation and soils. *In* "Air Pollutants and Their Effects on the Terrestrial Ecosystem" (A. H. Legge and S. V. Krupa, eds.). Wiley, New York.

Chandler, G., and Goosem, S. (1982). Aspects of rainforest regeneration. III. The interaction of phenols, light and nutrients. *New Phytol.* 92, 369–380.

Clayton, J. L. (1972). Salt spray and mineral cycling in two California coastal ecosystems. *Ecology* 53, 74–81.

Clements, R. G., and Colon, J. A. (1975). The rainfall interception process and mineral cycling in a montane rain forest in eastern Puerto Rico. *In* "Mineral Cycling in Southeastern Ecosystems" (F. G. Howell, J. B. Gentry, and M. H. Smith, eds.). U.S. Energy Research and Development Administration, Springfield, VA.

Coxson, D. S. (1991). Nutrient release from epiphytic bryophytes in tropical montane rain forest (Guadeloupe). *Can. J. Bot.* 69, 2122–2129.

Coxson, D. S., and Kershaw, K. A. (1983). Nitrogenase activity during chinook snow melt sequences by *Nostoc commune* in *Stipa–Bouteloua* grassland. *Can. J. Microbiol.* 29, 938–944.

Coxson, D. S., Webber, M. R., and Kershaw, K. A. (1984). The thermal operating environment of corticolous and pendulous tree lichens. *Bryologist* 87, 197–202.

Coxson, D. S., McIntyre, D. D., and Vogel, H. J. (1992). Pulse release of sugars and polyols from canopy bryophytes in tropical montane rain forest (Guadeloupe, French West Indies). *Biotropica* 24, 121–133.

Crittenden, P. D. (1983). The role of lichens in the nitrogen economy of subarctic woodlands: Nitrogen loss from the nitrogen-fixing lichen *Stereocaulon paschale* during rainfall. *In* "Nitrogen as an Ecological Factor" (J. A. Lee, S. McNeill, and I. H. Rorison, eds.). Blackwell, Oxford.

Davison, I. R., and Reed, R. H. (1985). The physiological significance of mannitol accumulation in brown algae: The role of mannitol as a compatible cytoplasmic solute. *Phycologia* 24, 449–457.

deBoois, H. M., and Jansen, E. (1975). Effects of nutrients in throughfall water and of litterfall upon fungal growth in a forest soil layer. *Pedobiologia* 16, 161–166.

Denison, W. C. (1973). Life in tall trees. *Sci. Am.* 228, 74–80.

Denison, W. C. (1979). *Lobaria oregana*, a nitrogen-fixing lichen in old-growth Douglas fir forests. *In* "Symbiotic Nitrogen Fixation in the Management of Temperate Forests" (J. C. Gordon and D. A. Perry, eds.), pp. 266–275. Oregon State Univ. Press, Corvallis.

Dudley, S. A., and Lechowicz, M. J. (1987). Losses of polyol through leaching in subarctic lichens. *Plant Physiol.* 83, 813–815.

Edwards, P. J., and Grubb, P. J. (1977). Studies of mineral cycling in a montane rain forest in New Guinea. I. The distribution of organic matter in the vegetation and soil. *J. Ecol.* 65, 943–969.

Edwards, R. Y., Soos, J., and Ritchey, R. W. (1960). Quantitative observations on epidendric lichens used as food by caribou. *Ecology* 41, 425–431.

Eickmeier, W. G. (1982). Protein synthesis and photosynthetic recovery in the resurrection plant, *Selaginella lepidophylla. Plant Physiol.* 69, 135–138.

Esseen, P. A. (1985). Litterfall of epiphytic macrolichens in two old *Picea abies* forests in Sweden. *Can. J. Bot.* 63, 980–987.

Evans, L. S., Gmur, N. F., and Dacosta, F. (1977). Leaf surface and histological perturbations of leaves of *Phaseolus vulgaris* and *Helianthus annuus* after exposure to simulated acid rain. *Am. J. Bot.* 64, 903–913.

Fitter, A. H. (1985). Functioning of vesicular-arbuscular mycorrhizas under field conditions. *New Phytol.* 99, 257–265.

Forman, R. T. T. (1975). Canopy lichens with blue-green algae: A nitrogen source in a Colombian rain forest. *Ecology* 56, 1176–1184.

Frangi, J. L., and Lugo, A. E. (1986). Ecosystem dynamics of a subtropical floodplain forest. *Ecol. Monogr.* 55, 351–369.

Frangi, J. L., and Lugo, A. E. (1992). Biomass and nutrient accumulation in ten year old bryophyte communities inside a flood plain in the Luquillo experimental forest, Puerto Rico. *Biotropica* 24, 106–112.

Fritz-Sheridan, R. P., and Coxson, D. S. (1988). Nitrogen fixation on a tropical volcano, La Soufrière (Guadeloupe): The interaction of temperature, moisture, and light with net photosynthesis and nitrogenase activity in *Stereocaulon virgatum* and response to periods of insolation shock. *Lichenologist* 20, 63.

Fritz-Sheridan, R. P., and Portecop, J. (1987). Nitrogen fixation on the tropical volcano, La Soufrière (Guadeloupe): I. A survey of nitrogen fixation by blue-green algal microepiphytes and lichen endophytes. *Biotropica* 19, 194–199.

Gaber, B. A., and Hutchinson, T. C. (1988). Chemical changes in simulated raindrops following contact with leaves of four boreal forest species. *Can. J. Bot.* 66, 2445–2451.

Galloway, C. M., and Black, C. C. (1989). Enzymes of sucrose metabolism in bryophytes. *Bryologist* 92, 95–97.

Gillis, A. M. (1990). The new forestry. An ecosystem approach to land management. *BioScience* 40, 558–562.

Golley, F. B., McGinnins, J., and Clements, R. (1971). La biomasa y estructura de algunos bosques del Darién, Panama. *Turrialba* 21, 189–196.

Grier, C., and Nadkarni, N. M. (1987). The role of epiphytes in the nutrient cycles of two rain forest ecosystems. *In* "People and the Tropical Forest. U.S. Man and the Biosphere Program" (A. E. Lugo, J. J. Ewel, S. B. Hecht, P. G. Murphy, C. Padoch, M. C. Schmink, and D. Stone, eds.), pp. 28–30. U. S. Gov. Printing Office, Washington, DC.

Grubb, P. J. (1977). Control of forest growth and distribution on wet tropical mountains: With special reference to mineral nutrition. *Annu. Rev. Ecol. Syst.* 8, 83–107.

Gupta, R. K. (1976). The physiology of the desiccation resistance in bryophytes: Nature of organic compounds leaked from desiccated liverwort, *Plagiochila asplenioides. Biochem. Physiol. Pflanz.* 170, 389–395.

Gupta, R. K. (1977). An artefact in studies of the responses of respiration of bryophytes to desiccation. *Can. J. Bot.* 55, 1195–1200.

Hale, M. (1967). "The Biology of Lichens." Edward Arnold, London.

Harvey, A. E., Jurgensen, M. F., Larsen, M. J., and Graham, R. T. (1987). Relationships among soil microsite, ectomycorrhizae, and natural conifer regeneration of old-growth forests in western Montana. *Can. J. For. Res.* 17, 58–62.

Henderson, A. (1993). Literature on air pollution and lichens. XXXVII. *Lichenologist* 25, 191–202.

Herwitz, S. R. (1991). Aboveground adventitious roots and stemflow chemistry of *Ceratopetalum virchowii* in an Australian montane tropical rain forest. *Biotropica* 23, 210–218.

Hilton, G. (1987). Nutrient cycling in tropical rainforests: Implications for management and sustained yield. *For. Ecol. Manage.* 22, 297–300.

Hofstede, R. G. M., Wolf, J. H. D., and Benzing, D. H. (1993). Epiphyte biomass and nutrient status of a Colombian upper montane rain forest. *Selbyana* 14, 37–45.

Hosker, R. P., and Lindberg, S. E., (1982). Review: Atmospheric deposition and plant assimilation of bases and particles. *Atmos. Environ.* 16, 889–910.

Hutchinson, T. C., Dixon, M., and Scott, M. (1986). The effect of simulated acid rain on feather mosses and lichens of the boreal forest. *Water, Air, Soil Pollut.* 31, 409–416.

Jacobs, M. (1988). "The Tropical Rain Forest." Springer-Verlag, Berlin.

Jenik, J. (1973). Root systems of tropical trees. 8. Stilt-roots and allied adaptations. *Preslia* 45, 250–264.

Jones, K., and Wilson, R. E. (1978). The fate of nitrogen fixed by a free-living blue-green alga. *Ecol. Bull.* 26, 158–163.

Jordan, C., Golley, F., Hall, J., and Hall, J. (1980). Nutrient scavenging of rainfall by the canopy of an Amazonian rain forest. *Biotropica* 12, 61–66.

Kellman, M., Hudson, J., and Sanmugadas, K. (1982). Temporal variability in atmospheric nutrient influx to a tropical ecosystem. *Biotropica* 14, 1–9.

Kelly, B. B., and Becker, V. E. (1975). Effects of light intensity and temperature on nitrogen fixation by *Lobaria pulmonaria, Sticta weigelii, Leptogium cyanescens* and *Collema subfurvum. Bryologist* 78, 350–355.

Kershaw, K. A., and MacFarlane, J. D. (1982). Physiological–environmental interactions in lichens. XIII. Seasonal constancy of nitrogenase activity, net photosynthesis, and respiration in *Collema furfuraceum* (Am.) Dr. *New Phytol.* 90, 723–734.

Klinge, H., Rodriguez, W. A., Bruning, E., and Fittkau, G. (1975). Biomass and structure in a Central American rain forest. *In* "Tropical Ecological Systems" (F. B. Golley and E. Medina, eds.), pp. 115–122. Springer-Verlag, New York.

Knops, J. M. H., Nash, T. H., III, Boucher, V. L., and Schlesinger, W. H. (1991). Mineral cycling and epiphytic lichens: Implications at the ecosystem level. *Lichenologist* 23, 309–321.

Kost, G. (1988). Interactions between Basidiomycetes and Bryophyta. *Endocytobiosis Cell Res.* 5, 287–308.

Lang, G. E., Reiners, W. A., and Heier, R. K. (1976). Potential alteration of precipitation chemistry by epiphytic lichens. *Oecologia* 25, 229–241.

Lang, G. E., Reiners, W. A., and Pike, L. H. (1978). Structure and biomass of epiphytic lichen communities of balsam fir forests in New Hampshire. *Ecology* 61, 541–550.

Larson, D. W. (1987). The absorbtion and release of water by lichens. *Bibl. Lichenol.* 25, 351–360.

Lechowicz, M. J. (1981). Adaptation and the fundamental niche: Evidence from lichens. *In* "The Fungal Community: Its Organization and Role in the Ecosystem" (D. T. Wicklow and G. C. Carroll, eds.), pp. 89–108. Dekker, New York.

Lechowicz, M. J. (1982). Ecological trends in lichen photosynthesis. *Oecologia* 53, 330–336.

Lesica, P., and Antibus, R. K. (1991). Canopy soils and epiphyte richness. *Natl. Geogr. R. Explor.* 7, 156–165.

Levings, S. C., and Windsor, D. M. (1984). Litter moisture content as a determinant of litter arthropod distribution and abundance during the dry season on Barro Colorado Island, Panama. *Biotropica* 16, 125–131.

Lewis, D. H., and Smith, D. C. (1967). Sugar alcohols (polyols) in fungi and green plants. *New Phytol.* 66, 185–204.

Lindberg, S. E., Lovett, G. M., Richter, D. D., and Johnson, D. W. (1986). Atmospheric deposition and canopy interactions of major ions in a forest. *Science* 231, 141–145.

Linzon, S. N. (1988). Tree decline in industrialized areas of North America. *GeoJournal* 17, 179–183.

Lodge, D. J., Scatena, F. N., Asbury, C. E., and Sanchez, M. J. (1991). Fine litterfall and related nutrient inputs resulting from Hurricane Hugo in subtropical wet and lower montane rain forests of Puerto Rico. *Biotropica* 23, 336–342.

Lousier, J. D., and Parkinson, D. (1976). Litter decomposition in a cool temperate deciduous forest. *Can. J. Bot.* 54, 419–436.

Lovett, G. M. (1981). Forest structure and atmospheric interactions: Predictive models for subalpine balsam-fir forests. Ph.D. Thesis, Dartmouth College, Hanover, NH.

Lovett, G. M. (1984). Rates and mechanisms of cloud water position in a subalpine balsam fir forest. *Atmos. Environ.* 18, 361–371.

Lovett, G. M., Lindberg, S. E., Richter, D. D., and Johnson, D. W. (1985). The effects of acidic deposition on cation leaching from three deciduous forest canopies. *Can. J. For. Res.* 15, 1055–1060.

Lugo, A., Farnworth, E., Pool, D., Jerez, P., and Kaufman, G. (1973). The impact of the leaf-cutter ant *Atta colombica* on the energy flow of a tropical wet forest. *Ecology* 54, 1292–1301.

Lyford, W. H. (1969). The ecology of an elfin forest in Puerto Rico. 7. Soil, root, and earthworm relationships. *J. Arnold Arbor., Harv. Univ.* 50, 210–224.

Mann, F. D., and Steinke, T. D. (1989). Biological nitrogen fixation (acetylene reduction) associated with blue-green algal (cyanobacterial) communities in the Beachwood Mangrove Nature Reserve. I. The effect of environmental factors on acetylene reduction activity. *S. Afr. J. Bot.* 55, 438–446.

Marrs, R. H., Proctor, J., Heaney, A., and Mountford, M. D. (1988). Changes in soil nitrogen-mineralization and nitrification along an altitudinal transect in tropical rain forest in Costa Rica. *J. Ecol.* 76, 466–482.

Maser, C. (1990). "The Redesigned Forest." Stoddart, Toronto.

Maser, C., Trappe, J. M., and Nussbaum, R. A. (1978). Fungal–small mammal interrelationships with emphasis on Oregon coniferous forests. *Ecology* 59, 799–809.

Matelson, T. J., Nadkarni, N. M., and Longino, J. T. (1993). Longevity of fallen epiphytes in a Neotropical montane forest. *Ecology* 74, 265–269.

Mooney, H. A., Vitousek, P. M., and Matson, P. A. (1987). Exchange of materials between terrestrial ecosystems and the atmosphere. *Science* 238, 926–931.

Muir, P. S., and McCune, B. (1988). Lichens, tree growth, and foliar symptoms of air pollution: Are the stories consistent? *J. Environ. Qual.* 17, 361–370.

Murach, D., and Ulrich, B. (1988). Destabilization of forest ecosystems by acid deposition. *GeoJournal* 17, 253–260.

Murphy, P. G., and Lugo, A. E. (1986). Structure and nutrient capital of a subtropical dry forest in Puerto Rico. *Biotropica* 18, 89–96.

Nadelhoffer, K. J., and Raich, J. W. (1992). Fine root production estimates and belowground carbon allocation in forest ecosystems. *Ecology* 73, 1139–1147.

Nadelhoffer, K. J., Aber, J. D., and Melillo, J. M. (1984). Seasonal patterns of ammonium and nitrate uptake in nine temperate forest ecosystems. *Plant Soil* 80, 321–335.

Nadkarni, N. M. (1981). Canopy roots: Convergent evolution in rainforest nutrient cycles. *Science* 214, 1023–1024.

Nadkarni, N. M. (1983). The effects of epiphytes on nutrient cycles within temperate and tropical rainforest tree canopies. Ph.D. Thesis, University of Washington, Seattle.

Nadkarni, N. M. (1984). Epiphyte biomass and nutrient capital of a Neotropical elfin forest. *Biotropica* 16, 249–256.

Nadkarni, N. M. (1988). Tropical rainforest ecology from a canopy perspective. *Mem. Calif. Acad. Sci.* 12, 57–67.

Nadkarni, N. M., and Longino, J. T. (1990). Macroinvertebrate communities in canopy and forest floor organic matter in a montane forest, Costa Rica. *Biotropica* 22, 286–289.

Nadkarni, N. M., and Matelson, T. (1989). Bird use of epiphyte resources in Neotropical trees. *Condor* 69, 891–907.

Nadkarni, N. M., and Matelson, T. J. (1991). Litter dynamics within the canopy of a neotropical cloud forest, Monteverde, Costa Rica. *Ecology* 72, 849–860.

Nadkarni, N. M., and Matelson, T. J. (1992). Biomass and nutrient dynamics of epiphytic litterfall in a Neotropical montane forest, Costa Rica. *Biotropica* 24, 24–30.

Nadkarni, N. M., and Primack, R. B. (1989). A comparison of mineral uptake and translocation by above-ground and below-ground systems of *Salix syringiana*. *Plant Soil* 113, 39–45.

Nadkarni, N. M., Matelson, T. J., Clark, K., Schaefer, D., and Solano, R. (1993). "Ecological Roles of Canopy Organic Matter in a Cloud Forest, Costa Rica." Association for Tropical Biology, San Juan, PR.

Nasrulhaq-Boyce, A., and Hajimohamed, M. A. (1987). Photosynthetic and respiratory characteristics of Malayan sun and shade ferns. *New Phytol.* 105, 81–88.

Nieboer, E., and MacFarlane, J. D. (1984). Modification of plant cell buffering capacities by gaseous air pollutants. *In* "Gaseous Air Pollutants and Plant Metabolism" (M. J. Koziot and F. R. Whately, eds.). Butterworth, London.

Nieboer, E., Richardson, D. H. S., and Tomassini, F. D. (1978). Mineral uptake and release by lichens: An overview. *Bryologist* 81, 226–246.

Niederer, M., Pankow, W., and Wiemken, A. (1989). Trehalose synthesis in mycorrhiza of Norway spruce: An indicator of vitality. *Eur. J. For. Pathol.* 19, 14–20.

Olson, J. (1963). Energy storage and the balance of producers and decomposers in ecological systems. *Ecology* 44, 322–331.

Ong, B. L., Kluge, M., and Friemert, V. (1986). Crassulacean acid metabolism in the epiphytic ferns *Drymoglossum piloselloides* and *Pyrrosia longifolia:* Studies on responses to environmental signals. *Plant, Cell Environ.* 9, 547–557.

Orndorff, K., and Lang, G. (1981). Leaf litter redistribution in a West Virginia hardward forest. *J. Ecol.* 69, 225–235.

Parker, G. G. (1983). Throughflow and stemflow in the forest nutrient cycle. *Adv. Ecol. Res.* 13, 57–133.

Pike, L. (1971). The role of epiphytic lichens in nutrient cycles and productivity in an oak woodland. Ph.D. Thesis, University of Oregon, Eugene.

Pike, L. H. (1978). The importance of epiphytic lichens in mineral cycling. *Bryologist* 81, 247–257.

Pocock, K., Duckett, J. G., Grolle, R., Mohamed, A. H., and Pang, W. C. (1985). Branched and swollen rhizoids in hepatics from montane rain forest in peninsular Malaya. *J. Bryol.* 13, 241–246.

Pocs, T. (1980). The epiphytic biomass and its effect on the water balance of two rainforest types in the Uluguru Mountains. *Acta Bot. Acad. Sci. Hung.* 26, 143–167.

Pocs, T. (1982). Tropical forest bryophytes. *In* "Bryophyte Ecology" (A. E. Smith, ed.), pp. 59–104. Chapman & Hall, London.

Proctor, J., Anderson, J. M., and Vallack, H. W. (1983). Comparative studies on forests, soils and litterfall at four altitudes on Guning Mulu, Sarawak. *Malays. For.* 46, 60–76.

Proctor, M. C. F. (1982). Physiological ecology: Water relations, light and temperature responses, carbon balance. *In* "Bryophyte Ecology" (A. J. E. Smith, ed.). Chapman & Hall, London.

Putz, F. E., and Holbrook, N. M. (1989). Strangler fir rooting habits and nutrient relations in the llano of Venezuela. *Am. J. Bot.* 51, 264–274.

Reiners, W. A., and Olson, R. K. (1984). Effects of canopy components on throughfall chemistry: An experimental analysis. *Oecologia* 63, 320–330.

Rhoades, F. W. (1977). Growth rates of the lichen *Lobaria oregana* as determined from sequential photographs. *Can. J. Bot.* 55, 2226–2233.

Rice, E. L., and Pancholy, S. K. (1974). Inhibition of nitrification by climax ecosystems: II. Additional evidence and a possible role of tannins. *Am. J. Bot.* 61, 1095–1103.

Richardson, D. H. S. (1988). Understanding the pollution sensitivity of lichens. *Bot. J. Linn. Soc.* 96, 31–43.

Richardson, D. H. S., and Nieboer, E. (1983a). Ecophysiological responses of lichens to sulphur dioxide. *J. Hattori Bot. Lab.* 54, 331–351.

Richardson, D. H. S., and Nieboer, E. (1983b). The uptake of nickel ions by lichen thalli of the genera *Umbilicaria* and *Peltigera*. *Lichenologist* 15, 81–88.

Rodin, L. E., and Bazilevich, I. (1967). "Production and Mineral Cycling in Terrestrial Vegetation." Oliver & Boyd, Edinburgh.

Roskowski, J. (1980). Nitrogen fixation (C_2H_2 reduction) by epiphylls on coffee, *Coffea arabica*. *Microb. Ecol.* 6, 349–355.

Ruinen, J. (1953). Epiphytosis, a second view on epiphytism. *Ann. Bogor.* 1, 101–157.

Runeckles, V. C. (1986). Photochemical oxidants. *In* "Pollutants and Their Effect on the Terrestrial Ecosystems" (A. H. Legge and S. V. Krupa, eds.). Wiley, New York.

Schofield, W. B. (1981). Ecological significance of morphological characters in the moss gametophyte. *Bryologist* 84, 149–165.

Schwarzmaier, U., and Brehm, K. (1975). Detailed characterization of the cation exchanger in *Sphagnum magellanicum* Brid. *Z. Pflanzenphysiol.* 75, 250–255.

Scott, M. G., and Hutchinson, T. C. (1987). Effects of a simulated acid rain recovery episode on photosynthesis and recovery in the caribou-forage lichens, *Cladina stellaris* (Opiz.) Brodo and *Cladina rangiferina* (L.) Wigg. *New Phytol.* 107, 567–575.

Scott, M. G., and Larson, D. W. (1985). The effect of winter field conditions on the distribution of two species of *Umbilicaria*. I. CO_2 exchange in reciprocally-transplanted thalli. *New Phytol.* 101, 89–101.

Scotter, C. W. (1962). Productivity of arboreal lichens and their possible importance to barren ground caribou (*Rangifer arcticus*). *Arch. Soc. Bot. Fenn.* "Vanamo" 16, 155–161.

See, L. S. (1990). The mycorrhizal association of the Dipterocarpaceae in the tropical rain forests of Malaysia. *Ambio* 19, 383–385.

Sehmel, G. A. (1980). Particle and gas dry deposition: A review. *Atmos. Environ.* 14, 973–1011.

Sengupta, B., Nandi, A., Samanta, R., Pal, D., Sengupta, P., and Sen, S. (1981). Nitrogen fixation in the phyllosphere of tropical plants: Occurrence of phyllosphere nitrogen-fixing microorganisms in eastern India and their utility for the growth and nitrogen nutrition of host plants. *Ann. Bot. (London)* [N.S.] 48, 705–716.

Sheridan, R. P. (1991). Epicaulous, nitrogen-fixing microepiphytes in a tropical mangal community, Guadeloupe, French West Indies. *Biotropica* 23, 530–541.

Sigmon, J. T., Gilliam, F. S., and Partin, M. E. (1989). Precipitation and throughflow chemistry for a montane hardwood forest ecosystem: Potential contributions from cloud water. *Can. J. Bot.* 19, 1240–1247.

Smith, A. J. E. (1982). Epiphytes and epiliths. *In* "Bryophyte Ecology" (A. E. Smith, ed.), pp. 191–227. Chapman & Hall, London.

Songwe, N. C., Fasehun, F. E., and Okali, D. U. (1988). Litterfall and productivity in a tropical rain forest, Southern Bakundu Forest Reserve, Cameroon. *J. Trop. Ecol.* 4, 25–37.

Steinhardt, U. (1979). Untersuchungen uber den Wasser- un Nahrstoffhaushalt eines andinen Wolkenwaldes in Venezuela *Gottinger Bodenkd. Ber.* 56, 1–185.

Steinke, T. D., and Naidoo, Y. (1990). Biomass of algae epiphytic on pneumatophores of the mangrove, *Avicennia marina,* in the St. Lucia estuary. *S.-Af. Tydskr. Plantkd.* 56, 226–232.

Strong, D. R. (1977). Epiphyte loads, tree falls, and perennial forest disruption: A mechanism for maintaining higher tree species richness in the tropics without animals. *J. Biogeog.* 4, 215–218.

Tanner, E. V. J. (1977). Four montane rain forests of Jamaica: A quantitative characterization of the floristics, the soils, and the foliar mineral levels, and a discussion of the interrelations. *J. Ecol.* 65, 883–918.

Tanner, E. V. J. (1980). Studies on the biomass and productivity in a series of montane rain forests in Jamaica. *J. Ecol.* 68, 573–588.

Tanner, E. V. J., Kapos, V., and Franco, W. (1992). Nitrogen and phosphorus fertilization effects on Venezuelan montane forest trunk growth and litterfall. *Ecology* 73, 78–86.

Terborgh, J. (1986). Keystone plant resources in the tropical forest. *In* "Conservation Biology: The Science of Scarcity and Diversity" (M. Soulé, ed.), pp. 330–343. Sinauer Assoc., Sunderland, MA.

Vance, E. D., and Nadkarni, N. M. (1990). Microbial biomass and activity in canopy organic matter and the forest floor of a tropical cloud forest. *Soil Biol. Biochem.* 22, 677–684.

Vance, E. D., and Nadkarni, N. M. (1992). Root biomass distribution in a moist tropical montane forest. *Plant Soil* 142, 31–39.

Veneklaas, E. (1991a). Litterfall and nutrient fluxes in two montane tropical rain forests, Colombia. *J. Trop. Ecol.* 7, 319–335.

Veneklaas, E. (1991b). Nutrient fluxes in bulk precipitation and throughflow in two montane tropical rain forests, Colombia. *J. Ecol.* 68, 573–588.

Veneklaas, E. (1992). Rainfall interception and aboveground nutrient fluxes in Colombian montane tropical rain forest. PH.D. Thesis, University of Amsterdam, Amsterdam.

Veneklaas, E., and Ek, R. V. (1991). Rainfall interception in two tropical montane rain forests, Colombia. *Hydrol. Processes* 4, 311–326.

Veneklaas, E., Zagt, R., Leerdam, A. V., Ek, R. V., Broekhoven, G., and Genderen (1991). Hydrological properties of the epiphyte mass of a montane tropical rain forest, Colombia. *Vegetatio* 89, 183–192.

Veres, J. S. (1990). Xylem anatomy and hydraulic conductance of Costa Rican *Blechnum* ferns. *Am. J. Bot.* 77, 1610–1625.

Vermes, J. F., and Myrold, D. D. (1992). Dentrification in forest soils of Oregon. *Can. J. For. Res.* 22, 504–512.

Vitousek, P. M. (1982). Nutrient cycling and nutrient use efficiency. *Am. Nat.* 119, 553–572.

Vitousek, P. M. (1984). Litterfall, nutrient cycling, and nutrient limitation in tropical forests. *Ecology* 65, 285–298.

Vitousek, P. M., and Denslow, J. S. (1986). Nitrogen and phosphorus availability in treefall gaps of a lowland tropical rainforest. *J. Ecol.* 74, 1167–1178.

Vitousek, P. M., and Matson, P. A. (1988). Nitrogen transformations in a range of tropical forest soils. *Soil Biol. Biochem.* 20, 361–367.

Vitousek, P. M., and Sanford, R. L., Jr. (1986). Nutrient cycling in moist tropical forest. *Annu. Rev. Ecol. Syst.* 17, 137–167.

Vitousek, P. M., Walker, L. R., Whiteaker, L. D., Mueller-Dombois, D., and Matsn, P. A. (1987). Biological invasion by *Myrica faya* alters ecosystem development in Hawaii. *Science* 238, 802–804.

Vogt, K. A., Publicover, D. A., and Vogt, D. J. (1991). A critique of the role of ectomycorrhizas in forest ecology. *Agric. Ecosys. Environ.* 35, 171–190.

Waring, R. H., and Schlesinger, W. H. (1985). "Forest Ecosystems: Concepts and Management." Academic Press, Orlando, FL.

Weathers, K. C., Likens, G. E., Bormann, F. H., Bicknell, S. H., Bormann, B. T., Daube, B. C., Jr., Eaton, J. S., Galloway, J. N., Keene, W. C., Kimball, K. D., McDowell, W. H., Siccama, T. G., Smiley, D., and Tarrant, R. A. (1988). Cloudwater chemistry from ten sites in North America. *Environ. Sci. Technol.* 22, 1018–1028.

Weaver, P. L., and Murphy, P. G. (1990). Forest structure and productivity in Puerto Rico's Luquillo Mountains. *Biotropica* 22, 69–82.

Weber, M. G., and Van Cleve, K. (1981). Nitrogen dynamics in the forest floor of interior Alaska black spruce ecosystems. *Can. J. For. Res.* 11, 743–751.

Welbourn, M., Stone, E., and Lassoie, J. (1981). Distribution of net litter inputs, with respect to slope and wind direction. *For. Sci.* 27, 651–659.

Westman, W. E. (1978). Inputs and cycling of mineral nutrients in a coastal subtropical eucalypt forest. *J. Ecol.* 66, 513–531.

White, C. S., Moore, D. I., Horner, J. D., and Gosz, J. R. (1988). Nitrogen mineralization–immobilization response to field N or C perturbations: An evaluation of a theoretical model. *Soil Biol. Biochem.* 20, 101–105.

Whitford, W., Steinberger, Y., and Ettershank, G. (1982). Contributions of subterranean termites to the "economy" of Chihuahuan Desert ecosystems. *Oecologia* 55, 298–302.

Wolf, J. H. D. (1993). Ecology of epiphytes and epiphytic communities in montane rain forests, Colombia. Ph.D. Thesis, University of Amsterdam, Amsterdam.

Yatazawa, M., Hamball, G., and Uchino, F. (1983). Nitrogen fixing activity in warty lenticellate tree barks. *Soil Sci. Plant Nutr.* 15, 285–294.

IV

Human Impacts on Canopy Research

21

Ethnobotany and Economic Botany of Epiphytes, Lianas, and Other Host-Dependent Plants: An Overview

Bradley C. Bennett

I. Introduction

A. Importance of Host-Dependent Plants

Forest canopies, the uppermost continuous vegetation layers (Lincoln *et al.*, 1982), contribute significantly to human welfare. Canopy-forming trees provide food, fuel, rubber, medicines, and multitudes of other products. Some consider the top strata of shrublands and grasslands to be canopies as well. Stature is no measure of a plant's importance; shrubs and grasses contribute significantly to human welfare.

Other life-forms depend on canopy plants for mechanical support. Epiphytes, lianas, and parasites live on species that form the upper strata of vegetation. They abound in tropical forests, particularly where annual precipitation exceeds 2000 mm. These host-dependent species are poorly represented in botanical surveys and in ethnobotanical studies. Many ethnobotanical studies fail to note a plant's habit, making it difficult to distinguish terrestrial species from their cogeneric epiphytic or climbing relatives. This inadequate description and sampling has underestimated the importance of host-dependent species for humans.

In this chapter, I examine the use of host-dependent plants by traditional and modern societies. I include epiphytes, lianas, hemiepiphytes, vines, and parasites, all species that grow on other plants. Life-form descriptions are not precise. Some authors use the term climber to describe scandent, herbaceous plants that share characteristics of hemiepiphytes and vines (e.g., many aroids). Benzing (1990) provided precise definitions of these host-dependent life-forms. Some host-dependent plants occur high in forest canopies or exposed sites (e.g., *Tillandsia usneoides*); others sprawl along the

Copyright © 1995 by Academic Press, Inc.
All rights of reproduction in any form reserved.

upper layers of shrubs or herbaceous vegetation (e.g., *Melothria pendula*). Some have an obligate relationship with a supporting host (e.g., *Cyrtopodium punctatum*); others are facultative epiphytes that grow both on trees and on terrestrial substrates (e.g., *Tillandsia ionochroma*).

How important are host-dependent species? Quantitative ethnobotanical studies of trees are rare; quantitative studies of mechanically dependent species are even rarer. The Siona–Secoya of Ecuador employ 46 species (47%) of lianas found in a 1-ha plot (Paz y Miño *et al.*, 1990). This is in the range of the estimated usage for all Amazonian plants, 30–50% (Bennett, 1992a). Though 90% or more of all tree species may be used by indigenous people (e.g., Balée, 1986; Boom, 1989; Bennett, 1992a), use of host-dependent species is considerable.

Plants that require a supporting host constitute 15% of the 670 documented plant species used by the Ecuadorian Shuar. They represent 10–22% of the species employed by other Amazonian ethnic groups (Bennett, 1992b). Araceae, Bignoniaceae, Cucurbitaceae, Ericaceae, and Orchidaceae are among the most important of the host-dependent taxa (six or more useful species each). These families provide significant resources for other indigenous people in Amazonia. Twenty-one epiphyte and liana species are reported in two or more studies from this region. Three lianas are among the most important nonfood plants in northwest Amazonia: *Lonchocarpus nicou* provides a fish poison, *Strychnos tomentosa* an arrow poison, and *Banisteriopsis caapi* the principal hallucinogen of northwest Amazonia (Bennett, 1992b,c). In India, several hemiepiphytic species of *Ficus* are sacred plants and important medicines (Jain, 1991).

Host-dependent plants contribute not only to the welfare of traditional peoples; several species are well known throughout the world. Vanilla and black pepper are among the world's most important spices (Purseglove, 1968). Sweet potatoes, yams, and many cucurbits are major food resources (Simpson and Conner-Ogorzaly, 1986). Species of *Dioscorea* provide precursors for synthetic hormones used in birth control pills. The muscle relaxant tubocurarine, derived from a tropical liana, revolutionized modern surgery. Another liana provides physostigmine, an antidote for tubocurarine (Morton, 1977a). Guaraná and grapes furnish popular drinks, and epiphytic orchids are the national flowers of many Central and South American countries (Lawlor, 1984).

B. Sources of Data

Much of the data are from Latin America, particularly Amazonia, for two reasons: the abundance of epiphytes reaches its zenith in the neotropics, and ethnobotanical studies are more extensive in the New World tropics. I draw data from my studies of the lowland Quichua, Shuar, and Chachi in Ecuador, published ethnobotanical studies from the neotropics, and general reviews. A list of the major ethnic groups and their locations is in

Table I   Major Ethnic Groups Discussed in the Text
and Their Geographic Locations

Name	Location	Source
Aztec	Mexico	Lawlor (1984)
Cofan	Ecuador	Pinkley (1973)
Guaraní	Brazil	Plowman (1969)
Huastec	Mexico	Alcorn (1984)
Kubeo	Colombia	Plowman (1969)
Mayan	Guatemala	Lawlor (1984)
Maina	Peru	(H. Wiehler, unpublished data)
Quechua	Peru	Bennett (1990, 1994)
Quichua	Ecuador	Alarcón (1988); Bennett (1992a)
Secoya–Siona	Ecuador	Vickers and Plowman (1984)
Shuar (Jívaro)	Ecuador	Bennett (1992a, 1994)
Tikuna	Colombia	(H. Wiehler, unpublished data)
Tsatchila (Colorado)	Ecuador	(H. Wiehler, unpublished data)
Waika	Venezuela	(H. Wiehler, unpublished data)
Waorani (Auca)	Ecuador	Davis and Yost (1983)
Zulu	South Africa	Lawlor (1984)

Table I and definitions of commonly used ethnobotanical and ethnomedicinal terms are in Table II.

General ethnobotanical studies, particularly those from Ecuador, provided primary data on the use of canopy-dwelling plants (e.g., Pinkley, 1973; Davis and Yost, 1983; Vickers and Plowman, 1984; Alarcón, 1988; Bennett *et al.*, 1994). Schultes and Raffauf's (1990) treatment of medicinal plants of the northwestern Amazon Basin yielded information on useful epiphytes and lianas. Alcorn's (1984) monograph on Huastec Mayan ethnobotany was another rich source of data, providing many non-Amazonian examples. Comprehensive reviews by Morton (1977a, 1982) were useful. Data on four of the most important epiphytic plant families are found in Plowman (1969) (Araceae), Lawlor (1984) and Benzing (1990) (Orchidaceae), Bennett (1994) (Bromeliaceae), and H. Wiehler (unpublished data) (Gesneriaceae); Bennett (1992b) was the first to document all the useful mechanically dependent species for a single ethnic group. Phillips (1991) reviewed the multitudinous use of vines and his careful study facilitated this compilation. Dictionaries of useful plants were valuable guides to plant use at the generic level and useful nomenclature tools (Uphof, 1968; Usher, 1974; Mabberly, 1987).

II. Results

A. Useful Host-Dependent Plants

I found data on 776 useful host-dependent species in 363 genera and 70 families of vascular plants. These are undoubtedly minimal figures. A com-

Table II   Commonly Used Ethnobotanical and Medical Terms [a]

Term	Definition
Abortifacient	Agent that produces abortion
Amebicide	Agent that causes the destruction of amoebas
Anodyne	Compound capable of relieving pain
Anthelmintic	Agent that destroys or expels intestinal worms
Antihypertensive	Drug or treatment that reduces blood pressure
Anti-inflammatory	Agent that reduces inflammation
Antipyretic	Agent that reduces fever
Antitussive	Agent that relieves coughs
Cauterant	Agent for scarring, burning, or cutting the skin by means of heat, cold, electric current, or caustic chemicals
Chicha	Fermented beverage made from many root and fruits, especially *Manihot esculenta* (cassava or yuca)
Cicatrazant	Agent that promotes scar formation
Diuretic	Agent that increases the amount of urine excreted
Emmenagogue	Agent that induces or increases menstrual flow
Erysipelas	Acute inflammation of the skin and subcutaneous tissues
Galactagogue	Agent that promotes the secretion and flow of milk
Hemostat	Agent that arrests the flow of blood from an open vessel
Leukorrhea	Vaginal discharge containing mucus and pus cells
Lumbago	Pain in the middle and lower back
Masticatory	Substance chewed, often for its stimulating effect
Mordant	Substance that increases the affinity of a dye for its substrate
Orchitis	Inflammation of the testis
Parturifacient	Agent that induces or accelerates labor
Vesicant	Agent that induces blistering

[a] Definitions from Hensyl (1990).

prehensive review is nearly impossible because epiphytes are poorly understood and ethnobotanical treatments of these resources are uneven.

The taxonomic distribution of major and minor families of useful host-dependent plants is in Tables II and III. For each of the 70 families, the number of useful genera and species by life-form (vines, lianas, epiphytes, climbers, and hemiepiphytes) is given. I will discuss 25 families in detail. Most of these families have more than ten useful species in published reports. Some have fewer representatives, but are of unquestionable value (e.g., *Strychnos*). For each family, I describe the most important genus and list other representative genera and their uses. A brief discussion of the minor families (those with ten or fewer useful species) follows. A list of the species discussed in the text and author citations for species names is in the Appendix.

A family use index, representing the sum of the number of use categories recorded for each species, is a measure of each family's utility (Table IV).

Table III Number of Useful Genera (G) and Species (S) of Major Families of Mechanically Dependent Canopy Plants

Family	Vines		Lianas		Epiphytes		Climbers		Hemi-epiphytes		Parasites		Number of species used
	G	S	G	S	G	S	G	S	G	S	G	S	
Apocynaceae	10	10	12	21	0	0	0	0	0	0	0	0	31
Araceae	0	0	1	1	2	30	7	41	0	0	0	0	72
Arecaceae	0	0	4	10	0	0	0	0	0	0	0	0	10
Aristolochiaceae	0	0	1	8	1	1	0	0	0	0	0	0	8
Asclepiadaceae	7	7	3	7	0	0	0	0	0	0	0	0	15
Bignoniaceae	0	0	19	25	0	0	0	0	0	0	0	0	25
Bromeliaceae	0	0	0	0	6	24	0	0	0	0	0	0	24
Cactaceae	0	0	0	0	3	5	3	5	0	0	0	0	10
Convolvulaceae	8	24	2	4	0	0	0	0	0	0	1	3	31
Cucurbitaceae	22	39	1	1	0	0	0	0	0	0	0	0	40
Dioscoreaceae	1	14	0	0	0	0	0	0	0	0	0	0	14
Fabaceae	15	28	10	14	0	0	0	0	0	0	0	0	42
Gesneriaceae	0	0	0	0	10	39	0	0	0	0	0	0	39
Gnetaceae	0	0	1	2	0	0	0	0	0	0	0	0	2
Loganiaceae	0	0	1	6	0	0	0	0	0	0	0	0	6
Loranthaceae	0	0	0	0	0	0	0	0	0	0	4	10	10
Malpighiaceae	0	0	9	13	0	0	0	0	0	0	0	0	13
Menispermaceae	0	0	18	23	0	0	0	0	0	0	0	0	23
Orchidaceae	0	0	0	0	51	89	0	0	0	0	0	0	89
Passifloraceae	2	13	0	0	0	0	0	0	0	0	0	0	13
Piperaceae	0	0	1	6	1	15	0	0	0	0	0	0	21
Rubiaceae	0	0	6	13	0	0	1	1	0	0	0	0	14
Sapindaceae	1	2	3	11	0	0	0	0	0	0	0	0	13
Smilacaceae	0	0	2	20	0	0	0	0	0	0	0	0	20
Vitaceae	0	0	7	16	0	0	0	0	0	0	0	0	16

Table IV Number of Useful Genera (G) and Species (S) of Minor Families of Mechanically Dependent Canopy Plants

Family	Vines		Lianas		Epiphytes		Climbers		Hemi-epiphytes		Parasites		Number of species used
	G	S	G	S	G	S	G	S	G	S	G	S	
Acanthaceae	1	2	1	2	0	0	0	0	0	0	0	0	4
Adiantaceae	0	0	0	0	1	1	0	0	0	0	0	0	1
Amaryllidaceae	1	4	0	0	0	0	0	0	0	0	0	0	4
Annonaceae	0	0	4	5	0	0	0	0	0	0	0	0	5
Aspleniaceae	0	0	0	0	1	2	1	1	0	0	0	0	3
Asteraceae	3	9	0	0	0	0	0	0	0	0	0	0	9
Basellaceae	1	2	1	4	0	0	0	0	0	0	0	0	6
Begoniaceae	0	0	0	0	1	6	0	0	0	0	0	0	6
Blechnaceae	0	0	0	0	1	1	0	0	0	0	0	0	1
Boraginaceae	0	0	1	3	0	0	0	0	0	0	0	0	3
Campanulaceae	0	0	0	0	1	1	0	0	0	0	0	0	1
Cecropiaceae	0	0	0	0	0	0	0	0	1	2	0	0	2
Celastraceae	0	0	3	5	0	0	0	0	0	0	0	0	5
Clusiaceae	0	0	0	0	0	0	0	0	2	6	0	0	6
Combretaceae	0	0	2	3	0	0	0	0	0	0	0	0	3
Connaraceae	0	0	3	5	0	0	0	0	0	0	0	0	5
Cyclanthaceae	0	0	0	0	4	6	0	0	0	0	0	0	6
Dilleniaceae	4	7	0	0	0	0	0	0	0	0	0	0	7
Ericaceae	0	0	0	0	7	7	0	0	0	0	0	0	7
Euphorbiaceae	0	0	3	4	0	0	0	0	0	0	0	0	4
Flagellariaceae	0	0	1	1	0	0	0	0	0	0	0	0	1
Hymenophyllaceae	0	0	0	0	1	1	0	0	0	0	0	0	1

Family																	
Icacinaceae	0	0	2	5	0	0	0	0	0	0	0	0	0	0	0	0	5
Lauraceae	0	0	0	0	0	1	1	0	0	0	0	0	0	0	1	1	1
Liliaceae	0	0	0	0	1	1	1	0	0	0	0	0	0	0	0	0	1
Marcgraviaceae	0	0	1	1	1	2	0	5	5	0	0	0	1	2	0	0	3
Melastomataceae	0	0	0	0	0	0	0	0	0	0	0	0	0	0	0	0	7
Monimiaceae	0	0	1	1	0	0	0	2	1	0	0	0	0	0	0	0	1
Moraceae	0	0	0	0	0	0	1	0	0	0	0	0	1	4	0	0	4
Myrsinaceae	0	0	0	0	1	2	0	0	0	0	0	0	0	0	0	0	2
Nepenthaceae	0	0	0	0	0	0	0	0	0	0	0	0	0	0	0	0	2
Nyctaginaceae	0	0	2	3	0	0	0	0	0	0	0	0	0	0	0	0	3
Oleaceae	1	1	0	0	0	1	1	0	0	0	0	0	0	0	0	0	1
Pandaceae	0	0	1	2	1	0	0	0	0	0	0	0	0	0	0	0	2
Phytolaccaceae	0	0	0	0	0	0	1	0	0	0	0	0	0	0	0	0	1
Polygonaceae	0	0	1	1	1	0	0	7	3	0	0	0	0	0	0	0	1
Polypodiaceae	0	0	0	0	0	0	0	0	0	0	0	0	0	0	0	0	7
Ranunculaceae	1	6	0	0	0	0	0	0	0	0	0	0	0	0	0	0	6
Rhamnaceae	0	0	3	7	0	0	1	0	0	0	0	0	0	0	0	0	7
Rosaceae	1	5	0	0	1	3	0	0	0	0	0	0	1	5	0	0	5
Schizaeaceae	0	0	0	0	0	0	0	0	0	0	0	0	0	0	0	0	3
Solanaceae	2	4	1	1	0	0	1	1	1	0	0	0	0	0	0	0	6
Tropaeolaceae	1	2	0	0	0	0	0	0	0	0	0	0	0	0	0	0	2
Ulmaceae	0	0	1	1	1	2	0	0	0	0	0	0	0	0	0	0	1
Urticaceae	0	0	0	0	0	0	0	0	0	0	0	0	0	0	0	0	2
Verbenaceae	0	0	2	3	0	0	0	0	0	0	0	0	0	0	0	0	3
Viscaceae	0	0	0	0	0	0	0	0	0	0	0	0	0	0	2	7	7

For example, a value of one means that one species has one use. A value of ten may mean that of two species, each has five uses, or that one species is employed in ten use categories. Five families rank high with this index: Araceae (103), Cucurbitaceae (98), Fabaceae (85), Orchidaceae (134), and Vitaceae (83).

B. Major Families of Host-Dependent Plants

1. Apocynaceae   One of the richest alkaloid-bearing families, Apocynaceae contains many tropical lianas (Schultes and Raffauf, 1990). *Strophanthus* is the most important host-dependent genus in the family. Nigerian natives treat gonorrhea, ulcers, and wounds with *S. gratus. Strophanthus sarmentosus,* another Nigerian medicinal species, provides an arrow poison and a source of cortisone (Phillips, 1991). Ouabain, a rapid-acting cardiac glycoside, is the main active principal in *S. gratus.* The related compound strophanthin comes from *S. kombé* (Morton, 1977a; Hensyl, 1990).

Allamanda cathartica—This common ornamental is used as an emetic and laxative in Suriname (Morton, 1981).

Fernaldia pandurata—Huastecs eat the leaves and stem as snack food and appetite stimulant. They also make medicines from the plant to treat constipation, edema, and vomiting (Alcorn, 1984).

Gonolobus barbatus—This provides a remedy for mouth sores (Morton, 1981). Huastecs eat the fruits and employ the latex of *G. niger* as a glue and an aid to removing thorns (Alcorn, 1984).

Landolphia spp.—Fruits are eaten in Tanzania. The latex is used as a cicatrizant and a rubber source in Africa and Madagascar (Uphof, 1968; Usher, 1974; Mabberley, 1987; Phillips, 1991).

Mandevilla steyermarkii—This is the source of an aphrodisiac. *Mandevilla vanheurckii* and *M. neriodes* provide antifungal medicines (Phillips, 1991). A venereal disease remedy is obtained from *M. subsagittata,* but the treatment causes female sterility (Morton, 1981).

Urceola and *Willughbeia* spp.—Some species in these genera bear edible fruits. They are also rubber sources in southeastern Asia (Uphof, 1968; Usher, 1974; Phillips, 1991).

2. Araceae   This large tropical family, characterized by acrid, milky latex containing calcium oxalate crystals, includes many epiphytic and climbing species. Many *Anthurium* and *Philodendron* species are employed by traditional cultures. The Shuar, for example, use at least seven *Anthurium* species. They eat the leaves of *A. acrobates* and *A. alienatum;* women scent their bodies with *A. apaporanum,* and they apply crushed *A. eminens* fruits to kill burrowing insect larvae in domesticated animals (Bennett, 1992b). Waorani also use this species medicinally (Davis and Yost, 1983). The Kubeo of Colombia blacken their teeth with *A. infectorium* during ceremonies. Other uses include fiber for guitar strings, baskets and lashings, and medicines for aphrodisiacs, contraceptives, and warts (Plowman, 1969; Bennett, 1992b).

The genus *Philodendron* is also employed widely. The Shuar cure dandruff with liquid from the petiole of *P. angustialatum.* They use the roots of this species and *P. bipinnatifidum* for cord (Bennett, 1992b). The Guaraní eat the fruits of the latter species and treat orchitis, rheumatism, and ulcers with the plant's caustic juice (Plowman, 1969). Natives of Colombia flavor chicha with *P. remifolium* leaves (Schultes and Raffauf, 1990). *Philodendron* provides basket material, fish poisons, pesticides, anthelmintics, and snake bite medicines (Plowman, 1969).

Heteropsis oblongifolia—Aerial roots are an important fiber source in northwestern Amazon (Figs. 1 and 2). (Davis and Yost, 1983; Bennett, 1992b).

Monstera deliciosa—The ripened spadix, which tastes like a mixture of pineapple and banana, is a popular tropical food. Considered a delicacy in

Figure 1 Jaimie Cerda, a Quichua man from Ecuador's Napo Province, collecting roots of an unidentified epiphyte (probably *Heteropsis oblongifolia,* Araceae, or perhaps a cyclanth). The roots are used to make baskets and to lash timbers in house construction.

Figure 2   House frame lashed together with roots of *Heteropis oblongifolia,* an epiphytic aroid.

Europe, it is used to flavor champagne. Baskets and a decoction to treat arthritis are made from the aerial roots (Plowman, 1969). *Monstera pertusa* also possesses an edible spadix. Juice from the crushed leaves is used as a vesicant and cauterant and also to treat snake bites. The stems are believed to ward off snakes. It contains the active principal monsterine, which supplies remedies for erysipelas, eczema, dandruff, hearing ailments, and ulcers (Plowman, 1969).

Syngonium podophyllum—Fruits are reported to be edible (Plowman, 1969). Fresh latex or pulverized ash is placed on wounds once a day until they heal. Latex also is applied as a hemostat, especially for machete cuts, but it irritates the skin (Bennett, 1992b). If a pregnant woman eats this plant, it is believed that her children will be promiscuous (Alcorn, 1984). *Syngonium podophyllum* also provides pig fodder, basket material, and medicines for dysentery and ant stings (Alcorn, 1984; Schultes and Raffauf, 1990).

3. Arecaceae   Palms are among the most important and versatile botanical resources in the world (e.g., Balick and Beck, 1990). Although most of its members are trees or shrubs, the family contains some lianas. Commonly called rattans, these lianas rank among the most economically valuable of all palms. The international furniture trade uses at least 25 species, mostly *Calamus* and *Daemonorops* (Dransfield, 1985). A single species, *C. manan,* dominates the furniture industry (Phillips, 1991). The value of rattan products may be as much as U.S. $4 billion per year (Myers, 1984).

Ancistrophyllum secundiflorum—This African palm is a source of rattan used for furniture and construction (Uphof, 1968; Usher, 1974; Phillips, 1991).

Calamus spp.—Several species of *Calamus* provide medicines used to treat malaria and to reduce menstrual and birth pain (Usher, 1974). *Calamus ornatus* var. *philippinensis* and *C. merrilli* have potential as fruit crops (Phillips, 1991). *Calamus scipionum* is the source of cane for walking sticks and umbrella handles (Mabberley, 1987).

Daemonorops spp.—Stems and young leaves are edible. The genus also has medicinal merit as sedatives and tonics (Phillips, 1991), and is a varnish source in southeastern Asia (Uphof, 1968; Usher, 1974; Mabberley, 1987).

Desmoncus spp.—These produce edible fruits and many are important fiber sources in the Amazon (Phillips, 1991). A furniture industry based on Neotropical palms of this genus may one day rival the importance of Asian rattans.

4. Aristolochiaceae Members of this family are famed for their therapeutic value in both temperate and tropical zones. The name *Aristolochia* is derived from a Greek word meaning "best birth," in reference to its parturifacient properties. *Aristolochia serpentaria* and many other species provide snake bite remedies (Mabberley, 1987). The Shuar treat kidney or back pain with *Aristolochia* (Bennett, 1992b). Other medicinal uses include treatment of chills, colds, cough, dysentery, fever, menstrual bleeding, and also as an abortifacient (Morton, 1981; Alcorn, 1984; Phillips, 1991). *Aristolochia petersiana* provides an arrow poison (Mabberley, 1987).

5. Asclepiadaceae This taxon is well represented in the world's pharmacopeia. The terrestrial genus *Asclepias* is named for the Greek god of medicine, Aesculapius. *Cryptostegia grandiflora,* a rubber source, is comparable in quality to *Hevea* (Mabberley, 1987).

Dischidia spp.—Species of this ant-inhabited epiphyte furnish remedies for boils, eczema, fish stings, goiter, gonorrhea, and herpes in Indonesia and the Philippines (Phillips, 1991).

Gymnema sylvestre—A compound found in this plant blocks sweet taste sensation and is used to treat diabetes (Usher, 1974; Mabberley, 1987).

Hoya spp.—Indonesians obtain remedies for asthma, coughs, digestive stimulants, gonorrhea, and poison fish stings from this genus (Usher, 1974; Phillips, 1991).

Periploca gracea—This supplies local medicines; other species are rubber sources (Mabberley, 1987; Phillips, 1991).

6. Bignoniaceae   This family contains many common lianas in the lowland tropics. Nearly every genus has a reported use (Phillips, 1991). *Arrabidaea chica,* one of the most widely employed species, provides an anti-inflammatory agent, tonic, and tooth-coloring agent (Phillips, 1991). The Shuar, Cofan, and Secoya–Siona extract a dye from the plant (Pinkley, 1973; Vickers and Plowman, 1984; Bennett, 1992b). Shuar women drink a medicine made from *Tynnanthus polyanthus* stems to change the sex of an unborn child (Bennett, 1992b).

Distictella pulverulenta—Coca chewers add an alkaline substance to their coca quid to liberate the plant's alkaloids. Many substances can be added, including the ashes of *D. pulverulenta. Distictella racemosa* is a curare ingredient (Schultes and Raffauf, 1990).

Macfadyeana uncata—The Shuar make a trap from this vine to capture vampire bats (Bennett, 1992b). *Macfadyeana* spp. contribute medicines to treat conjunctivitis and heavy menstrual bleeding (Phillips, 1991).

Mansoa standleyi—A medicine to treat bronchitis and coughs is made from the leaves (Bennett, 1992b).

Mussatia hyacinthina—Leaves are mixed with coca and also furnish a stimulant and medicine in Peru and Bolivia (Plowman, 1980).

Pachyptera alliacea—Stems smell strongly of garlic and are widely used as medicines in northwestern Amazonia. In Mexico, the plant supplies a condiment and medicines to treat chills, rabies, tumors, and spleen ailments (Alcorn, 1984). *Pachyptera hymenaea* provides an antipyretic medicine in Panama (Morton, 1981).

Tanaecium nocturnum—Native Brazilians make a hallucinogen from *T. nocturnum* (Mabberley, 1987).

7. Bromeliaceae   Nearly half the members of this large Neotropical family are epiphytic. *Tillandsia usneoides,* the most widely distributed bromeliad, has a myriad of medicinal applications, including treatments of coughs, fever, hemorrhoids, hernias, measles, mouth sores, rheumatic arthritis, external sores, and ailments of the lung, liver, kidney, and heart (Burlage, 1968; Soukup, 1970; Chávez Velásquez, 1977; Núñez Meléndez, 1982; Alcorn, 1984; Moerman, 1986; Bennett, 1995). Highland Quechua of Peru treat dandruff with a hair rinse made from this plant (Bennett, 1990). Huastec women drink a decoction made from a handful of *T. usneoides* as a contraceptive, but consider this practice to be dangerous (Alcorn, 1984). The plant was once an important low-grade fiber source in the southeastern United States with an annual production of 5000 tons (Bennett, 1990). *Tillandsia usneoides* serves as a Christmas ornament throughout Latin America, often in nativity scenes, where it forms the bed for the Christ figure. Tzeltal speakers hang the plant on doorways during celebrations (Berlin *et al.,* 1974), and the plant is employed in floral arrangements in North America. In the United States, *T. usneoides* is a quintessential symbol of the South.

Aechmea zebrina—The lowland Quichua of Ecuador use this epiphyte as medicine (Alarcón, 1988).

Catopsis spp.—Tzeltal speakers of Mexico decorate altars with several *Catopsis* species (Berlin *et al.,* 1974).

Guzmania spp.—The Shuar hunt monkeys where *Guzmania* spp. abound. Arboreal primates feed on *Guzmania* inflorescences. Ecuadorian Quichua make an aphrodisiac tea from the yellow flowers of *G. melinonis* (Bennett, 1992b, 1995).

Streptocalyx longifolius—Quichua occasionally eat the fruits of this species (Bennett, 1995).

Tillandsia benthamiana—This is used to cure anemia or kidney troubles (von Reis Altschul, 1973). *Tillandsia complanata* leaves are used to wrap tamales (von Reis Altschul, 1973). Highland Quechua of Peru decorate wedding ceremonies with *T. ionochroma,* which they call huicunto. The Peruvian Quechua adorn funeral ceremonies with *T. sphaerocephala,* which they call ayahuicunto (death huicuntu) (Bennett, 1990, 1995). Except for *T. usneoides, T. recurvata* is the most widely employed *Tillandsia* species. Its medicinal uses include treatment of hemorrhoids, gallbladder afflictions, leukorrhea, and menstrual irregularities (Morton, 1981; Núñez Meléndez, 1982). *Tillandsia recurvata* also provides edible shoot apices and treatments for coughs, fever, headache, and chest pain (Bennett, 1995).

Vriesea spp.—These are used as church ornamentals in Mexico (Berlin *et al.,* 1974).

8. Cactaceae   Although largely associated with xeric, terrestrial environments, this Neotropical family includes some epiphytes and climbers. The terrestrial taxa include several hallucinogenic species but none of the epiphytes is known to contain psychoactive principals. *Hylocereus* contains at least eight species with edible fruits, for example, *H. polyrhizus.* The leaves of this plant are placed on burns and tumors (Bennett, 1992b). *Hylocereus unundatus* also has edible fruits and an edible stem and provides a medicine for mouth sore erysipelas and childbirth (Williams, 1981).

Acanthocereus pitijaya and *A. pentagonus*—These bear edible fruits. The former is also used as a diuretic (Duke, 1986; Mabberley, 1987).

Epiphyllum phyllanthus var. *phyllanthus*—Fruits are edible (Alarcón, 1988; Bennett, 1992b) and the plant provides a cardiac tonic in Colombia (Morton, 1981). *Epiphyllum strictum* fruits are edible (Williams, 1981).

Pereskia aculeata—This is cultivated for its edible fruit (Usher, 1974; Mabberley, 1987).

Rhipsalis baccifera—Fruits are edible and the plant supplies medicines for dizziness, fever, headache, heartache, kidney trouble, nervousness, and rib pain (Alcorn, 1984).

Selenicereus grandiflorus—This is cultivated in Mexico for an antirheumatic drug and is used in Costa Rica as a heart stimulant (Mabberley, 1987). *Selenicereus* cf. *spinulosus* fruits are eaten in Mexico and the flowers are used medicinally for headache, pain, and obstetrics and to defend against bewitching shamans (Alcorn, 1984).

9. Convolvulaceae This predominantly tropical family includes *Ipomoea,* a terrestrial or climbing vine. The sweet potato (*I. batatas*) is one of the world's major foods. Medicinal uses of sweet potato include treatment of trauma from rabid dogs (Alcorn, 1984) and dysentery and diabetes (Phillips, 1991). *Ipomoea alba* is used to treat dandruff, gastrointestinal ailments, hair loss, and hiccups (Alcorn, 1984). The sap coagulates *Castilla* latex and the stems are used for fibers, and the young leaves are edible (Williams, 1981). *Ipomoea indica* provides pig fodder (Alcorn, 1984) and a purgative (Morton, 1981); *I. nil* has been used as an abortifacient, anthelmintic, insecticide, and purgative (Morton, 1981; Mabberley, 1987; Phillips, 1991); *I. orizabensis,* known as scammony root, is the source of a strong purgative (Mabberley, 1987); and *I. violacea,* commonly called heavenly blue, contains ergot alkaloids that have been employed as hallucinogens (Mabberley, 1987; Phillips, 1991).

Convolvulus floridus—Roots are the source of an essential oil (Mabberley, 1987). *Convolvulus scammonia* provides a drastic purgative in southwestern Asia (Mabberley, 1987).

Cuscuta americana—An infusion made from this species is used to treat jaundice and is also a purgative (Morton, 1981). Other uses include a love charm, diuretic, and laxative (Honychurch, 1980). *Cuscuta corymbosa* can be used to treat facial eruptions (Morton, 1981) and *C. tinctoria* is a dye source (Kuijt, 1969). Species of this parasitic climbing genus are used to make a shampoo and a purgative (Phillips, 1991).

Maripa spp—Fruits are eaten in French Guiana (Phillips, 1991).

Merremia dissecta—Cubans obtain a sedative from this plant (Morton, 1981). *Merremia quinquefolia* leaves are eaten and used medicinally (Alcorn, 1984), and *M. umbellata* provides cures for earache and mouth sores (Alcorn, 1984). Tubers and leaves of some *Merremia* species are edible and have value as anthelmintic, diuretics, purgatives, and snake bite cures (Phillips, 1991).

Turbina corymbosa—This provides a parturifacient and contraceptive and medicines to treat rheumatic pains (Morton, 1981). Its well-known hallucinogenic use is due to ergoline and lysergic acid amides (Mabberley, 1987).

10. Cucurbitaceae This family includes ancient domesticates and many important food species (squashes and melons). Secondary compounds abound, including the eponymous cucurbitacins. Among the most significant of the nondomesticated species is *Fevillea cordifolia*. Its seeds are burned as candle substitutes throughout the Amazon (Bennett, 1992b). The seeds also are used as a poison antidote and to treat colic, dermatitis, and leprosy (Morton, 1981).

Apodanthera spp.—These furnish an antisyphilitic medicine in Brazil (Phillips, 1991). *Apodanthera undulata* is a minor oil seed crop (Mabberley, 1987).

Cayaponia attenuata—Stems are used for scrubbing dirt and soap substitutes in El Salvador (Williams, 1981). The Cofan, Quichua, and Waorani of Ecuador eat *C. ruizii* fruits (Pinkley, 1973; Davis and Yost, 1983; Alarcón, 1988). *Cayaponia glandulosa* leaves are used as an insect repellent (Mabberley, 1987).

Gurania spp.—Macerated leaves are applied to snake bites (Bennett, 1992b) and provide medicines for infected cuts, parasites, boils, headaches, and diarrhea (Schultes and Raffauf, 1990). A decoction made from *G. spinulosa* is used to treat constipation (Morton, 1981).

Hodgsonia heteroclita—The inside of the seeds is eaten and opium is roasted in oil obtained from the seed (Usher, 1974; Mabberley, 1987).

Melothria pendula—Fruits are edible (Bennett, 1992b) and the plant furnishes medicines to treat vomiting, diarrhea, gonorrhea, and inflammation (Morton, 1981; Alcorn, 1984).

Momordica charantia—This climbing vine provides one of most popular folk remedies in the Caribbean. Its uses include the treatment of rheumatism, postpartum cleansing, blood tonic, and abortifacient (Morton, 1981). An extract from the plant has an insulinlike effect (Mabberley, 1987).

11. Dioscoreaceae This family of tropical vines includes many edible species and produces globally significant medicines. Many species of *Dioscorea* contain steroidal saponins. Diosgenin, derived from the genus, is a precursor to progesterone and cortisone. *Dioscorea composita* and *D. floribunda* are commercial sources of diosgenin (Phillips, 1991). Its annual retail sales exceed $1 billion (Mabberley, 1987). *Dioscorea alata* has edible tubers and is the source of a medicine for urinary ailments (Alcorn, 1984); *D. batatas, D. bulbifera, D. cayenensis, D. esculenta, D. macrostachys, D. matagalpensis,* and *D. trifida* are also edible (Alcorn, 1984; Phillips, 1991; Bennett, 1992b). Other medicinal uses of the genus include treatment of kidney trouble, syphilitic sores, arthritic and rheumatic pain, and insanity.

12. Fabaceae (sensu latu) The legumes are second in importance only to the grasses in terms of human use. Though shrubs, trees, and herbs predominate, the family also contains many vines and lianas. Alkaloids and other toxic secondary compounds abound within the family. Abrin from *Abrus precatorius* is one of the most toxic substances known. The plant is used to treat asthma, cough, conjunctivitis, fever, and sore throats and also provides a contraceptive and abortifacient in India (Morton, 1981; Mabberley, 1987). The seeds are used as beads and weights (Mabberley, 1987).

Bauhinia tarapotensis—The Shuar treat wounds with a decoction made from this species (Bennett *et al.,* 1995). Climbing species of *Bauhinia* provide cures for dysentery, fever, parasites, and venereal disease. Some are used as astringents, expectorants, and fish poisons (Phillips, 1991).

Caesalpinia bonduc—A medicine made from the toasted seeds is employed to treat vomiting. Dry stems are used to douse for water (Alcorn, 1984). Other medicinal products obtained from the plant include a diuretic, cardiac edema remedy, and quinine substitute in Europe. The seeds are also used as beads (Morton, 1981; Mabberley, 1987).

Derris spp.—Many species are widely employed as fish poisons. *Derris elliptica* is a commercial source of rotenone (Mabberley, 1987).

Entada spp.—These provide medicines for snake bite and hair treatments and are also used as a soap substitute (Phillips, 1991). *Entada gigas* has edible seeds and is eaten as a vegetable. Its stems provide fibers for nets and a saponin source (Mabberley, 1987). Children make tops and beads from the seeds (B. C. Bennett, unpublished data).

Lonchocarpus spp.—These are common sources of rotenone-based fish poisons. They also provide medicines to treat jaundice and venereal disease and an indigo dye (Phillips, 1991). *Lonchocarpus nicou* roots supply a fish poison in South America; the bark is toxic. Chickens are washed in a solution made from the roots to remove fleas. Shuar shamans cleanse the bodies of sick people or animals with *L. nicou* (Bennett, 1992b,c). The plant is cultivated in Amazonia as a commercial rotenone source. *Lonchocarpus rariflorus* is used to kill leaf-cutter ants (Phillips, 1991).

Machaerium riparium—Bark supplies toothache remedies (Alcorn, 1984). *Machaerium lunatum* provides a hemostat, venereal disease cure, and aphrodisiac in Surinam (Morton, 1981).

Mucuna argophylla—This is the source of medicines for nosebleed, headache, vomiting, and whooping cough (Alcorn, 1984). *Mucuna pruriens* supplies anti-inflammatory, anthelmintic, and diuretic remedies, and, in the southeastern United States, animal fodder (Mabberley, 1987). Other *Mucuna* species provide twine, food (seeds), medicines for hemorrhoids and intestinal parasites, black dye, and a source of L-dopa used in treating Parkinson's disease (Williams, 1981; Phillips, 1991).

Physostigma venenosum—This supplies an ordeal poison, administered to test one's guilt (those who survive are deemed innocent), in Nigeria. *Physostigma venenosum* is the source of physostigmine, an alkaloid antidote for tubocurarine, a derivative of another liana genus (see *Chondrodendron*). It also furnishes anti-inflammatory, antirheumatic, sedative, and ophthalmological medicines (Morton, 1977a; Mabberley, 1987).

Pueraria spp.—These provide a fodder and cover crop. *Pueraria lobata,* the infamous kudzu, is also a starch source (Mabberley, 1987; Phillips, 1991).

13. Gesneriaceae   Gesneriads comprise a large family of tropical herbs, shrubs, and many epiphytes. These common members of moist tropical forests have ubiquitous therapeutic value. *Dalbergaria* (often included in *Columnea*) is one of the most important medicinal genera. Many species have red-tipped leaves, suggesting their gynecological usage; the sanguine foliage, according to those who believe the Doctrine of Signatures, indicates the plant's utility in treating menstrual bleeding. Most species are employed to treat snake bites, including *D. asteroloma, D. consanguinea, D. crassa, D. difficilis, D. eburnea, D. ericae, D. medicinalis, D. picata,* and *D. villosissima.* The Chachi of western Ecuador use every *Dalbergaria* species that researchers have collected. They apply heated leaves to stop blood flow from snake bites and then drink a leaf decoction as an antidote (B. C. Bennett, unpublished data). Many species provide treatments for eczema, hemorrhages, headaches, heart problems, boils, and abnormal menstrual flow. The Shuar make a permanent contraceptive from one species (Bennett, 1992b). Others are employed as abortifacients, flea killers, and crab poisons (H. Wiehler, unpublished data).

Codonanthe calcarata—The Waika of the Venezuelan Orinoco heal wounds with this plant (H. Wiehler, unpublished data). *Codonanthe crassifolia* supplies medicines to treat colds and whooping cough (Morton, 1981). The Tikuna of Amazonian Colombia treat infections with *C. uleana* (H. Wiehler, unpublished data).

Columnea bilabiata—The Chachi heal coral and bushmaster snake bites with *C. bilabiata.* The Waunana of Colombia use the plant to relieve stomach pains (H. Wiehler, unpublished data).

Drymonia alloplectoides—This is an important Chachi snake bite medicine. *Drymonia coccinea* supplies a Maina anodyne for children (H. Wiehler, unpublished data), and *D. coriacea* provides a Siona mouthwash to treat toothaches and mouth ulcers. Chachi eat the sepals like candy. The Tsatchila (Colorado) treat snake bites with *D. macrophylla,* and *D. umecta* inflorescences are applied to a woman's nipple to increase milk flow (H. Wiehler, unpublished data).

Pentadenia ecuadorana—The Tsatchila treat snake bite with this species. *Pentadenia spathulata* is a Chachi snake bite medicine (H. Wiehler, unpublished data).

14. Gnetaceae   This group of primitive plants, related to *Ephedra,* includes *Gnetum,* a genus of lianas and climbers. At least 13 species of *Gnetum* possess edible seeds (Phillips, 1991). The young leaves and seeds of the cultivated *G. gnemon* are edible and the plant is a source of fiber (Mabberley, 1987). A seed decoction of *G. nodiflorum* is used to reduce swelling (Schultes and Raffauf, 1990).

15. Loganiaceae Neotropical members of the genus *Strychnos* are well-known curare sources. At least 19 species of the genus provide the arrow poison (Phillips, 1991). The Cofan use *S. darienensis, S. erichsonii, S. jobertiana, S. peckii,* and *S. toxifera* as hunting poisons (Pinkley, 1973). The Shuar employ *S. tomentosa* for the same purpose (Bennett, 1992b).

16. Loranthaceae and Viscaceae These widespread families of parasitic plants occur in both tropical and temperate zones. *Phoradendron serotinum,* often placed in Viscaceae, is the commercial source of mistletoe, a common Christmas ornamental in North America. *Viscum album* is the well-known mistletoe of Europe; its fruits are used as birdlime and medicine (Kuijt, 1969). The lowland Quichua treat broken bones with *P. crassifolium* (Alarcón, 1988) and crushed leaves are applied to wounds (Schultes and Raffauf, 1990). *Phoradendron piperoides* is employed as an antispasmodic (Morton, 1981) and to alleviate anemia (Schultes and Raffauf, 1990). Other species provide treatments for bruises, dizziness, headache, fractures, and a parturifacient (Morton, 1981; Alcorn, 1984; Mabberley, 1987).

Oryctanthus florilentus—This is used by the Ecuadorian Quichua for an undisclosed medicine (Alarcón, 1988). *Oryctanthus occidentalis* provides a remedy for high blood pressure in Jamaica (Morton, 1981).

Phthirusa pyrifolia—Macerated fruits are employed in an enema to treat diarrhea (Bennett, 1992b) and potentially could be a commercial rubber source (Usher, 1974). Bahamians treat colds with a tea made from *P. pauciflora* (Morton, 1981).

Psittacanthus cucullaris—A leaf decoction keeps skin free from wrinkles (Schultes and Raffauf, 1990); *P. schiedeanus* aids wound healing (Alcorn, 1984).

Struthanthus crassipes—This is the source of an ulcer medicine (Alcorn, 1984). *Struthanthus flexilis* macerated fruits are used in an enema to treat diarrhea (Bennett, 1992b); *S.* aff. *quercicola* supplies medicines for headache, dizziness, and heartache (Alcorn, 1984).

17. Malpighiaceae Vines from this family provide the principal hallucinogens of northwestern Amazonia and are among the region's most important plants for traditional cultures. *Banisteriopsis caapi,* known by many common names, including ayahuasca, caapi, natem, and yaje, is the most commonly used species. The Shuar preparation is typical. The stem is peeled, split, broken into small pieces, mixed with *Diplopterys cabrerana* and *Herrania* sp. fruit husk, and then placed in a receptacle, where it is allowed to ferment and turn red. Shamans drink the liquid (only at night) in order to converse with the spirit world and to diagnose illnesses. Although at least 20 additional species are added to ayahuasca, two are common components: *D. cabrerana,* employed by the Shuar, and *Psychotria viridis,* used by the Quichua [see Schultes and Raffauf (1990) and Bennett (1992c) for more detailed discussion of hallucinogens]. The Waorani of Ecuador use the related *B. muricata* similarly (Davis and Yost, 1983).

Diplopterys spp.—These are common additives to hallucinogenic beverages, especially *D. cabrerana* leaves, which are mixed with *Banisteriopsis caapi* (Schultes and Raffauf, 1990; Bennett, 1992c).

Heteropteris spp.—These provide medicines to treat gonorrhea and diarrhea and also are fiber sources (Mabberley, 1987; Phillips, 1991). *Heteropteris beecheyana* stems are used to make baskets (Alcorn, 1984).

Mascagnia castanea—This is a curare ingredient (Schultes and Raffauf, 1990).

Stigmaphyllon cordifolium—A tea made from this plant is used to treat flu and irregular menstruation (Honychurch, 1980). *Stimaphyllon sagraeanum* supplies cures for dandruff and gonorrhea (Morton, 1981); *Stigmaphyllon* spp. suppress vomiting, relieve toothaches, and are used in sorcery (Schultes and Raffauf, 1990; Phillips, 1991).

18. Menispermaceae   Lianas in this family, especially *Chondrodendron tomentosum,* are the preferred source of curare in northwestern Amazonia and provide tubocurarine, a muscle relaxant in modern surgery. Medicines made from the plant are used as antipyretics, diuretics, emmenagogues, and snake bite antidotes (Morton, 1977a; Hensyl, 1990; Schultes and Raffauf, 1990).

Abuta rufescens—This supplies a Cofan arrow poison (Pinkley, 1973) and urogenital medicines (Mabberley, 1987). *Abuta obovata* is considered one of the strongest curares (Schultes and Raffauf, 1990).

Anamirta cocculus—This supplies remedies for malaria and ringworm. It also is the source of picrotoxin, a central nervous system stimulant, an antidote for barbiturate poisoning, and a fish poison (Morton, 1977a; Mabberley, 1987; Hensyl, 1990). *Anamirta* spp., which have edible fruits, provide medicines to treat colds and malaria and to relieve toothaches (Phillips, 1991).

Anomospermum chloranthum—This is the source of a Siona arrow poison (Schultes and Raffauf, 1990).

Cissampelos grandifolia—This provides remedies for snake bite, malaria, and fever (Morton, 1981). *Cissampelos pareira* supplies medicines to treat edema, fever, fright, jaundice, pain, and snake bites. The plant also has edible fruits and is used for spiritual cleansing (Morton, 1981; Williams, 1981; Alcorn, 1984). *Cissampelos* spp. supply fish bait and abortifacient, antipyretic, astringent, colic, diarrhea, diuretic, ophthalmological, snake bite, and venereal disease medicines (Phillips, 1991).

Cocculus spp.—These supply treatments for snake and insect bites, fever, and a remedy to promote menstruation. Some species have edible fruits (Mabberley, 1987; Phillips, 1991).

Curarea tecunarum—This provides an oral contraceptive in Brazil and a common arrow poison (Mabberley, 1987; Phillips, 1991).

Dioscreophyllum cumminsii—This is the source of monellin, an artificial sweetener that is 9000 times as sweet as sugar (Mabberley, 1987).

Sphenocentrum jollyanum—Roots are used as chewing sticks, rendering food sweet afterward (Mabberley, 1987). *Sphenocentrum* spp. supply an aphrodisiac and medicines to treat coughs and wounds (Phillips, 1991).

Tinospora crispa—This is used to flavor cocktails (Phillips, 1991). *Tinospora* spp. supplies antipyretic, tetanus, and jaundice medicines.

19. Orchidaceae   Orchids constitute the largest family of flowering plants. Their name is derived from *orchis,* the Greek word for "testicle." Some species resemble male reproductive organs and much sexual lore is associated with the family. According to myth, orchids were the preferred foods of satyrs, who had an unquenchable sexual appetite. The generic name *Satyrium* and the term satyriasis (male equivalent of nymphomania) help preserve this lore (Lawlor, 1984; Hensyl, 1990). Except for ornamentals and vanilla, orchids are not considered to be of great human importance. This misconception is due in large part to the inaccessibility of orchids and their difficult taxonomy. However, studies have identified many beneficial orchid species. Lawlor (1984) compiled an impressive body of data on the uses of orchids. Nonetheless, scores of other reports mention the importance of orchids and much data awaits synthesis.

Orchids have long been known for their beautiful flowers and unusual morphology, but the family's most important product is *Vanilla,* cultivated since pre-Columbian times. The Mayans and Aztecs used the fruits as a perfume and to flavor chocolate. Today, only *V. planifolia, V. pomona,* and *V. taitensis* are of commercial importance (Lawlor, 1984). Vanillin, the compound responsible for the plant's culinary properties, is also added to medicines to treat bites from poisonous animals, convulsions, hysteria, impotence, parasites, rheumatism, and sterility. Its use as an aphrodisiac is also well known (Lawlor, 1984). Vanilla export generates between $60 and $80 million in foreign exchange for countries producing it (Smith *et al.,* 1992). The Shuar use the fruits of an undescribed *Vanilla* species as a perfume and to flavor sugarcane alcohol (Bennett, 1992b). Several species provide fibers, including *V. ovalis* (basket material in the Philippines), *V. grandiflora* (ropes for fishing nets in Gabon), and *V. crenualta* (guitar strings) (Lawlor, 1984).

Acampe pachyglossa—This provides an antimalarial medicine (Lawlor, 1984).

Ansellia gigantea and *A. humilis*—The Zulu use the roots to make unmarried girls temporarily sterile. These species also provide antimalarials, emetics, and aphrodisiacs (Usher, 1974; Lawlor, 1984).

Campylocentrum sp.—Juice from the pseudobulb is applied to skin ulcers (Bennett, 1992b).

Catasetum spp.—The sap from many species is employed as a glue (e.g., *C. cristatum* and *C. luridum;* Lawlor, 1984). The Shuar eat the stem of one species to reduce heart pain (Bennett, 1992b).

Cattleya skinneri—Guatemalans call the plant *flor de San Sebastían* and decorate churches with it on this saint's days. *Cattleya skinneri* is the national flower of Costa Rica, and *C. trianaei* is the national flower of Colombia (Lawlor, 1984).

Cyrtopodium punctatum—Sap from the stem serves as a glue substitute and a medicine made from the plant is believed to prevent baldness (Lawlor, 1984).

Dendrobium spp.—These are among the more widely used orchid species. In Papua New Guinea, the name *duruagle* is given both to a *Dendrobium* orchid and to a whore. According to legend, fair-skinned female ghosts stroll naked along the river at night, seducing young men. Failure to satisfy the female apparitions results in painful disease. *Dendrobium* stems are eaten in Australia and many species are employed for wicker and hat bindings. Dendrobine, from *D. nobile,* provokes uterine contractions and lowers blood pressure. *Dendrobium acinaciforme* is considered a love charm and good luck charm for hunters; *D. bigibbum,* the state flower of Queensland, also produces a yellow dye. Sri Lankans present *D. macartiae* to Buddha as a temple offering. In Indochina, *D. pulchellum* flowers are fed to dogs to make them better hunters. The Shuar and other Amazonian groups give hallucinogens to their dogs for the same purpose (Bennett *et al.,* 1994). *Dendrobium salaccense* flavors rice in Indomalaysia (Lawlor, 1984).

Epidendrum bifidum—This provides purgatives and anthelmintics. Mexican Huastecs treat dizziness and heartache with *E. difforme* (Alcorn, 1984); *E. radicans* is called *flor de Jesus; E. pastoris* supplies a mordant for painting, and diarrhea and dysentery treatments in Mexico (Lawlor, 1984); *E. atropurpureum* is used to treat coronary difficulties in Venezuela (Morton, 1981).

Lycaste virginalis—This is the national flower of Guatemala (Lawlor, 1984).

Oncidium cebolleta—Leaves are used to treat fractures and are employed as a peyote substitute in Mexico. This species may be hallucinogenic (Lawlor, 1984). *Oncidium carthagenense* furnishes medicines to treat headaches and dizziness (Alcorn, 1984).

Peristeria elata—Called *flor de muertos,* this is the national flower of Panama (Lawlor, 1984).

Phreatia spp.—Leaves are used for dress in Papua New Guinea (Lawlor, 1984).

Psygmorchis pusilla—The Cofan treat lacerations with this species (Pinkley, 1973).

Schomburgkia crispa—This supplies a medicine for the lowland Quichua

(Marles, 1988). *Schomburgkia tibicinis* hollow stems are used as musical instruments (Lawlor, 1984), and *S. thomsoniana* pseudostems provide pipe bowls and stems (Lawlor, 1984; Mabberley, 1987).

Sobralia spp.—These were used to decorate altars for human sacrifice. *Sobralia dichotoma* is known as *flor de paraíso* (Lawlor, 1984).

Tetramicra bicolor—This vanilla substitute is employed to flavor milk, sherbet, and ice cream (Lawlor, 1984).

20. Passifloraceae   This family of tropical and temperate vines contains several species cultivated for their edible fruits. Secondary compounds abound in Passifloraceae. The genus *Passiflora* contains at least 56 edible species, including *P. edulis, P. foetida, P. laurifolia, P. ligularis, P. mollisima, P. riparia, P. serratifolia,* and *P. vitifolia* (Pinkley, 1973; Alcorn, 1984; Alarcón, 1988; Phillips, 1991; Bennett, 1992b). Medicinal uses include treatment of broken bones, cuts, diabetes, earache, eye inflammation, fever, gastrointestinal pain, gonorrhea, headaches, hysteria, inflammation, postpartum problems, sores, vomiting, snake bites, and wounds (Morton, 1981; Alcorn, 1984; Alarcón, 1988; Schultes and Raffauf, 1990; Phillips, 1991).

Adenia spp.—These supply a fish poison and medicines for conjunctivitis, headache, lumbago, ringworm, snake bite, and stomach trouble (Phillips, 1991).

21. Piperaceae   Two large genera, *Peperomia* and *Piper,* dominate this tropical family of epiphytic and terrestrial herbs and shrubs. *Piper betel* is a popular masticatory used with the betel-nut palm *Areca catechu* of southeastern Asia. *Piper nigrum,* black pepper, is the world's most important spice. Native to India and Malaysia, black pepper vines are cultivated throughout the moist tropics. Long pepper comes from *P. chaca, P. longum,* and *P. retrofactum. Piper cubea* is used to flavor cigarettes, food, bitters, and throat lozenges. Other useful vines within *Piper* provide flavorings, medicines, stimulants, antipyretics, tonics, irritants, and abortifacients (Phillips, 1991).

Peperomia emarginella—This supplies medicines to treat colds, influenza, fever, and asthma in Venezuela (Morton, 1981). *Peperomia glabella* relieves swelling (Alcorn, 1984) and is used as a diuretic and treatment for conjunctivitis (Morton, 1981); *P. maculosa* leaves are bound to the head to cure headaches and the heated leaves are applied to swellings (Morton, 1981). Tea made from *P. rotundifolia* is used to treat colds in Dominica (Morton, 1981). The Shuar treat headaches with *P. urocarpa.* It also may be the source of a mild hallucinogen (Bennett, 1992b).

22. Rubiaceae   This largely tropical family of shrubs and trees also contains a few epiphytes and vines. *Manettia* spp. provide tooth-blackening agents and antipyretic medicines (Phillips, 1991). The Shuar relieve toothaches with *M. glandulosa; M. cordifolia* is an ipecac adulterant (Mabberley, 1987).

Chiococca alba—Fruits are eaten as a snack food and a root tonic made from the plant is used to treat joint aches. Huastecs use the stem for arrow shafts and loom parts (Alcorn, 1984). *Chiococca alba* also supplies diuretics, purgatives, and snake bite medicines (Mabberley, 1987; Schultes and Raffauf, 1990).

Paederia foetida—This is a commonly used medicine in India. *Paederia scandens* is used as a vegetable (Mabberley, 1987). *Paederia* spp. supply medicines to treat dysentery, eye infections, flatulence, gout, herpes, inflammation, malaria, ringworm, and toothache (Phillips, 1991).

Sabicea amazonensis—This is sometimes added to hallucinogenic ayahuasca mixtures. *Sabiacea villosa* provides an antimalarial medicine (Schultes and Raffauf, 1990).

Uncaria gambir—This supplies a tanning agent, printing dye, and clarifying agent for beer. It also is mixed with betel quids and is used to treat diarrhea (Phillips, 1991). *Uncaria africana* provides medicines to treat coughs, stomach pains, and syphilis (Phillips, 1991); *U. guianensis* is employed to treat dysentery (Schultes and Raffauf, 1990).

23. Sapindaceae   This family of trees, shrubs, and lianas commonly contains saponins. *Paullinia cupana* is the source of Brazil's famous beverage, guaraná. Prepared as a tonic or carbonated beverage, guaraná has several times the caffeine content of coffee. Preparations from the plant are also used as aphrodisiacs and analgesics (Phillips, 1991). The related *P. yoco* is a widely used stimulant in the northern Amazon. Many indigenous groups drink a cold-water infusion made from the rasped bark for a stimulant (Schultes and Raffauf, 1990). *Paullinia fuscescens* and *P. pinnata* are fish poisons (Williams, 1981). Medicinal uses of *Paullinia* spp. include antipyretics, contraceptives, emetics, and hemostats, and treatment of dysentery, snake bite, and toothache (Alcorn, 1984; Schultes and Raffauf, 1990; Phillips, 1991).

Cardiospermum halicacabum—Seeds are used to make necklaces (Bennett, 1992b). Medicinal uses include treatment of edema, ulcers, and wounds and the plant has diuretic properties (Morton, 1981). Wearing *C. halicacabum* beads is believed to ward off snakes (Schultes and Raffauf, 1990); *C. grandiflorum* seeds also are used for beads and the leaves are eaten (Mabberley, 1987).

Serjania inflata—This supplies a Quichua medicine in lowland Ecuador (Marles, 1988). *Serjania rubicaulis* stems provide a cord substitute (Bennett, 1992b); *S. mexicana* stems are chewed to relieve toothache (Morton, 1981). Seven species of *Serjania* spp. supply fish poisons and medicines for blood purification, malaria, postpartum problems, and vaginal washes. The plant stems supply fiber for baskets and cordage (Mabberley, 1987; Phillips, 1991).

24. Smilacaceae   This family of tropical and a few temperate vines is closely related to Liliaceae. It is best known as the source of sarsaparilla (*Smilax aristolochifolia, S. ornata, S. regelii,* and *S. spruceana;* Phillips, 1991). *Smilax aristolochifolia* shoots are eaten, the stems furnish basket material (Alcorn, 1984), and a venereal disease medicine is made from the plant (Phillips, 1991); *S. domingensis* is also used to treat venereal disease (Morton, 1981); *S. bona-nox, S. havanensis, S. macrophylla,* and *S. megacarpa* possess edible rhizomes (Morton, 1977b; Phillips, 1991). *Smilax* spp. also provide cordage, a pulque beverage, edible berries, and contraceptive, kidney pain, rheumatism, syphilis, and toothache medicines (Morton, 1981; Alcorn, 1984; Mabberley, 1987; Phillips, 1991). *Lapageria rosea,* known as copihue, is the national flower of Chile (Mabberley, 1987).

25. Vitaceae   This family is intimately related with Western culture and religion. One of the earliest cultivated plants, *Vitis vinifera,* is the source of wine grapes. Several thousand cultivars of this temperate plant exist. *Vitis rotundifolia* produces edible fruits, as do many noncultivated grapes; *V. mesoamericana* provides medicines to treat conjunctivitis, fever, headache, and toothache and to increase breast milk production (galactagogue). A diuretic and venereal disease medicine is made from *V. tiliifolia.*

Ampelocissus spp.—This genus contains 14 species with edible fruits and supplies medicines to treat boils, cuts, cholera, inflammation, lumbago, rheumatism, sprains, and swellings (Phillips, 1991).

Cayratia spp.—Three species possess edible fruits. The genus also furnishes cordage and medicine for severe headaches (Phillips, 1991).

Cissus gongylodes—This is a recently discovered cultivar of Káyapo people of Brazil (Phillips, 1991). *Cissus rhombifolia* provides a medicine to treat painful joints (Morton, 1981); *C. sicyoides* supplies cattle forage, cordage, a diuretic, an emmenagogue, and remedies for boils, sores, and swollen feet (Alcorn, 1984; Phillips, 1991). *Cissus* has 19 species with edible fruits. Its medicinal uses include treatment of boils, headaches, insect bites, skin infections, sore eyes, snake bites, and stomachaches. The stems of some species provide cordage (Bennett, 1992b; Phillips, 1991).

Rhoicissus spp.—The genus possesses six species with edible fruits and supplies medicines for leprosy and inflammation (Phillips, 1991).

Tetrastigma spp.—The genus has six edible species (e.g., *T. harmandii*) and provides cordage for fencing and footbridges, and medicines for headaches, fevers, boils, and dropsy (Usher, 1974; Mabberley, 1987; Phillips, 1991).

C. Minor Families of Useful Canopy-Dependent Plants

Acanthaceae—*Mendoncia aspera* and *M. pedunculata* provide fish poisons in Colombia (Schultes and Raffauf, 1990).

Actinidiaceae—*Actinidia chinensis* is the familiar kiwi.

Annonaceae—Some *Artabotrys* and *Popowia* species bear edible fruits (Mabberley, 1987).

Asteraceae—Indigenous people in South America treat poisonous snake bites with *Mikania guaco* (Bennett, 1992b).

Begoniaceae—The Shuar relieve stomachaches and inflammation with medicines made from *Begonia glabra* (Bennett, 1992b).

Celastraceae—*Celastrus paniculatus* is used as an opium antidote (Phillips, 1991).

Cyclanthaceae—The stems of several genera (e.g., *Asplundia, Eviodanthus,* and *Thoracocarpus*) are important fibers for basketry (Bennett, 1992b).

Dilleniaceae—Indigenous people drink water from the cuts stems of *Doliocarpus* spp. when other water sources are not available (Fig. 3) (B. C. Bennett, unpublished data).

Liliaceae—Several species of *Bomarea* possess edible tubers (Phillips,

Figure 3 Rocío Alarcón drinking water from the cut stem of *Doliocarpus* spp., which provides a good source of potable water.

1991). The highland Quechua of Peru use *Bomarea* stems for cord (B. C. Bennett, unpublished data).

Melastomataceae—The Chachi treat snake bites with *Clidemia epiphytica* leaves (B. C. Bennett, unpublished data).

Moraceae—Hindus in India consider the hemiepiphytic *Ficus benghalensis* and *F. religiosa* to be sacred trees (Mabberley, 1987).

Nepenthaceae—Stems of *Nepenthes* spp. serve as rope substitutes and basketry material (Phillips, 1991).

Nyctaginaceae—The large liana *Pisonia aculeata* is used to treat boils, fever, and snake bite (Morton, 1981; Alcorn, 1984).

Oleaceae—*Jasminum sambac* provides the flavor of jasmine tea (Phillips, 1991).

Polygonaceae—Though mainly cultivated for its pink flowers, *Antigon leptopus* possesses an edible tuber (Mabberley, 1987).

Polypodiaceae—*Phlebodium aureum* has much medicinal value, including use as an abortifacient, antitussive, and antihypertensive agent (Morton, 1981).

Rhamnaceae—*Gouania lupuloides* is a common chewstick and in Jamaica serves as a substitute for hops in making beer (Morton, 1977b, 1981).

Rosaceae—Blackberry and raspberry jams, familiar products made from *Rubus* spp., are found on many temperate tables.

Solanaceae—*Solandra* spp. provide hallucinogens in Mexico (Alcorn, 1984; Mabberley, 1987).

Verbenaceae—The commonly cultivated *Petrea volubilis* provides medicines to treat toothache, menstrual bleeding, asthma, and cough.

III. Discussion

Host-dependent plants are significant resources for indigenous and modern cultures. They contribute to all the major use classes, particularly fibers, foods, and medicines. *Anthurium, Asplundia, Dendrobium, Desmoncus, Eviodanthus, Heteropsis, Thoracocarpus, Tillandsia, Vanilla,* and others supply fibers for cordage, lashing material, baskets, and clothing. Many epiphytes and lianas possess edible fruits, leaves, or tubers, including *Dioscorea, Gnetum, Hylocereus, Ipomoea, Monstera, Passiflora, Smilax,* and several genera in Cucurbitaceae and Vitaceae. Domesticated animals consume the same genera plus *Cissus, Mucuna, Pueraria,* and *Syngonium.*

At least one liana is a fuel source. The oil-rich seeds of *Fevillea cordifolia* are burned as candle substitutes. Managed natural forests of *Fevillea* could rival the fuel produced on an oil palm plantation (Gentry and Wettach, 1986). By far, medicines form the largest use category of host-dependent plants. At least 529 species are used medicinally (Table V). Among the more

Table V Family-Use Index (FUI) for Major and Minor Host-Dependent Families[a]

Family	CO	CR	DP	FI	FH	FO	FP	FR	FU	ME	OR	RM	VE	MI	FUI
Acanthaceae	0	0	0	0	2	0	0	0	0	2	0	0	0	0	4
Adiantaceae	0	0	0	0	0	0	0	0	0	1	0	0	0	0	1
Amaryllidaceae	0	0	0	1	0	4	0	0	0	0	0	0	0	0	5
Annonaceae	0	0	0	0	0	2	0	1	0	4	0	0	0	2	9
Apocynaceae	0	6	0	6	4	7	0	0	0	25	1	0	0	3	53
Araceae	0	0	1	18	4	18	4	2	0	40	2	2	3	9	103
Arecaceae	0	2	0	7	0	7	0	0	0	2	0	0	0	0	18
Aristolochiaceae	0	0	0	0	1	0	0	0	0	8	0	0	0	0	9
Asclepiadaceae	0	4	1	2	2	2	0	0	0	14	0	0	0	4	29
Aspleniaceae	0	0	0	0	0	0	0	0	0	3	0	0	0	1	4
Asteraceae	0	0	0	0	0	0	1	0	0	9	0	0	0	0	10
Basellaceae	0	0	0	0	0	6	0	0	0	1	0	0	0	0	7
Begoniaceae	0	0	0	0	0	1	0	0	0	5	0	0	0	0	6
Bignoniaceae	1	1	1	7	3	0	1	0	0	23	0	6	0	5	48
Blechnaceae	0	0	0	1	1	1	0	0	0	0	0	0	0	0	3
Boraginaceae	0	0	0	0	0	1	0	0	0	4	0	1	0	0	6
Bromeliaceae	0	0	0	1	0	5	1	4	0	9	6	5	0	2	33
Cactaceae	0	0	0	0	0	18	0	0	0	9	0	0	0	0	27
Campanulaceae	0	0	0	0	1	1	0	0	0	0	0	0	0	0	2
Cecropiaceae	0	0	0	0	0	0	0	1	0	1	0	0	0	0	2
Celastraceae	0	0	0	1	0	2	0	0	0	3	0	0	0	0	6
Clusiaceae	0	1	0	1	0	0	0	0	0	5	0	0	1	2	11
Combretaceae	0	0	1	2	0	0	0	0	0	3	0	0	0	1	7
Connaraceae	0	0	0	0	2	0	0	0	0	3	0	0	0	2	7
Convolvulaceae	0	0	1	1	0	23	0	2	0	22	0	6	1	5	61
Cucurbitaceae	0	3	0	0	0	32	5	3	1	31	0	2	0	21	98

continues

Table V—*Continued*

Family								Use class							
	CO	CR	DP	FI	FH	FO	FP	FR	FU	ME	OR	RM	VE	MI	FUI
Cyclanthaceae	0	2	0	2	0	2	1	0	0	2	0	1	0	0	10
Dilleniaceae	0	0	0	0	0	1	0	0	0	3	0	0	0	4	8
Dioscoreaceae	0	0	0	0	0	8	0	0	0	10	0	0	0	0	18
Ericaceae	0	0	0	0	0	3	0	1	0	4	0	0	0	2	10
Euphorbiaceae	0	0	0	1	1	2	0	0	0	1	0	0	0	1	6
Fabaceae	0	1	2	4	6	18	0	6	0	31	0	2	1	14	85
Flagellariaceae	0	0	0	0	1	0	0	0	0	0	0	0	0	1	2
Gesneriaceae	0	0	0	0	1	0	0	0	0	38	0	0	0	2	41
Gnetaceae	0	0	0	1	0	13	0	0	0	1	0	0	0	0	15
Hymenophyllaceae	0	0	0	0	0	0	0	0	0	1	0	0	0	0	1
Icacinaceae	0	0	0	0	0	5	0	0	0	0	0	0	0	0	5
Lauraceae	0	0	1	0	0	0	0	0	0	0	0	0	0	0	1
Liliaceae	0	0	0	0	0	4	0	0	0	2	0	0	0	2	8
Loganiaceae	0	0	0	0	19	0	0	0	0	2	0	0	0	2	23
Loranthaceae	0	0	0	0	1	0	0	4	0	16	1	2	0	0	24
Malpighiaceae	0	0	0	4	2	0	0	0	0	12	0	8	0	1	27
Marcgraviaceae	0	1	0	1	1	1	0	1	0	5	0	0	0	1	11
Melastomataceae	0	1	0	1	0	2	0	0	0	5	0	0	0	1	10
Menispermaceae	0	0	2	0	18	5	0	0	0	23	0	1	0	3	52
Monimiaceae	0	0	0	0	0	0	0	0	0	1	0	0	0	0	1
Moraceae	1	0	0	2	0	1	0	1	0	0	0	2	0	0	7
Myrsinaceae	0	0	0	0	0	2	0	0	0	1	0	0	0	0	3

Family	CO	CR	DP	FH	FI	FO	FP	FR	FU	ME	OR	RM	VE	MI	Total
Nepenthaceae	0	1	0	3	0	0	0	0	0	0	0	0	0	0	4
Nyctaginaceae	0	0	0	0	0	0	0	0	0	3	1	0	0	0	4
Oleaceae	0	0	0	0	0	1	0	0	0	0	0	0	0	0	1
Orchidaceae	0	14	0	5	4	14	1	1	0	48	13	22	1	11	134
Pandanaceae	0	0	0	1	0	1	0	1	0	0	0	0	0	0	2
Passifloraceae	0	0	0	0	1	56	0	0	0	5	0	2	0	0	64
Phytolaccaceae	0	0	0	0	0	1	0	0	0	0	0	0	0	1	2
Piperaceae	0	0	0	0	0	0	0	0	0	17	0	1	1	1	19
Polygonaceae	0	0	0	0	0	1	0	0	0	1	0	0	0	1	2
Polypodiaceae	0	0	0	0	0	1	0	0	0	6	0	0	0	1	8
Ranunculaceae	0	0	0	2	0	0	0	0	0	6	1	0	0	0	11
Rhamnaceae	0	1	1	1	0	2	0	0	0	3	0	2	0	2	10
Rosaceae	0	0	0	1	1	3	0	0	0	1	0	0	0	0	5
Rubiaceae	0	1	0	0	1	3	0	0	0	12	0	1	0	3	26
Sapindaceae	0	3	0	3	10	3	0	0	0	11	0	3	0	1	34
Schizaeaceae	0	3	0	3	0	0	0	0	0	0	0	1	0	0	7
Smilacaceae	1	1	1	2	0	10	0	0	0	10	1	1	1	0	27
Solanaceae	0	0	0	0	0	1	0	0	0	3	2	1	0	2	9
Tropaeolaceae	0	0	0	0	0	2	0	0	0	0	0	0	0	0	2
Ulmaceae	0	0	0	0	0	0	0	0	0	1	0	0	0	0	1
Urticaceae	0	0	0	0	0	0	0	0	0	2	0	0	0	2	4
Verbenaceae	0	0	0	0	0	0	0	0	0	3	0	0	0	0	3
Vitaceae	0	0	0	15	0	52	0	4	1	12	0	0	0	0	83
Total	3	46	17	100	86	348	14	31	1	529	28	73	8	115	1399

[a]The value of the index represents the sum of the scores for each host-dependent species in the family. The score for each species can range from 0 (no uses) to 14 (employed for all 14 use classes). CO = construction, CR = craft, DP = dye-paint, FH = fishing/hunting, FI = fiber, FO = food, FP = food processing, FR = forage, FU = fuel, ME = medicine, OR = ornamental, RM = ritual/mythical, VE = veterinary, and MI = miscellaneous.

important taxa are *Anthurium, Aristolochia, Dioscorea, Peperomia, Philodendron, Smilax,* and most species of Bignoniaceae, Gesneriaceae, and Loranthaceae. These contribute therapeutic products, ranging from amebicides to emollients.

Epiphytes are best known for their ornamental value, particularly Bromeliaceae, Gesneriaceae, and Orchidaceae. Use of many epiphytes as national flowers illustrates their aesthetic value. The ornamental category often merges with ritual/mythical uses. Indigenous people decorate shrines and altars with many colorful orchids and bromeliads. Peruvian Quechua-speakers adorn weddings and funerals with *Tillandsia* species. Passifloraceae is named for the "passion of Christ," as symbolized by the family's floral structure.

Banisteriopsis caapi and other species of Malpighiaceae are the most important hallucinogens in Amazonia (Fig. 4). Several other lianas (*Ipomoea violaceae* and *Turbina corymbosa*) and even one orchid (*Oncidium cebolletta*) are also hallucinogenic. Canopy plants yield many other beneficial products, most notably medicines and curare arrow poisons from Menispermaceae and fish poisons from Fabaceae and Sapindaceae. Many epiphytes and lianas are purported aphrodisiacs, including *Anthurium infectorium, Guzmania melinonis, Machaerium lunatum, Mandevilla steyermarkii, Paullinia cupana, Sphenocentrum* spp., and *Vanilla planifolia.*

Scores of host-dependent plants reach local markets and many are exchanged internationally. Among the most common are a ubiquitous spice

Figure 4   Peeled and cut stems of *Banisteriopsis caapi* (Malpighiaceae), the principal hallucinogen in northwestern Amazonia.

and flavoring (black pepper and vanilla), two beverage plants (guaraná and grape), several foods (kiwis, passion fruits, sweet potatoes, and yams), and major medicines (*Chondrodendron, Dioscorea, Strophanthus,* and *Physostigma*). The economic value of host-dependent plants is considerable. Black pepper is the world's most important spice. Producing countries earn up to $80 million per year through the export of vanilla (Smith *et al.,* 1992). Annual retail sales of diosgenin from *Dioscorea* exceed $1 billion (Mabberley, 1987). The value of furniture produced from rattan each year exceeds several billion dollars (Myers, 1984).

The plants mentioned in this chapter represent only a subset of all useful epiphytes, lianas, and parasites. Considering their importance to traditional, developing, and modern societies, greater emphasis should be placed on the documentation, management, and utilization of host-dependent species. First, we need ethnobotanical studies of epiphytes and lianas used by rapidly disappearing indigenous cultures. This research requires collaboration of ethnobotanists, taxonomists, and field researchers with technical climbing abilities.

Second, we need studies on the ecology of individual species. Any of the taxa mentioned in this chapter could be the subject of a dissertation. Particular attention should be paid to the management of these resources. Do indigenous people knowingly manage canopy-dwelling plants? Do they harvest epiphytes and lianas in a sustainable manner? In Ecuador's Esmeraldas Province, nonindigenous colonists have harvested an epiphyte called piquihua excessively. This important fiber resource has been extirpated throughout much of its former range. Because of its rarity and growth high in the canopy, we have not yet identified this species. After establishing its botanical identity, scientists can attempt to propagate piquihua and reintroduce it to forests where it once occurred.

A third need is evaluation of the present and potential economic value of canopy-dwelling plants. The value of nontimber forest products is considerable, often exceeding the value of all other uses of the same parcel of forest (Peters *et al.,* 1989; Balick and Mendelsohn, 1992; B. C. Bennett, unpublished data; Grimes, unpublished data). Most studies have focused on tree species but lianas and epiphytes also have value (e.g., *Petrea maynensis;* Grimes, unpublished data). Sustainable propagation and legal export of ornamental epiphytes (in accord with CITES regulations) also offer economic alternatives for indigenous and rural people.

A fourth topic in need of study is the secondary chemistry of lianas and epiphytes. Why are so many hallucinogens, fish poisons, and curares derived from lianas? Phytochemical analyses would not only complement ethnobotanical field studies but would also help address interesting theoretical questions. Doest the patchy distribution of epiphytes have any influence on their secondary chemistry? Do the quality and quantity of secondary compounds differ in canopy-dwelling and canopy-forming life-forms? Do epiphytes and parasites share similar chemical defense strategies with their hosts?

Host-dependent species are widely used by indigenous people. Why do they use epiphytes and lianas when other life-forms are more abundant and more accessible? Some are used because they are effective (e.g., *Chondrodendron tomentosum*); others because they share taxonomic affinities with effective taxa (perhaps *Melothria pendula*). Other taxa are used because their morphology or anatomy suggests a particular use (e.g., *D. consanguinea*). Myth and tradition also reinforce the utilization of certain plant species (e.g., *Dendrobium* spp.).

IV. Summary

Despite their diversity, epiphytes, lianas, and parasitic plants are often overlooked by ethnobotanists. Yet these life-forms contribute many useful products for indigenous peoples throughout the world. Some, such as vanilla (*Vanilla planifolia*) and passion fruit (*Passiflora edulis*), are globally well known. Others, such as yoco (*Paullinia yoco*) and ayahuasca (*Banisteriopsis caapi*), are important regionally. Three of the most culturally significant plants in northwestern Amazonia are lianas; many other epiphytes and vines provide foods, medicines, and fibers. Indigenous people collect, protect, and sometimes intentionally plant host-dependent species. They are also responsible for the present distribution of some taxa (e.g., *P. yoco* and *B. caapi*). Some host-dependent species contribute signifncantly to regional and national economies. The sustainable harvest of selected epiphytes and lianas may provide alternatives to destructive uses of the forest. Studies on the use, management, economics, and secondary chemistry of these host-dependent species are needed.

V. Appendix

Host-dependent species and genera cited in text (including author citation).

Acanthaceae
  Mendonica aspera (Ruíz & Pavón) Nees, *M. pedunculata* Leonard
Actinidiaceae
  Actinidia chinensis Planchon
Annonaceae
  Artabotrys
  Popowia

Apocynaceae
Allamanda cathartica L.
Fernaldia pandurata (DC) Woodson
Gonolobus barbatus HBK, *G. niger* (Cav.) R.Br.
Landolphia
Mandevillia neriodes Woodson, *M. steyermarkii* Woodson, *M. subsagittata* Woodson, *M. vanheurckii* (Muell. Arg.) Markgraf
Strophanthus gratus Franchet, *S. sarmentosus* DC., *S. kombé* Oliver
Urceola spp.
Willughbeia spp.
Araceae
Anthurium acrobates Sodiro, *A. alienatum* Schott, *A. apaporanum* Schultes, *A. eminens* Schott, *A. infectorium* R.E. Schultes
Heteropsis oblongifolia Kunth
Monstera deliciosa Liebm., *M. pertusa* (L.) de Vreise
Philodendron angustialatum Engler, *P. bipinnatifidum* Schott, *P. remifolium* R.E. Schultes,
Syngonium podophyllum Schott
Arecaceae
Ancistrophyllum secundiflorum Mann et Wendl.
Calamus manan Miq., *C. merrilli* Becc., *C. ornatus* Blume ex Schultes var. *philippinensis* Becc., *C. scipionum* Lour.
Daemonorops
Desmoncus
Aristolochiaceae
Aristolochia serpentaria L., *A. petersiana* Klotzsch
Asclepiadaceae
Cryptostegia grandiflora R.Br.
Dischidia
Gymnema sylvestre (Retz.) R. Br. ex Schultes
Hoya
Periploca gracea L.
Asteraceae
Mikania guaco Humbl. & Bonpl.
Begoniaceae
Begonia glabra Aublet
Bignoniaceae
Arrabidaea chica (HBK) Verlot
Distictella pulverulenta Sandwith, *D. racemosa* Bur. et K. Schumann ex Mart.
Macfadyeana uncata (Andrews) Sprague & Sandwith
Mansoa standleyi (Steyermark) A. Gentry
Mussatia hyacinthina (Standley) Sandwith

Pachyptera alliacea (Lam.) A. Gentry, *P. hymenaea* A. Gentry
Tanaecium nocturnum (Barb. Rodr.) Bur. & Schumman
Tynnanthus polyanthus (Burret) Sandw.

Bromeliaceae

Aechmea zebrina L.B. Smith
Catopsis
Guzmania melinonis Regel
Streptocalyx longifolius (Rudge) Baker
Tillandsia benthamiana Klotzsch, *T. complanata* Benth., *T. ionochroma*
Andé ex Mez, *T. sphaerocephala* Baker, *T. recurvata* (L.) L., *T. usneoides*
(L.) L.
Vriesea

Cactaceae

Acanthocereus pitijaya (Jacq.) Dugand ex Croizat, *A. pentagonus* (L.) Britton & Rose
Epiphyllum phyllanthus L. Haw var. *phyllanthus*, *E. strictum* (Lemaire)
Britt. & Rose
Hylocereus polyrhizus (A. Weber) Britton & Rose, *H. unundatus* (Haw.)
Britt & Rose
Pereskia aculeata Miller
Rhipsalis baccifera (J. Miller) Stearn
Selenicereus grandiflorus (L.) Britton & Rose, *S.* cf. *spinulosus* (DC) Britton
& Rose

Celastraceae

Celastrus paniculatus Willd.

Convolvulaceae

Convolvulus floridus L.f., *C. scammonia* L.
Cuscuta americana L., *C. corymbosa* Ruíz & Pavón, *C. tinctoria*
Ipomoea batatas (L.) Lam., *I. alba* L., *I. indica* (Burm.) Merrill, *I. nil* (L.)
Roth., *I. orizabensis* (J. Pellet.) Steudel, *I. violacea* L.
Maripa
Merremia dissecta Hallier f., *M. quinquefolia* (L.) Hallier f., *M. umbellata*
(L.) Hallier f.
Turbina corymbosa (L.) Raf.

Cucurbitaceae

Apodanthera undulata A. Gray
Cayaponia attenuata (Hook. & Arn.) Cogn., *C. ruizii* Cogn., *C. glandulosa*
(Poeppig & Endl.) Cogn.
Fevillea cordifolia L.
Gurania spinulosa (Poeppig & Endl.) Cogniaux
Hodgsonia heteroclita (Roxb.) Hook.f. & Thomson
Melothria pendula L.
Momordica charantia L.

Cyclanthaceae
Asplundia
Eviodanthus
Thoracocarpus
Dilleniaceae
Doliocarpus spp.
Dioscoreaceae
Dioscorea alata L., *D. batatas* Decne., *D. bulbifera* L., *D. cayenensis* Lam.,
D. composita Hemsley, *D. esculenta* (Lour.) Burkill, *D. floribunda* Mar. &
Gal., *D. macrostachys* Benth., *D. matagalpensis* Uline, *D. trifida* L.f.
Fabaceae (*sensu latu*)
Abrus precatorius L.
Bauhinia tarapotensis Bentham
Caesalpinia bonduc (L.) Roxb.
Derris elliptica (Wallich) Benth.
Entada gigas (L.) Fawcett & Rendle
Lonchocarpus nicou (Aublet) DC, *L. rariflorus* Mart.
Machaerium riparium Brandegee, *M. lunatum* Ducke
Mucuna argophylla Standley, *M. pruriens* (L.) DC
Physostigma venenosum Balf.
Pueraria lobata (Wild.) Ohwi
Gesneriaceae
Codonanthe calcarata (Miquel) Hanstein, *C. crassifolia* Morton, *C. uleana*
Fritsch
Columnea bilabiata Seeman
Dalbergaria asteroloma Wiehler, *D. consanguinea* (Hanstein) Wiehler,
D. crassa (Morton) Wiehler, *D. difficilis* Wiehler, *D. eburnea* Wiehler,
D. ericae (Mansfield) Wiehler, *D. medicinalis* Wiehler, *D. picata* (Kar-
sten) Wiehler, *D. villosissima* (Mansfield) Wiehler
Drymonia alloplectoides Hanstein, *D. coccinea* (Aublet) Wiehler, *D. coriacea*
(Oersted ex Hanstein) Wiehler, *D. macrophylla* (Oersted) H.E. Moore,
D. umecta Wiehler
Pentadenia ecuadorana Wiehler, *P. spathulata* (Mansfield) Wiehler
Gnetaceae
Gnetum gnemon L., *G. nodiflorum* Brongn.
Liliaceae
Bomarea
Loganiaceae
Strychnos darienensis Seemann, *S. erichsonii* R. Schomb., *S. jobertiana* Bail-
lon, *S. peckii* B.L. Robinson, *S. tomentosa* Bentham, *S. toxifera* R. Schomb.
ex Bentham
Loranthaceae (including Viscaceae)
Phoradendron serotinum (Raf.) M.C. Johnston, *P. crassifolium* (DC) Eich-
ler, *P. piperoides* Nutt.

Oryctanthus O. florilentus (Rich.) van Tieghem, *O. occidentalis* Eichler

Phthirusa pyrifolia (Kunth) Eichler, *P. pauciflora* Wr. ex Sauv.

Psittacanthus cucullaris (Lam.) Blume, *P. schiedeanus* (Schlect. & Cham.) Blume

Struthanthus crassipes (Oliv.) Eichl., *S. flexilis* (Rusby) Kuijt, *S.* aff. *quercicola* (Cham. & Schl.) Blume

Viscum album L.

Malpighiaceae

Banisteriopsis B. caapi (Spruce ex Grisebach) Morton, *B. muricata* (Cav.) Cuatrec.

Diplopterys D. cabrerana (Cuatrec.) B. Gates

Heteropteris beecheyana Juss.

Mascagnia castanea (Cuatrec.) Anderson

Stigmaphyllon cordifolium Niedenzu, *S. sagraeanum* A. Juss.

Melastomataceae

Clidemia epiphytica (Triana) Cogn.

Menispermaceae

Abuta rufescens Aublet, *A. obovata* Diels

Anamirta cocculus Wight & Arn.

Anomospermum chloranthum Diels

Chondrodendron tomentosum Ruíz & Pavón

Cissampelos grandifolia Triana & Planchon, *C. pareira* L.

Cocculus

Curarea tecunarum Barneby & Krukoff

Dioscreophyllum cumminsii (Stapf) Diels

Sphenocentrum jollyanum Pierre

Tinospora crispa (L.) Hook.f. & Thomson

Moraceae

Ficus benghalensis L., *F. religiosa* L.

Nepenthaceae

Nepenthes

Nyctaginaceae

Pisonia aculeata L.

Oleaceae

Jasminum sambac Aiton

Orchidaceae

Acampe pachyglossa Rchb.f.

Ansellia gigantea Rchb.f., *A. humiilis* Bull

Campylocentrum

Catasetum cristatum Lindl., *C. luridum* Lindl.

Cattleya skinneri Bateman, *C. trianaei* Linden et. Rchb.f.

Crytopodium punctatum (L.) Lindl.

Dendrobium acinaciforme Roxb., *D. bigibbum* Lindl., *D. nobile* Lindl., *D. macartiae* Thw., *D. pulchellum* Roxb., *D. salaccense* Lindl.

Epidendrum atropurpureum Willd., *E. bifidum* Aublet, *E. difforme* Jacq., *E. pastoris* La Llave et Lex, *E. radicans* Pav. ex Lindl.

Lycaste virginalis Linden

Oncidium cebolleta Sw., *O. carthagenense* (Jacq.) Sw.

Peristeria elata Hook.

Phreatia

Psygmorchis pusilla (L.) Dodson et Dressler

Schomburgkia crispa Lindl., *S. tibicinis* Bateman, *S. thomsoniana* Rchb.f.

Sobralia dichotoma Ruíz & Pavón

Tetramicra bicolor Rolfe

Vanilla crenualta Rolfe, *V. grandiflora* Lindl., *V. ovalis* Blanco, *V. planifolia* G. Jackson, *V. pompona* Schiede, *V. tahitensis* J. W. Moore

Passifloraceae

Adenia

Passiflora edulis Sims, *P. foetida* L., *P. laurifolia* L., *P. ligularis* Juss., *P. mollisima* (Kunth) L. Bailey, *P. riparia* Mart., *P. serratifolia* L., *P. vitifolia* HBK

Piperaceae

Peperomia emarginella C.DC., *P. glabella* (Sw.) Dietr., *P. maculosa* Hook., *P. rotundifolia* (L.) Kunth, *P. urocarpa* Fischer & Meyer

Piper betel L., *P. cubea* L., *P. chaca* Hunter, *P. longum* L., *P. nigrum* L., *P. retrofactum* Vahl

Polygonaceae

Antigon leptopus Hook. & Arn.

Polypodiaceae

Phlebodium aureum (L.) J. Sm.

Rhamnaceae

Gouania lupuloides (L.) Urban

Rosaceae

Rubus

Rubiaceae

Chiococca alba (L.) Hitch.

Manettia glandulosa Poeppig & Endl., *M. cordifolia* Mart.

Paederia foetida, P. scandens (Lour.) Merr.

Sabicea amazonensis Wernham, *S. villosa* Roemer et

Uncaria africana G. Don., *U. gambir* Roxb., *U. guianensis* (Aublet) Gmelin

Sapindaceae

Cardiospermum halicacabum L., *C. grandiflorum* Sw.

Paullinia cupana HBK, *P. fuscescens* HBK, *P. pinnata* L., *P. yoco* R.E. Schultes & Killip

Serjania inflata Poeppig & Endl., *S. mexicana* Willd., *S. rubicaulis* Bentham ex Radlkofer

Smilacaceae
  Lapageria rosea Ruíz & Pavón
  Smilax aristolochifolia Mill., *S. bona-nox* L., *S. domingensis* Willd., *S. havanensis* Jacq., *S. macrophylla* Roxb., *S. megacarpa* DC., *S. ornata* Lam., *S. regelii* Killip, *S. spruceana* A.DC
Solanaceae
  Solandra spp.
Verbenaceae
  Petraea volubilis L.
Vitaceae
  Ampelocissus
  Cayratia
  Cissus gongylodes, C. rhombifolia Vahl, *C. sicyoides* L.
  Rhoicissus
  Tetrastigma harmandii Planchon
  Vitis rotundifolia Michaux, *V. mesoamericana* Rogers, *V. tiliifolia* Humb. & Bonpl., *V. vinifera* L.

Acknowledgments

I thank B. Boom, R. E. Schultes, M. Lowman, and N. Nadkarni for their thoughtful comments on this manuscript. The U.S. Agency for International Development Grants 518-0023-G-SS-4110-00-605 and LAC-0605-G-SS-7037-00; the Rockefeller Foundation, Program for Economic Botany in Latin America and the Caribbean; the U.S. Agency for International Development Sustainable Uses of Biological Resources Project; and the Edward John Noble Foundation supported fieldwork in Ecuador.

References

Alarcón, R. (1988). "Etnobotánica de los Quichuas de la Amazonia Ecuatoriana," Misc. Antropol. Ecuatoriana, Ser. Monogr., Vol. 7. Museos del Banco Central del Ecuador, Guayaquil.

Alcorn, J. B. (1984). "Huastec Mayan Ethnobotany." Univ. of Texas Press, Austin.

Balée, W. (1986). Análise preliminar de inventário florestal e a etnobotânica Ka'apor (Maranhao). *Bol. Mus. Para. Emílio Goeldi, Bot.* 2, 141–167.

Balick, M. J., and Beck, H. T. (1990). "Useful Palms of the World: A Synoptic Bibliography." Columbia Univ. Press, New York.

Balick, M. J., and Mendelsohn, R. (1992). Assessing the economic value of traditional medicines from tropical rain forests. *Conserv. Biol.* 6, 128–130.

Bennett, B. C. (1990). The ethnobotany of bromeliads: The use of *Tillandsia* species in the highlands of southern Peru. *J. Brom. Soc.* 40, 64–69.

Bennett, B. C. (1992a). Plants and people of the Amazonian rainforests: The role of ethnobotany in sustainable development. *BioScience* 42, 599–607.

Bennett, B. C. (1992b). Uses of epiphytes, lianas, and parasites by the Shuar people of Amazonian Ecuador. *Selbyana* 13, 99–114.

Bennett, B. C. (1992c). Hallucinogenic plants of the Shuar and related indigenous groups in Amazonian Ecuador and Peru. *Brittonia* 44, 483–493.

Bennett, B. C. (1995). Ethnobotany and economic botany of Bromeliaceae. *In* "The Biology of the Bromeliaceae" (D. H. Benzing, ed.). Cambridge Univ. Press, London. (in press).

Bennett, B. C., Baker, M., and Gómez, P. (1995). Ethnobotany of the Untsuri Shuar. *Adv. Econ. Bot.* (in press).

Benzing, D. H. (1990). "Vascular Epiphytes." Cambridge Univ. Press, New York.

Berlin, B., Breedlove, D. E., and Raven, P. H. (1974). "Principles of Tzeltal Plant Classification." Academic Press, New York.

Boom, B. M. (1989). Use of plant resources by the Chácobo. *Adv. Econ. Bot.* 7, 78–96.

Burlage, H. M. (1968). "Index of Plants of Texas with Reputed Medicinal and Poisonous Properties." Henry M. Burlage, Austin, TX.

Chávez Velásquez, N. A. (1977). "La Materia Medica en el Incanato." Editorial Mejia Baca, Lima, Peru.

Davis, E. W., and Yost, J. A. (1983). The ethnobotany of the Waorani of eastern Ecuador. *Bot. Mus. Leafl., Harv. Univ.* 29, 159–217.

Dransfield, J. (1985). Prospects for lesser known canes. *In* "Proceedings of the Rattan Seminar" (K. M. Wong and N. Manokaran, eds.), pp. 107–114. Rattan Information Centre, Kepong, Malaysia.

Duke, J. A. (1986). "Isthmian Ethnobotanical Dictionary." Pawan Kumar Scientific Publishers, Jodphur, India.

Gentry, A. H., and Wettach, R. H. (1986). *Fevillea*—A new oil seed from Amazonian Peru. *Econ. Bot.* 40, 177–185.

Hensyl, W. R., ed. (1990). "Stedman's Medical Dictionary," 25th ed. Williams & Wilkins, Baltimore, MD.

Honychurch, P. N. (1980). "Caribbean Wild Plants and Their Uses." Macmillan, London.

Jain, S. K., ed. (1991). "Contribution to Indian Ethnobotany." Pawan Kumar Scientific Publishers, Jodhpur, India.

Kuijt, J. (1969). "The Biology of Parasitic Flowering Plants." Univ. of California Press, Berkeley.

Lawlor, L. J. (1984). Ethnobotany of Orchidaceae. *In* "Orchid Biology: Reviews and Perspectives" (J. Arditti, ed.), Vol. 3, pp. 27–149. Cornell Univ. Press (Comstock), Ithaca, NY.

Lincoln, R. J., Boxshall, G. A., and Clark, P. F. (1982). "A Dictionary of Ecology, Evolution, and Systematics." Cambridge Univ. Press, Cambridge, UK.

Mabberley, D. J. (1987). "The Plant-Book: A Portable Dictionary of the Higher Plants." Cambridge Univ. Press, Cambridge, UK.

Marles, R. J. (1988). The ethnopharmacology of the lowlands of eastern Ecuador. Ph.D. Dissertation, University of Illinois at Chicago.

Moerman, D. E. (1986). "Medicinal Plants of Native America," Tech. Reps, No. 19 (Res. Rep. Ethnobot., Contrib. 2), Vol. 1. University of Michigan Museum of Anthropology, Ann Arbor.

Morton, J. F. (1977a). "Major Medicinal Plants: Botany, Culture and Uses." Thomas, Springfield, IL.

Morton, J. F. (1977b). "Wild Plants for Survival in South Florida." Fairchild Tropical Garden, Miami, FL.

Morton, J. F. (1981). "Atlas of Medicinal Plants of Middle America." Thomas, Springfield, IL.

Myers, N. (1984). "The Primary Source: Tropical Forests and Our Future." Norton, New York.

Núñez Meléndez, E. (1982). "Plantas Medicinalis de Puerto Rico." Editorial de la Universidad de Puerto Rico, San Juan.

Paz y Miño, G., Balslev, H., and Valencia, R. (1990). Aspectos etnobotánicos de las lianas utilizadas por los indígenas Siona–Secoya de la Amazonía del Ecuador. *In* "Las Plantas y el Hombre" (M. Ríos and H. Pedersen, eds.), pp. 105–118. Ediciones Abya-yala, Quito, Ecuador.

Peters, C. M., Gentry, A. H., and Mendelsohn, R. O. (1989). Valuation of an Amazonian rainforest. *Nature (London)* 339, 655–656.

Phillips, O. (1991). The ethnobotany and economic botany of tropical vines. *In* "The Biology of Vines" (F. E. Putz and H. A. Mooney, eds.), pp. 427–475. Cambridge Univ. Press, Cambridge, UK.

Pinkley, H. V. (1973). The ethnoecology of the Kofán. Ph.D. Dissertation, Harvard University, Cambridge, MA.

Plowman, T. (1969). Folk uses of new world aroids. *Econ. Bot.* 23, 97–122.

Plowman, T. (1980). Chamairo: *Mussatia hyacinthina*—An admixture to coca from Amazonian Peru and Bolivia. *Bot. Mus. Leafl., Harv. Univ.* 28, 253–261.

Purseglove, J. W. (1968). "Tropical Crops: Dicotyledons." Longman Scientific and Technical, Essex.

Schultes, R. E., and Raffauf, R. F. (1990). "The Healing Forest: Medicinal and Toxic Plants of the Northwest Amazonia." Dioscorides Press, Portland, OR.

Simpson, B. B., and Conner-Ogorzaly, M. (1986). "Economic Botany: Plants in Our World." McGraw-Hill, New York.

Smith, N. J. H., Williams, J. T., Plucknett, D. P., and Talbot, J. P. (1992). "Tropical Forests and Their Crops." Cornell Univ. Press (Comstock), Ithaca, NY.

Soukup, J. (1970). "Vocabulario de los Nombres Vulgares de la Flora Peruana." Colegio Salesiano, Lima, Peru.

Uphof, J. C. T. (1968). "Dictionary of Economic Plants." Cramer, New York.

Usher, G. (1974). "A Dictionary of Plants Used by Man." Constable, London.

Vickers, W. T., and Plowman, T. (1984). Useful plants of the Siona and Secoya Indians of Eastern Ecuador. *Fieldiana, Bot. [N.S.]* 15, 1–63.

von Reis Altschul, S. (1973). "Drugs and Foods from Little Known Plants: Notes in the Harvard University Herbaria." Harvard Univ. Press, Cambridge, MA.

Williams, L. O. (1981). The useful plants of Central America. *Ceiba* 24, 1–381.

22

The Collection and Preservation of Plant Material from the Tropical Forest Canopy

Stephen W. Ingram and Margaret D. Lowman

From all over the earth—tundra, desert, steppe, mountain, ocean island, and lake—we gather the threads of plant life on land, and we trace them in the canopy of the forest, which first fitted plants for their life in the desert and on the mountain. . . .
Too little is known of these places . . .
—E. J. H. Corner, "The Life of Plants" (1964, p. 284)

I. Introduction

The forest canopy harbors a diverse assemblage of plants of many different growth habits. The reproductive parts of tall trees, many epiphytes, and lianas may exist only in the upper canopy. Access to these upper regions is essential to identify canopy plants and to collect and document voucher specimens for a particular forest region. The majority of botanists, however, have relied on collecting techniques that are opportunistic (i.e., "reach and grab") rather than on methodically conducting more comprehensive samples of forest composition. Reproductive parts (essential materials for proper identification and botanical study) and leaves from the upper canopy (whose physical and physiological structure may differ strikingly between the canopy and understory) are not accurately represented in herbaria because collection of these specimens has historically posed an insurmountable challenge. This limits our ability to classify forest plants and our understanding of their taxonomy and ecology.

Why collect plant material from the canopy when it is more easily ob-

Copyright © 1995 by Academic Press, Inc.
All rights of reproduction in any form reserved.

tained from the understory? Fallen branches and entire treefalls sometimes enable a botanist to collect canopy specimens with ease, but information associated with the plant's specific habitat is often lost in a tangle of broken branches. Collecting an identifiable plant fragment may be sufficient for the botanist who is interested only in recording the presence of a certain species in a region, but data on plant habit and environment are necessary for ecological and taxonomic treatments. Systematic surveys should be assessed not only in terms of their breadth of taxa, but also in terms of their inclusion of both canopy and understory materials that are essential for an accurate description of whole forests.

In this chapter, we review the most commonly used methods for collecting plants from tropical forest canopies, discuss collection techniques necessary for different plant groups, give information on collecting and preserving specimens from the canopy, and summarize the coverage of canopy plant material in herbaria throughout the world.

II. Collecting Methods

A. Collecting Methods Used in Early Expeditions

Plant collecting became popular as both a hobby and a vocation over two hundred years ago. Although early plant collectors contributed a great deal to the scientific record, some also inadvertently contributed to the demise of canopy species (e.g., orchids). To the "plant hunters" of the eighteenth, nineteenth, and early twentieth centuries, tropical forests represented an endless bounty of exotics available for collection. Tropical plants were sought for both their scientific curiosity and horticultural value. Many species, especially colorful epiphytic orchids, were collected and transported by ship to plant fanciers in England and Europe, where they were grown in greenhouses or Wardian cases (large terrariums).

Three methods were used by Europeans to collect plants from tropical forest tree canopies prior to the 1920s: collecting from fallen limbs, hiring indigenous forest-dwelling people to climb, and cutting down trees. Botanical expeditions were typically very elaborate affairs. Among Joseph Hooker's entourage for collecting plants in India were "an attentive valet, coolies, seed gatherers, runners, cooks, tree climbers, plant dryers and pressers, an instrument mender and taxidermist, and, naturally, sepoys to guard the lot" (Whittle, 1970). Albert Millican's (1891) adventures in the Colombian Andes were typical of many who sought out canopy plants for the commercial trade: "After two months' work we had secured about ten thousand plants (of *Odontoglossum crispum*), cutting down to obtain these some four thousand trees, moving our camp as the plants became exhausted in the vicinity" (p. 151).

The chronicles of early botanical collectors illustrate a wealth of innovative, sometimes amusing, techniques for sampling plants from forest canopies. For example, good fortune played a role for the Czech collector Benedict Roezl in South America in 1871. While canoeing down a swollen river with a local guide after an unsuccessful expedition on the Pacific slope of the Andes, he came upon a large tree floating downstream. The recently fallen tree was "festooned . . . with epiphytic orchids, ferns and mosses of every shade of green." Roezl and his guide, taking advantage of this rare opportunity, "moored themselves to the tree and collected plants from its branches as they traveled downriver." Their "rich haul" offset the previous disappointments of the collecting trip (Whittle, 1970).

It was not until the Oxford University Expedition in 1929 to former British Guiana (now Guyana) that a Western botanist ascended into the rain forest canopy (described in Mitchell, 1986). Ladders were placed in the canopy of a huge "morubukea" tree by two local tribesmen, and then an armchair was hoisted up using ropes and pulleys. Since then, both simpler and more elaborate methods have been used to reach the upper forest canopy (Lowman and Moffett, 1993; Chapter 1).

B. Collecting Techniques Currently Used

Early methods of free-climbing and felling trees are still used by modern plant collectors, though less frequently than in the past. However, the free-climbing expertise of local inhabitants, chicleros, and other intrepid individuals should not be underestimated. In Peru, while two botanists were preparing to climb, their Bora assistant ascended a tree and cut specimens from branches 25 m high before they had their spikes ready (B. Bennett, personal communication). In Sarawak, Asah Amak Unyang free-climbed into the canopy and spent most of the day making collections from 35 canopy-level trees (P. Ashton, personal communication).

Technological advances in the twentieth century have led to improved climbing techniques that cause less tree damage from the sampling process (Chapter 1). Felling trees for specimen collection should be confined to those trees that are in imminent danger of destruction from road-building crews, dam projects, and other human activities. Conservation priorities make nondestructive collecting techniques essential for future maintenance of many forest regions.

We conducted a survey of contemporary field botanists to quantify which methods are used most often to collect canopy-level plants in tropical rain forests (the forest type considered most complex and subsequently most challenging for botanical surveys). We sent an eight-question survey to 39 botanists in nine countries; respondents' experience comprised 606 collective years of fieldwork and 437,000 collection numbers. Our questions related to their collection of canopy plant material: trees, epiphytes, herbaceous vines/woody lianas, and other. The respondents addressed two topics:

1. field methods that were used most commonly for plants of different habits (some botanists also related stories of special interest owing to the idiosyncracies of their collecting techniques) and
2. ranking of herbaria according to quality and usefulness of collections of the six major groups of canopy plants.

Field methods were separated into the categories of ground-based and canopy-based sampling. The more conventional techniques of ground-based collection reported by the respondents included searching through natural treefalls, using a fishing line with a weight to pull down vegetation, and clipping branches with extendible pole-pruners (Table I). Unconventional ground-based techniques have included harassing monkeys so they would retaliate by breaking branches (R. Foster, personal communication) and collecting foliage that was strewn on the forest floor by raucous gangs of galahs (M. Lowman, personal communication). The use of boats to collect plants overhanging water was included with ground-based collecting methods.

Ground-based techniques were used approximately twice as often as canopy-based sampling methods. The ground-based technique of searching

Table I Ground-Based Collecting Techniques and Their Proportional Use by Field Botanists

	Percentage class for those surveyed who used method for:[a]			
Collection method	Trees	Epiphytes	Vines	Unspecified
Using naturally fallen trees and branches	3	4	4	1
Searching along road cuts	2	4	3	0
Pruning pole	3	4	3	0
Rope or line with weighted end	2	2	2	0
Rope with pruning saw	1	2	1	0
Throwing stones	1	2	1	0
Using a boat to reach plants over water	3	3	3	2
Using "legally" felled trees	0	0	0	3
Rifle to shoot branches down	0	0	0	1
Felling trees	0	0	0	1
Using stick to knock plant down	0	1	0	0

[a] Classes 0–4 correspond to the percentage of time this field technique was employed by botanists surveyed: 0 = never; 1 = 1–25%; 2 = 26–50%; 3 = 51–75%; 4 = 76–100%.

among naturally fallen trees and limbs was the most common technique, followed by the use of a pruning pole and searching along road cuts (mainly used for facultative epiphytes). Seventy percent of respondents had collected "canopy" material from boats (Table I). Approximately two-thirds of those surveyed had collected specimens from trees recently felled by road crews, and most of the collectors had followed crews specifically for the purpose of making plant collections.

Despite their common use, ground-based methods do not enable the collector to make comprehensive surveys of upper-canopy plants. Collection of material that has fallen to the forest floor is opportunistic and often does not provide important ecological information, for example, it is often impossible to ascertain the height and environment from which a specimen originated. Ground-based sampling creates three shortcomings for herbarium collections: (1) inaccurate depiction of the canopy foliage of tall trees with polymorphic leaves where collectors harvest predominantly understory leaves; (2) inadequate classification of plant specimens (as well as undiscovered species) from the upper canopy levels; and (3) bias toward understory species and plant material, which may cause disproportionate representation of forest canopy structure and diversity.

Canopy-based methods described by respondents include climbing trees with single-rope techniques (Fig. 1); ladders, the peconha method (climbing with a loop of webbing held around the trunk), and tree grippers/Swiss "tree bicycles" (Fig. 2); climbing irons with spikes/*patas de loro* (Fig. 3), or freehand; cherry pickers, booms, platforms or walkways (Fig. 4); and innovative methods such as cranes, dirigibles, and the canopy raft (Chapter 1). The collection of plants from tree canopies requires a greater investment in time and equipment than from the forest floor, but the use of both canopy-based and ground-based techniques results in a more accurate and complete depiction of the flora.

Several botanists reported the discovery of unexpected plants in the canopy after climbing into trees. One advantage of using canopy-based techniques is the potential discovery of flowers, fruits, and even foliage that is not visible from the ground. Another advantage is to facilitate more comprehensive sampling through the vertical stratification of forests. Just as SCUBA diving enhances one's understanding of coral reef life below the ocean surface, the view from above the ground gives the botanist a better perception of forests and their canopy components.

Although canopy-based methods are more comprehensive than ground-based sampling, they have limitations in terms of suitability for certain types of fieldwork. For example, French climbing spikes are only useful for ascending trunks with a diameter between 10 and 50 cm (Mori and Prance, 1987). Spikes will damage trees after frequent, long-term use, but probably

Figure 1   Karen Ferrell-Ingram using Jumar ascenders and climbing rope in Costa Rica. (Photo by Stephen Ingram).

do little damage to trees climbed only once or several times (S. Mori, personal communication). Hence, they are more suitable for floristic than for ecological studies. Swiss "tree bicycles" can be used for ascending trees with diameters of 18 to 72 cm that are free of most lianas, epiphytes, and climbing plants (Mori, 1984), but their design makes it awkward to move onto branches (T. Croat, personal communication). Single-rope techniques offer access primarily to a vertical transect near the rope, where most foliage is beyond reach. However, a "web" of ropes or nylon webbing secured to branches increases access within the canopy (Perry, 1986).

Combinations of several canopy-based methods (Tables I and II) offer

Figure 2 Scott Mori using Swiss "tree bicycles" in French Guiana. (Photo courtesy of Scott Mori, New York Botanical Garden.)

the most comprehensive sampling opportunities. For example, single-rope techniques plus the use of an extendible pruning pole are a good combination, especially with the help of an assistant on the ground. Extendible pruning poles, typically considered a ground-based method (Table I), may also be used from canopy walkways, platforms, or the canopy raft. Ascending suitably sized neighboring trees with either climbing spikes or tree bicycles and an extendible pruner can be very effective for collecting from the canopy of large-diameter trees (S. Mori, personal communication). Relative costs and ease of different canopy access techniques are summarized in Chapter 1.

Figure 3   Scott Mori collecting tree canopy material with the use of French climbing spikes (also called *patas de loro* or "parrot feet") and extendible pruner in French Guiana. (Photo by Carol Gracie, New York Botanical Garden.)

General safety precautions are necessary for any tree-climbing technique, and specific precautions may be necessary for certain techniques. In general, one should check knots and equipment on the ground, avoid climbing partially dead trees, climb with a spotter on the ground, wear a hardhat, and climb only when fully alert. Make sure the rope and branch will support the full weight of at least two people before ascending via single-rope techniques or arborist techniques. One needs to be alert to the potential dangers of wasps, biting ants, and venomous snakes, especially in low- and middle-elevation tropical forests. Dripping sap and plant spines are nuisances that are best avoided.

Figure 4   Meg Lowman using walkway constructed to access forest canopy in Belize.

C. Biological Differences between Canopy and Understory Plant Material

One of the greatest shortcomings of early herbarium specimens of canopy plants was the lack of information about the habitat of the collections. Information on the height, light conditions, and identity of the supporting host plant (for epiphytes and vines) was often omitted, in part because such information was not known for specimens collected opportunistically from fallen branches. Identification of tropical host trees in many regions did and still does pose a difficult challenge.

Canopy and understory plant material exhibit physical and biological differences, which makes each important to collect. Some epiphyte species are limited to a certain vertical range in the forest (Johansson, 1974; Kelly, 1985; S. Ingram, K. Ferrell-Ingram, and N. Nadkarni, unpublished data). Sampling among different crown heights may yield ecological information about the canopy processes of a forest (Lowman, 1984). The size differ-

Table II Canopy-Based Collecting Techniques and Their Proportional Use by Field Botanists

Collection method	Percentage class for those surveyed who used method for:[a]			
	Trees	Epiphytes	Vines	Unspecified
Climbing/scrambling to 4 m	2	4	3	0
Free-climbing higher than 4 m (or asking someone else to do so)	3	3	2	0
Using climbing irons/spikes/tree grippers	2	1	1	0
Using tree-climbing ladders	1	1	1	0
Climbing ropes with Jumar ascenders or similar gear	1	1	1	0
Permanently rigged cable and basket system	1	1	1	0
Semipermanent platforms or canopy walkway	1	1	1	0
Canopy raft	1	1	1	0

[a] Classes 0–4 correspond to the percentage of time this field technique was employed by botanists surveyed: 0 = never; 1 = 1–25%; 2 = 26–50%; 3 = 51–75%; 4 = 76–100%.

ences between understory and canopy foliage may vary by 10-fold; leaf area, toughness, longevity, chemistry, photosynthetic capacity, and other physical attributes differ significantly throughout the canopies of many tree species (Lowman, 1992).

III. Application of Methods for the Collection of Canopy Plants

A. General Flora Surveys

When plant collectors make general surveys, such as for a regional flora, they usually begin with opportunistic collecting. For example, branches overhanging rivers may have fertile material that can be collected from boats (Table II); recent treefalls may provide flowering lianas, epiphytes, and foliage from the host tree; low-growing branches and epiphytes can often be reached along forest trails or ridges. Many species of lianas that inhabit the upper crowns of mature forests also occur in gaps and forest edges (Putz, 1984), where they may be easier to observe and collect. If an area has been collected over many decades, then opportunistic sampling will probably produce an adequate (although incomplete) description of the canopy flora. The addition of canopy-based techniques to collect plants may yield previously undocumented lianas, epiphytes, and trees.

B. Canopy-Level Trees

Several methods can be used for collecting specimens in the crown after opportunistic approaches are exhausted. Shotguns or rifles are used to harvest branches of tall trees (although the specimens may be damaged from their descent to the ground and the foliage may be riddled with shot). A slingshot used with nylon line can snag a branch up to 25 m high (depending on the amount of foliage interference), and a crossbow can shoot a line as high as 45 m (Perry, 1986). It may then be possible to cut limbs using a rope with a pruning saw or chainsaw blade attached, or to cut some foliage by tugging on the line, but such tasks are time-consuming and often unproductive. The advantages of these ground-based methods are their relatively low costs, light weight for field transport, and ability for even an acrophobe to collect canopy flora. Some botanists prefer collecting tree specimens with the rope-and-saw method instead of using more bulky and expensive climbing gear (M. Grayum, personal communication), and others prefer the relatively lightweight, inexpensive French climbing spikes, also called *patas de loro* (A. Gentry, S. Mori, and M. Nuñez-Vargas, personal communication).

One of us (M.D.L.) spent 15 years collecting and measuring ecological attributes of canopy foliage in rain forests. Leaves in the upper canopy were significantly tougher, smaller, thicker, and more sclerophyllous, and differed in their levels of tannins and phenolics. For example, the leaves of *Doryphora sassafras* in the understory of a subtropical rain forest in Australia were up to 4 times longer in length, 10 times greater in surface area, and 25 times softer, and an average of up to 26% surface area was defoliated (as compared to an average of 7% in the understory) (Lowman, 1992). In lowland tropical rain forests of Cameroon, the leaves of *Sacoglottis gabonensis* were 2 times longer and 4 times greater in surface area in the understory than in the upper canopy, and suffered 25 times the level of herbivory, 0.5% versus 12.9% (M. Lowman, M. Moffett, and B. Rinker, unpublished data).

The challenges of conducting statistically replicable ecological sampling throughout different crown regions of the forests in this study were daunting, and even most canopy-based techniques were inadequate for comprehensive ecological canopy research. To measure a canopy process such as herbivory, two types of canopy measurements were made that used a range of access techniques. "Long-term measurements" involved frequent access for monthly measurements of foliage over many years, without harvesting foliage (Lowman, 1984). A second method, called "discrete measurements," involved destructive sampling of multiple branches from measured heights and light environments in the crown. Almost every type of canopy-based and ground-based sampling was used to measure herbivory as a canopy process: single-rope techniques (sometimes combined with pole-pruners and/or swinging across on horizontal vines), shotguns, rifles, slingshots, treefalls and branchfalls, following behind logging operations, exploiting destructive galahs who broke off leaves, cherry pickers, leaning over cliffs and edges, trams, ladders, canopy raft, dirigible and sled, canopy crane, scaffolding, platforms, and walkways. Although these methods vary in effectiveness, it was necessary to use many techniques to accumulate adequate ecological information about canopy foliage.

C. Lianas and Vines

Lianas are the most undercollected major plant group (Gentry, 1991). Because lianas often grow between trees, they do not tend to descend with treefalls as readily as epiphytes do, and serendipity becomes a more important factor in their collection. Their flowers and fruits may be found on the ground, but foliage and branches from lianas of mature forest canopies are often impossible to observe or obtain, except by canopy cranes or other costly methods. In Panama, the canopy crane erected by the Smithsonian Tropical Research Institute was used to survey vines on over 80 trees within several days (M. Lowman, unpublished data). With any other method of access, such efforts would require several months.

Canopy-based techniques may enable a collector to reach vines in the upper canopy, although even ropes are not successful in reaching the outer crown, where vines tend to congregate. More intrepid botanists in our surveys confessed to free-climbing upward or outward while anchored to the tree with nylon webbing, despite obvious dangers. Using extendible pruning poles from a secure perch in the canopy is probably the safest low-cost method for collecting vines from the outer canopy. One innovative ground-based technique involved the use of a hand winch to literally wind vines back down to ground level (Putz, 1990). We predict that using a combination of ground-based and canopy-based techniques for studies of vine abundance and diversity in tropical forests would be more successful than ground-based techniques alone (e.g., Putz, 1984).

D. Epiphytes

At The Marie Selby Botanical Gardens, where the herbarium collection contains predominantly epiphytes, fewer than 10% of specimen labels include information about habitats or host trees. Labels indicate that most collectors have obtained specimens opportunistically from the ground, although others have used single-rope techniques (e.g., Ingram and Nadkarni, 1993). Ecologists who collect and study epiphytes most commonly use felled trees (e.g., Kelly, 1985), ladders with extendible pruners (Bøgh, 1992), rope ladders (Hofstede *et al.*, 1993), and single-rope techniques (e.g., Wolf, 1992).

One of us (S.W.I.) compared ground-based with canopy-based collection techniques to calculate the percentage of a known local vascular epiphyte flora via these two collecting methods. Canopy-based collection techniques for this ecological and floristic study were crucial for two reasons: (1) accurate information about the type of microhabitat occupied by each epiphyte species was quantified and (2) a more complete epiphyte flora could be found by including single-rope climbing techniques. Another advantage of single-rope techniques was the ability to repeatedly climb individual trees at different seasons to collect canopy plants in flower. Eighty-five percent of the nearly 250 epiphyte species from a 4-ha site in Monteverde, Costa Rica, were collected using ground-based techniques (i.e., from fallen trees, branches, and lower tree trunks). Although ultimately 15% of the epiphyte flora was collected with canopy-based methods only, the inclusion of these 38 species made the final checklist more comprehensive and valuable (S. Ingram, K. Ferrell-Ingram, and N. Nadkarni, unpublished data). In addition, the specimens made from standing tree canopies have more useful habitat information than those from fallen branches. As with lianas and foliage from canopy trees, we advocate a combination of ground-based and canopy-based collection techniques to survey vascular epiphytes.

IV. Preservation of Field Collections

The accession of plant specimens from tropical forests, whether ground-based or canopy-based access techniques are used, requires three steps: (1) collecting and numbering specimens, (2) recording data, and (3) processing the specimens. Plant collections should be fairly representative of the local population and should contain as much information as possible. The ultimate goal is to make a more or less two-dimensional representation of a three-dimensional part with associated information.

The plant itself is the botanist's most important source of data, but recording collection information is crucial for a useful specimen. Each collection should be given an unambiguous number. Ideally, the final specimen with the label should provide information regarding the plant species' important anatomical features, its habitat, associated species or vegetation, relative abundance, geographical location and elevation, date of collection, and collector(s) name and number. Ephemeral characteristics, such as flower color and odor, should be noted at the time of collection and included with the basic collection data. Plant parts of certain taxonomic groups (such as some orchid flowers) should be preserved separately in a solution of alcohol, water, and glycerin. Recording certain features of the live plant, and making specific preparations before pressing specimens, is necessary for identification of some plant families (Bridson and Forman, 1992).

The preservation method used for plant collections depends on the climate, available resources for drying, and the preferences of the collector. The specimens can either be dried in the field or kept saturated in alcohol within plastic bags (the Schweinfurth method). In the humid tropics, most vascular plants will grow mold within several days if not dried sufficiently. Pressed plants can be suspended over kerosene lanterns to dry; a bed of hot embers is a more effective heat source. The greatest advantage of field-drying is that the dried specimens keep their color and do not become as brittle as when they are dried after being stored in alcohol. The Schweinfurth method is advantageous where mold presents a threat, or during collecting trips with continuous travel. This method requires less material, takes less time, and requires no further care for weeks until specimens are ready to dry (Bridson and Forman, 1992).

V. Herbaria with Important Collections of Canopy Plant Material

We surveyed botanists to identify herbaria with good representations of canopy material in six major plant groups (see Section II, B). Most botanists listed the herbaria they were most familiar with as being among the herbaria with the most useful collections. Most of the herbaria listed in Table III can be considered to belong to either of two groups: large or

Table III   Plant Groups and Herbaria, in Alphabetical Order, with Notable Collections of Canopy Plants [a]

Epiphytes: AMES, B, BR, COL, CR, F, G, HNT, K, MO, NY, SEL, U, US, W
Hemiepiphytes: A, GH, MICH, MO, NY, S
Herbaceous vines/climbers: K, MO, NY, SEL, US
Woody vines (lianas): CR, F, K, KEP, MO, NY, P, U, US
Paleotropical forest trees: A, BM, DUKE, F, K, L, MO, P, US, W, WAG
Neotropical forest trees: A, AAU, CUZ, F, GH, K, MO, NY, P, QCA, UC, US

[a] Key to abbreviations of herbaria (Holmgren *et al.* 1990): **A,** Herbarium, Arnold Arboretum (Harvard University Herbaria), MA, U.S.A.; **AAU,** Herbarium Jutlandicum, University of Aarhus, Denmark; **AMES,** Orchid Herbarium of Oakes Ames (Harvard University Herbaria), MA, U.S.A.; **B,** Herbarium, Botanischer Garten und Botanisches Museum Berlin-Dahlem, Berlin, Germany; **BM,** Herbarium, The Natural History Museum, London, England; **BR,** Herbarium, Nationale Plantentuin van Belgïe, Meise, Belgium; **COL,** Herbario Nacional Colombiano, Bogotá, Colombia; **CR,** Herbario Nacional de Costa Rica, San José, Costa Rica; **CUZ,** Herbario Vargas, Universidad Nacional San Antonio Abad del Cusco, Cuzco, Peru; **DUKE,** Herbarium, Duke University, NC, U.S.A.; **F,** Herbarium, Field Museum of Natural History, IL, U.S.A.; **G,** Herbarium, Conservatoire et Jardin botaniques de la Ville de Genève, Genève, Switzerland; **GH,** Harvard University Herbaria, MA, U.S.A.; **HNT,** Herbarium, Huntington Botanical Gardens, CA, U.S.A.; **K,** Herbarium, Royal Botanic Gardens, Kew, England; **KEP,** Herbarium, Forest Research Institute of Malaysia, Kuala Lampur, Malaysia; **L,** Rijksherbarium, Leiden, Netherlands; **MICH,** Herbarium, University of Michigan, MI, U.S.A.; **MO,** Herbarium, Missouri Botanical Garden, MO, U.S.A.; **NY,** Herbarium, New York Botanical Garden, NY, U.S.A.; **P,** Herbier, Muséum National d'Histoire Naturelle, Paris, France; **QCA,** Herbario, Pontificia Universidad Católica del Ecuador, Quito, Ecuador; **S,** Herbarium, Swedish Museum of Natural History, Stockholm, Sweden; **SEL,** Herbarium, Marie Selby Botanical Gardens, FL, U.S.A.; **U,** Herbarium, State University of Utrecht, Utrecht, Netherlands; **UC,** University Herbarium, University of California (Berkeley), CA, U.S.A.; **US,** United States National Herbarium, Smithsonian Institution, Washington, DC, U.S.A.; **W,** Herbarium, Naturhistorisches Museum Wien, Wien, Austria; **WAG,** Herbarium Vadense, Agricultural University, Wageningen, Netherlands.

specialized. Herbaria in North America and Europe with large collections (more than two million specimens, such as BM, F, G, GH, K, L, MO, NY, S, US, and W) were recommended by respondents for their large, representative collections of most of the six plant groups listed in Table III (abbreviations follow Holmgren *et al.,* 1990; Table III). Herbaria that are not large (fewer than one million specimens) often have specialized collections of certain plant taxonomic groups, ecological growth habits, or regions. For example, herbaria in Bogotá, Colombia; San Jose, Costa Rica; and Quito, Ecuador, have significant collections of different plant groups from their respective countries.

The herbaria listed indicate that the most useful collections are housed at large herbaria. However, there are hundreds of small herbaria around the world with extensive collections from the local flora (e.g., Cuzco, Peru) or from other regional floras (e.g., the University of Aarhus' Ecuadorian collections). Large collections of nonvascular Neotropical epiphytes can be found in Bogotá, the University of Florida (Gainesville), and the State University of Utrecht, The Netherlands. Significant collections of vascular Neotropical epiphytes of several families are housed at Selby Botanical Gardens and Huntington Botanical Gardens, in addition to the larger herbaria cited in Table III.

VI. Future Directions in Plant-Collecting Methodology

The goals of specific research projects and resources available will dictate, to a large extent, the collection methods that field botanists use. For example, the collection and observational methods used by Conservation International's Rapid Assessment Program (RAP) team are aimed at the quick evaluation of an area's biodiversity (Parker and Carr, 1992). Because of the urgency and magnitude of their field work, RAP botanists collect juvenile lianas and canopy-level trees from branches or sucker shoots at ground level, note observations of canopy species out of reach, and rely on their knowledge of plants to assess the species diversity of certain tropical forests (R. Foster, personal communication). If complete habitat data are essential for canopy vine or epiphyte collections, or for ecological studies of canopy vegetation, then it is imperative to use one or more canopy-based techniques. If the goal of a project is a complete florula over the time span of a short funding period, then it is advisable to employ both canopy-based and ground-based collecting methods. Although the most cost-effective approach depends on the goals of the project and funds available, in most cases a combination of ground-based and canopy-based methods yields more complete results than either method alone. For ecological studies, or for returning to a certain tree repeatedly, single-rope techniques or ladders are probably best. For floristic studies, or for collection of canopy-level trees, climbing spikes probably offer the easiest, least-expensive access (at least in forests without lush epiphytic vegetation).

We advocate more comprehensive collection of canopy plants with careful documentation of their taxonomic and ecological attributes. With the current rapid destruction of tropical rain forests at the rate of 142,000 square kilometers per year (Wilson, 1992), the description of canopy plant species becomes more urgent to implement conservation strategies.

VII. Summary

We reviewed the methods used for the taxonomic collection of forest canopy plants. Collectors from a range of institutions were surveyed to quantify the diversity of collecting techniques. The most commonly used methods for the collection of herbarium plant specimens are ground based, such as opportunistically gathering epiphytes from fallen trees or cutting tree foliage with extendible pruners. However, canopy-based methods offer better information about the plant's environment and, in many cases, offer the only access to plant reproductive parts. For more careful documentation of the taxonomic and ecological characteristics of different canopy plants, and for more comprehensive floral surveys of different forest regions, we encourage the combined efforts of both canopy-based and ground-based collection techniques.

Acknowledgments

Many thanks to the botanists who responded to our survey. We also thank T. Croat, C. M. Taylor, and an anonymous reviewer for helpful criticism of an earlier draft of this manuscript.

References

Bøgh, A. (1992). Composition and distribution of the vascular epiphyte flora of an Ecuadorian montane rain forest. *Selbyana* 13, 25–34.

Bridson, D., and Forman, L., eds. (1992). "The Herbarium Handbook." Royal Botanic Gardens, Kew.

Corner, E. J. H. (1964). "The Life of Plants." Univ. of Chicago Press, Chicago.

Gentry, A. H. (1991). The distribution and evolution of climbing plants. *In* "The Biology of Vines" (F. E. Putz and H. A. Mooney, eds.), pp. 3–42. Cambridge Univ. Press, Cambridge, UK.

Hofstede, R. G. M., Wolf, J. H. D., and Benzing, D. H. (1993). Epiphytic biomass and nutrient status of a Colombian upper montane rain forest. *Selbyana* 14, 37–45.

Holmgren, P. K., Holmgren, N. H., and Barnett, L. C., eds. (1990). "Index Herbariorum. Part I: The Herbaria of the World." New York Botanical Garden, Bronx.

Ingram, S. W., and Nadkarni, N. M. (1993). Composition and distribution of epiphytic organic matter in a neotropical cloud forest, Costa Rica. *Biotropica* 25, 370–383.

Johansson, D. R. (1974). Ecology of vascular epiphytes in West African rain forest. *Acta Phytogeogr. Suec.* 59, 1–136.

Kelly, D. L. (1985). Epiphytes and climbers of a Jamaican rain forest: Vertical distribution, life forms and life histories. *J. Biogeogr.* 12, 223–241.

Lowman, M. D. (1984). An assessment of techniques for measuring herbivory: Is rainforest defoliation more intense than we thought? *Biotropica* 16, 264–268.

Lowman, M. D. (1992). Leaf growth dynamics and herbivory of five canopy species in Australian rain forests. *J. Ecol.* 80, 433–447.

Lowman, M. D., and Moffett, M. (1993). The ecology of tropical rain forest canopies. *Trends Ecol. Evol.* 8, 104–108.

Millican, A. (1891). "The Travels and Adventures of an Orchid Hunter." Castle and Company, Ltd., London.

Mitchell, A. W. (1986). "The Enchanted Canopy." Macmillan, New York.

Mori, S. A. (1984). Use of "Swiss tree grippers" for making botanical collections of tropical trees. *Biotropica* 16, 79–80.

Mori, S. A., and Prance, G. T. (1987). A guide to collecting Lecythidaceae. *Ann. Mo. Bot. Gard.* 74, 321–330.

Parker, T. A., III, and Carr, J. L., eds. (1992). "Status of Forest Remnants in the Cordillera de la Costa and Adjacent Areas of Southwestern Ecuador." Conservation International RAP Working Papers 2.

Perry, D. (1986). "Life above the Jungle Floor." Simon & Schuster, New York.

Putz, F. E. (1984). The natural history of lianas on Barro Colorado Island, Panama. *Ecology* 65, 1713–1724.

Putz, F. E. (1990). Growth habits and trellis requirements of climbing palms (*Calamus* spp.) in north-eastern Queensland. *Aust. J. Bot.* 38, 603–608.

Whittle, T. (1970). "The Plant Hunters." Chilton Book Co., Philadelphia.

Wilson, E. O. (1992). "The Diversity of Life." Harvard Univ. Press, Cambridge, MA.

Wolf, J. H. D. (1992). Species richness patterns of epiphytic bryophytes and lichens along an altitudinal gradient in the northern Andes. *Selbyana* 13, 166–167.

23

Tourism, Economics, and the Canopy: The Perspective of One Canopy Biologist

Donald Perry

A new century of environmental reality is about to dawn. Under the pressure of explosive human population growth, the planet's natural communities are shriveling rapidly. They are being constricted on all sides by the expansion of agriculture, for example, corn, beans, coffee, bananas, pineapple, rice, chocolate, and beef.

As a tropical biologist who has worked in the Central American country of Costa Rica since 1974, I have witnessed the impact of a major tourist boom in what was once a little-known paradise. I have watched the devastating effects of rampant logging that threaten the integrity of the Costa Rican National Park system. It appears likely that, in the not too distance future, most wild places will be "natural zoos" embedded in seas of domesticated land.

As the planet's populations of wildlife become increasingly rare, more and more people want to see what is left. In Nepal, hikers have stripped the landscape bare for fuel, and they leave their trash to spoil the experience for future visitors. In the Galapagos, increasing numbers of visitors and the support facilities to accommodate them have created an onerous strain on these sensitive islands.

Often, the influx of too many visitors into natural habitats has caused wildlife to retreat to more remote areas. This has more than likely occurred in every corner of the globe, and I have witnessed it in Costa Rica's Manuel Antonio National Park, Carrara Nature Reserve, and the Monteverde Cloud Forest Reserve. Animals that were once regularly encountered on trails are

Copyright © 1995 by Academic Press, Inc.
All rights of reproduction in any form reserved.

now hard to find. Tour guides invade deeper and deeper into the forest to search out subjects for camera-ready tourists. The stress on plants and animals in these areas is apparent. Uncontrolled nature tourism, so called eco-tourism, can in some conditions be considered eco-destruction.

Many believe the current tidal wave of travelers threatens to inundate nature's survivors, putting the final nail in the coffin of earth's great ecosystems. But in this essay I advocate that tourists may be environmentally important. Why wave a flag for tourism? This question is frequently asked by staunch European and American conservationists who attend my slide shows about tropical rain forests. They visibly pale when I mention a project I am starting that will take people into jungle treetops to educate them about conservation of rain forests. They do not think that conservation and economics can be compatible. But I think they can be.

The late Dian Fossey, who studied mountain gorillas (and wrote *Gorillas in the Mist*), was my mentor, even though we never met. By all accounts, she was an ardent preservationist who did not want the mountain home of her gorillas to be made into a gorilla theme park. She felt that the animals had a right to live without being disturbed by meddlesome tourists. Dian strived for this purist idea even though tourism could have aided in the conservation of her beloved gorillas. Dian taught us a valuable lesson: when nature becomes an island amid people who are impoverished, nature is never safe.

Many see Costa Rica as a shining example of conservation, but it is neither better nor worse than North America or Europe. My perspective is that of a biologist who once studied Costa Rica's vast lowland rain forests. Banana cultivators, lumbermen, and farmers are stripping those lowlands of every last rain forest tree. At this time, Costa Rica has proportionally less rain forest left than many other tropical countries. In fact, only a few islands of relatively untouched lowland Caribbean rain forest remain.

Countless documentaries and articles report statistics of lost acreage and dwindling species. Is anyone getting the message? Certainly, a lot of funds have been accumulated in temperate countries in the name of saving the rain forest. But the truth is, little land has been conserved to attest to all these temperate-based efforts. Less than 10% of Costa Rica's forests are national parks. Moreover, the national park service says there is not enough money to protect all park borders. FUNDACORE, a nonprofit organization devoted to the conservation of national parks, also says there is not enough money to protect the national parks. Still, many people naively believe that the national parks are safe from destruction.

While Costa Rica is busy cutting the final 10% of the forest outside its parks, many people fear that Costa Rica will then begin to harvest wood inside the parks. At the present rate, nonpark forests will last only 7–10 years in this country that supposedly represents the bastion of conservation in Latin America.

Tourists come to Costa Rica to admire a country that appears to have performed a conservation miracle. They arrive believing that rain forests are being saved. They bring affluence and jobs to a broad spectrum of Costa Ricans who represent a potent political force. Tourism does not poison the ecosystem, the estuaries, rivers, or the oceans; tourism has the potential to spread wealth more evenly and more deeply into the population. As tourism grows, the determination of Costa Ricans to protect national parks would also grow. If Costa Rica does not ensure the preservation of all the forest that remains, its conservation image will decline drastically.

Many Costa Ricans are unaware that their country possesses one of the most complex communities of life on our planet. My research has shown that their canopy may be three times richer in species than forests in Africa and Asia. It is frightening to think what would happen if someday all Costa Rican forests were allowed to be selectively logged. Canopy communities would be virtually dead. Costa Rica's single most important tourist attraction might disappear.

Who should pay for a massive educational campaign for Costa Ricans to learn about rain forest conservation? How could a project that must reach a large number of Costa Ricans be made cost-effective? There are no obvious short-term economic returns from educational projects in tropical countries. After careful consideration of these problems, I have developed a plan that may make a difference: The Rain Forest Aerial Tram involves scientists, tourists, Costa Ricans, and students in a project of conservation, tourism, and science.

The Aerial Tram had its origin in my efforts to devise tree-climbing techniques for scientific investigation. After experiencing the limitations of single-rope techniques, I realized that to study the canopy effectively, researchers needed a vehicle for access.

By 1982, the development of this vehicle became my prime objective. In 1983, I began designing the vehicle with the engineering expertise of John Williams. The Automated Web for Canopy Exploration (AWCE) is located at Rara Avis, in Costa Rica, and the facility is now open for general scientific use. It is composed of a power and winch station, support and control cables, and a radio-controlled, steel platform that holds up to three people. The support cable spans a forested canyon and is about 300 m in length. The platform is suspended from the support cable and can move along its length. It can also carry scientists from ground level to above the treetops through approximately 22,000 cubic meters of forest. This was made at a cost of about $1.82 per cubic meter of access. Because the support cable is stationary, the Automated Web is a linear system. AWCE is a prototype for a canopy vehicle to investigate the treetops.

For Phase II of AWCE, I continue to seek the additional funding that is necessary to give the support cable a sweeping capacity. It will then be able to move over all the tree crowns in a given area at the touch of a radio-control. Access to forest volume will be increased to 728,000 cubic meters, covering an area about the size of a football field.

The 1.3-km route of the tram itself has been surveyed and will be situated on 338 ha close to San José. There will be 24 cars with a capacity to hold up to six people. Each car will be separated by about 100 m, giving considerable privacy to its occupants. The cars are attached to a cable that rotates around two end stations—in actuality, the system will be a converted ski lift. The system will carry up to 70 people per hour through the canopy. The maximum annual capacity is about 300,000 round-trips, but we expect only around 40,000 visitors per year. This surplus capacity will be directed to Costa Rican students. Up to 10,000 free rides per year will be donated to students. We are also developing the educational materials for these students. Costa Rican citizens will receive ticket discounts.

Displays will be made to educate people on important aspects of the world environment—from waste management to ozone depletion. Sponsors will be solicited to fund these displays and to provide us with environmentally safe products. The Aerial Tram will establish a nonprofit organization called the Center for Canopy Exploration (CCE). CCE will assist in monitoring and implementing conservation programs within the site. A primary role of this organization will be to manage the land as a nature and research preserve.

CCE will work in cooperation with scientists and institutions worldwide. Canopy researchers will be appointed as advisors to the scientific board of CCE, some of whom have already agreed to participate. This will bring a broad range of research experience, which will allow us to catalog canopy communities and train para-taxonomists and others in techniques of canopy access. We expect these scientists to proceed in the exploration of other tropical regions as well. CCE will also raise money to purchase additional property to enlarge the nature-science reserve and to protect additional national park borders.

It is expected that the Aerial Tram combined with CCE will be a model that demonstrates how people can visit nature with minimal impact on the wildlife and the environment. I hope that it will become a model that is used in sensitive habitats throughout the world.

24

Canopy Science:
A Summary of Its Role
in Research and Education

Nalini M. Nadkarni and Margaret D. Lowman

The time has come when scientific truth must cease to be
the property of the few, when it must be woven into the common
life of the world.
—Louis Agassiz (1863)

So, what next, we wonder? These chapters attest to the enormous accomplishments gained in the frontiers of canopy biology—most in the last five years. We see many new directions of research to respond to the unanswered questions that arise. Perhaps more significant, we face the challenges of translating the complexities of canopy biology to applied situations and to the people living in or near forest ecosystems. Education on the scientific processes of the canopy is a critical priority, especially because it can be used as an effective tool for forest conservation.

One goal of assembling this book was to allow readers to evaluate the "state of the art" of canopy science. Research accomplishments, gaps in our knowledge of the canopy, and avenues for future investigations are now evident. These authors have documented that the forest canopy is a region of great ecological importance. Forest canopies contain a major portion of the diversity of organisms on Earth and constitute the bulk of photosynthetically active foliage and biomass in forest ecosystems. Members of canopy communities contribute substantially to the dynamics of the forest ecosystems they inhabit.

Studies of canopy phenomena have shifted from earlier, primarily descriptive work. Historically, canopy studies were dominated by people seek-

Forest Canopies

Copyright © 1995 by Academic Press, Inc.
All rights of reproduction in any form reserved.

ing the thrill of climbing and the excitement of discovering a new arboreal species. Early quantitative estimates of the complex nature of the canopy were restricted to ground-based surveys or to short-statured trees. More recently, however, the innovation of high-strength, low-cost climbing equipment has made more detailed canopy study a viable option for scientific research. With the improvement of effective technological climbing methods, the canopy raft, the canopy crane, and ground-based methods such as fogging, researchers can now spend less time figuring out how to work in the treetops and more time recording, analyzing, and interpreting meaningful data and results.

Within the past decade, the amount of information on canopy biology has burgeoned. The number of scientific publications on canopies has grown at a disproportionately rapid pace relative to the general field of biology (Fig. 1). Other evidence for the growing interest in the canopy are recent symposia on canopy biota, books, popular articles, and documentaries. This attention is a consequence both of new techniques for canopy access and of growing concern for conservation issues such as biodiversity, global atmospheric change, and conservation of tropical rain forests.

The research summarized here documents that both the types and amounts of canopy data are changing rapidly. In the last decade, the simplicity of rope climbing generated studies by scientists who worked solo or in small groups and produced fairly small data sets. The more recent development of access techniques that permit teams of scientists to work within

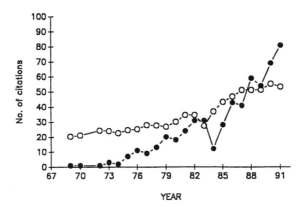

Figure 1   Indication of the rate of scientific literature published on canopy structure compared to the rate of literature published in the general biological literature. Data points are the number of citations with keywords related to canopy structure tallied in a bibliographic search of the data base BIOSIS (solid circles) and the total number of all citations indexed in BIOSIS for that year ($\times 10,000$) (open circles). Note that the rate of publications concerning "canopy structure" greatly exceeds the rate of "call citations" after 1984, indicating the explosion of interest and study of forest canopy structure in recent years.

the same volume of the canopy has led to complex data sets whose collecting expense warrants collaborative use.

Scientists are notorious for their independent (at times idiosyncratic) ways of taking, storing, and analyzing data. Because the data from access tools such as construction cranes are expensive and difficult to obtain, canopy researchers will need to deal with (1) new types of data, (2) much more data, and (3) the necessity of sharing data among scientists who have different research questions. We need to plan for developing the means to collect, process, store, analyze, portray visually, interpret, and distribute these types of data. In many ways, the current situation is analogous to what planetary scientists faced when anticipating data from the *Voyager* space probe: large quantities of new types of data that must be shared by a variety of scientists asking different questions.

The field of canopy science is ready for research that will (1) quantify the distribution and abundance of the major canopy components and the processes that link them; (2) formulate protocols that will allow other researchers to gather comparable data on canopy processes, thereby building the available data on canopies for all scientists; and (3) lead to the generation of specific hypotheses that can be rigorously tested.

One approach to integrate canopy research is the Canopy Research Network (CRN), which has received funding from the National Science Foundation (Database Activities Program). It brings together forest canopy researchers, quantitative scientists, and computer scientists to establish methods to collect, store, analyze, and interpret three-dimensional spatial data relating to tree crowns and forest canopies. The CRN mandate is to (1) compile an array of research questions and needs from the canopy research community, (2) examine potentially applicable information models and software tools that are in use in allied fields, and (3) develop conceptual models and recommendations for the types and format of information and analyses necessary to answer research questions posed by forest canopy researchers.

A second approach is to integrate the results of canopy research into the same publications rather than have disciplines appear in an array of disjointed journals as has historically been the case. This volume is a first attempt to unify canopy research results for the scientific audience. Similarly, the journal *Selbyana* has taken the lead in publication of canopy research via the proceedings of the epiphyte and canopy symposia held at The Marie Selby Botanical Gardens in the 1980s and 1990s.

A third approach is to facilitate the communication of research ideas and inspirations, access techniques, and results within and outside of the multifaceted canopy research, education, recreation, and conservation communities. Although much of this communication has occurred among the few canopy scientists through informal "networking" and word-of-mouth exchanges, the growing size and diverse interests of current and future canopy workers demand a more formal structure for information transfer. In 1993, the CRN initiated an electronic mail bulletin board for those interested in the canopy, with a first-year subscribership of 250 scientists. A newsletter to summarize and disseminate information to non-computer-linked canopy scientists will be initiated in 1994.

Another conspicuous feature of canopy science highlighted in this book is that few comparative data exist at the ecosystem level. One goal of many canopy researchers is to develop standardized protocols to measure canopy organisms and processes and to compare them directly between ecosystems. In the longer term, canopy scientists envision the incorporation of canopy measurements into many forest studies in a variety of locations: parks, field stations, forest plantations, and university campuses. Ultimately, canopy scientists will be able to generate testable hypotheses based on comparable data that will contribute to our understanding of forest structure and function.

The image of canopy science is generally poor among traditional scientists. Because early canopy biology focused on techniques of getting to the canopy and documenting its poorly known biota, canopy science was viewed by many academics as "Tarzan stuff" or a throwback to nineteenth-century descriptive biology. There is a clear need to continue to explore canopy access methods and to document unknown species. However, work described in this book demonstrates that experimental approaches and theoretical issues are well integrated into canopy science.

How might we improve research and communication in canopy science? First, we must increase communication among canopy scientists. One scenario would be to formalize the existing informal network by establishing an international canopy network, which could publish a regular newsletter and/or journal, stage annual meetings, and produce a directory. Regular international meetings would help to keep colleagues abreast of progress and facilitate personal interactions that lead to collaborative studies.

Second, canopy scientists should be made aware that a wealth of canopy workers exists outside the realm of scientific journals, symposia, and granting agencies. These are the arborists, who are currently more closely allied to horticultural rather than academic worlds. They have their own organizations, publications, and meetings. Major strides may be made if we cross-pollinate ideas and techniques with professional tree climbers, tree surgeons, and tree pathologists.

Third, there are few direct outlets by which scientifically sound information can be shared with the general public. Most research results are presented and discussed at closed scientific meetings. We need to increase the flow of canopy information to the general public. The few outlets of this information to nonscientists have been extremely well received. For example, Ben Shedd's OMNIMAX film *Tropical Rainforest,* which featured segments on biologists working in the canopy, has received a great deal of positive critical acclaim. In 1994, the Jason Project (an international science education program that broadcasts live from field sites) featured a canopy biologist studying the ecology of rain forest tree crowns in Belize. The success of these popular media treatments of the canopy attests to the inherent interest of the general public in canopy biology.

In the same breath that we advocate greater interest of the general public in forest canopies and encourage the "person on the street" to vicariously take to the treetops, we recognize the need to develop and enforce wise methods of canopy access. There is still no overarching moral ethos for climbing. Anyone can buy a rope and sling it over any tree in any manner. Safety considerations are up to the individual. As with rock climbing and caving, there is a general sense that ethical climbing must be espoused by canopy visitors. This is currently being discussed among arborists, canopy scientists, and science educators.

In this volume, canopy research has been shown to be relevant to both basic biology and applied ecology. One example of how canopy research relates to a major environmental issue is the potential use of epiphytes as indicator species of global atmospheric changes. Human-influenced changes to the atmosphere include the increasing concentration of particular nutrients. Because many epiphytes are partially or entirely dependent on atmospheric sources for their nutrients and water, some species may be highly sensitive to changes in atmospheric chemistry, which in turn would affect those organisms directly or indirectly associated with them. "Vulnerable" epiphytes and their associated biota could serve as biological indicators in the canopy for particular attributes of atmospheric conditions, the proverbial "canaries in the coal mine." The documentation of canopy ecology in "healthy" forests provides a benchmark foundation for future investigations on the effects of atmospheric changes on canopy communities.

Relating basic ecological processes to solving environmental problems is an important, albeit underutilized, application for current biological research. The "Sustainable Biosphere Initiative," generated by the Ecological Society of America in 1991, emphasizes that the exploitation of natural ecosystems may have an irreversible and detrimental effect on the ability of ecosystems to regulate themselves naturally. Ecological understanding of the canopy lags far behind the current knowledge of most other components of terrestrial ecosystems. Understanding the canopy as part of whole-ecosystem processes is an obvious priority if we are to responsibly manage and conserve forests in the future.

Index

Physiological Ecology
A Series of Monographs, Texts, and Treatises

Continued from page ii

F. S. CHAPIN III, R. L. JEFFERIES, J. F. REYNOLDS, G. R. SHAVER, and J. SVOBODA (Eds.). Arctic Ecosystems in a Changing Climate: An Ecophysiological Perspective, 1991

T. D. SHARKEY, E. A. HOLLAND, and H. A. MOONEY (Eds.). Trace Gas Emissions by Plants, 1991

U. SEELIGER (Ed.). Coastal Plant Communities of Latin America, 1992

JAMES R. EHLERINGER and CHRISTOPHER B. FIELD (Eds.). Scaling Physiological Processes: Leaf to Globe, 1993

JAMES R. EHLERINGER, ANTHONY E. HALL, and GRAHAM D. FARQUHAR (Eds.). Stable Isotopes and Plant Carbon–Water Relations, 1993

E.-D. SCHULZE (Ed.). Flux Control in Biological Systems, 1993

MARTYN M. CALDWELL and ROBERT W. PEARCY (Eds.). Exploitation of Environmental Heterogeneity by Plants: Ecophysiological Processes Above- and Belowground, 1994

WILLIAM K. SMITH and THOMAS M. HINCKLEY (Eds.). Resource Physiology of Conifers: Acquisition, Allocation, and Utilization, 1995

WILLIAM K. SMITH and THOMAS M. HINCKLEY (Eds.). Ecophysiology of Coniferous Forests, 1995

MARGARET D. LOWMAN and NALINI M. NADKARNI (Eds.). Forest Canopies, 1995

BARBARA L. GARTNER (Ed.). Plant Stems: Physiology and Functional Morphology, 1995